GREEN
MEDIA
GROUP

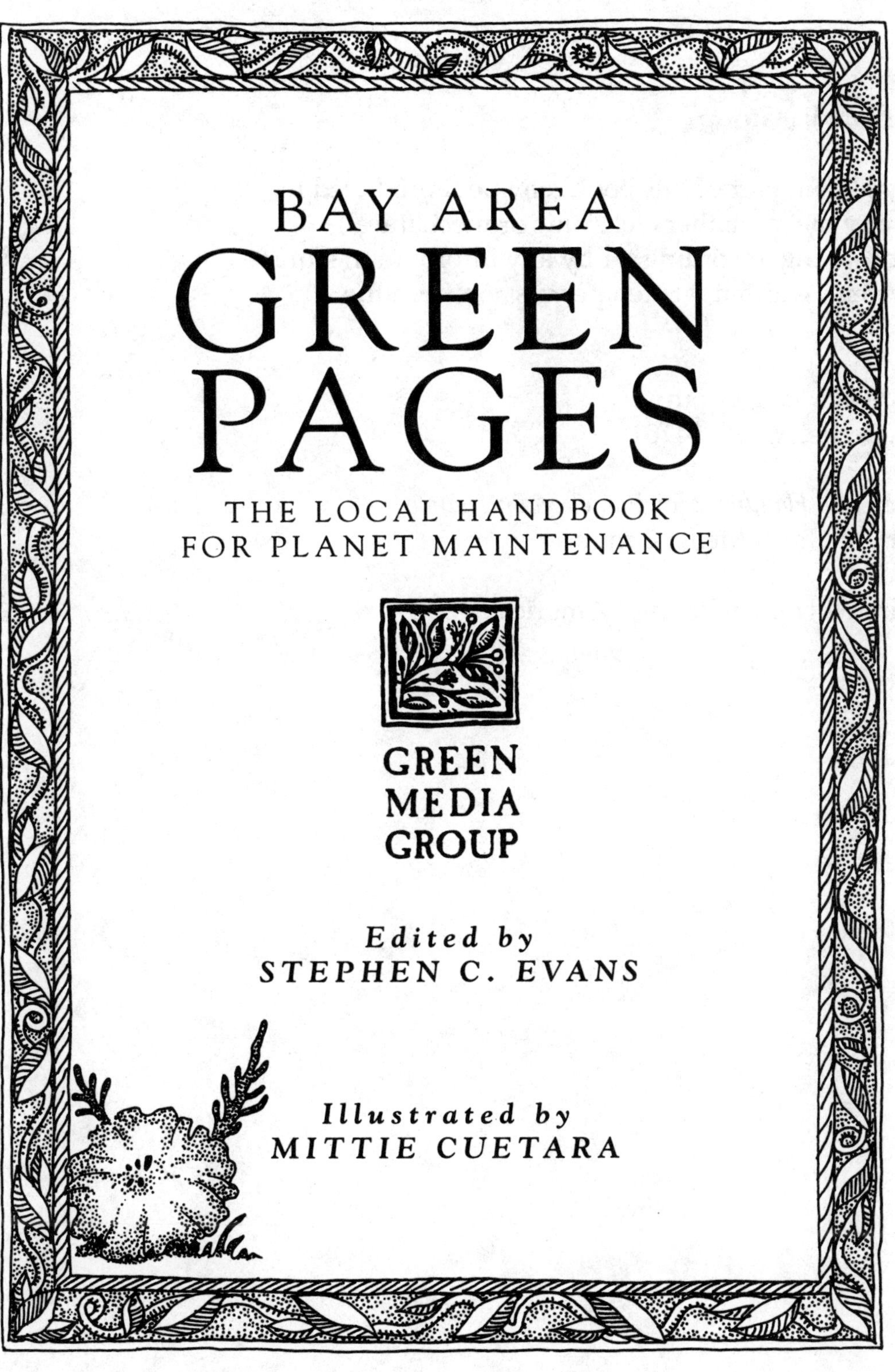

BAY AREA

GREEN PAGES

THE LOCAL HANDBOOK
FOR PLANET MAINTENANCE

GREEN
MEDIA
GROUP

Edited by
STEPHEN C. EVANS

Illustrated by
MITTIE CUETARA

Published by Green Media Group,
P.O. Box 11314, Berkeley, CA, 94701

Manufactured in the United States of America

ISBN 0-9627358-0-9

First edition.

BAY AREA GREEN PAGES

THE LOCAL HANDBOOK FOR PLANET MAINTENANCE

GREEN MEDIA GROUP

PUBLISHER
Eric Ingersoll

EDITOR
Stephen C. Evans

RESEARCH DIRECTOR
Monica J. Fletcher

ASSOCIATE EDITOR
Susan E. Davis

COVER & ILLUSTRATIONS
Mittie Cuetara

PUBLISHER'S ASSISTANT
Willa Eastman

PRODUCTION
Bryan Hammond

DESIGNERS
Susan Redding
John Nguyen

ADDITIONAL LAYOUT
Nick Braun

TYPIST
George Russell

MAPS
Eureka Cartography

ADVERTISING MANAGER
Gloria Lucia Viana

ADVERTISING SALES
Eileen Duffy
Andrew Getz
Kate Littleboy
Debra Nash
Bruce Rinehart
Wendy Roberts
Carrie Sealine
George Spies
Gloria Lucia Viana

DATABASE DEVELOPMENT
Michael Schumann

INTERNS
Liz Brooks
Bill Buck
Matthew Gauland
Jacob Lieb
Lynn Schached
Alex Zulauf

RESEARCHERS
Karen Kho
Lisa Levy
Christine McKenna
Scott Morehouse

ACCOUNTS
Suzanne Pettigrew

ADVISORY BOARD
Seth Adams
Carl Anthony
Ernest Callenbach
Tyrone Cashman
Debra Lynn Dadd
Claire Greensfelder
Denis Hayes
Huey Johnson
Daniel Knapp
John Knox

If you have any comments, criticisms, or suggestions for future editions, please contact the Green Media Group, P.O. Box 11314, Berkeley, CA 94701.

Acknowledgements

This book went from dream to reality in about eight months, a feat that would have been impossible had we not been blessed with an extraordinary staff and the generous support and advice of friends, relatives, and the many people we came to as perfect strangers who gave freely of their time, experience, and wisdom.

It also helped that much of our work had already been done for us. Many non-profit organizations and professional trade associations were very helpful in providing resources, lists of members, and other invaluable information.

The Ecology Center in Berkeley and the Data Center in Oakland both proved to be virtually bottomless wells of information on the subjects that interested us. Perhaps the most valuable resource, one that we began to call "the bible," was the *Harbinger File*, a listing of most of the environmental organizations in California, published by Harbinger Communications in Santa Cruz. Very special thanks go to Bill Leland for permission to use descriptions from the *Harbinger File*, and for having put it together in the first place.

Dale Sanders of the UC Berkeley planning office deserves special mention for his expert guidance through the intricacies of the California government's environmental bureaucracy. Dan Knapp and Mary Lou van Deventer of the Northern California Recycler's Association answered countless questions about recycling and the Bay Area's landfills. Debra Lynn Dadd's books on environmentally sound consumer goods were also incalculably valuable resources, as was her advice.

John and Zinta Bastin of the Bastin Steamship Control Company deserve special accolades. Not only did they lease office space to us at Depression prices, but they patiently endured months of Green Pagers traipsing through their office to send and receive faxes, pour coffee, and use their refrigerator. These trespasses were only matched by those against the sailmakers at Hogin's Sails, whose work loft has our front door at its far end. Rather than complain about all of the traffic, these wonderful people gave us organically-grown food from their home garden.

Jonathan Maslow kept us on course, shared his prodigious knowledge of the publishing business, constantly urged us to meet deadlines, and came in to help control the damage on the several occasions when we failed to do so. Invaluable advice about the publishing trade was also provided by Ben Bianco, Warren Jones, Arnie Kotler, Bill Dodd, Jerry Mander, Joel Makower, Malcolm Margolin, Julie Bennett, and the people at Publisher's Group West. As for keeping us to our schedule, nobody applied loving pressure better than our publicists, Carol Butterfield and Associates.

If it hadn't been for the loan of fast computer equipment by Rain Burns of Apple Computers, Jerry Lander at Lapis Technologies, and Chris McKenna, we probably would have been here for another year slaving away at our sluggish old equipment. We were the happy recipients of many such gifts, loans, and "at-cost" deals. We are particularly grateful to Megan Keelaghan, John Parsons and the others at Eureka Cartography for the beautiful maps; Bill Parks, Don Abbott, and the people at Recreational Publications for the big breaks on camera work; Alan Davis, Janis Brewer, and Mark Lesser of Conservatree for the advice and good deals on recycled paper; Jim Duffy and Alonzo Printing for being the best eco-printers around; Ken Munro and Zellerbach for shipping supplies; and Ryan Best and Mark Allen of Direct Language for the late nights and weekends helping spew forth the final pages. Special thanks go to Jerry Ingersoll for his work in obtaining decent phone equipment for us, and to Peter Trueblood for all the free consultations.

Claire Greensfelder, Tyrone Cashman, Susan Alexander, Ernest Callenbach, Peter Barnes, Beth Weinberger, Denis Hayes, John Knox, Jamie Levin, Ken Maley, Gil Friend, everyone on our advisory board, and too many others to list here were all very helpful in suggesting avenues of research and providing us with important contacts.

Whole Earth Access and Esprit helped tremendously by agreeing to buy quantities of the book sight unseen, before we went to press.

Thanks also go to our families, particularly our parents, for all of the moral and material support.

We will never be able to show enough appreciation to Arthur Evans and William E. Evans, who were the first to demonstrate their belief in the project (and us) by loaning the seed money to get it started. Later, in one of those dark periods that strikes at the middle of many projects, Jack Anderegg came through with another loan, salvaged our spirits, and quite possibly saved the book. Theo Gund was also a tremendous help in this respect

Finally, and foremost, our biggest debt of gratitude goes to the great group of staffers who worked on the book for next to no pay, sometimes for very long hours, out of the goodness of their hearts and their love for the planet. We couldn't have done it without you.

-Stephen C. Evans & Eric Ingersoll

TABLE OF CONTENTS

TABLE OF CONTENTS

TABLE OF CONTENTS

Preface

About this Book

My friend Eric Ingersoll came back from a trip to Europe in February. On one of those extraordinary warm winter days, brought in such abundance by this long drought (is it global warming?) we sat out by the little weed-choked pond I had dug in my back yard, watching dragonflies and talking about the environment.

At the time, I was planning to go to Alaska to shoot a documentary about the aftermath of the Exxon Valdez disaster, and Eric was looking for work as an eco-activist. He hadn't found the right job yet, but the search had given him a wonderful idea. His research had revealed that, despite being renowned as a hotbed of environmentalism, the Bay Area lacked a good comprehensive directory to all of that activity. - This was something that each of us would have found quite useful right at that moment.

I needed such a guide for a raft of little household problems, including pests in the kitchen and dying plants in the garden. (My wife and I have stubbornly refused to use the standard poisonous remedies.) There was also the question of what to do with the shelves stacked with old paint and strange chemicals our predecessors had left stacked on shelves in the basement when they moved out.

Above all, Eric and I were both tired of feeling glum and powerless about the mounting multitude of messes our ways of living had gotten us into, and we wanted to know what we could *do* to extricate ourselves. More specifically, we wanted to know what we could do *locally*.

Together, we started to dream and scheme about a book that would answer all of our questions, that would put everything that had to do with environmentalism in the Bay Area— issues, organizations, public agencies, products, services — in one handy guide. We would make a comprehensive "green" directory for the region. Nothing like this had ever been done before, as far as we knew.

We got very excited. The afternoon by the pond became two, then five. We moved inside to the computer and started working on spreadsheets and charts. Planning for the Valdez documentary was pushed to one side and finally supplanted entirely. A gang of the greatest friends in the world got enthusiastically involved. A saint rented us office space in the Alameda Marina at a price that would have been considered reasonable in 1960. The Green Media Group was born.

Now, six months later, the first edition of the *Bay Area Green Pages* is ready to go to press. It is the product of the democratic process at its best. In those first animated days, the whole staff of 15 hashed out the details of the Consumer Guide in afternoon-long bull sessions that led to some pretty heated discussions.

One of the hottest was about automobiles, those arch-enemies of the environment upon which we also rely for almost all of our transportation. Should we advertise the evil, polluting machines at all? If so, which ones? Only the most gas-efficient one? Or only those that run on alternative fuels? How about used cars? We are, after all, firm advocates of re-use. Yet older cars tend to pollute far more than newer models.

Some of us firmly held that gas-powered vehicles of any sort had no place in the *Green Pages*. Others responded, practically spitting, that there was no way the *Green Pages* was going to pry Bay Area drivers from behind the wheel by choosing to ignore them. Someone made a pitch for including used cars, saying that it cost as much fuel to build a car as it would be likely to burn in its entire lifetime. Someone else responded that the actual energy break-even point came after something like 10,000 miles for the average car, a figure that an energy expert later confirmed. The debate raged on for days. In the end we decided to list the ten least polluting conventional autos, plus whatever sound alternatives we could find.

Similarly, we debated whether or not to sell advertising to companies that provided one good service or product, but which otherwise had poor environmental records. We ultimately decided in favor of permitting such advertisements, as long as only the beneficial product or service was mentioned in the ad. This decision was based on our wish to promote the development of markets for "green" goods, and to encourage companies to move in greener directions. We believed that forbidding, say, a paper company that once polluted a river from advertising their recycled paper line in our publication would neither punish the company nor serve our readers.

We certainly would like to help all paper companies sell more recycled paper, however, and we would like to see increased demand push those companies to make it more readily available. At present, high quality, fully recycled book stock is virtually impossible to find on the open market. (In our long search for paper for this book, for example, the only 100 percent "post-consumer waste" recycled paper we were able to find was a thin newsprint that was, we finally decided, unacceptable. The paper we decided to use instead is "50 percent recycled," although only ten percent of the content has actually been used before. The rest is mill scrap, which is normally thrown back into the hopper anyway. That was the best we could do — in our opinion, not good enough.)

Other decisions were easier to make, and required no discussion — just lots of research. For instance, we knew that we wanted to put in every gardener and landscape architect with expertise in pesticide-free or drought tolerant gardening techniques we could find. But finding them was the problem.

Our five researchers ferreted out trade associations, interest groups, and knowledgeable individuals that provided leads not only to the gardeners, but to organic farmers, solar energy firms, builders and architects who specialize in energy-efficient and non-toxic building techniques, recyclers, purveyors of recycled goods, and much more.

But enough about the Consumer Guide. There is much more to this book, and we hope our readers will make use of all of it. In particular, we'd like you to wear thin the pages describing the many wonderful local non-profit organizations that tirelessly work for a broad range of important causes, and without which our world would most certainly be in much sorrier circumstances than it is today.

The Bay Area is blessed with as fine a brain trust and as dedicated a group of "green" activists as can be found anywhere. People are fighting to get the local refinery to reduce toxic emissions, cruising the Bay watching for illegal dumpers, working to restore a nearby stream or to get a bit of open space declared a park, trying to save the rain forest in Brazil, and lobbying for better mass transit. Whatever problem you can think of, somebody is probably working on it not far from home.

All of these groups rely heavily on volunteers to be their foot soldiers; if it weren't for volunteers, they couldn't exist. It is not enough to simply improve our consuming habits and take out the recycling once every couple of weeks. We must all become activists.

In order to be good activists, we need to know what is happening. Few of us really understand the Bay Area's diverse ecosystem, or the patchwork of public agencies that oversees it. We have therefore put together a diverse collection of articles, reviewed literature on topics from agriculture to waste disposal, and described educational opportunities and public agencies' responsibilities.

The rest is up to you.

-Stephen C. Evans

A Call to Action

By Denis Hayes

Twenty years after the first Earth Day, those of us who set out to change the world are poised on the threshold of utter failure. Measured on virtually any scale, the world is in worse shape today than it was then. Despite a string of spectacular successes since that first show of strength on April 22, 1970, environmental threats now vie with nuclear war as the preeminent peril to our species.

How could we have fought so hard, and won so many battles, only to find ourselves now on the verge of losing the war?

The answers are complex. Occasionally we were blind-sided by 'miracles of technology' gone astray. We relied on government to solve all of our problems. We looked the other way while that same government made mistakes and steered us into water more troubled than that we left behind. All too often, we pawned off responsibility on others and looked for techno-fixes when we should have been changing our own behavior.

We possess only a rudimentary understanding of the complex interactions of life in the biosphere and of the myriad subtle effects of human action upon long-established processes. For instance, if in 1970 a poll had been taken of industrial chemists, asking each to name ten triumphs of modern chemistry, most probably would have listed chlorofluorocarbons (CFCs) as one. These compounds had an array of beneficial uses, and they appeared to have no undesirable side effects. They are not toxic, carcinogenic, or mutagenic. They do not corrode materials; they are not flammable; they don't explode.

Not until 1974 did Professor Sherwood Rowland and his colleagues at the University of California at Irvine discover that CFCs posed a danger to the stratospheric ozone layer that protects the earth from ultraviolet radiation. And it was not until 1985 — just five years ago — that a British team discovered the huge seasonal thinning of the ozone layer over the Antarctic.

Another example of scientific uncertainty, one that has gained widespread public attention only in the last year, is the debate over the health effects of non-ionizing electromagnetic radiation. Until recently, there was a consensus among mainstream scientists that electromagnetic fields (EMFs) had no biological effects. New evidence indicates that EMFs can promote childhood leukemia, fetal deformities, learning disabilities, depression, miscarriages, and cancer.

Asbestos, EMFs, and CFCs have given us a degree of humility. When yesterday's "triumph of modern chemistry" turns into today's deadly threat to the global environment, we can legitimately ask, "What else don't we know?"

It is always easier to tackle urgent problems than distant threats — even when the distant threats are more important.

The U.S. is what boxers call a counterpuncher. What we do best is respond. Bomb Pearl Harbor and America will pull out all the stops. Launch Sputnik and America will have NASA functioning overnight.

What we do not do well is anticipate and avoid problems. Unfortunately, many environmental phenomena involve thresholds that, when passed, cause irreversible damage. If we wait until the damage occurs and then respond, it will be too late.

We face many such thresholds in the years ahead. Some, such as rain forest destruction, are already causing irrevocable harm. Every area cleared is lost forever. Others, such as global warming, could eventually result in rising oceans covering huge tracts of land, including the rice-producing river deltas of East Asia. These are not problems to which we can respond after they occur. They are problems we must avoid.

Environmentalists must force the political system to assign high priority to distant, but dire, threats. We must draw a line in the political sand on this side of each irreversible threshold. The public intuitively understands this. Environmental victories are always carried on the shoulders of a mobilized public.

Time and again, the environmental movement has relied too heavily upon government. The government has aggressively promoted unsustainable agricultural practices, an unbalanced transportation system, the nuclear power quagmire, park barrel dam projects, and logging policies reminiscent of Paul Bunyan. Government is the nation's largest polluter, and it frequently exempts itself from the rules it applies to industry. Nuclear weapons facilities may be the most contaminated sites in the world. The estimated price tag to clean them up is more than $150 billion.

Our excursions into what I call lemon socialism — having the government fund projects that the private sector is too shrewd to finance — produced a notable collection of gold-plated turkeys. The Synfuels Corporation, the federal government's answer to the gas crisis, was America's biggest bust before Star Wars. Despite an initial budget of $88 billion — more than the space race, the Marshall Plan, and the interstate highway program combined — the synfuels program yielded no net energy at all. In the nuclear sector, the Clinch River Breeder Reactor was a classic example of Cheop's Law: Nothing ever gets built on schedule or within budget.

In both cases, failure was fortuitous. A successful synfuels program could have increased America's contribution to global warming threefold, and a successful breeder program would have confronted us with the need to manage thousands of tons of bomb-grade plutonium.

Our national energy program has been a bust, but this is not to suggest that the environmental movement should ignore the government. On the contrary, governments (local, state, federal, and international) must be a major focus of our efforts. Governments must set the rules and establish the framework if we are ever to build a sustainable society.

However, ninety-nine cents of every environmental dollar raised today — other than funds earmarked for land acquisition — is spent trying to influence government. Such single-mindedness has caused us to ignore other vitally important opportunities.

All the most successful movements, and all the world's major religions, have succeeded in part because they ask people to improve their behavior. The civil rights movement and the women's movement, for example, ask their supporters for heroic changes in their personal lives. Environmentalists, on the other hand, have often tried to convince the public that we could all eat our cake and have it, too. People were encouraged to believe that if we could effect the necessary changes in government and industry, people would not have to change their habits at all.

The answer to air pollution was claimed to be catalytic converters on tailpipes and scrubbers on smokestacks. We have pursued this strategy, at enormous cost, for twenty years. Yet the sky today at times resembles split pea soup. We have been spectacularly unsuccessful at cleaning up automobile exhaust. Our cities have grown larger. People have moved farther away from their jobs and they drive more miles; their cars idle more at stop lights, drive-through windows, and traffic jams.

It was not sufficient to scrub pollutants out of exhaust. We must also begin using cleaner fuels and more efficient engines. We should create incentives for people to live closer to their workplaces to reduce commuting and congestion. We should encourage widespread use of, and improvements in, public transportation and promote bicycle riding (and bicycle lanes) wherever possible.

No behavior is more ripe for change than our waste habits. Sending our natural resources on a one-way trip from the mine to the dump makes no sense, no matter how you look at it. We throw away valuable resources, eliminate jobs, waste embedded energy, and destroy the environment — all because people can't be bothered to put glass in one trash container and aluminum in another.

Some of our landfills are now richer in resources than some of our mines. Regulatory and tax systems designed to promote exploration and exploitation in a pioneer society have acquired their own inertia and their own vested interests. So we mine virgin ore instead of reducing use and repairing, reusing, and recycling substances that have already entered the stream of commerce. To take, perhaps, the most obscene example, the federal government is currently selling 300-year-old trees in the Tongass for less than the price of a Big Mac.

Comprehensive recycling is essential. At even a 50 percent recycling rate, after just five cycles, only three percent of the original material is left in the economy. We need to do much better than that. Comprehensive recycling and composting will require significant government involvement — to end its bias in favor of raw materials, landfills and incinerators,and to provide curbside pickups, set standards, and provide near-term markets for recycled goods.

But the necessary first step is to do our part. We must comprehensively recycle all used items, and we must purchase recycled goods whenever possible.

The environmental movement must broaden.

The most dangerous environments are in communities that are the least powerful. Poor people and minorities are downwind from the most toxic incinerators and adjacent to the most hazardous waste dumps. They are in the fields when pesticides are sprayed from planes. They work in factory jobs having the highest exposure to dangerous pesticides. Yet poor people are not well represented in the ranks of the environmental movement. In communities racked by the devastation of drugs, plagued with violent crime, suffering rising problems of homelessness and malnutrition, environmental issues are not considered a priority. But they should be. The right to lead healthy, productive lives means that environmental values should be of great importance to those communities most deeply scarred by environmental degradation.

Right now, there are probably no more than 10 million dues-paying environmentalists in the country. They are powerful beyond their numbers because they tend to be highly educated, well paid, and politically active. That is enough to pass some good, narrowly-tailored legislation.

However, it was not enough to withstand the full frontal assault of the Reagan administration. Let me cite a particularly painful example. In 1980, the United States led the world in every renewable energy technology. Today we lead in none. Most of our photovoltaic industry has been sold to foreign companies or abandoned. Renewable energy sales have declined by more than 95 percent.

This was not an accident. The destruction of the U.S. renewable energy industry and the abandonment of solar research by many of our most prestigious scientists were the result of explicit governmental policies that served powerful economic interests — the conventional energy industry — Those interests were colorfully characterized during this period by budget director David Stockman as "pigs in a trough." For solar advocates, the Reagan years were like Dunkirk without the boats. The environmental movement, despite all its rhetoric about the absolute necessity for a renewable energy future, was helpless to reverse the assault. It simply didn't have enough troops or enough clout.

We will face the same hard battles again and again in the years ahead. Global warming, for example, requires that we move swiftly off fossil fuels and on to renewable fuels. It demands an explosive growth of photovoltaics, a swift transition from oil and gas to solar and hydrogen, a universal use of passive solar architecture, and perhaps a trillion-dollar investment in energy efficiency. Such a transition will necessarily entail winners and losers. Conventional energy producers will be among the losers. They are some of the richest and most powerful institutions in the country, and they will fight like hell to avoid being phased out of existence.

There are no powerful economic institutions on the solar side. The transition will only be achieved if it enjoys enthusiastic backing from a broad cross-section of society.

It will not be possible to build a sustainable society without confronting some controversial, emotional issues. Many environmental organizations have avoided issues that should be of central concern; two are of paramount concern.

First, the world cannot build a sustainable future so long as it spends $1 trillion annually — virtually all the discretionary capital — on military ends. The proposed U.S. military budget for next year is $305 billion — all ostensibly to defend our national security. Far more vital threats to our security — global warming, ozone destruction, the ecological undermining of agricultural productivity, mounting dependence upon foreign oil, the crack epidemic, the creation of a permanent urban underclass, and the erosion of much of the national infrastructure — all cry out for more money. These threats cannot be averted as long as 75 percent of all federal research and development is devoted to military research.

Even as Martin Luther King was told to stay out of the war issue, we have well-intentioned friends and allies cautioning environmentalists to stay out of the defense debate. We must firmly reject that advice. We cannot save the planet while we are still spending $1 trillion annually on instruments designed to destroy it.

Similarly, we are frequently urged to sidestep the population issue. Environmental advocacy of family planning, it is said, will alienate major religions, certain racial and ethnic leaders, and some heads of state from the environmental cause. Some leaders feel we should focus all of our attention upon matters over which we can build a consensus.

Again, we must ignore the advice. Current population levels are undermining the biological basis for our future. Water tables are plummeting far faster than they are recharged. Topsoil is eroding five times faster than it is replaced. Deserts are on the march across Africa, Asia, Australia, and America. There is not a single important problem facing the planet that could not be more easily solved with a population of under five billion.

The claim is sometimes made that the United States is overpopulated, but India is not. The purported explanation is that the average American consumes twenty times more resources than the average Indian. The core assumption underpinning such an argument is that India will never develop, and that the average Indian's impact on the earth will remain negligible. While the environmental destruction caused by contemporary Americans is unconscionable and should not be continued or replicated, it is similarly unconscionable to consign the majority of the world's population to perpetual poverty.

Global population growth must be addressed with substantial family-planning assistance and provisions for social mechanisms (for example, old age insurance) to undercut the motivations for large families while advancing social justice. For $4 billion per year, family planning could be provided to all who want it. It might be the single most cost-effective investment available to the world.

We have the power to choose our future.

A common feature of all the problems we have been discussing is that none are the result of forces beyond human control. None are caused by sunspots or the gravity pull of the moon or volcanic activity. All are the result of conscious human choices. All can be cured by making other choices.

First, we need to make our own lives congruent with our values. For most of us, there is room for improvement in virtually all spheres. We should conserve energy with easy things, such as replacing incandescent light bulbs with folded flourescents (which are five times as efficient), insulating our water heaters, and doing laundry in cold water. Then we should do the more expensive and difficult things, such as superinsulating our dwellings and buying more efficient furnaces and appliances.

We should take every possible step to minimize driving, and pledge not to purchase another new car until we can buy one that meets our needs while getting at least fifty miles per gallon. We should install flow restrictors in our faucets and showers and dams in our toilets. We should plant indigenous vegetation, search out environmentally sensible soaps and cosmetics, and look for recycled paper and other products.

We should eat lower on the food chain and develop a preference for fresh organic products grown nearby. We should carry our own, reusable string bags to the supermarket, and search out ways to eliminate other, unnecessary packaging. We should recycle our metals, glass, paper, and plastics and compost all organic waste.

In the aggregate, such life-style changes make a huge difference. If everyone used the most efficient refrigerators available, we could save an amount of energy equivalent to that produced by twelve large nuclear power plants. Using the most efficient cars with the same internal dimensions as our current vehicles would cut gasoline consumption in half. Every year, we send more iron and steel to our dumps than we use in the entire automobile industry. The aluminum we throw away every three months could replace the country's entire fleet of airplanes.

What should an individual do to make a difference? Leading a life that is congruent with your values is a necessary and important first step, but it does not discharge your responsibilities. Next you need to explore what you can do as an employee, an investor, a parent, and a member of your church and civic clubs. You should be alert to ways you can lessen the environmental impact of your job, from avoiding styrofoam coffee cups to suggesting modifications in industrial processes. You should ask your pension fund trustees to adhere to the Valdez Principles (see text on page 42) in choosing investments.

Integrating your values into your job and your other activities is another important step, but it still does not discharge your responsibilities. Next, join local and national organizations that share your goals and your philosophy, and proselytize on their behalf. Give gift memberships for Christmas; display their publications on your coffee table; support their campaigns financially and with your volunteer efforts.

Finally, become actively involved in politics. Support candidates who share your vision; vigorously oppose those who do not. Invest the time, energy, and financial support needed to win elections. Play the sort of role that causes political friends and foes alike to view you as a person with whom to be reckoned. Communicate your environmental goals and values to your candidate, and make clear that there are narrow limits on how much compromise is acceptable.

The *Bay Area Green Pages* provides the information and leads to the local resources you need to take action. Use it to find the products and services that will help minimize your personal impact on the planet; to locate the groups in which you may want to become active; to get information from, and lobby, the government; and to inform yourself further on the issues.

We have, at most, ten years to embark on some undertakings if we are to avoid crossing some dire environmental thresholds.

Individually, each of us can do only a little. Together, we can save the world.

Denis Hayes was the international chairman of Earth Day 1990 and is the current chairman of Green Seal, the Palo Alto-based environmental labelling program for consumer products.

How to Use This Book

Using the information gathered here will help Bay Area residents lead "greener" lives and become effective, informed champions of the planet. Think of it as a local operating manual for the environment.

The times call upon us to change the way we live. Some of the changes, like switching to organic food and recycled paper, should be painless. Others — like curtailing our driving habits or donating ten hours of work a week to a non-profit organization — require more commitment. To make any kind of change, though, we need information. We need to know *what* products and services are environmentally sound, *where* to find them locally, *who* is working on the issues we care about, and *how* to make the changes that matter most.

All of that information is here in an easy-to-access form. And the information the *Bay Area Green Pages* gives you is the most useful kind — local information.

The *Consumer Guide* and the *Business-to-Business Guide* offer comprehensive listings of hundreds of local businesses which provide environmentally sound products and services. You can find almost anything you really need here — just look in the index at the front of the *Consumer Guide*.

But being a good consumer is only one step in the right direction. We must also be informed and active. That's why we've given you more than 140 pages of useful articles on topics ranging from planting dwarf fruit trees to nuclear transportation around the Bay. Many of these articles are followed by a "resource box," which suggests titles for further reading on the subject or lists organizations active with the issues discussed. If you want to go deeper into a particular subject, check the main index for other organizations and literature. Then find the appropriate descriptions in the *Reviews* or *Organizations* sections.

When you need to get information or help from a government agency, check the *Public Agencies* section. We explain what they do and how to get in touch with them. If you or your children want to learn more about the environment, turn to the *Education* section for a full listing of regional educational resources for everyone from preschooler to post-graduate.

If you are looking for a career working for the environment in the non-profit sector, be sure to check the article *Working for Change* on page 135, and the job resource listings on page 139.

We have tried to be as comprehensive as possible. Some of the resources we need to make the change to a sustainable way of life are just emerging, and we realize there are some stones we may have left unturned. Please let us know. The criticism and information you share will make the next edition of the Bay Area Green Pages an even better resource. Meanwhile, we hope this one will inspire and inform you.

Section One

THE GREEN DIGEST

Chapter One

Home & Garden

There is a tendency to blame everything on the other guy. Environmental problems, we tell ourselves, are the fault of the big corporations, or of inadequate governmental regulation, or of greedy land barons in the Third World.

But the problems that are closest to home often have their roots right there — at home. Each day every one of us contributes something to environmental degradation. We spray our roses with pesticides and our lawns with fungicides. We don't insulate our houses because we think the weather is too nice here to really make the energy savings worthwhile. We get lazy about recycling.

Many of us do things we know are bad, because we don't know the alternatives. In this way, for instance, an estimated equivalent of half an Exxon Valdez in motor oil gets dumped straight onto the ground or down storm drains in the Bay Area each year. That's more than any of the big refineries have ever spilled here.

This chapter tells you how you can make a real difference in the world, starting right on the home front.

The Integral Urban House Re-visited

By Sim Van Der Ryn and Therese Peffer

Our society's habits are taking us down the road towards extinction, but an older conscience beckons us to fit back into the picture. Attitudes are changing, and people are making commitments to change their lifestyle. We need good models to show us a sustainable existence in an urban environment.

Sustainability means a better balance between using resources and renewing them through natural systems. Since the home is a major consumer of resources—such as energy, water, and food—redesigning your home towards sustainability not only saves you money but helps planetary ecology and health.

The Integral Urban House (IUH) was one such model of a sustainable home. As a project of the Farallones Institute, an older Berkeley house was redesigned in the mid-1970's to respond to the energy crisis and serve as an example of urban-scale appropriate technology. From its inception in 1975 until its completion in 1985, the IUH provided thousands of Bay Area residents with ideas and inspiration to redesign their homes towards sustainability. (The project is chronicled in *The Integral Urban House: Self-Reliant Living in the City*, Sierra Club Books).

Fifteen years later, the house is closed, but the idea of an integral urban house serves as a model for living in a city, reconnecting with the basics of life, and "treading lightly on the earth." The House implemented new design and encouraged new attitudes towards conserving energy, reducing water consumption, raising plants and animals in the backyard for food, and recycling.

The redesign of the older home entailed creating structural additions, room furnishings, and appliances that moderated climate and conserved energy resources, so that the least amount of supplemental energy would be needed for us to live comfortably within the house and reuse the greatest possible amount of materials.

The first step was weatherization; caulking and sealing of doorways, windows, and other entryways for drafts. Then the sun's energy was harnessed to heat the indoor space and water for the household. Simple use of solar panels and a heat-storing bottle wall provided energy to heat both the indoor environment and tap water. Opening the window curtains on the sunny side of the house during the day and closing them at night, and using insulated shutters held the heat in. The reverse procedure kept the heat out. Strategic placement of outdoor landscaping provided windbreaks from the winter weather and shade in the summer months.

Appliances such as the refrigerator, dishwasher, stove, and oven are major consumers of energy. (Goldbeck's *The Smart Kitchen* lists qualities to look for when choosing kitchen appliances. Efficiency guides for major appliances are available through the American Council for an Energy Efficient Economy. Top rated household appliances are also listed in the Consumer Guide at the back of this book.) Modification of ingrained habits can also save you money. Simple things like labelling the contents of your refrigerator to reduce time spent searching for items with the refrigerator door wide open, and using the "air dry" option on your dishwasher can reduce your energy bill. Electronic ignitions for gas appliances can be installed as a safety and energy-saving feature. Energy-efficient fluorescent light fixtures, especially the new compact "warm white" fluorescent lights, lack the annoying flicker and glare of older versions. A few brands are available that do not contain radioactive elements, and are thus less of a concern in disposal.

Another goal of the House was to conserve water by carrying it from source-point (rain, well, municipal delivery system) through the system, reusing it as often as possible before it left the property. The typical toilet system, which uses many gallons of water to transport feces to sites unseen, was replaced by a Clivus Multrum composting toilet which uses no water and recycles human waste to supplement the garden soil. (Low-flush toilets and water-saving attachments for toilets, or just putting a couple of bricks in the tank to displace some of the water, are other alternatives). Aerators and flow restrictors were used on kitchen and bathroom faucets to reduce the amount of water used without compromising water pressure.

Water was further recycled in the House by diverting greywater from the bathroom fixtures to the garden. This had a positive side-effect; using greywater to water plants required us to limit use of chloride bleach and boron. Of course, much water can be conserved by merely using a cup when brushing your teeth, or washing dishes and vegetables in a tub instead of letting the water run. Outdoors, watering deeply and infrequently instead of sprinkling every day makes better use of water. Trickle or drip-irrigation systems are the best for this purpose. They put the water in the ground, where you want it, rather than into the air to evaporate.

SOLAR ENERGY
SOLAR COLLECTORS capture the sun's radiant energy for heating household water. A flat-plate collector on the southern roof daily heats 120 gallons of water to temperatures above 140°. A small electric water heater provides a back-up on cloudy days
A "BOTTLE WALL" in the southern window of the bathroom employs the principle of "thermal lag" to store the sun's energy to moderate internal temperatures
A SOLAR OVEN warms and cooks food produced in the vegetable garden, in the animal yard, and breads from the kitchen
FOOD RAISING
DOMESTICATED BEES produce honey and pollinate vegetable crops and fruit trees. An observation hive provides an inside view of the honey bee's life
A VEGETABLE GARDEN, based on labor saving and environmentally sound techniques of food raising, yields produce enough for the family of four
AN AQUACULTURE POND tests the feasibility of raising fish as a supplementary protein source
A ROOF TOP GARDEN utilizes otherwise nonproductive space for raising vegetables in light weight planters
A GREENHOUSE provides a warm protective environment for germinating seeds and raising tomatoes and cucumbers in the winter
ORNAMENTAL CROPS demonstrate how food can be produced by planting a landscape of dwarf fruit trees, herbs, and edible flowering foliage
INSECTS are controlled using biological and cultural methods of pest management. No synthetic pesticides are used
RABBITS AND CHICKENS, housed in sanitary pens on the cool north side of the house, provide a dependable source of high quality protein. Much of their diet is raised on the premises, and their wastes are recycled to the soil
A KITCHEN PANTRY provides storage for garden surpluses preserved by canning, pickling, and drying
WASTE RECYCLING
HUMANS
THE CLIVUS MULTRUM waterless toilet converts human excrement into a pathogenically safe soil conditioner for use on fruit trees and ornamental crops. The process conserves water and recycles nutrients
SOLID WASTES, such as glass, aluminum, tin, and newspaper, are sorted into bins and delivered to neighborhood recycling centers
HOUSEHOLD WASTE WATER from wash basins and the shower is filtered, mixed with human urine, and reused as a garden irrigation water rich in nutrients
COMPOSTING wastes and returning their nutrients to the soil is a central theme in the Integral House. A variety of biological systems transform garden, animal and kitchen wastes into a valuable soil amendment
AT

INTEGRAL URBAN HOUSE

Another goal of the Integral Urban House was to increase self-reliance by backyard food production. Outdoor areas were designed to maximize use of space in producing food with the least labor input, using shaded areas for compost bins, animal cages, and storage of compostable materials until processed, and providing space for drying clothes. Rabbit cages were placed above chicken coops so rabbit dropping could be picked over by the chickens. Alfalfa, instead of the typical lawn, was raised for rabbit fare. A bee hive provided honey and was situated over the aquaculture pond so the dead bees could be consumed by the fish. Fresh vegetables, herbs, and fruit trees were raised in the garden and organized so that only 3-4 hours a week were needed for maintainance. The kitchen was designed to accomodate the cleaning and preparation of garden vegetables, and the storage of foods in bulk.

Both organic and inorganic materials were recycled. According to Garbage Magazine (July/August 1990), 20-30 percent of residential trash is kitchen scraps and garden debris—organic materials that can be composted to be used as fertilizer on the plants and trees in the garden. The kitchen must be designed to make the sorting of aluminum, "tin" cans, paper (newspaper, cardboard), glass, plastics, and the depositing of organic compostables convenient. Other discards—furniture, appliances, rugs, batteries—were given away for reuse.

Today the ideas of the Integral Urban House have been advanced by the increasing attention paid to the use of natural materials. New forms of the family unit, created by cultural changes and by economic necessities, have entered the picture.

The concept of sustainability has been broadened by these concerns, but the original issues addressed by the project—energy conservation, water use, and the importance of a closer relationship to food—remain as crucial as they ever were.

Although the Integral Urban House is no longer functioning as a model open to the public, Farallones Institute is moving to provide design workshops for homeowners who wish to remodel for sustainability. Look for announcements for these Home Design Workshops, to begin this winter at Fort Mason Center in San Francisco.

Sim Van der Ryn is an architect and a founder of the Farallones Institute, a non-profit educational organization located in Occidental, California. The Institute provides classes and workshops in subjects ranging from renewable energy technologies to edible landscaping. Therese Peffer works with Mr. Van der Ryn.

The Hassles of Household Hazardous Waste

By Stephen C. Evans

Every home has some. It fills the cabinet under the kitchen sink, clutters shelves in the back closet, and takes over in the basement and the gardening shed. More and more of it seems to accumulate each year.

"It" is household hazardous waste (known in the trade as HHW), and it comes in many forms: old paint, used motor oil, solvents, strippers, cleaners, pesticides, fungicides — anything, in fact, marked with signal words such as "caution," "danger," or "warning." It is the stuff of our day-to-day lives, and it poses a great threat to the environment.

As long as you have some use for these substances, you can always store and use them up carefully. But what happens when you want to remodel the kitchen, put a darkroom in that back closet, or another bedroom in the basement? In other words, what do you do when you just want to get rid of the stuff?

The law says you can't dump anything with those signal words on the label in the trash, down the sink, or onto the ground. Nonetheless, that is exactly what happens to huge amounts of it each year.

Although exact figures are impossible to come by, the state's Waste Management Board estimates that a city the size of San Jose (population 700,000) discharges about 21 tons of toilet bowl cleaner, 96 tons of liquid household cleaners, and 24.5 tons of used oil into the sewage system each year. The quantities of hazardous waste being surreptitiously or unwittingly thrown into the trash undoubtedly amount to much more. As for the amount of used motor oil that is illegally dumped, estimates vary from the equivalent of 24 to 36 Exxon Valdez's nationwide each year.

In other words, the worst polluters may not be the big oil companies and manufacturing corporations after all. The worst polluters may well be the rest of us.

The reason we are all breaking the law so routinely is simple: there are few practical alternatives. The situation in the Bay Area is a real case in point. There is only one permanent free household hazardous waste drop-off point in the entire region, and it is exclusively for the use of San Francisco residents. The only business that regularly accepts household hazardous waste for pay is Bay Area Environmental in Richmond, where you pay $10 per gallon to dispose of it, with a $50 minimum

Most of the counties and many cities do run occasional HHW collection days, but in many places their scheduling is erratic and they are usually poorly publicized (see the sidebar for the number to call about collection dates in your county). This is no accident; often only those citizens who request notification are told about them. Collecting, treating, and disposing of the waste is terrifically expensive. Counties and cities already have trouble footing the bill for their present modest efforts. The most recent HHW collection day in San Jose, for instance, was not widely publicized but was nonetheless "totally swamped," according to county program analyst Paula Stoner. The day ended up costing about $335,000 — approximately $150 per car load. That's about average for collection days in the Bay Area. The high costs have made some jurisdictions decide to go years without scheduling a collection day; the last county-wide drop-off day in Alameda County, for example, was in 1987, although various cities in the county have held their own days since then.

This picture should change soon. A new state law, AB939, requires counties to come up with a plan for handling household hazardous wastes by the end of 1991. These plans must include a source reduction and recycling element that will handle 25 percent of the waste by 1995. However, that's about all the new law requires. It does not mandate that drop-off centers or even collection days be part of the plans; it is entirely up to the local jurisdictions to determine what their needs are and how to deal with them.

Most of the planners charged with coming up with the new HHW program are looking closely at the San Francisco drop-off facility, located at 501 Tunnel Avenue in San Francisco, as a model. Originally opened on an experimental basis in January, 1988, the facility was last year deemed a success and given the green light for permanent operation.

The Sanitary Fill Company, which runs the San Francisco center, has succeeded in recycling or reusing 65 percent of the material brought to it. Most of what is recycled is house paint, which amounts to a full 60 percent of the waste deposited. Latex paint is sent to a Standard Brands factory, where it is thrown together and reconstituted into an institutional beige mixture. The city is under contract to buy this product back from Standard Brands, and it is ultimately used to paint city buildings and donated to various organizations for "graffiti abatement."

Oil-based paints and solvents are sent to the California Solvent Recycling Company in Palo Alto. There they are converted into an alternative fuel which is used to fire a cement kiln in southern California.

What happens to more toxic materials that can't be recycled is somewhat more complex and a lot more expensive. Such materials must be specially packed in absorbent materials ("lab-packed"), sealed in steel drums, and shipped to the lone surviving hazardous waste dump in all of California, the Kettleman Hills landfill in the Central Valley. Things that must be treated this way include flammable solids, pesticides, fungicides, aerosols, dioxin precursors, acids, and heavy metals.

Several Bay Area counties, including Alameda, Contra Costa, San Mateo, Santa Clara, and Santa Cruz are seriously considering plans to set up similar drop-off centers. Contra Costa and Santa Clara counties are both considering making such a center part of the regular transfer station, so that going to the dump can be a one-stop errand. Alameda County officials want to create three HHW sites, strategically located in the northern, southern, and eastern parts of the county.

One of the stipulations of AB939 is that localities provide a way of disposing of household hazardous waste free of charge, because charging the actual disposal costs would deter people from using the service. Costs therefore will be spread to everyone via garbage bills or other forms of taxation. Alameda County officials estimate that the rate hike will amount to about five dollars per year.

Providing a cheap and easy way of dumping our hazardous wastes doesn't really solve the problem. In fact, making dumping easier will very likely result in more toxins entering the waste stream.

That's why many of the agencies involved in planning the new HHW programs are actively seeking ways of reducing household hazardous waste at the source — that is, all of us. Here are some things you can do to avoid being part of the problem:

• Don't buy hazardous products in the first place. Seek safer alternatives, and buy them even if they cost a bit more. One of the cheapest alternatives to many toxic products is good, old-fashioned elbow grease. Remember that the ultimate costs are a lot lower if you avoid damaging the environment. Absolutely spurn poisons — pesticides, fungicides, etc. — as even their normal use puts toxins into the environment.

• Most household hazardous waste becomes waste only when you decide that's what it is. If there is any chance you will have a use for a hazardous material in the future, don't throw it away. Store it in a safe place, and use it another day.

• If you must get rid of it, try to give it to a friend or organization who needs it, rather than throwing it away. If it can be recycled, make sure that is what happens to it. Paints,uncontaminated solvents, motor oil, and automobile batteries can all be recycled.

County Household Hazardous Waste Programs

ALAMEDA COUNTY
Department of Environmental Health, (415) 271-4300
Developing an HHW program which will include 3 drop-off points by late 1991, but no program at present. Numerous cities within the county sponsor periodic collection days.

CONTRA COSTA COUNTY
Department of Health, (415) 646-2286
Periodic collection days for used oil, latex paint, and lead acid batteries. Call above for locations and times.

MARIN COUNTY
County Planning Department, (415) 499-6269
Periodic collection days. Call for exact locations and times.

NAPA COUNTY
Hazardous Materials Section (707) 253-4269
Program limited to latex, car batteries and waste oil. Car batteries can be deposited at Clover flat near Calistoga or at the American Canyon landfill Oil recycling is available at Clover Flat or Napa Recycling. Planning oil-based paint recycling at the American Canyon landfill.

SAN FRANCISCO COUNTY
Sanitary Fill Company/NORCAL, (415) 468-2442
Permanent HHW collection facility at 501 Tunnel Ave. open only to residents of San Francisco. Open Thursday, Friday, and Saturday, 8AM to 4PM.

SAN MATEO COUNTY
Environmental Health DIvision, (415) 363-4718
Two or three collection days per month. Call to make an appointment, available between 8:30am and 3:00pm.

SANTA CLARA COUNTY
Hazardous Waste Management Program (408) 441-1195 or (408) 299-6930 for unincorporated areas
Periodic collection days. Call for locations and times. Planning permanent facility and a mobile collection unit.

SANTA CRUZ COUNTY
Environmental Health Services, (408) 425-2341
Periodic collection days. Planning permanent collection facility.

SOLANO COUNTY
Department of Environmental Management, (707) 421-6770
No HHW collection program. Contact above for information on where to dispose of HHW materials and for educational information. Limited permanent facility for recyclables — used motor oil, latex paint, and lead acid automotive batteries — open to residents of Solano County. Operated by Pleasant Hill Bayshore Disposal, 441 N. Buchanan Circle, Pacheco. Phone (707) 745-3242.

SONOMA COUNTY
Environmental Health Department (707) 525-6565
No HHW collection program. Call to find out about plans.

Household Toxics Guide

PRODUCT	SAFE DISPOSAL	BETTER ALTERNATIVES
Antifreeze Ethylene glycol	If local sewage plant can break it down (check first) dilute and pour down drain.	None. Some cars don't need antifreeze unless temperatures drop below freezing
Batteries - automotive lead, sulfuric acid	Recycle at service station, collection event, or with a recycler.	None
Batteries - household cadmium, lithium, mercury	Take to a collection event or retailer who collects them for re-processing.	Use rechargeable batteries or house current.
Disinfectants Diethylene or methylene glycol, cresol, phenols, sodium hypochlorite	Dilute greatly and flush down drain if served by a sewer system	One-half cup borax in a gallon of hot water.
Drain openers hypochloric acid, lye, petroleum distillates	Dilute and flush down drain. With septic tanks, dilute and flush over several days.	A quarter-cup of baking soda mixed with a half-cup of white vinegar; plunger.
Finishes & stains glycol ethers, halogenated hydrocarbons, ketones, naptha, petroleum distillates	Donate to someone who can use them or take to a collection event or site.	Natural earth-pigment finishes or water-based paint.
Furniture & Floor Polish Diethylene glycol, niotrobenzene, petroleum distillates	Give away or take to a HHWcollection event or site.	Mix two parts vegetable oil to one part lemon juice.
Metal primers methylene chloride, petroleum distillates, toluene	Give away or take to a HHW collection event or site.	Unknown
Mothballs Paradichlorobenzene	Keep in air-tight container until collection event.	Store in a cedar chest, or use cedar chips.
Motor oil hydrocarbons such as benzene, heavy metals	Recycle	None at present; there are some promising experiments with vegetable oils.
Oven cleaners ammonia, potassium or sodium hydroxide	Can be diluted with water and slowly flushed down drain.	Baking soda in water and a scouring pad.
Paints, oil-based aliphatic hydrocarbons, ethylene, fungicides, pigments	Take to an HHW collection site or event for recycling.	Milk or latex-based paint; plant and wood-oil based paints.
Pesticides Broad range of toxins, including carbamates, chlorinated hydrocarbons, and organophosphates	Take to an HHW collection site or event for final disposal at special landfill.	Physical barriers and biological controls. See article on integrated pest management in the Home and Garden chapter.
Strippers Acetone, alcohols, methylene chloride, methyl ethyl ketone, toluene, xylene	Donate to furniture refinisher or take to collection event or site.	Heat gun, sand paper, water-based strippers.
Thinners acetone, ketone, methyl isobutyl, N-butyl alcohol, petroleum distillates	Let sludge settle and funnel off liquid for reuse. Sludge must be taken to collection site or event.	Use water-based paints, which can be diluted with water.
Wood preservatives arsenic compounds, creosote, chlorinated phenols	Take to collection site or event.	Rot-resistant wood, water-based preservatives.

Some Tips for Energy Efficiency at Home

By Sunday Oliver

Consuming energy is something most of us do most of the time without even being aware of it. But a few simple changes in home fixtures and personal habits can lead to significant energy savings.

• Water heating is one of the areas where being energy-conscious is less expensive and more convenient than being an energy hog. Even if you wrap a blanket around your conventional water heater, it's still heating water 24 hours a day—whether you're there to use it or not. And it can only give you as much hot water as there is in the tank.

Tankless heaters, such as those used in Western Europe and Japan where energy costs are much higher, ignite only when you turn on the hot water tap. Water passes through heated tubing and arrives at your faucet steaming hot. Their best feature is the ready , steady supply of hot water as long as you want it.

These highly efficient units can cut your water-heating bill 20 to 50 percent. Since a water heater is the second biggest energy consumer in your house, that can mean a big difference in the total bill. If your present heater is electric, a tankless heater can pay for itself in two years in savings alone. In the average house, a tankless heater can save more than $100 a year over a conventional gas heater; more than $150 over an LP heater; and more than $450 over an electric heater.

• When it starts to get cool, it's time to make sure your house will keep you warm with minimal artificial heating. In the fall, take down the screens and wash windows. This will give you 30 percent more light, as well as passive solar heat. Close attic vents, seal off swamp coolers, and cover them up. If you have roof vents, cover them too.

Draw curtains or shades in the evening to keep heat from leaking out of windows. Change the heater thermostat to a programmable time, so it's on low when people aren't around. If you have a gas heater, re-light the pilots, and check the furnace filters. If you rely on woodstove heat, be sure it's getting its draft from the outside—or it will pull cold air into the room through door jambs, floors, or any other available crack. (Now is a good time to clean out the creosote in the chimney, too).

Check the local utility company for credits on attic insulation, double-paned windows, and other heat savers.

• You probably don't need any air conditioning to keep cool in the summertime. All it really takes is an understanding of basic solar design principles, and a little common sense.

Make sure your house is "breathing" properly. Keep doors and windows closed during the day; in the evening, open them up to flush out hot air (cross ventilation makes the job more thorough). Use table, window, and ceiling fans if your house doesn't breathe well. If you go to bed late, close the house up again to seal in the cool. If you get up very early, close it up then.

Make sure gable or attic vent fans are free and clear—and if you don't have one, get one. These alone can make your house 20 percent cooler. Then make sure the hot air is going where you want it to (out), try this trick: light a stick of incense and watch which way the smoke goes.

If your house doesn't have a good overhang on the south side, shades or awnings will make a big difference. Shades and awnings all around will help reduce daytime temperatures significantly. Don't forget shades for skylights.

• Evaporative coolers (better known as swamp coolers) work very well in a dry climate—and they're about ten times as efficient as air conditioning. (In dry areas, there's usually a nighttime temperature drop which makes cooling without air conditioning especially easy.) If your house doesn't have any trees to shade it, a cooler may be the best solution, along with vents and shades.

• Trees are one of the best evaporative coolers around. Before you cut down trees for that beautiful view, consider these facts: one good shade tree can make as much cool air as ten room-size air conditioners running twenty hours a day. A good-size tree also makes enough oxygen for four people for a year—and every tree planted is part of the fight against the greenhouse effect.

Plant on the west and south sides of your house for best cooling in summer. The best shade trees drop their leaves in the fall, so you can take full advantage of solar heat in winter. Good fast-growing shade trees for this area are catalpa, fruitless mulberry, silver maple, birch, and red-flowering plum. Fruit trees—apple, plum, peach, pear—are also good choices.

Once trees are thoroughly established, they become drought-resistant. Trees need to be watered thoroughly, slowly, and infrequently. Getting professional advice on trees for your particular situation helps avoid frustration. And it gives you more time for lying in the shade.

Sunday Oliver is a home energy-effieciency expert and alternative energy consultant living in Nevada City, CA.

The Changing of the Bulb

The Advantages of Compact Fluorescent Lightbulbs

By Lisa Schiffman

For the most part, people are still in the dark about compact fluorescents. They hear the word "fluorescent," and think of the flicker, the buzz, the harsh light in their fourth-grade classroom. Tired of the barrage of eco-savings statistics, they raise their eyebrows at the $20 average price tag and the acclaimed $50 savings in energy one can expect over the life of the bulb.

What are the facts? According to Chris Calwell, research associate for the Natural Resources Defense Council, substituting a compact fluorescent for a standard bulb saves the atmosphere from a half-ton of carbon dioxide (CO_2), and other so-called greenhouse gases emitted during the bulb's lifespan. Coal-burning electric power plants expel the pollutants into the atmosphere, amplifying the problems of acid rain and global warming.

The standard incandescent hasn't changed much from the device Edison invented in 1879. Producing light by heating the thin wire inside the bulb, incandescents are ecologically unsound. Almost 90% of the electricity in Edison's bulb goes towards generating heat. Plus, it burns itself out after 750 hours. A compact fluorescent lasts for up to 10,000 hours, and uses one fourth the energy of incandescents — without sacrificing the quality of the light.

In the early eighties, the compact fluorescents produced were heavy, odd-shaped glass bulbs that often didn't fit properly anywhere. Like small torpedoes, they jutted out from standard fixtures. Now the major manufacturers, including Osram, Philips, and Panasonic, have honed the technology and housed it in sleek, aesthetically-pleasing bulbs.

These energy-efficient fluorescents rely on electronic, instead of electromagnetic, ballasts to start and operate the lamps. In the past, all ballasts were electromagnetic, consisting of a core of steel laminations surrounded by two copper or aluminum coils. The vibrations from the electromagnetic field were created by the core and coil, and they emitted a steady drone. Electronic ballasts use electronic components, virtually eliminating the hum. They weigh up to 50 percent less than electromagnetic ballasts and give off less heat.

Some utility companies are providing customers with discounts on the bulbs, and other energy saving incentives as well. According to the Rocky Mountain Institute, more than 60 utilities now offer rebate programs to encourage efficient use of energy. Some propose giveaways. Southern California Edison, for example, has handed out more than 800,000 compact fluorescents. When customers reduce energy usage, utilities benefit. They're able to reduce operating costs. They can save billions of dollars in construction costs for the building of new plants.

Utilities, recognizing that the greatest economic and environmental gains would come from the businesses, have mostly concentrated their efforts in the commercial sector. Owners of supermarket chains with hundreds of ceiling lights and whopping energy bills can see the advantages of buying bulbs that last for years and save on electricity costs.

What about residential light users? They gain by installing fluorescents in fixtures which are turned on for at least two hours a day. Still, consumers hesitate. Who wants to spend more than two dollars for a household bulb, and then wait around for a rebate? Old purchasing habits die hard, and only an educated consumer will pass up the familiar incandescent and find out which eco-catalogue or retailer that sells compact fluorescents.

At present, compact fluorescents are not easy to find. Many retailers cite low turnover, low demand, and a concern for sales per square foot as reasons not to carry them(See "lighting" in the Consumer Guide for a local list). Who wants to stock a bulb that lasts for roughly 6 years? Consumers need to be vocal, and pressure retailers to carry fluorescents and other energy efficient appliances. Stores will react if they know the demand exists.

Beyond using compact fluorescents, individuals and industries must reduce their energy appetite. Nationwide, government agencies, utilities, and consumers need to design and follow strategies for energy efficacy. A recent *Scientific American* article by John P. Holdren, professor of energy and resources at the University of California, Berkeley, reports that if people outside of the United States used energy the way we do, global energy use in 1990 would be more than four times as large as it is now.

Individuals can start small, by being open to new lighting and other super efficient technologies. They can contact their local utility, and ask about their energy saving recommendations and rebates. Finally, when that old bulb in the table lamp blows, they can screw in a compact fluorescent.

Lisa Schiffman is a researcher for the Natural Resources Defense Council and a freelance writer.

The Why and How of California Native Plants

By Mia Amato

Oh! That's so beautiful! What is it?" she shrieked, taken aback by the beauty of a towering, flowering plant. I thought to myself, "that's a common enough reaction from people I take to visit the Berkeley Rose Garden." It's a favorite spot, a terraced hillside that's one grand, sweeping swath of roses in every color.

But my visitors were from Manhattan and not easily impressed. And what caused my friend to drop her jaw was not a rose at all.

She was staring at a common weed, too far back in the bramble of rose hedges for the caretaker to remove. I recognized it as *Romneya coulteri*, a native to California, sometimes called Matilija poppy. Its white flowers are spectacular, up to nine inches wide, with shiny, silken petals topped by a dense, round pompom of golden stamens — the characteristics which evoke its much more common name — the fried-egg plant.

Native California plants and flowers have come into vogue in recent years, largely because of the long drought. In the absence of rain, the search has been on for garden plants that will take care of themselves without regular watering. Natives fit the bill.

Plants that 'grew up' adapting to the environmental cycles of our region are more than just water-savers, though. Some (like the wiry manzanitas and succulents) retard the spread of wildfires; others (like the fried-egg plant) have strong, spreading root systems that hold eroding soil on hillsides. Many (like the ubiquitous anise) are necessary for birds and insects specific to the region.

And the colors! The famed British borders of herbaceous perennials seem dim once you realize what blooms under blazing West Coast sunshine. Sage (*Salvia*) in bright reds, blues, purple and white. Drooping pink panicles of our local flowering currant (*Ribes sanguineum*). The neon-orange glow of California Poppy (*Eschscholzia*). One could go on, but better to see them yourself. Large displays of flowering native plants are fixtures at both the University of California Botanical Garden in Berkeley and the Strybing Arboretum in San Francisco's Golden Gate Park.

From Wilds To Backyard

Finding native California plants for your own garden isn't as easy as heading for the wilderness and digging up some pretty flowers. That's a practice not only frowned upon (if everybody did it there'd be nothing left!) but is in fact illegal.

Neither is the process as haphazard as simple digging. Universities and commercial growers are actively researching outstanding varieties of native plants. One goal: hybridizing the useful ones to create "garden variety" versions with all the vigor and good traits of the homeboys. Several Bay Area nurseries now specialize in native species, and conveniently label each plant with directions for cultivation.

The Saratoga Horticultural Foundation acts as a bridge between the plant-finders and the landscape industry. The Foundation has two ten-acre sites in the Santa Cruz area, plus innumerable test spots throughout California, where native plants are "developed" for eventual sale in garden centers.

"It takes about a year to seven years for a new plant to catch on," says Lowell Cordas, director of the Foundation. Plants with potential are offered to nursery owners, who buy them in small lots to try them out. Using methods suggested by the foundation, they'll also propogate the plants in their own greenhouses, perpetuating the supply as the demand for certain new species increases.

About 40 of the several hundred varieties of native plants propagated by the Foundation have become mainstays

of the nursery trade in California, according to Cordas. Two hybrids have gone on to become real stars. They are the mounding manzanita (*arctostaphylos densiflora*), named 'Howard McMinn,' and a pretty blue-flowered Pacific lilac (*ceanothus*) named 'Yankee Point.'

Just about every commercially available variety of the graceful landscape tree known as liquidambar (also called sweet gum) was developed by the Foundation, says Cordas. "'Burgundy,' 'Palo Alto,' 'Festival,' Autumn Gold' — they're all ours."

Saratoga's offerings are wholesale and to the trade only. But twice a year, members of the Foundation flock to a plant sale where they can buy unusual and exciting plants individually. Other good places to find unusual varieties are Strybing Arboretum and U.C.'s Botanical Garden, which hold seasonal sales of their excess plants to raise funds.

Caring For Natives

Plants native to the California Coast thrive best in our natural climate: semi-arid high desert, with a touch of Mediteranean balminess, and a good dose of nightly fog. No rain from May to October is — well, natural to them.

The worst thing you can do for a native plant is water it too much, especially in the summertime. The very features that make a plant drought-resistent also make it susceptible to disease when the ground is too wet.

This is why 'Tam' junipers, and other evolved members of the local juniper family, die a gruesome death in many a foundation planting. They can't take the water when a well-meaning homeowner aims the hose at them. They get root rot — a fungus disease that's so hard to eradicate the only solution is digging up the Tams and putting something else there instead.

A better idea is to identify your landscape plants and look up their culture in a book like Sunset's *Western Garden Book*. Then all you may have to do is leave them alone.

One of my favorite professors at the University of California is a tree expert. During the course year, he likes to liven up a seminar on tree care with this series of slides:

The first slide shows two large, live oak trees on a U.C. campus, near a building framed by a sweeping lawn. The tree on the right doesn't look so good.

The next slide, taken the next year, illustrates that the tree that seemed to do poorly is obviously dying. And now the oak on the left isn't looking so chipper either.

The third slide, taken the third year, shows only one tree, on the left..

"The one on the right died and had to be removed, at some cost," says the teacher. "What caused the oaks to decline?" the students want to know. Diagnosis: the trees are too near a lawn irrigation system. The same amount of water that keeps the campus lawn so lush and green is rotting the roots of the native oaks.

The last slide, taken in the fourth year, shows the remaining oak — hale, hearty, and bursting with green leaves. Strangely enough, the lawn doesn't look so good.

What happened?

"It's simple," says the teacher. "That was the first year of our drought. The University was forced to save water so the lawn irrigation system was turned off for the summer."

That anecdote, I think, tells you all you need to know about native California plants. They're smarter than we are. After all, they were here first.

Mia Amato is a certified Master Gardener from the UC Cooperative Extension in Berkeley and a freelance writer.

Sources and Resources:

Western Garden Book, Sunset Publishing, Menlo Park, CA. 94025

Strybing Arboretum
Golden Gate Park
9th Avenue & Lincoln Way
San Francisco, CA 94122 (415) 661-0668

University of California
Botanical Gardens
Centennial Drive
Berkeley, CA 94720 (415) 642-0849

Saratoga Horticultural Foundation
15185 Murphy Avenue
San Martin, CA 95046 (408)779-3303

Nurseries That Specialize In Native Plants:

Wildwood Farm
10300 Sonoma Highway
Kenwood, CA (707)833-1161

Mostly Natives Nursery
27115 Highway One
Box 258
Tomales, CA 94971 (707)878-2009

California Flora
Somers & D Street (P.O. Box 3)
Fulton, CA 95435 (707) 528-8813

See the Garden section of the Consumer Guide for more..

"Heirloom" Vegetables:

Maintaining Genetic Diversity in Food Plants

By Mia Amato

Earnest environmentalists active in saving whales and preserving rain forests have begun to turn their attention to another group of endangered species: vegetables and grains. The genetic diversity of food plants, especially in the U.S., has dwindled dramatically in the past century as big agribusiness has come to dominate the seed marketplace.

In the search for bigger and "better" vegetables, commercial seed companies have focused on development of hybrids. A big motivator for the seed companies is that hybrid seed, the product of carefully cross-breeding specific plants for particular qualities, usually produces infertile plants and must therefore be bought anew by the farmer each year.

Since the hybrids are bred for high productivity and good appearance, among other factors, using them has made sense to farmers as well.

But almost universal reliance on hybrids has created a dangerous uniformity in the genetic make-up of major food plants; the potential for a single new pest or disease wiping out entire crops without encountering resistant strains is correspondingly high.

Meanwhile, many older varieties of food plants have been forced into smaller and smaller niches, leaving many today perilously close to extinction. Once such varieties are lost, any potential beneficial attributes they may have had — whether it be resistance to frost or disease or a delicious taste — are lost with them.

Before hybrids, farmers relied on "open-pollinated" vegetable varieties, whose seeds could be economically saved and replanted from one year to the next. A few endangered vegetable varieties have survived because gardeners and small truck farmers stuck to open-pollinated types they had special luck with, often handing down the seeds from generation to generation.

Today, these are often called "heirloom" seeds. Prized antiques of the plant kingdom, heirloom vegetables are often hardier and taste better than commercial hybrids.

Much of the work on preserving heirloom seeds in the U.S. is done by members of the Seed Saver's Exchange. The Exchange helps alert people to the endangered status of old vegetable types, and catalogs an inventory of more than 5,000 kinds of rare vegetable seeds. In 1990, Exchange founders Kent and Diane Whealy, who run the network from their home and small office in Decorah, Iowa, were honored with a MacArthur Grant, the so-called "genius grant" from the John D. and Catherine T. MacArthur Foundation.

California claims one of the five national "curators" of the Seed Savers Exchange. Suzanne Ashworth, who grew up on a farm and "remembers when tomatos tasted like tomatos," gardens near Sacramento and is in charge of cole crops (kales and cabbages), eggplants, and beans.

A visit to Ashworth's garden is like entering a vegetable wonderland. There are blood-red carrots and gleaming white ones, vines hung with passionately purple hyacinth beans, and Italian kales that grow in five-foot-high, fountain-like plumes. One of her Bay Area connections is Alice Waters, the owner of Chez Panisse. Ashworth often supplies unusual vegetables for tastings and use in the Berkeley restaurant, which is known for its inventive use of produce; she recalls once sending an emergency shipment of tiny, jewel-like red currant tomatos, in a box of dampened paper towels, to the Bay Area by Greyhound Bus.

"Most gardeners like their heirlooms because they are so beautiful," says Ashworth. "Much too beautiful to become extinct."

Candy-striped eggplants such as *'Listada de Gandia'* and the marble-white 'Ghostbusters' occasionally seen in specialty groceries are one direct result of the Seed Saver's Exchange work in promoting antique eggplants. Pale shades once were common, says Ashworth, until a fad for purple eggplant prompted commercial seed companies to "select out" [breed for] the color typically found in supermarkets.

When grocery owners discovered that dark eggplant bruised less when handled, and hence could be held in store bins much longer, pale eggplants disappeared.

"The original colors of eggplant were white and green, and these are milder and much less bitter than purple ones," she notes. "If you've never liked purple eggplant, you should try a white or green kind." These do not have to be salted and left to "sweat" before cooking, she adds.

An unusual eggplant in her garden is a large vine studded with tiny, hard, button-like fruits of a very dark green. The seeds came from Southeast Asian immigrants who have settled in the Sacramento area.

"The fruits are very bitter and would taste terrible to most people," Ashworth points out, but the plant is Exhibit A in her efforts to preserve a diverse eggplant gene pool. The vine appears to be tolerant in cold weather, even standing up to light frosts, which would shrivel and kill most Mediterranean varieties of eggplant. Perpetuating this Southeast Asian type for cross-breeding might eventually produce a cool-weather eggplant that could be grown, say, in San Francisco fog.

The San Francisco League of Urban Gardeners (S.L.U.G.) began experimenting in 1987 with heirloom tomato seeds in various community gardens around the city. S.L.U.G. member Pam Peirce, who is just about to go to press with a new book on Bay Area gardening (see below), supervised the tomato testing program. She says one goal of the project was to find out which tomato variety would do best in the area's chilly summers.

"One we tried we called 'Visitacion Valley,' because the seed came, I think, from someone who had been planting it for years in the community garden in that part of the city," says Peirce. "It was similar to [the commercial variety] Early Girl and probably had been selected from Early Girl at some point in the past."

In S.L.U.G. tests in 1987-1990, 'Visitacion Valley' was "slightly rated better on flavor" than Early Girl, Peirce notes, "but it did not have as quite as large a tomato or as many tomatos as Early Girl."

Her full report, published in S.L.U.G.'s Spring 1989 newsletter, suggests 'Visitacion Valley' may not be so superior as to warrant widespread distribution; its chief merit is that as a non-hybrid, open-pollinated variety, the seed can be saved by the self-sufficient gardener for use from year to year. Some S.L.U.G. members still grow the seed for fun.

Joining the Seed Savers Exchange is a good way for the average gardener to get involved in the important task of maintaining genetic diversity among food plants while having good fun at the same time. A $15 annual membership fee pays for a newsletter and the omnibus catalog that supplies names and addresses of gardeners who have heirloom seeds.

"You can buy seed from any member, grow it out, save the seed and re-offer the seed on the Exchange the next year," says Ashworth. Once in the "loop," she adds, members can then exchange seeds for the price of postage only. Sometimes the organization puts out a call for help to gardeners to grow out seeds that are especially in danger of becoming extinct.

Another, more informal network of heirloom seed exchange is *National Gardening*, a magazine which publishes a monthly list of requests for seed and offers of seed from gardeners across the country.

Whether you have only a few potted plants on a patio or a veritable victory garden out back, growing heirloom vegetables is a marvelous way to benefit not only yourself, your family, and your friends by putting fresh and unusual produce on the table; it is also a sort of insurance for the future of all humanity. Genetic diversity means strength in the face of calamity. Worldwide, many varieties of food plants are quickly disappearing. You can play an important part in reversing the trend.

Who knows? Perhaps the seeds that will save the day against some plant disease of the future will come from your own back yard

Mia Amato is a certified Master Gardener from the UC Cooperative Extension in Berkeley and a freelance writer.

Sources and Resources:

ORGANIZATIONS:

Seed Saver's Exchange
(Membership: $15 a year)
RR 3, Box 239
Decorah, Iowa 52101

READING:

The Joys of Home Hybridizing
Horticulture, August 1990.
(20 Park Plaza, Suite 1220,
Boston, MA 02116

Seed Swap
various issues of *National Gardening*
180 Flynn Avenue
Burlington, VT 05401.

The Heirloom Gardener
by Carolyn Jabs. 1984
Sierra Club Books.

Artichokes and Beyond,
by Pam Peirce, 1991
Heyday Press
(CALL (415) 549-0993 for release date and availability.)

Seed Saver's Exchange: The First 10 Years
edited by Ken Whealy and Arllys Adelman, 1986.
Seed Saver's Exchange. Contains many articles, including the Seed Saving Guide. Also available: *Seed To Seed*, by Suzanne Ashworth, 1990.

SOURCES OF HEIRLOOM SEEDS:

Ornamental Edibles
(catalog $1)
3622 Weedin Court
San Jose, CA 95132 (408) 946-SEED

Johnny's Selected Seed
Foss Hill Road
Albion, ME 04901

Shepherd's Garden Seeds
(call for catalog)
6116 Highway 9
Felton, CA 95018 (408) 335-3638

Seeds Blum
(catalog $3)
Idaho City Stage
Boise, ID 83706

Abundant Life Seed Foundation
(catalog $l)
P.O. Box 772
Port Townsend, WA 98368

Territorial Seed Co.
(Catalog $1)
PO Box 27
Lorane, OR 97451

Planting a Dwarf Fruit Tree Orchard

By Mia Amato

There are plenty of reasons to grow your own fruit in an in-home orchard. Fruit trees are lovely in the garden, with pretty flowers in spring, leafy branches for shade, and bright yellow and red globes of fruit in the fall. You know that the fruit your family picks and eats can be totally organic and free from pesticides, and your children can learn and enjoy seeing how their favorite fruits come into being. You can make your own cider, dry your own figs, and — in the Bay Area — enjoy the taste of fresh-picked fruit every day of the year.

How much fruit can you grow? If garden space is limited to tubs and window boxes, you can still enjoy a summer's worth of fresh strawberries and blueberries and perhaps the bounty of a potted dwarf apricot — enough for a special treat. If you have a backyard, you can grow as many as eight to ten dwarf fruit trees in a 20′ by 20′ space.

Fruit Tree Selection

A true dwarf fruit tree is made by grafting a nicely fruiting variety onto the dwarfed rootstock of a similar type of tree. The result is the tree will never grow more than ten or twelve feet high. Some genetic dwarfs never get taller than five feet. Advantages to dwarf (as opposed to semi-dwarf or standard) is that the tree starts giving fruit in two or three years, instead of five or six, and the fruit is low enough so the lazy gardener can easily look after it and pick it when ripest. Dwarf trees, pruned up to six feet, do well in containers on patios and decks.

In catalogs and at your local nursery, you'll find dwarf fruit trees that run the gamut from asian pears to pomegranates to tangelos. But be forewarned: some types sold locally may simply not grow or fruit for you in the Bay Area.

The key factor is *chilling hours*. Simply put, this is roughly the number of hours in winter with an average temperature of 45 degrees Fahrenheit or below. Apples, pears, and most of the plum/apricot family need a certain minimum of chilling hours or the fruit won't set. Since the Bay Area has warm winters, varieties that require the least number of chilling hours tend to do best. (see chart) Your local extension office can tell you how many chilling hours are expected in your particular microclimate.

Don't despair if your neighborhood's microclimate precludes growing apples and pears. Gardeners in warm zones enjoy success with all the citrus fruits — oranges, tangerines, lemons, grapefruits — and can grow exotic tropicals such as avocado, pineapple guava, and dwarf banana.

The other factor for fruit set is pollination. Most dwarf apricots, peaches, and nectarines are self-pollinating, so you only have to buy one of each. The Alberta peach, however, requires cross-pollination, so you'll have to plant at least two. Apples, cherries and most

CHILLING CHART

Neighborhood	Chill hrs/yr
San Francisco downtown	197
San Francisco Sunset district	484
San Mateo	575
Richmond	614
Woodside	1281
Livermore	1296

Fruit tree	Hrs req.
Figs	
any type	100
Apricots	
Gold Kist	200
Blenheim	500
Plums	
Santa Rosa	300
Greengage	700
Peaches	
Early Amber	250
Bonanza	300
Apples	
Anna	300
Delicious	600
Granny Smith	600
Macintosh	1000

Source: UC Cooperative Extension

plums need at least two plants for cross-pollination. Sometimes you can mix and match varieties that flower at the same time.

Planting the Hole

Buy and plant bare-root trees in December and January; leafed-out trees in containers are best set out February and March. If trees are ordered by mail, dig the planting hole soon after ordering, so you'll be ready when the plant arrives. Once you have your tree, soak the root system for 24 hours in a bucket of water and set it out the next day.

Select a sunny spot that is well-drained (tip: make sure there are no standing puddles in rainy season) and dig a *big* hole. Five feet deep and a few feet wide is about right . Put loose topsoil in the bottom of the hole. If you are planting a bare-root tree, mound this loose soil into a cone shape, place the plant on top of the mound and then arrange the roots around it.

For leafed-out trees, remove the container and backfill with loose soil till the junction of the graft (seen as a knob on the stem) is 2-3 inches above the soil line.

The latest research suggests you do a young tree no favor if you add fertilizer or compost instead of original soil as you continue to backfill the hole. Use the soil already there; the tree will have to get used to it eventually. Gently pat the soil into place and water the tree well.

Resist the urge to fertilize for the first year. It may be beneficial to cut back the tree about two feet from the ground, right after planting, to enourage good branching, if this hasn't been done at the nursery.

In the second year, trim the new growth to three main limbs with a spread of about 120 degrees apart.

In the third year, prune back about a third of the new growth; by the third or fourth year you'll start to see flowers and fruit.

Container plants can be treated the same way Make sure there is an adequate drainage hole, andthat the pot is at least 20 inches in both dimensions(wide and deep).

Tree Care

Water your tree deeply at least once a week (more often for containers) after planting. An established tree in the ground should get a nice, deep soaking once a month to help develop fruit.

Don't be alarmed if fruits start dropping off your tree in the early green stages. This is a natural process of "thinning" you can help along. Removing some of the small unripe fruits allows the tree to put more energy into developing the ones that are left. Result: better, bigger fruit. The University of California recommends thinning apples and pears to one fruit every six inches; plums and apricots can be thinned to three inches apart.

Common insect pests on fruit trees are certain types of scale and aphids, which can be controlled with a dormant oil spray just before the leaf buds pop in late winter. Sticky traps will keep down populations of whitefly and codling moth; other caterpillars can be controlled with *Bacillus thuringiensis* (Bt). A typical problem is a tree full of fruits with tiny holes. Birds are the culprits. Netting or hanging light-reflecting aluminum pie pans will keep them away.

More serious problems are fungus diseases which spread to the flowers by mold spores on wind. These are very, very common in the Bay Area.

Fireblight on apples and pears is first noticed as scorched-looking leaf tips that may be blackened or sticky. It will spread and kill all branches if not checked. The cure is to cut back the infected branches to new growth.

Brown rot attacks trees in the plum/apricot family. It withers entire branches to a gummy mess, leaving brown "mummy" fruit on the plant through the winter. The solution is to prune off all the dead branches and spray with a fungicide (one containing Benylate will be most effective) twice when the tree begins to flower.

About Berries & Citrus

Raspberries and blackberries grow to rampant vines in our climate and are best for waste areas. You'll have to prune vigilantly to keep them in bounds, and eradicate any "volunteers" which can quickly grow into thorny backyard barbed wire. Shallow-rooted blueberries work well in containers with the addition of peat moss and a bit of fertilizer yearly. Passion-fruit and kiwi, berrylike fruits that grow on vines, are easy to grow in warm microclimates, and can be trained on a trellis or fence. Strawberries may be treated like biennials: remove all flowers the first year to get a hefty big crop the next.

Planting citrus is the same as for deciduous fruits. For maximum output of oranges, lemons, grapefruit and the like, remove sucker branches that grow at the base of the plant. Thinning fruit isn't necessary. Yellowing or splotched leaves signal a deficiency in trace minerals such as zinc and manganese. This is a common problem, but can be prevented with a monthly feeding of a fertilizer such as Miracle-Grow that contains trace elements.

Pests common to citrus are aphids (spray off with jets of water), a barnacle-like scale (scrape off trunks with fingernail) and snails, which will chew big holes in the leaves. A collar of four-inch copper strip, wrapped around the trunk or around the container, is a simple, organic method of snail control.

Winter Chores

Having dwarf citrus around is a good way to have your own fresh fruit throughout the winter. Navel oranges and tangerine varieties like Dancy, Satsuma, and Fremont are ready for picking December through March.

Pruning is done on deciduous trees from December to February. Evergreen citrus is pruned from March to May. Pruning fruit trees is too specific to the species to go into here, but there are many books on the subject. If you're uneasy about cutting, you can always hire a professional.

Winter tasks to maintain tree health include spraying dormant oils or Benylate (if brown rot was a problem) and clearing dead leaves or debris from the ground beneath the tree so it won't be a haven for disease and mold spores. Put the debris in outgoing trash instead of the compost pile. And don't forget to water trees on a regular schedule through the winter if we have another drought year.

Mia Amato is a certified Master Gardener from the UC Cooperative Extension in Berkeley and a freelance writer.

Common Questions About Fruit Trees

My apricot tree has black-brown gunk and its branches are dead. What is it?

Probably brown rot. Cut away the infected branches to new growth, dipping your pruning tool into a solution of 1% household bleach and 99% water after each cut. Discard branches — don't put in compost pile. Next spring, spray flower buds with Benomyl, copper bordeaux, or similar fungicide just before they open and 2-3 times more while the branches are flowering.

I grew an avocado tree from seed. If I put it in the yard, will it bear fruit?

Odds are one in a million, since the avocados we buy are hybrids that have been perpetuated by grafted cuttings for decades. Trees grown from pits are happiest as houseplants, but if your climate is warm enough, you can put it in the yard to enjoy its foilage . For a fruiting avocado, try the Dwarf Mexicola variety as a pot plant.

The temperature's slipped below freezing and I have citrus trees. Help!

Use a cloth to cover small trees on frosty nights, erecting poles so the cloth does not touch the foilage. For large trees, Don Dillon of Four Winds Nursery suggests decking the inner branches with Christmas lights (the waterproof outoor type) to keep the heart of the tree warm during cold spells.

Sources and Resources:

REFERENCE BOOKS:

Pruning Simplified,
byLouis Hill
Pownal Storey Communications Inc.
1986

Designing and Maintaining Your Edible Landscape Naturally
by Robert Kourik
Santa Rosa Metaphysical Press
Santa Rosa, CA, 1986

TREE SOURCES:

Four Winds Nursery True Dwarf Citrus
Box 3538
Fremont, CA 94539 (415)656-2591

Stark Bro's.
Louisiana, MO 63353 1-800-325-4180

Raintree Nursery Catalog
Fruits, Nuts and Berriess for the Pacific Northwest
391-Butts Road
Morton, WA 98356 (202)496-6400

ORGANIZATIONS:

Home Orchard Society
PO Box 776
Clackamas, OR 97015

California Rare Fruit Growers
2209 McGee Avenue
Berkeley, CA 94703 (415)843-1657

Setting up a Home Compost Heap

By Stephen C. Evans

Home composting is recycling at its best. While cutting down on trips to the dump (about a quarter of the material sent to landfills is compostable), it provides the best fertilizer and soil conditioner available — eliminating the need for chemical substitutes. What's more, it's easy to do, and many seasoned composters find great pleasure in the damp sweet smell and the curl of steam from a freshly turned pile on a cool morning.

To start your compost heap, select a shady, out-of-the-way spot in your garden. Shade is important because you want to keep the pile damp (not wet) at all times, optimizing conditions for the billions of bacteria responsible for breaking the organic wastes into forms plants can use.

While not absolutely necessary, most gardeners build some kind of frame to contain the compost. The walls of such a frame should be made of wire mesh, or constructed in a way that will allow free air flow through and around the pile, to provide the bacteria with oxygen. It is a good idea to build a base for your frame of sturdy steel mesh about a foot above the ground. This allows oxygen to get to the bottom of the heap, and the small bits of finished compost that fall through the bottom can be shovelled out and used. Leave an opening in one side of the frame for working the pile with a pitchfork or shovel.

Build your pile with alternating layers of dry plant material, which provide carbon (no more than half the heap), moist or green organics (about 35%), and soil or old "finished" compost (the final 15%). Kitchen scraps and the greens organics contribute nitrogen and oxygen to the mix. If you wish to increase the nitrogen component, you can also add some manure to the pile.

The larger the pieces of plant material you put in your compost heap, the longer they will take to decompose. Chop everything into small bits. Although there are some electric or gas-powered chippers and mulchers available, we recommend using a hand tool, such as a hedge-trimmer or hatchet, in the name of good exercise and saving fossil fuels.

You can put most yard wastes and kitchen scraps (including moldy bread, coffee grounds and filters, egg shells, orange rinds, tea bags, and corn cobs) into your compost pile.

Do not use meat or dairy products, as they will putrefy and attract animals such as rats and flies. Plant detritus which is very slow to decompose, such as eucalyptus leaves, pine needles, or redwood sawdust, are also best avoided. Salt is soil's biggest enemy; avoid getting any in your compost heap.

Use a pitchfork or shovel to turn the compost about twice a week. This gives it air, speeding the decay process. Remember to dampen the pile with a garden hose regularly, but don't drench it.

Note that decomposition releases quite a bit of heat — temperatures of 160 degree Fahrenheit are normal.

When the pile no longer heats up and is brown and crumbly with a pleasing smell of rich soil, it is ready to use. Mix it liberally with soil for new plantings, or spread it up to two inches thick on the surface around old plantings.

Besides fertilizing the soil, compost used as a mulch deters weeds, guards soil against erosion, and holds moisture, thus aiding water conservation. So use it freely.

Compost problems & solutions

Problem	Solution
Pile doesn't heat up	This happens when the pile is too wet, too dry, too small, or contains too much carbon. The pile should be kept damp, but not soaking wet. Be sure there are plenty of nitrogen sources - greens, manure, kitchen scraps. Turn the pile regularly to keep it aerated.
Pile smells rotten, attracts flies	Eliminate meat, grease, bones, dairy products, and dog or cat feces. The pile may also be too wet — add dry materials and turn the compost to air it out.
Pile emits a strong ammonia odor	The pile is either too rich in nitrogen or too alkaline. Add carbon sources such as straw, dry leaves or sawdust and acid materials like coffee grounds and oak leaves.

Kicking the Habit:

Non-toxic alternatives to pesticides for the home and the garden.

By Christine McKenna

Fleas infest our pets, roaches rule our kitchens, lice nestle in our children's hair, and aphids suck the life from our garden plants. What are we to do? For most of us, poisoning everything in sight is too often the answer. According to a National Academy of Science survey, 90 percent of Americans are reaching for flea collars, roach bombs, lice shampoos, and other pesticide products when the bothersome beasties make an appearance.

Sixty-five million pounds of herbicides, insecticides and fungicides are used in homes and gardens each year according to EPA estimates. Yet, unlike agricultural applications, where safety suits and even gas masks may be used, often the only precaution taken by homeowners before dousing their children, gardens, and pets is a cursory scan of warning labels.

The result? Children, pets, and parents are poisoned. Poison Control Centers across the country received 41,499 pesticide and 10,626 rodenticide related calls in 1988. The threat to children's health can be particularly severe. A study done at the University of Southern California showed that children living in homes where household and garden pesticides were used had as much as seven times the chance of developing childhood leukemia.

While there have been many attempts to reduce the use of dangerous pesticides in agriculture, urban use of similar poisons seems to have largely been overlooked. Considering that up to 15 times more pesticides are applied in urban than in rural areas, it seems only logical that homeowners should follow the example of farmers using alternative pest management methods and institute non-toxic programs in their homes and gardens.

Nontoxic alternatives

Unlike chemical pesticides, which rely heavily on broad spectrum poisons that kill indiscriminately both harmful and beneficial organisms, natural alternatives offer much more specific means of control. Instead of using chemical panaceas, like Raid™, to destroy all six-legged kitchen intruders, the first step of a non-toxic program is to become more familiar with pests — to carefully identify each pest species and research its eating, sleeping, and breeding habits. Reference books and local Cooperative Extension Services are two sources that may be helpful in this initial diagnosis. (See list of Cooperative Extensions on page 30)

In contrast to chemical programs, where achieving zero population levels is held as a reasonable expectation, non-toxic (or "integrated") pest management emphasizes that complete eradication of pests is unlikely.

The second phase of a non-toxic program is to carefully examine the actual damage done or health hazards posed by a pest and then to reconsider tolerance levels; to ask whether these warrant going to extremes to remove them. Is honeysuckle on car windows so inconvenient as to justify spraying Carbaryl, or other suspected carcinogens, on all neighborhood shade trees? Do dandelions detract so much from lawns that application of toxic herbicides is neccessary?

Ideally, this reevaluation will result in higher tolerance levels for less offensive or destructive pests. Finding a non-toxic means to control the remainder is then relatively simple. Alternatives are as varied and bountiful as the pests themselves. While some are commercial products, many natural pest controls can be made using simple home recipes, or simply involve understanding the pest in question and doing things to make your house or garden uninviting or inaccessible to them.

Pests in the home

Listed below are some commonly used household pesticides, their possible health effects and safer methods of control. Compiled primarily from a publication by the UC Berkeley Office of Environmental Health and Safety, a few tips were provided by Debra Lynn Dadd's book *Nontoxic and Natural* (Jeremy P. Archer, Inc, 1984).

Ant & Roach Killer:

These are poisons which contain chloropyrifos and petroleum distillates. They can impair central nervous system function and are harmful if swallowed, inhaled, or absorbed through the skin. Alternative solutions include keeping food areas clean, removing water sources such as as leaky plumbing or water in houseplant trays, caulking or filling cracks to block entry or hiding spots, pouring a line of cream of tartar, dried peppermint, red chilli powder, or paprika at ant entry points, using boric acid to control roaches in inaccessible areas under the house or in cracks and crevices under the sink or refrigerator where roaches hide. Bay leaves or cucumber skins also can be placed in cracks of roach-infested areas.

Houseplant Insecticide:

These contain pentachlorophenol, toxaphene, malathion, lindane and petroleum solvents, and can be poisonous if ingested, inhaled, or absorbed through the skin. An alternative to these chemicals is washing leaves with mild soapy water and rinsing with plain water.

Flea Bombs:

These poisons contain carbaryl, benzene hexachloride-3, napthalene, phenols, and occasionally arsenic. They are toxic if ingested, inhaled, or absorbed through the skin and can cause nausea, diarrhea, tremors, convulsions and respiratory collapse, still-births, and fetal abnormalities. Alternative controls include vacuuming often and disposing of the bag outside, heating your home to 122F for several hours, and using a flea comb. Preventative measures include getting rid of your pets (just kidding), feeding your pet brewer's yeast (25 milligrams/per 10 pounds body weight) or rice-based B-complex vitamins, or applying external herbal repellants such as ground cloves, eucalyptus oil, strong wormwood tea or citrus oil.

Rat Poison:

Rat poisons contain arsenic, strychnine, phosphorous, or anti-coagulants. They can poison children, pets, and wildlife and create resistant rat populations. Alternative solutions include using mechanical traps, covering drains with screens, or sealing off entryways around pipes with steel wool. Cats are also quite effective rat killers, although they tend to hunt birds and other desirable species as well.

Pests in the Garden

Making your garden an unpleasant place for pests is the best way to avoid an infestation. Cultural techniques are often successful in detering pests. Careful plant selection, organic amendments, crop rotation, and companion planting all discourage pests by strengthening a garden's natural resistance. Like human bodies, unhealthy gardens, which lack proper nutrients or sufficient beneficial organisms, have weakened immune systems and are more vulnerable to pest invasions.

Selecting pest or disease resistant plants and creating a healthy environment for these plants to thrive are the first steps to a hardy garden. Rodale's gardening guides or books such as *Strawberries in November* (Judith Goldsmith, Heyday Books, 1987), which are area-specific, may be helpful in finding resistant varieties.

Healthy soil for plants can be maintained by applying organic amendments. Chemical amendments, although often a successful quick fix, weaken a garden's immune system by depleting its natural supply of beneficial organisms. In contrast, organic amendments such as compost, manure, bone and blood meal, or mulch work slowly and steadily in the soil and and serve to enhance this supply of microorganisms. Unlike their chemical counterparts, these additions are relatively inexpensive and some, such as composts, can be made at home using very simple recipes. (See the article on composting, page 25.)

Using alternative or irregular planting methods can also deter pests. Rotating plants both prevents nutrient depletion of soil and disrupts pests likely to settle on particular crops. Knowledge of plant families is crucial to this method, for alternating related plants, such as cauliflower and broccoli, defeats the purpose of rotation.

Companion planting can pool a garden's pest resistance resources. In this process, plants with tastes or characteristics pests find appealing are planted next to other varieties which they find repellent. Confused and disgusted, pests are often discouraged and leave the garden for better feeding grounds. Thus, radishes are interplanted with cucumbers to repel cucumber beetles, potatoes with collards to reduce flea beetle damage, and onions or garlic are scattered throughout the garden as they repel many pest species. Of course, as cauliflower and broccoli attract the same pests, they are kept separated. Suggested rotations and companion plantings can be found in guides such as *Rodale's Organic Gardening Encyclopedia* .

Removing old, diseased, or rotting plants can also considerably reduce a garden's appeal as they often serve as hiding places or foods for pests. Next season's pest population can also be reduced as eggs are often hidden in this debris.

Beneficial Organisms

If preventative measures fail and pests invade, more aggressive non-toxic controls may be necessary. Perhaps the best way to decrease pest populations is to increase the beneficial organisms in a garden. Although beneficial predators, parasites, and microorganisms often occur naturally, the existing population may need to be augmented. Beneficial organisms should be used carefully as they are often pest-specific and must be applied at a particular stage of growth. For example, *Bacillus thuringiensis* (Bt), a bacterium commonly unleashed on garden invaders, is only deadly to caterpillars and then only to those unlucky enough to consume it.

Insects such as lacewings, lady-beetles, and parasitoid wasps that devour or sting a variety of pests can be purchased from garden suppliers and can often be persuaded to remain in a garden if a source of food and water is provided. Other microbials, such as Bacillus popillae,"milky spore disease", and Nc nematodes, insect eating microscopic worms, are also effective controls and are available commercially.

Finally, manual efforts such as hand picking, hosing off plants, and pulling up weeds, although time-intensive, are often the most effective means of non-toxic control. Traps offer another alternative to spraying broad spectrum poisons and can be used to confuse, lure, and remove pests. Black light traps and other light traps, however, should be avoided as they also kill beneficial insects like lacewings.

Compiled below is a list detailing which non-toxic methods can be used in controlling particular garden pests.

Aphids:

These tiny black, gray, or green pear-shaped bugs suck plant juices by piercing plant tissues. Maintaining a soil's organic matter with composts and manures and interplanting mint or repellents such as garlic, chives, coriander, and anise is one way to discourage aphid infestation. Aphids can be removed by spraying plants with soap solution, dusting plants with lime or lime solution, trapping aphids in a yellow bowl of soapy water, or introducing natural enemies of the aphid such as ladybeetles and praying mantises.

Cabbage Loopers:

This caterpillar loops it's body when it crawls and attacks members of the cabbage family, peas, lettuce, and tomatoes. Planting strong smelling herbs such as mint, catnip, thyme, and rosemary may discourage butterflies or moths from laying eggs. Caterpillars can be removed by hand, hosing with soaps, dusting wet plants with lime or lime solution, or spraying with Bt.

Corn Borers or Earworms:

Although attracted to corn, these pests also infect beans, potatoes, peas, peppers, and tomatoes. Maintaining nitrogen levels through interplanting legumes, rotating crops, and plowing in the fall to destroy pupae should discourage infestation. Borers and worms can be removed by hand picking, releasing lacewings or spraying Bt on leaves and silks.

Cutworms:

These worms feed on a multitude of plants, including cabbage, beans, tomatoes, and corn and are particularly attracted to young transplants. They can be discouraged by wrapping individual foil or paper collars around plants and by placing sandpaper or roofing shingles on the ground near plants, as cutworms will not cross rough surfaces. Plant repellants include tansy and marigolds. Worms can be removed by handpicking, spreading beneficial nematodes, or by spraying Bt.

Earwigs:

More beneficial than not, these brown wingless insects eat the larvae of many harmful insects. Providing mulch as an alternative food supply can reduce damage done by these pests. Earwigs can easily be trapped by placing a hollow tube such as a rolled up newspaper or a piece of bamboo in the garden. Earwigs will crawl inside at night or when seeking a cool spot on hot days and can then be removed.

Flea Beetles:

These jumping beetles feed on eggplant, corn, tomatoes, spinach, and potatoes. Crop rotation, careful weeding, and interplanting strong-smelling herbs such as mint or garlic can discourage these pests. Beetles can also be controlled by handpicking, importing beneficial nematodes, or applying soap or lime solutions, or home-made sprays of garlic and water.

Mites:

As these pests can only be seen with magnification, the first indication that a garden is infected will be a yellowing or speckling of plant leaves. Often attacking eggplant, beans, squash and strawberries, they can be removed most effectively by other native predator mites. Other methods of control include introducing lacewings or ladybeetles and spraying with water, soap, or lime solutions.

Slugs and Snails:

These slippery pests have voracious appetites and will eat most anything in a garden. Removal methods include several means of trapping and collecting pests: placing a pan of stale beer in the infected area, overturning clay pots where snails will seek shelter from the heat, and laying boards between rows of planted vegetables where snails will attach themselves to the undersides during the day.

Thirps:

Plants that drop their blossoms early or whose leaf tips turn brown may be infected with thirps. These tiny pests suck the juices from plants and flowers and spread plant diseases. Control methods include interplanting marigolds, introducing predatory mites or dragonflies, and spraying with soap solutions.

Tomato Hornworms:

These horned green caterpillars devour tomatoes, eggplants, potatoes, peppers, and dill. Plowing pupae under in the fall and interplanting marigolds, borage, or nicandra can discourage them. Control methods include handpicking (unless there are wasp cocoons on the worms backs), supplementing natural enemies such as lacewings or ladybug beetles, and spraying with soap, lime or Bt.

Whiteflies:

More like moths than flies, these pests suck juices from leaves. Interplanting nasturtium and marigolds can discourage whiteflies. Control measures include supplementing natural enemies such as ladybeetles or lacewings, pruning or removing badly infected plants, spraying with soap solution and vacuuming plants.

Lawn Pests

Would you spray your child's playground with a known carcinogen? Line your doghouse with Dursban, a chemical shown to cause chronic kidney damage? You may be doing so without knowing it if you are one of the millions of Americans who apply nearly 5-10 pounds of pesticide per acre on their lawns each year. Most of the twelve pesticides most frequently used on lawns, such as Captan, Dursban or 2,4,D, are thought to pose severe long-term health hazards.

Just as one puff can lead to a pack a day, the innocent spread of a few bags of nitrogen fertilizer can lead to a dependence on lawn chemicals. Fertilizers stimulate not only grass but weeds to grow to new heights. At the same time, they discourage the existence of naturally occuring beneficial organisms in the soil that both enhance grass growth, and fight off harmful pests, weeds, and turf diseases. As a result, when a myriad of pests descend, when fertilized weeds begin to thrive and fungi begins to spread, the lawn is incapable of combating these threats with its compromised immune system.

At this point, many desperate lawnowners reach for synthetic pesticides to rescue their ailing turf. Although increasing chemical inputs may relieve some of these problems, the relief is fleeting as such products only exacerbate more serious concerns like the creation of pesticide-resistant pest populations, and the depletion of beneficial organisms.

The "greenest" solution to this dilemma is not to have a lawn at all — to replace turf with native drought resistant plants. Aside from the difficulty of maintaining a lawn without pesticides and herbicides, lawns are notorious water wasters. The East Bay Municipal Utility District has recently produced a book, *Water-Conserving Plants & Landscapes for the Bay Area*, which is an excellent resource for those ready to make the shift away from lawns.

For those not yet ready for such a transition, there are many alternatives to chemical care programs. The following steps can help to wean a lawn from synthetic fertilizers and pesticides:

- Aerate compacted soil by cutting narrow plugs out of the sod to allow oxygen to penetrate soil and to restore proper drainage.

- Use organic fertilizers: composts, mulches, manures, bloodmeals, seaweeds, and peat moss.

- Choose compatible drought or pest resistant grasses: tall fescue, perennial rye grass, or Kentucky bluegrass.

- Use Integrated Pest Control techniques: monitoring, picking, and hoeing.

- Reduce stress on grass by keeping mower blades sharp. Maintain maximum photosynthesizing area on grasses by mowing high.

When a lawn becomes infected with diseases or weeds invade, opt for one of the non-chemical alternatives listed below. Many of these suggestions were provided by *How to Get Your Garden Off Drugs*, a guide written by Carole Rubin for the Friends of the Earth.

Brown patch:
This fungus, which appears as round or crescent shaped brown patches, prefers high nitrogen levels, heavy thatch, and humid areas. Reducing nitrogen, dethatching and pruning shade trees and shrubs will help remove it.

Dollar Spot:
A fungus that appears as round bleached dead spots on lawns in spring or fall, it thrives on excessive water above the soil and not enough nitrogen. Aerating soil, watering in the morning, de-thatching in the fall, mowing high and amending soil to augment nitrogen (by spreading compost, for example) can correct this problem.

Mildew:
This fungus likes shady, wet areas and is found on the top part of grass blades. It appears in a grey cobweb-like or white powdery form. Aerating soil, watering in the morning, decreasing shade cover, and mowing high can discourage mildew.

Mushrooms:
Like mildew, this fungus also likes shady wet areas. The best way to control mushrooms without damaging a lawn is to remove them by raking.

Weeds:
Pulling weeds out by their roots after watering is usually the best means of control. The presence of weeds is often a sign that a lawn is suffering from nutritional deficiencies. Many weeds, such as clover and dandelions, may actually be more of an asset than a liability for your lawn. Clover attracts beneficial insects, fixes nitrogen in the soil, and can be planted as a green manure. Dandelions attract beneficial wasps and, as they are rich in vitamins, iron and potassium, can be used as a dietary supplement.

Getting Started

Help instituting a non-chemical lawn care program can be found through UC Cooperative programs and in many community organizations such as S.F. Urban Gardeners (SLUG), BioIntegral Resource Center (BIRC), Biological Urban Gardens (B.U.G.S.) or the Ecology Center in Berkeley. Periodicals such as *Organic Gardening* and guides published by the Rodale Press are also excellent resources. (Please see organization listing and consumer guide listing for suppliers and landscapers.)

Christine McKenna is a researcher and staff writer for the Bay Area Green Pages.

Sources and Resources:

FURTHER READING

Home

The Healthy Home
By Linda Mason Hunter
Rodale Press, 1989.
33 East Minor Street, Emmaus, PA, 18098

Nontoxic & Natural,
By Debra Lynn Dadd
Jeremy P. Tarcher, Los Angeles, 1984.

The Green Consumer,
By John Elkington, Julia Hailes & Joel Makower
Penguin Books, New York, NY, 1990.

Garden & Lawn

Growing Vegetables West of the Cascades,
By Steve Solomon
Sasquatch Books, Seattle, WA, 1989.

How to Get Your Lawn & Garden Off Drugs,
ByCarole Rubin
Friends of the Earth, Ottawa, ON, 1989.

Rodale Guide to Composting
By Jerry Minnich and Marjorie Hunt
Rodale Press, Emmaus, PA, 1979.

Shepherd's Purse: Organic Pest Control Handbook
Pest Publications, 1987.

The Organic Gardener's Complete Guide to Vegetables and Fruits
Rodale Press Inc., 1982.

Water Conserving Plants and Landscapes for the Bay Area
East Bay Municipal Utility District
Dept. GM, PO Box 937, Alamo, CA, 94507.

Periodicals

B.U.G.S. Newsletter
PO Box 76, Citrus Heights, CA 95610
Describes horticultural research and new products and techniques.

Common Sense Quarterly, The IPM Practitioner
BIRC, PO Box 7414, Berkeley, CA, 94707
How to control pests at home and professional news.

Organic Gardening
Rodale Press,(see address above)
This monthly describes organic methods and offers gardening tips.

RESOURCES

Agricultural Information and Publications, University of California, Davis, CA, 95616, (916) 757-8930
A catalogue of reasonably priced publications is put out by cooperative extension offices. Topics include: gardening, landscaping, pest control, livestock management, and fruit and nut tree planting tips.

Gardening Help Courtesy of the University

By Christine McKenna

If you know how to go about getting it, government agencies are a treasure trove of information on almost any subject you can imagine. When it comes to gardening advice, Californians are fortunate to have ready access to the very latest research, as well as general information through the auspices of the University of California Cooperative Extension.

Working with county, state, and federal governments, the University of California Cooperative Extensions provide agricultural assistance and educational programs throughout California. With research support and staff experts in agriculture and natural resources, the Extension services can be a valuable asset to home gardeners. Master Gardeners are volunteers trained to answer questions regarding urban gardening and small home orchards. They can offer advice on urban horticulture, turf management and landscaping, and fruit and nut trees. If you have pest or plant disease problems, you can bring a specimen in to the local Cooperative Extension office to have it identified, and to get suggestions for remedies. When looking for solutions to problems, however, you should be sure to ask about biological or other non-polluting controls.

The Cooperative Extension's Master Gardener training program is also an excellent way to learn a great deal about horticultural techniques in a short period of time. The intensive three-week courses are offered each summer. For more information, call one of the numbers listed below.

Cooperative Extension Offices by County

Alameda

(415) 670-5200
Fax (415) 670-5231

224 W. Winton Ave.
Rm. 174
Hayward 94544-1298

Contra Costa

(415) 646-6540
Fax (415) 646-6708

1700 Oak Park Blvd., Bldg. A-2,
Pleasant Hill 94523

Marin

(415) 499-6352

1450-A Lucas Valley Rd.,
San Rafael 94903

Napa

(707) 253-4221 or 944-2006
Fax (707) 253-4434
1436 Polk St.
Napa 94559-2597

San Francisco

(415) 586-4115

Mail:
P.O. Box 34066
San Francisco 94134

Location:
Cow Palace, South Hall
Geneva St. & Santos Ave., Daly City

San Mateo

(415) 726-9059
Fax (415) 726-9267

Mail:
P.O. Box 37,
Half Moon Bay 94019

Location:
625 Miramontes St., Suite 215
Half Moon Bay

Santa Clara

(408) 299-2635
Fax (408) 246-7016

2175 The Alameda
San Jose 95126

Santa Cruz

(408) 761-4056
Fax (408) 761-4106

1432 Freedom Blvd.
Watsonville 95076

Solano

(707) 429-6381
Fax (707) 429-5532
2000 W. Texas St.
Fairfield 94533-4498

Chapter Two

Food and Agriculture

Living in cities we somehow forget the intimate ways we are linked to the land. Food is perceived as just another commodity; it comes straight out of a package off a supermarket shelf. The soil from which it sprang is not even a dim memory.

The small family farm that was once the basis of American life has been replaced by mega-farms operated by large corporations whose only interests reside on the bottom line. In short, even the farmers have lost touch with the land — with disastrous results. Years of chemical farming techniques have left many of our soils depleted of organic content. Worse, the steady, heavy doses of poisons administered to crops to guarantee high yields of cosmetically-perfect produce have contaminated water supplies and come home in our food.

There are older, better ways of farming, and some farmers in the Bay Area are returning to them. It is the fact that we must eat to live that gives us our strongest natural stake in the health of the environment. Food is our tether to the web of life. Where it comes from and how it is produced should therefore be of the utmost concern to all of us.

The Perils of Pesticides

The Environmental and Health Costs of Spraying our Food

by Christine McKenna

It took Meryl Streep thirty seconds to do what environmentalists and consumer groups have been trying to do for years — alert the public to the health threats posed by agricultural pesticides.

Streep, appearing in a television commercial as a spokeswoman for Mothers and Others for a Livable Planet, disclosed the findings of the Natural Resource Defense Council's review of Alar: young children were being exposed to an "intolerable risk" by carcinogenic pesticides applied to apples and other produce. In fact, the NRDC report suggested, 5,000 to 6,200 of the current population of preschoolers could develop cancer solely from their exposure, before the age of six, to eight pesticides used on fruits and vegetables.

The NRDC report, *Intolerable Risk: Pesticides in Our Children's Food*, has been criticized by many groups, including the EPA, who hold that health threats posed by pesticides are minimal when compared to biological hazards like salmonella or carcinogens such as radon or x-rays. What's more, the 6,200 extra cancers, they maintain, represent an increased risk of only .025 percent per person.

Nevertheless, these statistics and the Alar media blitz accompanying the NRDC report have led consumers to doubt the safety of the American food supply. No longer complacent, they are asking questions about pesticide regulation. Who is regulating pesticide use? How thorough are inspections for dangerous chemical residues? Exactly what pesticides are used on produce and what are their possible health effects?

There are three federal agencies (the EPA, FDA and USDA) and two state agencies (the CDFA and County Agricultural Commission), that regulate pesticide use in California. The EPA works in concert with the CDFA to approve and register pesticides used on agricultural products. Testing and monitoring of imported produce or of items shipped interstate falls under the jurisdiction of the FDA. Inspection of meat and dairy commodities is done by the USDA. Other policing of use, and investigation of violations and pesticide-related illness, is done by County Agricultural Commissioners.

With the combined efforts of these five agencies, it would seem that Californians could feel confident that their produce was not contaminated with carcinogenic chemicals. Unfortunately, the ability of these agencies to provide the state with a safe, clean food supply has been questioned not only by consumer groups like the NRDC, but by the Senate Office of Research and even by employees of these agencies who feel severely limited by insufficient resources and staff. Concerns over food safety seem warranted in light of the following facts:

- Less than 1 percent of produce consumed by Americans is tested, and even that is tested only for 50 percent of the pesticides currently on the market.

- Of the 496 pesticides likely to leave residues, only 203 are detectable by the FDA's routine tests. Of the remainder, many are known to be oncogenic (causing tumor growth.)

- A report released by the Senate Office of Research in February, 1990 stated that, in 1987, CDFA officials ignored warnings from their own toxicologists, altered reports, and approved up to 20 potentially dangerous pesticides. The agency has acknowledged that at least eight of these should never have been approved.

Consumer groups claim that the inadequacy of pesticide policing can be traced to flaws in the regulatory procedures of the EPA, the limited scope of FDA detection tests, and outdated national legislation.

The EPA regulatory process most often criticized is determination of pesticide "tolerance levels," the maximum amount of residues allowed on food. The EPA makes these determinations despite the fact that many of these chemicals have never been tested for long-term effects. Even when EPA special reviews are initiated for some potentially carcinogenic pesticides, delays are inevitable, and can take four to eight years to complete. Although Congress has directed the EPA to reconsider up to 600 chemicals, the reviews will not be complete until 1997. The pesticides will continue to be used in the interim.

Exacerbating significant gaps in health data, the EPA sets tolerance levels based on a evaluation of pesticides in isolation, despite the fact that the effects of pesticides can be additive. Thus, possible synergistic reactions in which chemicals from different pesticides combine to form other possibly more toxic, compounds are often not considered.

Insufficient data on the health effects of inert ingredients in pesticides represent another flaw in the evaluation process. Disclosure of these chemicals is not required by the EPA as they are considered "trade secrets" by the pesticide industry. This only multiplies the possible number of unknown synergistic reactions that may be occuring. Although these "secret" ingredients aren't actively involved in pest control, the EPA has determined that as many as 50 of these inert ingredients pose serious health hazards.

Finally, the EPA has set tolerance levels for many older pesticides with unrealistic consumption levels in mind, failing to allow for excessive intake of certain produce or the greater effect of pesticides on children.

Although tolerance levels for newer pesticides may more accurately reflect the average American diet, the EPA-determined levels for many earlier pesticides using a diet survey estimate that the average American consumes only 7.5 ounces per year of almonds, avocadoes, blueberries, garlic, pecans, nectarines, and mushrooms.

Clearly this is hardly a realistic projection because many individuals, particularly vegetarians, can easily surpass these amounts in a matter of days.

Compounding the problems resulting from the EPA's questionable regulation is the limited scope of FDA and CFDA tests and the minimal enforcement of pesticide regulations.

Multi-residue tests, those most commonly used to monitor residues, can detect less than half of the 300 pesticides used on foods. EBDC's, fungicides widely used on produce, are just a few of the potential carcinogens which slip through this screen.

Enforcement of pesticide limits appears equally slippery. According to the General Accounting Office, 73 of 164 cases in which illegal residues were detected on imported produce in 1986 were allowed to reach the public.

Furthermore, in only eight of these cases did the FDA penalize importers responsible for the contaminated produce.

Health threat or media hype?

Are pesticides so prevalent in our foods that we should be worried? The number of different pesticides registered for use on fruits and vegetables alone is phenomenal: over 100 pesticides for apples, 100 for tomatoes, and more than 80 for grapes and grapefruit.

Although farmers certainly do not use every registered product on their crops, the residues left by those they do apply is significant. Between 1982 and 1985, the FDA detected pesticide residues in 48 percent of frequently-consumed fruits and vegetables. This statistic does not include possible residues from the 50 percent of pesticides applied to food that cannot be detected by standard FDA tests.

Many pesticide residues are quite benign. However, of the more than 60 pesticides the EPA has identified as pos-

Reducing Your Pesticide Intake

Ultimately the message to be taken from this discussion of pesticides is not to reduce vegetable intake — the health benefits to be gained still outweigh the risks — but rather to work at ways of minimizing and eventually eliminating unnecessary exposure. The following suggestions can help:

- Take the extra time in preparation to wash items with warm (100F) water and mild soap. There are also several vegetable washes on the market. (See our consumer guide section). Some produce, such as tomatoes and cucumber may require additional scrubbing as they are coated with waxes that make the pesticides difficult to remove with water alone. Peeling is always an alternative, especially with produce that has several layers such as lettuce. Removing the skins of other vegetables such as carrots or potatoes may reduce pesticide residues, but will result in nutrient loss as well.

- Cook produce if possible, as this aids in the breakdown of remaining pesticide residues.

- Avoid excessive consumption of particular produce. Remember the EPA has determined pesticide tolerance levels using mean consumption rates, ie. 7.5 ounces of mushrooms or avocadoes per year.

- Beware of cosmetically perfect fruits and vegetables, particularly apples, peaches, strawberries, and other produce grown in the soil. Their pesticide contamination may be greater.

- Avoid imported produce by eating in season and shopping at groceries that sell local produce or at farmer's markets. Better yet, visit a local farm and buy your produce directly. Some farms allow you to pick the fruits and vegetables for a reduction in the price. (See consumer guide listing for nearest farmers' market or call Farmer's Market Hotline for the latest update. 1(800) 952-5272. See Farm Trail listings at the end of this chapter to find farms in your area.)

- Buy organic produce whenever possible. Visiting a local organic farm could make buying organics less costly.

- Grow your own produce using alternative pest control methods. Community gardens are an option for apartment dwellers and those seeking a more social horticultural experience. (See consumer guide listing for nurseries and seed companies.)

- Write to state and federal congress people to support pending pesticide restrictions.

- Join pesticide watch groups in your area like the Bay Area Coalition for Safe Food or the Consumer Pesticide Project. (See our Organizations section for further group listings.)

sible carcinogens, 55 are likely to leave residues.

Although the Delancy Clause of the Federal Food, Drug, and Cosmetic Act (FIFRA) prohibits the use of carcinogenic pesticides in processed foods in which they concentrate, there currently exists no such legislation for raw products or for processed foods in which pesticides don't tend to concentrate. Carcinogens may be applied to these products as long as the EPA determines that the economic benefits of doing so outweigh potential health or environmental risks.

Some, such as UC Berkeley professor Bruce Ames, estimate that the risks from man-made pesticides found in water and food are minimal and account for nearly "zero" cases of cancer or birth defects. The EPA, however, holds that up to 6,000 cases of cancer each year can be linked to just one-third of the pesticides that have been tested.

Health threats posed to consumers have received a great deal of media attention. This does not necessarily mean they are the most pressing dangers of agricultural pesticides. The most serious hazards are posed to those who apply deadly pesticides and live near or on lands contaminated with them— mainly, farm workers and their families. Despite California's relatively progressive worker safety standards, many farm workers are still faced with unreasonable risks from pesticides. Even when precautions are taken, given the large gaps in human exposure data, there is no way of knowing exactly what amount or form of exposure might be dangerous.

Unlike those who manufacture pesticides, farm workers are not subject to careful monitoring for health effects from these chemicals. This is especially unsettling in light of the fact that 12 years old is the minimum age for California farm workers.

An increased incidence of childhood and other cancers, birth defects, and diseases of the central nervous system have been documented in farm families. In recent years, two Southern California towns, Earlimart and McFarland, have seen an unusually high number of incidents of cancer. These towns had six and fourteen cases respectively.

Farm workers worldwide are victims of American pesticides. Pesticides such as DDT, heptachlor, chlorodane, and 2,4,5,T that have been banned in the United States continue to be manufactured here and then sold to countries like Mexico. Both foreign farm workers and American consumers are exposed to these extremely dangerous chemicals as they are used on produce later exported to the US.

The dimensions of this problem, detailed by David Weir and Mark Schapiro in their expose *Circle of Poison*, become clear when one considers that at least half of the produce eaten in the U. S. during the winter months comes from Mexico and is twice as likely to be contaminated with pesticides as crops grown in the U.S. The FDA inspects only one percent of the produce entering the country.

Agricultural pesticides have been implicated in other problems ranging from threats to endangered species to groundwater contamination. According to EPA estimates, over half of the 450 endangered species are threatened by pesticide use across the country.

Some contamination lingers for years. DDT, banned in 1972 because of its destruction of endangered species of birds and fish, has been found as recently as 1987 in mussels in the Richmond Inner Harbor.

Agriculture has also been determined by the EPA to be the primary non-point source of groundwater pollution in the country. Drinking water supplies have been seriously affected as 50 percent of most urban supplies are drawn from groundwater sources. The first discoveries of pesticide contamination were in 1979, when DBCP (Dibromochloropane is a pesticide that causes sterility) was detected in 2500 wells in the San Joaquin Valley. To date, over 57 different pesticides have been detected in California ground water.

Aldicarb and atrazine are two of the most commonly detected compounds. Aldicarb, one of the most toxic pesticides registered by the EPA, was responsible for over 1000 illnesses in 1985, when it was illegally applied to watermelon. Atrazine is an herbicide currently pending EPA status as oncogenic (tumor producing).

A Hot Debate

EPA Administrator William K. Reilly once wrote that "...the nation's pesticide laws are seriously flawed and the EPA's ability to cancel a problem pesticide is inexcusably burdened by procedural requirements that can entail years of delay." The EPA is currently taking measures to correct some of the flaws in its regulatory process: the agency subjects new pesticides to much more stringent regulation and review, it has requested that the National Academy of Sciences conduct a study on the effects of pesticides on children, and it is supporting a presidential initiative which will expedite hazardous pesticide removal, increase enforcement penalties, and replace the Delancy Clause with a negligible risk standard for both processed and raw produce.

Nevertheless, Reilly says, "At times, we won't have all the scientific data we would like to have, but we will not be able to wait for it to come in before we take action." He adds that the EPA will continue to "make many decisions on public health in the face of uncertainty, based on less than complete information." For consumer groups and environmentalists, this simply is not acceptable. They are calling for a complete ban of potentially dangerous pesticides.

A California ballot initiative on the November, 1990 ballot includes a provision for such a ban. The initiative, known as "Big Green," has ignited a furious debate in California. Many, particularly large growers and representatives of agrichemical companies, argue that the use of synthetic pesticides, carcinogenic or otherwise, is crucial to the livelihood of California farmers. They estimate that a ban, even one limited to those pesticides causing cancer or birth defects as suggested by "Big Green," would increase food prices 40 to 50 percent, cost 189,000 jobs, and result in a 6 billion dollar loss from the state's economy.

Farmers and agricultural experts who support the ban say a severe curtailment of select pesticides could result in a small food price hike, but that prices would decrease again as consumers adjusted to imperfect produce.

Pesticides are not only unnecessary, they claim, but quickly become an expensive habit which can lead to a farm's economic downfall. Farmers who begin using only small amounts of chemicals soon find themselves on a pesticide treadmill: pest resistance to poisons increases while beneficial insects are destroyed, so farmers find it necessary to apply larger and larger quantities to crops to produce the same yields.

While the relative merits of organic and chemical agriculture are hotly disputed, there is one point of common ground in the pesticide debate — cosmetic use. Farmers apply huge amounts of herbicides, insecticides, and fungicides in order to meet rigorous federal cosmetic and insect-part standards, and to offer discriminating customers unblemished produce. Pesticides applied for purely cosmetic purposes, growers and grocers agree, are expendable. Their elimination depends solely on consumer support.

Such support seems to be growing. A 1989 Harris poll found that 84 percent of consumers would consider purchasing bruised or otherwise imperfect produce if it were grown with fewer pesticides, and that 49 percent would be willing to pay more for it. Clearly, the next step is to demonstrate this willingness to the federal government and to California farmers by rewarding growers already using decreased chemical inputs.

Farmers' reluctance to make the switch has less to do with their fondness for costly chemical inputs than with what they see as a lack of alternatives. Increased public, political, and financial support could remedy this, as it would pressure the federal government to fund sustainable agricultural research and to provide educational resources for farmers. While it's true that the success of organic farming or of reducing chemical inputs can vary from farm to farm, public demand for organics will provide the best incentive for farmers to attempt this transition.

Christine McKenna is a staff researcher and writer for the Bay Area Green Pages.

A Short List of Pesticides

The list that follows is a sample of some of the most dangerous pesticides routinely used in food production. For a more comprehensive list, see *Pesticide Alert*, a guide written by NRDC scientists Lawrie Mott and Karen Snyder.

Alachlor:

Tolerance levels for this herbicide were set before it was known to be a probable carcinogen. Its tolerance doesn't cover a break-down product that may cause cancer. Often used on corn, peanuts, and soybeans, it is not easily detected by FDA lab tests.

Aldicarb:

An insecticide that is commonly detected in potatoes and citrus fruit, minute doses can be toxic. It is one of the compounds most frequently detected in California ground water supplies. The EPA has restricted its use but it is still used on banana and citrus crops.

Captan:

A fungicide widely used on fruits and vegetables. Residues are found on apples, cherries, grapes, and strawberries. Use of this fungicide has been restricted by the EPA because it is a probable human carcinogen.

EBDCs:

These are fungicides used to control mold, mildew, and other fungal diseases on fruits and vegetables. Not regularly detected by FDA tests, they have recently been subject to curtailment by EPA, as they have been shown to cause cancer.

Parathion:

Developed initially as a nerve gas, this insecticide is a probable carcinogen commonly detected on broccoli, carrots, cherries, oranges, and peaches.

Resources

FURTHER READING

Circle of Poison
By David Weir & Mark Schapiro
Institute for Food and Policy
San Francisco, 1981

Intolerable Risk: Pesticides in Our Children's Food
Robin Whyatt and Bradford Sewell
NRDC, February 27, 1989

Pesticide Alert: A Guide to Pesticides in Fruits and Vegetables
Lawrie Mott and Karen Snyder
Sierra Club Books, S.F., 1987

PERIODICALS

Journal of Pesticide Reform,
Northwest Coalition for Alternatives to Pesticides, Box 1393, Eugene, OR 97440

REPORTS & RESOURCES

Citizen's Guide to Pesticides and *Citizen's Guide to Drinking Water*
U.S. Environmental Protection Agency, 401 M Street SW, Washington D.C., 20460, Free

Safety on Tap: A Citizen's Drinking Water Handbook
League of Women Voters Education Fund, 1730 M Street NW, Washington, D.C. 20036. 1987, $7.95.

EPA Pesticide Hotline
1-800-858-7378

National Pesticide Telecommunications Network
1-800-858-7378(PEST)
Fax 1-806-743-3094
Impartial information about pesticides for anyone in the contiguous US, Puerto Rico, and the Virgin Islands. Operates 24 hrs. a day, 365 days a year. Answers by phone or mail. Provides info on pesticide products and poisoning, referrals for investigation of pesticide incidents, emergency treatment, safety info, health and environmental effects and clean-up and disposal procedures.

So You Want To Go Organic

by Christine McKenna

Going or-ganic is not as simple as deciding to pay higher, sometimes much higher, prices for tastier chemical-free produce. As increased demand for natural foods is making the "organic" label more lucrative, "organic" claims have multiplied. The result? A labeling nightmare.

On a trip to the market, you can find "organic" lettuce labeled "Grown in accordance with California Health and Safety Code 26569.11", "organic" tomatoes that are certified by the California Certified Organic Farmers, and "organic" bananas which are "In Transition." Other organic produce may claim to have "No Detected Residues," "No Spray," or to be "Pesticide-Free."

What is California Code 26569.11? Who are the CCOF? What kind of transition are those bananas going through?

Confusion caused by this irregular labeling can often lead even those with the best intentions to become suspicious of, and apathetic towards, organics – particularly as they can sometimes cost twice as much as commercial produce.

Certification and labeling are currently being done by a hodgepodge of private and state agencies, and "organic" is defined differently by each. This is likely to change in the future, pending legislation such as the California Organic Food Law and the National Organic Foods Production Act of 1990, which could clarify "organic" status. In the meantime, the following descriptions should help the hapless find their way out of the organic labeling maze.

Produce labeled "Grown in accordance with California Health Code 26569.11" or simply "Organic" is produced, harvested, distributed, stored, and packaged without synthetically-compounded fertilizers, pesticides, or growth regulators.

One year of transition is required for farmers changing from chemical farming before they can make this claim. A "transition period" is defined as the time a farmer must keep land free of synthetic pesticides and fertilizers. (Crops grown on soil during this period will bear the Transitional label.) This produce must not contain pesticide residues which exceed 10 percent of the levels allowed by the Federal Food & Drug Administration. Growers are required to keep records of any chemical applications they make to plants, soil, and irrigation water. Consumers can obtain this information from the DHS for a small copying fee.

As the regulation and enforcement of these measures by the Department of Heath Services is minimal, and fraudulent labeling rarely prosecuted, it may be wise to look for other third-party certification such as the CCOF label. California Certified Organic Farmers is the largest organization of organic farmers in the state, with a membership that has expanded from 40 members in 1973 to 640 grower members in 1990. To obtain certification, CCOF growers must meet more stringent criteria than those required by California civil code. They must provide a 10 year land-use history, and are subject to inspections, anywhere from one to seven visits annually, by CCOF officials. The transition time, now one year, will be increased to three years in 1992.

Although CCOF is the primary certification agency in California, there are several other labels you may encounter. Among these are individual state agencies such as the Oregon Tilth, the Natural Organic Farmers Association (NOFA), the Demeter Association, and two international groups, Farm Verified Organic (FVO) and the Organic Crop Improvement Association (OCIA). Criteria used to certify produce as organic varies by organization and is available by request. (See addresses below).

A NutriClean seal is often one that accompanies a claim of "No Detected Residues." NutriClean, a firm in Oakland, is often hired by local supermarket chains and growers to monitor organic produce for pesticide residues. This private laboratory requires farmers to report pesticide use; the lab then tests crops for residues. However, produce with this label has been found to have no detected residues of the reported pesticides only. It does not mean they have been tested for all possible pesticides.

"No Spray," "Unsprayed," or "Pesticide-Free" are the most loosely defined and the least-regulated of organic labels. The first two usually indicate that no pesticides were sprayed directly on the crops. This is no guarantee, however, that they were not applied to the soil beforehand or through the irrigation process. A "Pesticide-Free" label is simply a grower's claim that his crop was produced without pesticides. It does not mean that the soil in which it was grown had no chemical residues.

This is not to say that such produce should be rejected. There are many reasons, economic, political and otherwise, that farmers choose to work independently of the certification process. Rather than shunning crops backed only by personal guarantees, ask your grocer for proof of a farmer's organic claims. Before marketing these products, retailers usually request some form of documentation to verify its organic status. Thus, they should be able to clear up any uncertainties you may have about a farmer's growing practices. Alternatively, shopping at farmers' markets or going on farm tours can offer excellent opportunities for you to chat with a grower about his farming practices. (See list of farmer's markets and list of organic farms offering direct farmer-to-consumer marketing.)

All of this may sound like a bit of a chore. However, as you are now armed with the means to decipher organic labelese, "going organic" should be a bit more straightforward. As tough organic labeling legislation is passed by both state and federal governments, purchasing chemical-free produce will become easier .

Finally, although it is tempting to patronize only those farmers who have been certified, it's important to encourage others who are still in the process of doing so. By purchasing transitional produce you can support farmers who are often taking serious financial risks in shifting from conventional, heavily chemical-based techniques to a more sustainable form of agriculture. Your support will help them through this rocky period. You may inspire others to follow suit as well.

Certification Organizations

CCOF
P.O. Box 8136
Santa Cruz, CA 95061
(408) 423-2263

Demeter Association
4214 National Ave
Burbank, CA 91505
(818) 843-5521

Farm Verified Organic
P.O. Box 45
Redding, CT 06875
(203) 544-9896, FAX (203) 544-8409

National Organic Farmers Association
RFD #2, Sheldon Road
Barre, MA 01005
(508) 355-2853

Organic Crop Improvement Association
3185 Township Road 179
Bellefontaine, OH 43311
(513) 592-4983

FARM TRAILS OF THE BAY

Farm Trail Maps

County farm trails are maps of local growers that sell directly to consumers. A visit to one of these farms can offer both the freshest in farm produce—at some farms the crops can be picked by the consumer for a reduction in prices— and a chance to talk with farmers about their growing practices and pesticide use. Some farms even provide tours, petting zoos, refreshments and demonstrations. To receive a map for your area, send a stamped, self-addressed envelope to the the regional farm trail organization in your county.

Alameda County Farm Trails
638 Enos Way
Livermore, CA 94550
(415) 670-5200
or
224 West Winton Avenue
Hayward, CA 94544
(415) 449-1677

Coastside Harvest Trails
765 Main Street
Half Moon Bay, CA 95353
(209) 726-4485

Country Crossroads
1368 North 4th Street
San Jose, CA 95112
(408) 453-0100
or:
600 Main Street #2
Watsonville, CA 95076
(408) 724-1356

Harvest Time in Brentwood
P.O. Box 773
Brentwood, CA 94513
(415) 634-3344

Napa County Farming Trails
4075 Solano Avenue
Napa, CA 94558
(707) 224-5403

Sonoma County Farm Trails
P.O. Box 6032
Santa Rosa, CA 95406
(707) 544-4728

Suisun Valley Farm Trails
2000 West Texas Street
Fairfield, CA 94533
(707) 429-6381

Northern California Organic Farms Which Sell Direct To Consumer

ALAMEDA

Finn's Farm
4400 Mines Road
Livermore 94550
(415) 447-9652
Open May 20-June 20
(Call first.)
Pick-your-own and pre-picked organic ollallaberries.

CONTRA COSTA

Brentwood Ranch Eggs
Route 2, Box 179E
Brentwood 94513
(916) 634-0404
(Sales on Balfour Road Westend.)
June-August, Daily 9am-5pm
September-May, Thurs-Sat, Noon-5pm
(Call first)
Tours, brochures, & refreshments are available. The farm sells organic almonds and walnuts in September and October.

Marsh Creek Farms
P.O. Box 284
Clayton 94517
(415) 672-8102/0101
(Sales on Marsh Creek Road.)
Open all year
(Call first and bring containers.)
Farm sells all organic -fed dairy goats & kids, poultry & eggs, weaned pigs, calves, lambs & wool, rabbits, angora wool, & turkeys. *(continued next page)*

Farms, continued

Rancho de Sueno Organic Farms
4525 Discovery Point
Byron 94514
(415) 634-2409 / 5598
(Sales on Marsh Creek Road at Sellers Ave, Brentwood)
(Call first.)
Farm sells all organic vegetables. Produce is pre-picked, or pick your own.

NAPA

Bella Vista
1250 Lokoya Road
Napa 94559
(707) 255-5464
Year-round, Sat. & Sun., 8am-6pm.
(Call first.)
Farm sells organic geese, ducks, turkeys, & baby chicks.

Hoffman Farm
2125 Siverado Trail
Napa 94558
(707) 226-8938
June-Dec., Thurs.-Mon., Noon-Dusk
(Call first and bring containers.)
June: organic boysenberries.
August to mid September: french prune plums and organic sugar.
Pick your own produce.

Napa Valley Grapevine Wreaths
P.O. Box 2286
Yountville 94599
(707) 963-0399
(Sales at 1796 St. Helena Hwy South St. Helena, CA 94574)
Open all year. Mail order also available.
Farm sells organic pumpkins.

SAN FRANCISCO

Hoskins Pure Honey
120 Eastwood Drive
San Francisco 94112
(415) 587-6222
All year by appointment
(Call first.) Bees are delivered throughout the Bay Area. Farm sells organic honey, liquid, and comb.

SAN MATEO

Arata Pumpkin & Animal Farm
P.O. Box 15
San Gregorio 94074
(415) 726-4359
(Sales 6 miles south of Half Moon Bay on Verde Rd at Hwy 1.)
Sept 25-Oct 31, daily, 7am-7pm
(Call first.)
The farm sells organic pumpkins in October, and offers tours for school groups. A picnic area, farm animals and pony rides are available.

Pescadero Creek Orchards
5901 Pescadero Road
Pescadero 94060
(415) 879-0136
Sept. 24-Nov. 6, weekends only
Orchard sells organic red & golden delicious apples.
Oct. 8-Nov. 6: organic pippin apples.
Pick your own produce.

Shamrock Ranch
South End Peralta Road
Pacifica 94044
(415) 359-1627
Mon., Thu., Fri.: 1pm-7pm
Sat 9am-5pm
(Call first and bring containers.)
Sells CCOF-certified organic produce: Beets and swiss chard all year;
Aug.-Nov.: peas, lettuce, carrots, broccoli, cabbage, kolhrabi, summer squash, green beans, cucumbers, sugar pie pumpkins and winter squash.

The Farmer's Daughter
P.O Box 370219
Montara 94037
(Sales on hwy across from Half Moon Bay Airport.)
March-Sept., Sat & Sun, 10am-dusk; Oct., Daily, 10am-dusk
Organic beets, Swiss chard, and eggs.

SANTA CLARA

Barker's Produce
10705 Center Avenue
Gilroy 95020
(408) 842-0658
Aug.-Sept.
(Call first and bring containers.)
Organic apricots, peaches, apples, and Kadota figs as in season.

Glendenning Ranch
16891 Stevens Canyon Road
Cupertino 95014
(408) 867-2969
(Sales at Stevens Canyon Road and Palo Alto Farmer's Market.)
Aug.-Oct., Wed.-Fri., 3pm-6pm, Sun, noon-5pm
(Call first and bring containers.)
Bartlett pears, prunes, black figs, and Pippin apples in season.

J.R. of California
1475 Rodeo Gulch Road
Soquel 95073
(408) 475-1727
(Sales at 1475 Mt. Messiah Grade)
Open all year
(Call first.)
Wheat, rye, and red kidney beans.
Health food seminars available.

Lynnde Farms
390 Leland Avenue
Palo Alto 94306
(415) 328-5022
Open all year.
(Call first and bring containers.)
Organic lettuce
May - Oct.: squash, basil, and Italian parsley
June - Oct.: Greek oregano

Redwood Hill Farm
23115 Summit Road
Los Gatos 95030
(Summit Road & Old Santa Cruz Hwy)
Aug. - Dec ., Thursdays 2pm-6pm
Weekends, 10am-6pm
Sweet corn, tomatoes, beans, squash, cucumbers, pumpkins, potatoes, melons, peppers, and eggplant.

Van Dyke Ranch
7665 Crews Road
Gilroy 95020
(408) 842-5423 / 8689
All year by appointment.
(Call first.)
All year: dried apricots and cherries.
June - Aug.:apricots & cherries.
November: persimmons.

SANTA CRUZ

Greensward
1255 Hames Road
Aptos 95003
(408) 728-4136
All year, Tues-Thurs, 11am-4pm
(Call first.)
Bay Area delivery is available
Wheat grass, herbs, alfalfa sprouts, chives, buckwheat lettuce, water-cress, and sunflower greens.

La Pajarosa
2010 Pleasant Valley Road
Aptos 95003
(408) 724-2044
Open all year.
Organic Christmas trees,
4-H feeder lambs, and wool.

Santa Cruz Orchards
P.O. Box 1510
Freedom 95019
408) 728-0414
Call for location.
Tours are offered.
March: peas.
Aug.: Organic squash.
Sept. - Dec.: apples.

Swanton Berry Farms
P.O. Box 308
Davenport 95017
(408) 425-8919
(Sales at 5221 Coast Road, 5 miles north of Santa Cruz on Hwy 1.)
May-Aug., Wed.-Sun., 11am-5pm
Farm offers varietal tastings and a self-guided tour of their demonstration plot and sell organic strawberries, tomatoes and other fruits and vegetables.

The Farm
5555 Soquel Drive
Soquel 95073
(408) 476-5613
All year daily, 9am-9pm
Restaurant.
Also sells organic vegetables.

Webb's Organic Farm
5381 Old San Jose Road
Soquel 95073
(408) 475-1020
All year daily, 8am-5pm
(Call first and bring containers.)
July: cattle, calves, hogs, apricots, grapes, nectarines peaches, Bartlett pears, plums, prunes honey.
Sept.: lemons and variety apples.
Oct.: Indian corn, squash, snap beans, beets, cabbage, carrots, peppers, potatoes, tomatoes, figs, almonds, English walnuts, pecans, pumpkins, flowers, and persimmons.

SOLANO

Old MacDonald's Farm
250 North Orchard Avenue
Vacaville 95688
(707) 448-3301
Jan. - Feb., and May - Early Fall
(Call first and bring containers)
Farm sells all organic produce
Jan.- mid-Feb.: oranges & lemons.
Early May: Royal Ann Cherries.
Summer - early Fall: cherry tomatoes, crookneck squash, lemon cucumbers and zucchini.

Star Valley Farm
2985 Mix Canyon Road
Vacaville 95688
(707) 448-5303
(Call first and bring containers)
Mail order is available
Sells all certified organic produce
May - Mid June: cherries.
June: apricots, figs.
June - Aug.: red beauty and Santa Rosa plums.
Sept - Oct: Fuyu and Hiachia persimmons, and walnuts.
Oct.: pomegranates.

SONOMA

Appleseed Orchards
1834 High School Road
Sebastopol 95472
(707) 829-1121
Open all year.
(Call first and bring containers)
Mail-order and picnic area are available.
Apple juice and honey.
July - Dec.: organic apples, vegetables and applesauce.

Buzzards Roost
1778 Facendini Lane
Sebastopol 95472
(707) 823-2799
Open all year
(Call first.)
Tours by appointment.
Picnic deck available.
All year: organic sugar, snow peas, spinach, Swiss chard and lettuce .
Angora and mohair wool and yarn products are also available.

Chicken Crossing
1100 Ozone Drive
Santa Rosa 95401
(707) 575-8917
March - Oct.
(Call first.)
Organic chicken, duck and goose eggs from range-fed birds; garden vegetables.

Figs A'Plenty
20301 - 5th Street East
Sonoma 95476
(707) 996-0375
June - July and Sept. - Oct.
(Call first and bring containers)
Pick your own figs.

Krout's Pheasant Farm
3234 Skillman Lane
Petaluma 94952
(707) 762-8613
May - Sept., by appointment
Sept. 15 - Jan. 15, Tues. - Sat, 1pm-5pm.
(Call first and bring grape containers)
Tours by appointment only.
May - July: pheasant eggs.
September: organic concord grapes.
Oct - Feb and May - June: pheasants.
Organic veal as available.

Farms , continued

Lucky Duck Christmas Tree Farm
2310 Magnolia Avenue
Petaluma 94952
(707) 763-9710
Thanksgiving - Christmas
10am - 5pm daily
Organic Christmas trees

Oak Hill Farm
(707) 996-6643
14805 Sonoma Highway
Glen Ellen 95442
(Sales at 15101 Sonoma Hwy)
July-October
(Bring your own containers.)
Farm sells all organic produce.
Aug. - Oct.: melons.
July - Oct.: vegetables & flowers.

Paul Orchards
(707) 823-7907
3561 Gravenstein Hwy North
Sebastopol 95472
Aug.-Dec. 22, daily, 10am-5pm
Mail order and free samples of apple juice, fresh cider, and jams.

Petaluma Mushroom Farm
(707) 762-1280, (707) 795-1260
782 Thompson Lane
Petaluma 94952
Open all year.
Mon.-Fri., 9am-5pm; Sat., 9am-12pm
(Call first.)
Organic mushrooms.

Puff Lane Organic Farm
(707) 539-8147
6969 St. Helena Road
Santa Rosa 95404
(Sales at Santa Rosa & San Rafael farmers' market.)
(Bring containers.)
Farm sells all-organic produce.
As in season: apples, pears, figs, rhubarb, cherries, peaches, eggs, walnuts, artichokes, asparagus, beans, beets, broccoli, cabbage, corn, carrots, cauliflower, cucumbers, eggplant, garlic, lettuce, melons, peas, peppers, radishes, spinach, Swiss chard, squash, tomatoes, herbs, flowers, and berries.

Reverie Orchards
(707) 823-3624
4815 Thomas Road
Sebastopol 95472
Oct.-Nov.
(Call first.)
Farm sells organic Rome apples.

River Valley Organic Garden
(707) 857-3868
21040 Railroad Avenue
Geyersvile 95441
(Sales off Hwy 128, one block from the town of Geyserville.)
All year daily, 8:30am-7:30pm
Picnic areas are available and children can collect eggs and pet rabbits.
Farm sells all-organic produce:.
All year: tomatoes, corn walnuts brown eggs, Muscovy ducks, French & English Angora Rabbits, Angora wool.
Aug: pears.
Oct: pomegranates.

Serpentine Garden
(707) 823-4366
1210 Furlong Road
Sebastopol 95472
Open all year
(Call first and bring containers.)
Farm sells organic vegetables and offers macrobiotic consultations.

Sonoma Antique Apple Nursery
(707) 433-6420
4395 Westside Road
Healdsburg 95448
Jan. 15- March 31: Tue., Wed., Fri & Sat, 9am-4:30pm; all year Wed 9am-4:30pm
(Call first and bring containers)
Mail order is also available.
Farm is a member of CCOF .
Jan. - March: 100 varieties of unusual & heritage varieties of apples, pears, and other fruit varieties.
Sept.: four varieties of wine grapes.

The Earthworm Co.
(707) 539-6335
3675 Calistoga Road
Santa Rosa 95404
All year, Mon.-Fri., 10am-5pm; weekends by appointment.
(Call first.)
Mail-order and consultations are available. Tours by appointment. Farm sells earthworms for soil correction & organic earthworm castings all year. Instructions provided with purchase. Consumers pick worms.

Vater, Roy
(707) 823-9566
5889 Lone Pine Road
Sebastopol 95472
June-Oct., 8am-6pm.
(Call First.)
June - July: boysenberries .
Late Aug. - Sept.: raspberries.
Sept .- Oct.: concord grapes
June - Oct.: honey.
July: peaches.

Via Verde Organic Gardens
(707) 527-8443
650 Irwin Lane,
Santa Rosa 95401
All year by appointment.
(Call first and bring containers.)
Farm sells all organic produce:
Canning tomatoes, Roma tomatoes, basil, seven varieties of summer squash, pickling cucmbers, elephant garlic, fresh and dried flowers as in season.

Voge Ten-Fruit Ranch
(707) 823-0485
1430 Hollman Lane,
Sebastopol 95472
June 15-Nov. 15, Fri.-Sun., 9am-6pm
Picnic area is available.
July - Sept.: plums.
Sept. - Oct.: grapes.
July - Oct.: various vegetables.

Western Skies Ranch
(707) 545-8079
1156 West Ave. , Santa Rosa 95407
(Sales in rear, behind the barn.)
July-Oct., Wed, 8am-6pm; Sat. & Sun., 11am-5pm.
(Call first.)
Recreation tours, swimming and yard swing available. Farm sells all organic produce: Golden, Red Delicious apples, & Rome Beauty apples, grapes, pears plums, prunes, figs, berries, Swiss chard, comfrey, cherry tomatoes, mints, as in season;
Nov.- Dec.: English walnuts .
Nov. - Jan.: pineapples and guavas.

Wheeler Ranch
(707) 874-3029
P.O. Box , Occidental 95465-253
(Call first.)
Mail order and brochures available.
July: organic garlic.

Chapter Three

Workplace

Most of us spend two thirds of our waking hours at work or getting there and back. Our workplaces, whether office or factory, are often highly destructive to the environment. That's why it's just as important to develop sound day-to-day environmental practices at work as it is at home.

Every business can recycle, conserve energy, and get rid of obvious toxics on-site. In addition, businesses have unparalleled opportunities for promoting green causes and creating green markets. By publicizing the things it is doing right, a company puts pressure on other businesses to follow suit. Such pressure is already being exerted by the market; recent surveys show a majority of American consumers consider the environment in making purchasing decisions, and many are willing to pay more for products that minimalize environmental impacts.

Many of the things individual businesses can do are specific to themselves; some things any business can do. This chapter covers only the more general things. Please let us know about the other things you think up yourself.

Greening Your Business

By Stephen C. Evans

Contrary to common expectations, converting a business to sound environmental practices is neither expensive nor particularly difficult. In fact, substantial savings can result from following the simple principles of reduce, reuse, and recycle.

There are many things any individual can do within the work setting to effect the kind of changes discussed below. But systematic structural changes reap the largest rewards. It is therefore very important that both management and staff be involved in assessing environmental policies on an ongoing basis. Keep channels of communication open, and don't become complacent or satisfied that the job is done. There is always room for improvement.

1) Set up an office recycling program.

Put all doubts behind you; recycling is not just a nice idea — it saves money, too. Hewlett Packard estimates a savings of $20,000 per year in one facility from recycling just white paper, computer paper, and corrugated cardboard.

To recycle right, you should first analyze your office's waste production. Be thorough at this stage and avoid the temptation to trivialize what appear to be small things. For instance, just one copy of the Wall Street Journal per day for a year requires four trees to print.

Once you've figured out what kind of garbage you're producing, and where, put recycling containers where they will do the most good: paper bins next to the copy machine and computer printers, beverage container bins in the lunch room, and so on. Place two small bins next to each desk: one for white paper, the other for mixed. Avoid the false economy of skimping on bins and containers; if a worker has to go too far to recycle an aluminum pop can, it may not get recycled. Frequent trips to the next room or down the corridor can also amount to a substantial waste of time (and hence money).

If there is no curbside recycling program in your area, lobby city politicians and the county solid waste disposal agency to start one. Curbside service is spotty at best in the Bay Area. This can be changed with the right number of phone calls and letters.

Until you get curbside service, you will need some way of getting your recyclables to a recycling center. Some areas have recycling businesses that make a specialty out of picking up recyclables from businesses and cashing them in at a buy-back center. The Palo Alto recycling program staff has estimated it takes about 200 office workers to create enough waste paper to be worth a commercial recycler's effort. (Check the Consumer Guide to see if there is one operating near you.)

Failing this, an employee may be willing to make the run to the recycling center in return for the buy-back fees. Whether it's your own employee or someone else's, though, do your best to make it worth their while. If there are several businesses in your building or complex, try to get others involved in a cooperative recycling program with large centralized bins (or even dumpsters) that several businesses can share. Remember, the more material the hauler can take per trip, the more profitable is his or her enterprise.

2) Persuade employees to leave their cars at home.

Air pollution, congestion, global warming — by now we are all painfully familiar with the rash of problems caused by the automobile.

Nonetheless, each morning hundreds of thousands of Bay Area residents get into their cars to face longer and longer commutes to work — in traffic that grows ever more sluggish.

The work day starts when the employee arrives at work, and ends upon departure for home. Traditionally employers have not concerned themselves with the logistics of their employees' comings and goings.

How different things might be if standard hourly wages included time spent sitting in traffic jams. Proximity of home to workplace would quickly take a high place in qualifying candidates for jobs.

In the absence of such sensible criteria, however, there are still several things employers can do at minimal cost to help alleviate the woes caused by all those cars. At almost no cost, for instance, a commute bulletin board can be set up in a central location to help people who live near each other set up car pools. Use of the board can be encouraged at staff meetings, via memos, and by awarding prizes to frequent ride sharers.

Provide a secure bicycle parking area and other incentives for using this most efficient of vehicles. The Alza Corporation of Palo Alto pays employees one dollar for each day they ride a bike to work. The city of Palo Alto requires large office buildings to provide secure bike parking and showers for workers who ride bikes.

If you have on-site parking, consider charging fees for its use. Money thus generated can be used to offer discounts on mass transit tickets.

If your company offers employees the use of company cars as a benefit, consider offering free mass transit passes instead.

3) Conserve, conserve, conserve.

Water, electricity, and fuel oil all have vastly higher commercial usage than residential usage. So if you have weatherstripped your doors and windows and installed compact fluorescent bulbs and low-flow shower heads at home, don't stop until you've seen to it that similar measures are taken at work.

It is not necessary to bathe the entire office in bright light. Only reading and work areas need be brightly-lit. In less-used areas, use lower wattage bulbs, or replace some fluorescent tubes with burned out ones.

For brightly-lit areas, it is better to use one large bulb than several small ones. By locating most of our desks near windows here at the Green Media Group, we use no electric lights whatsoever during the day.

Besides saving energy, evidence suggests that natural lighting (daylight) reduces depression, irritability, illness and fatigue. If you don't have windows in every room, install full-spectrum compact-flourescent lighting in the rooms that don't. The extra initial expense will certainly be compensated by the improvement in employee morale and increased energy.

A tremendous amount of water is routinely used in various manufacturing processes, for purposes ranging from cleaning to transport of material. If you are in a position to do so, figure out ways to do the same things better with less, or no, water. In the same vein, constantly seek ways to reduce the use of electricity or fuels in manufacturing. Electric motors, if not properly maintained, can be huge wasters of electricity, for example. (See "The Nega-Watt Revolution, p. 47.)

4) Provide a healthy workplace.

There are many small things in the workplace that can combine to create health problems for workers. For instance, adhesive tape, rubber cement, some correction fluids, and many cleaning agents contain toxic substances. Use paper clips or staples, white glue, tape that lifts off or covers errors, and non-petroleum based cleaners instead.

Encourage employees to bring indoor plants to work, or supply some. There is more than a purely psychological benefit to having house plants around; recent NASA studies show that plants actually act as natural air filters, cleansing the environment of toxic gases. Philodendrons were found to be particularly effective at absorbing carbon monoxide and formaldehyde (which is present in several common building materials). Spider plants, golden pothos, chrysanthemums, and gerbera daisies were also found to be very effective.

Better yet, avoid putting toxins into your work environment in the first place. Paint is a common, though often overlooked, source of interior air contaminants. It is conventional wisdom that water-based paints are less toxic. Though partially true, these paints in fact usually contain potentially harmful fungicides. If possible, use natural paints made from ingredients like chalk, pine resin, and beeswax. Other building and decorating materials, from particle board to carpeting can also contain toxic chemicals (see following article.)

5) Treat paper as a precious commodity.

The modern office couldn't exist without paper, from paper cups at the water fountain, to the stuff that spews out of our computer printers in endless ribbons, to the cartons and mailers we send all over the world. American offices dispose of 4.5 million tons of office paper every year. If it were all recycled, we would save 75 million trees, 30 billion gallons of water, and enough energy to heat 2.3 million homes, according to Earth Day, Inc. research. The Franklin Research Institute says paper makes up a whopping 41 percent of our garbage.

Every step you can take to reduce, recycle or reuse paper is a step in the right direction. Use the back of copier paper either for more copies or as scratch paper before throwing it in the recycling bin. Similarly, don't throw a piece of paper away if you make a mistake; use the back for something. Reuse manila envelopes and folders.

Don't use colored papers, as they require more bleaching when they are recycled. Also in the "don't" department are plastic window envelopes. Ask for glassine windows, made from biodegradable cellophane, or use open-windowed envelopes instead.

Above all, use recycled paper and recycled paper products whenever possible. Recycled paper costs manufacturers three percent less than ordinary paper to produce, and the only reason it costs more at retail is low sales volume. As soon as demand picks up, the prices should drop.

6) Consider your packaging habits.

Garbologists estimate that packaging in its myriad forms makes up about 65 percent of what we throw out every day. In a land of conspicuous waste, packaging is the most conspicuous of all.

A task force appointed by the Coalition of Northeastern Governors has come up with a short list of "preferred packaging guidelines." Listed from best to worst, they are:

1. *No packaging at all.* Many products that require no packaging — from produce to hardware and tools — are nonetheless shrink-wrapped, blister-packed, bundled, or boxed in a dizzying variety of styles. If it doesn't need it, save money and don't do it.

2. *Minimal packaging.* Use absolutely the minimum amount of material required to do the job. Given the opportunity, design your product in such a way as to require very little or no packaging. Pack larger quantities into a single container. If possible, reduce the weight of materials used.

3. *Packaging that can be consumed, returned, refilled, or reused.* Sterilizing and refilling bottles, as one example, is far less energy-intensive than recycling the glass into new bottles. Another good example is the jelly jar that doubles as a drinking glass when it is empty.

4. Recyclable and recycled packaging. If a package can't both be made of recycled materials and be readily recycled, it should at least be made to be recyclable so it doesn't have to be landfilled. If recycling facilities are not close enough to make it viable (as is the case with most kinds of plastic in the Bay Area) the packaging is not recyclable. A package is much more easily recycled if it is made entirely from one material — seals, labels, and closures. Packages that are made entirely from recycled materials, but which are not recyclable, will end up in the dump — leaving such packages at the bottom of the task force's preference list.

7) Publicize what you're doing right

There are many day-to-day opportunities for businesses to raise the environmental awareness of other businesses and the public. If your stationary is printed on recycled paper, put the recycling logo on it. If you package your goods in reusable containers, print the words "reusable shipping container" on the outside. Tell your colleagues about the green programs you've instituted, and encourage them to follow suit.

Recent polls show American consumers to be very concerned about environmental issues. So if you are doing things right, advertise it. It will set an example and create market pressures that will bring other businesses to follow.

8) Take a pledge: sign the Valdez Principles

The Coalition for Environmentally Responsible Economies, a project of the Social Investment Forum, has devised a broad set of principles, known as the Valdez Principles, as guidelines to help businesses operate in accordance with sound environmental practices (see the sidebar at the right). Companies that sign this oath make a long-term commitment to the goal of environmental sustainability. For more information about signing the Valdez Principles, contact the Social Investment Forum, 711 Atlantic Ave., Boston, MA, 02111. Phone (617) 451-0927, or FAX(617) 482-6179.

THE VALDEZ PRINCIPLES

By adopting these Principles, we publicly affirm our belief that corporations and their shareholders have a direct responsibility for the environment. We believe that corporations must conduct their business as responsible stewards of the environment and seek profits only in a manner that leaves the Earth healthy and safe. We believe that corporations must not compromise the ability of future generations to sustain their needs.

We recognize this to be a long-term commitment to update our practices continually in light of advances in technology and new understandings in health and environmental science. We intend to make consistent, measurable progress in implementing these Principles and to apply them wherever we operate throughout the world.

1. Protection of the Biosphere We will minimize and strive to eliminate the release of any pollutant that may cause environmental damage to the air, water, or earth or its inhabitants. We will safeguard habitats in rivers, lakes, wetlands, coastal zones and oceans and will minimize contributing to the greenhouse effect, depletion of the ozone layer, acid rain, or smog.

2. Sustainable Use of Natural Resources We will make sustainable use of renewable natural resources, such as water, soils and forests. We will conserve nonrenewable natural resources through efficient use and careful planning. We will protect wildlife habitat, open spaces and wilderness, while preserving biodiversity.

3. Reduction and Disposal of Waste We will minimize the creation of waste, especially hazardous waste, and wherever possible recycle materials. We will dispose of all wastes through safe and responsible methods.

4. Wise Use of Energy We will make every effort to use environmentally safe and sustainable energy sources to meet our needs. We will invest in improved energy efficiency and conservation in our operations. We will maximize the energy efficiency of products we produce and sell.

5. Risk Reduction We will minimize the environmental, health and safety risks to our employees and the communities in which we operate by employing safe technologies and operating procedures and by being constantly prepared for emergencies.

6. Marketing of Safe Products and Services We will sell products or services that minimize adverse environmental impacts and that are safe as consumers commonly use them. We will inform consumers of the environmental impacts of our products or services.

7. Damage Compensation We will take responsibility for any harm we cause to the environment by making every effort to fully restore the environment and to compensate those persons who are adversely affected.

8. Disclosure We will disclose to our employees and to the public incidents relating to our operations that cause environmental harm or pose health or safety hazards. We will disclose potential environmental, health or safety hazards posed by our operations, and we will not take any action against employees who report any condition that creates a danger to the environment or poses health and safety hazards.

9. Environmental Directors and Managers We will commit management resources to implement the Valdez Principles, to monitor and report upon our implementation efforts, and to sustain a process to ensure that the Board of Directors and Chief Executive Officer are kept informed of and are fully responsible for all environmental matters. We will establish a Committee of the Board of Directors with responsibility for environmental affairs. At least one member of the Board of Directors will be a person qualified to represent environmental interests to come before the company.

10. Assessment and Annual Audit We will conduct and make public an annual self-evaluation of our progress in implementing these Principles and in complying with applicable laws and regulations throughout our worldwide operations. We will work toward the timely creation of independent environmental audit procedures which we will complete annually and make available to the public.

Non-Toxic Building Design

by Sim Van der Ryn
and Therese Peffer

In addition to their impact on the environment, energy, and resource use, buildings we work and live in may be dangerous to our health. The health professions are tracing occupant health problems which were previously attributed to "flu" or "common cold" or even "hysteria" to real problems in the design and operation of building systems.

Since Americans spend up to 90% of their time indoors, it's no wonder that building-related illnesses (or sick building syndrome) have become increasingly common. The combination of the emphasis on energy-efficiency and the increased use of synthetic materials, as well as the trend towards centrally-controlled air systems, has led to tighter buildings with less exchange of air and correspondingly higher levels of contamination.

Research is just beginning on the sources of health hazards in construction materials and furnishings. Asbestos, radon, and formaldehyde have come under public scrutiny as health hazards, but for the vast majority of organic chemicals, no standard exists to determine what level of exposure threatens the health of the inhabitants. The research that has been done addresses indoor air quality, yet water-supply contaminants and electromagnetic radiation deserve study as well.

The effects on humans range from annoyance (fatigue, headache from smell of fresh paint and platicizers in auto interiors), to irritation of the mucus membranes (dry throat, eyes, and nose, allergies), to cancer (periodic exposure to asbestos and radon) to fatal heart attacks (exposure to methyl chloride) and other forms of death (Legionnaires' disease).

There are more questions about the effects of various chemicals than answers. Formaldehyde, for example, is prevalent in most building materials, from particle board to draperies to adhesives, and is considered a mutagen and animal carcinogen, and a possible human carcinogen. In extreme cases, sensitive persons suffer a breakdown of their immune systems from exposure to various chemicals (known as CAIDS — Chemically Acquired Immunity Deficiency Syndrome).

Major material contributors to the contamination of indoor air are:

• Particles and fibers (asbestos from insulation, and fiberglass).

• Outgassing of finishes, floor coverings, furnishings (volatile organic compounds such as formaldehyde from particle board).

• Adhesives (simple organic compounds such as toluene and xylene).

• Combustion products from gas stoves and appliances, wood burning stoves, attached garages, kerosene heaters (CO, CO2, NO, NO2, formaldehyde, benzene, benzo-a-pyrene).

• Pesticides and wood preservatives (heptachlor, chlordane; pentaclorophenol).

• Airborne biological contaminants (bacteria, molds, spores).

Contributors to water pollution include contaminants from asbestos, radon, and organic compounds from PVC (polyvinyl chloride), PCB (polychlorinated biphenyl), or PE (polyethylene) pipe, pipe dope, soldering flux, and adhesive for PVC pipe. Lead can leach into water supplies from solder, and lead particles from soldering residue can contaminate drinking water over years after the original plumbing work.

Part of the problem lies in assessing health risk. For example, the source needs to be isolated and measured accurately. This can be problematic since many factors influence indoor chemical concentration, such as temperature, humidity, air exchange rate, and "wall effects" (adsorptive properties of some materials).

Some materials emit large quantities of contaminants at first, which decrease with age. There are immediate effects (symptoms that surface with initial exposure) and long-term effects (symptoms that appear after chronic exposure). The sensitivity of the population must be taken into account (15 percent considered chemically sensitive). We presently lack a database of sources of contaminants and allowable levels, thus making risk assessment a subjective art.

For existing buildings, corrective action may involve removal, substitution, and/or modifications of the pollution source. In new construction, new design guidelines should play a role. The cost-effectiveness of using alternative materials (non-toxic paints and finishes, formaldehyde-free plywood), versus other techniques ("flushing," baking the environment, or sealing the offending material) should be assessed. If the material serves as a direct contributor or as a sink for other contaminants, perhaps it should be removed. Increased ventilation may be the solution in some cases, but not for formaldehyde, asbestos, or continuously evaporative substances. Ventilation systems need to be addressed; air to air exchangers (HRV's) have been suggested as an alternative. Continuous filtration of either air or water is another option.

Several public agencies and professional associations are currently assessing the health effects of toxics in building materials and attempting to define standards and alternatives. The Air Resources Board has launched a five year study of the sources and effects of indoor air pollution, and is develolping monitoring techniques. The American Institute of Architects will soon publish a summary of recent research on the subject in the form of an Environmental Resource Guide catering to architects and those in the construction trade.

In terms of regulations and guidelines in the area, the EPA is beginning an extensive study of emissions and health effects of many building materials to be produced as a series of guidebooks and a computerized database. Under a congressional mandate, the EPA is also taking a role in research and the development of mitigation technology.

Asbestos, UFFI, and lead paint have already been banned and the regulation of other materials is under consideration

If you suspect indoor air pollution as a cause of illness in your workplace, call the Occupational Safety and Health Administration (OSHA). It is the only federal agency with the authority to require alterations in the building — if it finds any violations of existing regulations, which unfortunately are not at present particularly stringent.

Sim Van der Ryn is an architect and a founder of the Farrallones Institute. Therese Peffer works with Mr. Van der Ryn.

Building Materials Guide

MATERIAL/ PROBLEMS	ALTERNATIVES
Particle Board Up to 10% by weight urea-formaldehyde; may emit fumes for years.	Pine boards and other solid wood boards.
Hardwood Plywood Urea-formaldehyde, although at lower levels than in particle board.	Pine panelling and other solid wood boards.
Softwood Plywood Interior grades contain urea-formaldehyde. Exterior grades contain Phenol-formaldehyde	Pine boards and other solid wood boards, or exterior grades.
Fiberglass insulation Glass fibers irritate eyes, skin and respiratory tract- particularly installer	Natural cork or Air Krete, nontoxic magnesium based foam.
Paints, Sealants, Stains and Preservatives May contain: Toxic solvents, mildewcides, and fungicides.	Nontoxic finishes, or no finish at all.
Carpets and Rugs May contain a wide assortment of stain repellents, biocides and fungicides.	Natural wood, ceramic tile or polished concrete floors, or old or handmade area rugs of natural fibres.
Countertops Usually used over particle board.	Sealed exterior grade plywood, or Corian
Drapes, furniture,upholstery May contain formaldehyde and other chemicals	Untreated natural materials: cottons, wool, linen, feathers, horeshair, down.
Furnaces, stoves, fireplaces and heaters Fuel burning appliances can give off a variety of toxic fumes that can leak into living areas.	Solar gain, heat pumps or electricity
Cellars and crawl spaces Possibility of infiltration of dampness and radon gas.	In new construction, build on a concrete slab with a moisture barier.

An excellent source for further information on toxins in building materials - and many other products with which we come into contact on a regular basis is *Nontoxic & Natural: How to Avoid Dangerous Everyday Products and Buy or Make Safe Ones*, by Debra Lynn Dadd (1984, Jeremy P. Tarcher, Inc).

The Nega-Watt Revolution

By Amory Lovins

Ron Perkins' colleagues at Compaq Computer Corporation were incredulous. Perkins, manager of facilities resource development, believed Compaq's next Houston office building could be designed to use a fourth less electricity per square foot than the previous one. "Why make it so hard on yourself?" they asked. "Our designs are already excellent. Savings of 5 or 10 percent would be pushing your luck." But a year later, when construction was completed, the building's use of electricity proved a third less than previous designs, thanks to the new electricity-saving technology that Perkins had harnessed. The payback period on the energy-saving equipment was only a few years. Perkins has since designed another building in which he expects to slash electric use by still another fifth.

The innovative equipment that Perkins uses in his buildings — chiefly advanced lighting fixtures, electronic ballast, high-efficiency lamps, and lighting controls — are part of a flood of astonishingly cheap and powerful electricity-saving techniques. They have the potential to add many tens of billions of dollars a year to business' bottom line.

Some companies are already enjoying the rewards that come from saving electricity. For example, Southwire, the largest independent rod, wire, and cable business in the United States, found itself facing hard times in the early 1980s as the energy-intensive heavy industry was squeezed between market prices and manufacturing costs. The company responded by cutting its total energy use per pound of product by half in eight years. The energy reductions — about a 60 percent savings in gas and 40 percent in electricity — yielded virtually all of the company's profits during the tough years of 1980 to 1986, and may have saved 4,000 jobs at 10 plants in six states. Southwire continues to make further improvements, which still generally pay back investment in fewer than two years.

Even in less energy-intensive industries, savings from energy-efficiency can be dramatic. A couple of years ago, a large company had an energy manager at one of its plants who was achieving annual energy savings of $3.50 per square foot. "That's nice — a few million extra on our bottom line," said one of the company's executives. He then added, in the same breath, "But I can't really get excited about energy. It's only a few percent of our cost of doing business." Such thinking is stalling energy improvements throughout the corporate world. Installing energy-efficient equipment may not be sexy, but the savings are real. If this executive's company had achieved similar savings at all its facilities worldwide, its total net would have gone up by 56 percent.

Industry has already seen major savings from its fuel-conserving programs initiated during the Arab oil embargo of the early '70s. Yet companies spend more than twice as much on electricity as on oil. Unbeknownst to many in industry, in the past few years there have been tremendous advances in electric efficiency. Electricity-saving technology is evolving so quickly that most of the best options now on the market didn't exist last year. Today, you can save twice as much electricity as you could five years ago, at only a third the real cost. Practically every building, however modern, can be made much more efficient.

American companies have a $93 billion annual electric bill, with 25 to 45 percent of the total going towards lighting — about three fourths of it directly and a quarter to counteract the heat generated by the lights. In most commercial buildings, lighting consumes more than a third of the electricity used—upward of half when the cooling load is considered. Yet according to studies by Lawrence Berkeley Laboratory (the leading national lab on saving energy in buildings) and Rocky Mountain Institute, 80 to 90-plus percent of this lighting energy could be saved by fully converting to today's most efficient lighting equipment.

The vast range of efficient lighting hardware now available fits almost any need, providing unchanged lighting levels with less glare, more pleasant and accurate color, no flicker, and no hum. Upgrading a typical office fluorescent lighting system can be accomplished by installing computer-designed reflectors, which deliver virtually the same light from half as many lamps, new lamps that give off more light per watt and nicer color, sophisticated high-frequency electronic ballasts, which start and regulate the current of the lamp and can now power four lamps instead of two, and several kinds of controls. As a result, a company will need only half as many lamps and a quarter as many ballasts which should save it 70 cents per square foot on maintenance costs-nearly half the total cost of the upgrade. Typical direct energy savings are about 70 to 90 percent, and including the 35 to 40 percent "bonus" for saved space-cooling, most paybacks are well under two years.

Even juicier savings come from converting incandescent lamps, such as the ubiquitous floodlamps in can fixtures, to compact fluorescent lamps.

These lamps can cut lighting bills by 75 to 85 percent, and they last 4 to 13 times as long, thereby more than paying for themselves just by reducing maintenance costs on replacement bulbs and the labor needed to install them.

Other improvements can boost lighting savings by another third or more, including better maintenance, lighter-colored finishes and furnishings to distribute light better, top-silvered blinds and glass-topped partitions to bounce sunlight three times as far into buildings, polarizing lenses that make reading easier by almost eliminating glare, half-watt electro-luminescent panels to replace 30 to 50 watt EXIT signs, and miniature tungsten-halogen spotlights for displays.

Together, these commercially available lighting innovations have the potential to save about a fourth of all the electricity in the country, at a net cost somewhat less than zero. In fact, Rocky Mountain Institute estimates that because the amount saved on maintenance costs would be more than the cost of the electricity-saving devices, the average cost of replacement will be about minus 1.4 cents per kilowatt-hour. In the United States this would displace 120 Chernobyl-size power plants costing about $200 billion and eliminate more than $30 billion a year in utility operation costs. This may be the biggest gold mine in the whole economy.

Opportunities nearly as dramatic abound in every other electricity-consuming device. Together, they can cut U.S. electricity consumption by another half.

After lights, motors are probably the next fattest opportunity. Motors use at least two thirds of industrial electricity and some 53 to 60 percent of all the electricity in the country—more than $90 billion a year's worth, or about 2 percent of our gross national product. In fact, making the electricity to run U.S. motors now uses more fuel than is consumed by all U.S. highway vehicles.

A typical big industrial motor consumes electricity costing some 10 to 20 times its own total capital cost per year. Over a motor's life, a 1 percentage point gain in efficiency typically adds at least $10 per horsepower to the bottom line. Direct efficiency gains averaging about three-and-a -half percentage points are currently possible, which for a motor-intensive company, such as a paper mill, can create enough savings to turn around a foundering firm.

Two measures that have gained wide acceptance are buying only high-efficiency new motors, which can now save twice the electricity that they could a decade ago, and using electronic speed controls. Immediately replacing a standard induction motor with a high-efficiency model has many advantages. In addition to cutting electricity costs, the replacement will last twice as long because it runs cooler and has better bearings, will need fewer capacitors to boost the motor's "power factor" (the fraction of electricity fed into the motor that actually turns it rather than heating it) and will work better with adjustable-speed drives.

Electronic speed controls have become popular because many machines, especially pumps and fans, need to vary their speed to match production needs. Before electronic adjustable-speed drives became widespread and affordable, output was usually varied by running the pump or fan at full speed while "throttling" its output with a partly closed valve or damper-like driving with one foot on the accelerator and the other on the brake. Today, electronic speed controls can eliminate this waste. When you need only half the flow from a pump, you can save almost seven-eighths of the power because its energy needs vary as roughly the cube of its flow. In all, electronic adjustable speed drives can save 14 to 27 percent of total U.S. motor energy, with paybacks of a year or two. Only a few percent of this opportunity has yet been grasped.

Currently, most engineers consider just these two measures, ignoring the other half of the total electricity-saving potential in motor systems. At Rocky Mountain Institute we have identified 33 kinds of further improvements that could be made to motor systems, comprising the choice, maintenance, sizing, and controls of motors and the systems that supply electricity and transmit torque from the motor to the driven machine. Implementing all 35 of the improvements can cut the motor systems use of electricity in half, for a potential national saving equivalent of 80 to 190 giant power plants. (This figure doesn't even take into account the potential for another 50 percent savings on the remaining electricity bill from improving the machinery that the motors are driving.) Because you pay for only seven of the 35 improvements-the rest are cost-free by-products-the average payback on the doubled efficiency is only about 15 months.

Capturing the savings depends on simultaneously doing many things right. For example, to double a motor's lifetime, the motor's shaft must be kept precisely aligned with the shaft it's driving or the bearing will fail prematurely, and the bearings themselves must be lubricated by someone with clean hands to prevent dirt from getting into the grease and eating the bearing-both simple steps that frequently aren't taken in American industry.

Companies should switch from V-belts, which stretch, slip, and require so much tension to stay in place that they harm the bearings, to "synchronous" belts, which have teeth that engage sprocket lugs so the belt doesn't slip, and fiberglass or Kevlar bands inside so it doesn't stretch. Not only would such a conversion save about 5 to 15 percent of the transmitted energy, but it would save about a dollar per kilowatt-hour because of immense maintenance savings.

Maintenance itself must also be improved. Poor maintenance ruins costly motors, wastes energy, and needlessly incurs downtime costs that can exceed $10,000 per hour.

Measurements by General Electric Company suggest that in the United States between $1 billion and $2 billion worth of electricity is wasted each year by the damage done to the iron cores of motors as a result of poor repair practices used to remove old windings. An alternative technique using only gentle warmth to loosen old windings for removal causes no damage and is faster and cheaper, but is known to relatively few motor repairmen. Another step to improve the state of motor maintenance would be to make lubrication and other motor upkeep a white-lab-coat profession, with a "motor doctor" who makes house calls and administers precisely metered dosages of special medicine to motors.

When it comes to energy efficiency, details matter. Jim Clarkson, the mastermind behind Southwire's dramatic savings, found that before executives toured plants, motors were often given a coat of shiny new paint. Over the years, so many coats built up that the heat couldn't get out. Today, you can't repaint a Southwire motor without first stripping off the old paint. Clarkson also discovered that of the typical 5 percent power loss between the meter and the machinery, three fourths could be saved, with a payback of around two years, just by installing wire twice as fat. The wire in most big buildings, it seems, is chosen by low-bid electricians told to meet the local building code, which is meant only to prevent fires, not to save money.

Many energy-saving techniques require no investment. Clarkson found he could save Southwire 10 percent of its motor electricity bill by turning off idling motors. At some machines he installed a red light that went on when high peak-period utility charges approached, and told the operators that if they would take a coffee break when the light went on, the company's overall profits would be more than if they kept working.

Almost every other electricity-consuming device holds potential savings as well. Replacing a desktop computer with an equally sophisticated laptop model can save up to 95 percent in electricity-enough to pay for the difference in cost for the laptop-while improving safety, portability, ergonomics, and space use, and eliminating the need for a costly uninterruptible power supply. Simple improvements to such common office devices as laser printers and photocopiers can save most of their energy and help avoid multimillion-dollar investments to expand air-conditioning capacity to handle machine heat in older buildings. For both these devices, for example, "cold fusing"-setting the toner onto the paper with a cold compression roller rather than a hot drum-saves 90 percent of the electricity, eliminates fumes and warm-up time, and gives twice the life with half the maintenance. Further savings can be gained by installing controls that turn such machines off or into a standby mode when not in use.

Making windows more efficient also saves money. Most buildings use plain glass, so you're hot in summer and cold in winter. But new "superwindows" provide year-round comfort. Some let in 60 percent of the visible light, thereby displacing electric light and the heat it produces, while admitting only two percent of the sun's heat. Other windows, designed for cold climates, can insulate up to six times as well as double glazing, and can even gain more heat than they lose in the winter while facing in any direction, including north. At Rocky Mountain Institute's research center, we have no furnace in a climate that goes down to -47 degrees Fahrenheit. Our cold-climate windows not only permit us to do without a furnace, cutting winter heating bills by $1,000 a month, but also have reduced our building's net capital cost. The reason: We saved more by eliminating the furnace and ductwork than it cost us to install the superwindows and superinsulation. Fully used, superwindows could save the United States four million barrels worth of oil and gas per day, at costs of a few dollars per barrel-far cheaper than drilling for more.

Increasingly popular super-efficient appliances are another fountainhead of savings. There are refreigerators and freezers on the market that consume 10 to 20 percent of the usual amount of energy, commercial refrigeration systems that save more than 50 percent, and televisions and high-performance showerheads that save 75 percent. The collective results can be astounding. My 4,000 square-foot home's lights and appliances cost only $5 a month to run—a 90 percent savings over normal bills. Installing new technology has also resulted in a more than 99 percent savings in space and water heating and a 50 percent reduction in water usage. Best of all, the payback period for my home's improvements was only 10 months, and that was with 1983 technology.

What do all these opportunities add up to nationwide? A comprehensive study by Rocky Mountain Institute suggests that if the thousand or so best electricity-saving innovations now on the market were fully installed in U.S. buildings and equipment, they'd save about three fourths of all electricity now used, at an average payback of slightly more than one year, while providing unchanged or improved services.

Some of these innovations are now becoming popular. Sales of many kinds of electricity-saving devices are more than doubling every year. Advanced windows, for instance, have gone from 1 percent to more than 60 percent of the insulated-glass market in just a few years. More than 20 million compact fluorescent lamps are expected to be sold this year.

Yet progress in converting to electricity-saving technologies has so far been much slower than it should be. A major obstacle to efficiency is the indifference or outright opposition of about a third of the utility industry. Some utilities have exemplary (and highly profitable) programs to help their customers use electricity more efficiently, but others are still trying to sell more electricity, not less. This reflects a basic misunderstanding of their business. Customers don't want kilowatt-hours, they want services such as hot showers, cold beer, lit rooms, and spinning shafts, which can come more cheaply from using less electricity more efficiently. Good programs to save commercial and industrial electricity cost only about a half cent per kilowatt-hour, which is severalfold cheaper than just operating a coal or nuclear plant, and 10 to 20 times cheaper than building a new one.

Many utilities, conditioned by a century of rising sales and revenues, still forget that, like any other business, they can make money on margin instead of volume. This is true even for utilities with overcapacity. If it's cheaper to save electricity than to make it, then a utility should save it regardless of how much capacity it has, because capacity is a sunk cost, whereas marginal variable costs can still be saved. New regulations in California, New York, and Massachusetts are encouraging such choices by decoupling utilities profits from their sales and letting them keep part of the savings as extra profit, thereby directly rewarding efficient behavior. A dozen more states are developing similar incentives for their utilities.

A second obstacle to efficiency is that many electricity-using devices are purchased by people who won't be paying their running costs and thus have little incentive to consider efficiency when comparing prices. Furthermore, most customers don't know what the best efficiency buys are, where to get them, or how to shop for them. Business customers have trouble conveniently buying integrated packages of efficient equipment; only a handful of companies can do everything to your lighting systems and do it right, and nobody as yet offers such a service for completely overhauling your motor systems.

Perhaps the most critical obstacle to overcome is the "payback gap" between consumers and utilities. If you invest your own money to save energy in your business or home, you'll probably want it back within a couple of years, implying a real discount rate upward of 60 percent a year. In contrast, if a utility has to build or expand a power plant to meet increased demand, it'll probably use a 20-year payback horizon, or about a 5 or 6 percent real annual discount rate. The utility's great technical and financial strengths, low information costs, diversified risk portfolio, and steady cash flow allow it to take a more relaxed view of investments than consumers can.

Although these respective discount rates are rational for each party, for the American economy their tenfold payback gap makes us invest too little in efficiency and too much in new power plants, misallocating some $60 billion a year.

Many utilities are seeking to equalize the disparity in discount rates between them and their customers by financing efficiency via concessionary loans, rebates, and even gifts. Southern California Edison Company, for instance, has given away more than 800,000 compact fluorescent lamps because it's cheaper than operating the company's existing power plants. Utilities are also beginning to explore leasing electricity-efficient lamps and motor systems to consumers. For example, a 20-cent-per-lamp-per-month charge on a consumer's electric bill lets him pay for the efficiency improvement over time, exactly as he now pays for power plants.

Rocky Mountain Institute has come up with an innovative way to foster such efficiency gains, creating negawatt markets. Negawatt markets would treat saved electricity as a commodity, just like copper, wheat, and pork bellies. Negawatts (saved watts) would be subject to competitive bidding, arbitrage, and secondary markets. Some entrepreneurial utilities even want to become "negawatt brokers" and create spot, future, and options markets in saved electricity. Such markets could be highly profitable. Arbitrageurs make money on spreads of a fraction of a percent, but the spread in discount rate between utilities and their customers is closer to 1,000 percent.

Perhaps the strongest incentive to create negawatt markets is their win-win solution to many environmental problems. Because it's now generally cheaper to save fuel than to burn it, global warming, acid rain, and urban smog can be reduced not at a cost but at a profit. A 1989 Swedish State Power Board study found that by using electricity twice as efficiently, Sweden could fulfill the electorate's mandate to phase out the nuclear half of the nation's power supply while simultaneously supporting 54 percent growth in real gross national product, reducing the utilities carbon dioxide output by a third, and cutting the total cost of electrical services by nearly $1 billion per year. This finding is all the more encouraging because Sweden has a severe climate, a heavily industrialized economy, and perhaps the world's highest aggregate energy efficiency to start with.

Today, the best energy investments provide the most environmental protection. A study by Rocky Mountain Institute found that a dollar spent on nuclear power will displace less than a seventh as much coal-fired electricity as would spending the same dollar on efficient use of electricity. This means that each dollar spent on nuclear energy will result in the release of at least six units of extra carbon that would not have been released if it had been spent instead to improve electric efficiency. From this perspective, nuclear power makes global warming worse. Most of global warming, Rocky Mountain Institute analysts believe, can be abated by advance energy-saving techniques at a net profit of about $200 billion per year.

Given the negawatt market's profit opportunities, why are a market-oriented Administration and many in the business community opposing aggressive abatement of energy-related pollution such as acid rain and global warming? Probably because they think abatement will cost extra. Eminent economists running computer modeling studies have shown costs running into the trillions of dollars for reducing fossil fuel combustion by, say, 25 percent by 2005. But these models base costs on economic theory, while ignoring the results of real-life efficiency programs. What those economists presumably have in mind is that because fossil-fuel use declined when energy prices quadrupled after the Arab oil embargo of 1973, a decline today in fossil-fuel use must be accompanied by similar price increases. As a result, their models only ask how high energy prices need go, based on historical elasticities, to reduce fossil-fuel use by a given amount. They forget that major efficiency gains are cost-effective at well below current fuel prices. It is a national tragedy that a few noted economists' ignorance of the empirical costs of energy efficiency has so widely spread the myth of costly environmental protection that it threatens to paralyze energy-efficiency and anti-pollution programs, thereby blocking major profit opportunities for the private sector.

Energy efficiency ultimately represents a trillion-dollar-a-year global market. American companies have at their disposal the technical innovations to lead the way. Not only should they upgrade their plants and office buildings, but they should encourage the formation of negawatt markets. And they should let the United States Government know that the best energy policy for the nation, for business, and for the environment is one that focuses on using electricity efficiently — for it's the only policy that makes economic sense.

Physicist Amory B. Lovins directs research at the Rocky Mountain Institute, a non-profit resource policy center in Snowmass, Colorado. This article was originally printed in Across The Board.

Chapter Four

The Green Marketplace

Our material culture is at the root of most of our environmental ills; the denuding of wilderness in pursuit of raw materials, pollution from manufacturing, and the heaps of waste from all of our discards. The sheer volume of 'goods' (a misnomer if ever there was one) purchased, consumed, or discarded by each of us daily is dizzying. While good environmental sense calling on us to reduce consumption, business depends on getting us to consume more.

As long as a healthy, growing economy means accelerated plundering of natural resources, the goals of business and the needs of the environment will remain at odds. But there are ways the economy can grow in a sustainable direction, like by increasing reliance on recycled as opposed to virgin materials, or using non-toxic alternatives to toxic products.

As consumers we have a great deal of power to influence the transition to a more sustainable society. It's the power of the pocketbook. If we make our investments and purchases wisely, rewarding those who consider the environment and spurning those who don't, we may yet see a better tomorrow.

Green Product Blues

The Need for Standard Eco-Labeling

By Stephen C. Evans

Ever since polls showed in late 1989 that consumers were willing to pay more for products that somehow minimized environmental impacts, manufacturers and Madison Avenue ad executives have been busy developing a bewildering array of claims for their products: "recyclable" disposable diapers and plastics; "ozone-friendly" aerosols; "biodegradable" trash bags.

These efforts to cash in on public concern for the environment have often been little more than just that. Many of the claims have been misleading or meaningless. The profusion of self-proclaimed "eco-friendly" goods on the shelves of the supermarket have left environment-conscious consumers feeling confused, skeptical and frustrated.

A number of large retail and supermarket chains — including Wal-Mart, Safeway, Giant, and Loblaw — have jumped in with their own environmental labeling initiatives, or have set aside aisles for green products. But these labels come with no independent verification or testing programs, and in the absence of real standards they have simply added to the confusion. The need for an independent body that sets environmental standards for products and then tests them to see who comes up to snuff has become obvious.

Two new Bay Area non-profit organizations are independently trying to do exactly that. Green Seal of Palo Alto and the Green Cross Certification Company of Oakland both recently launched projects that could eventually lead to much easier choices for the environmental shopper.

Unfortunately, it may be a long while before a wide range of products bears either green mark. There are a lot of thorny issues that need to be worked through first. Different products will need different sorts of criteria, and there are a number of difficult questions to answer. Must a given product contain recycled components to qualify? If so, how much recycled material is enough? Should consideration be given to a manufacturer's overall environmental record, or should certification be strictly product-by-product? What if the manufacture of an item involves the creation of toxics, or the exorbitant use of energy, even though the final product is not harmful in itself?

Green Cross and Green Seal have markedly different approaches to these issues.

Green Cross is taking the narrow path of simply checking manufacturers claims and certifying specific achievements. Their mark is then accompanied by a short explanation of the basis for the approval: "These RENEW bags are certified to be made from at least 80% recycled plastics," or "This Start brand bath tissue is certified to be made from 100 percent recycled paper fibers."

"We're not the police," said Green Cross Vice President Linda Brown in a phone interview. "It's not practical to try to be a total environmental assessment program. We're just trying to distinguish products from each other within certain categories." Within this narrow scope, Green Cross has already awarded its emblem to about fifty products, most of them for their content of recycled materials.

Green Seal, on the other hand, is attempting to develop an ambitious set of criteria based on a product's full life cycle, from the mining of raw materials to its ultimate recycling or disposal. Under the direction of an environmental standards council appointed by Earth Day 1990 organizer and Green Seal founder Denis Hayes, Green Seal product analyzers will look at energy use, water use, impact on natural areas and wildlife, toxic and radioactive pollution, use of material resources, and effect on the atmosphere in evaluating products and companies for the seal.

Specific criteria for particular products are still on the drawing boards, and Hayes says it will probably take until spring 1991 for the Green Seal to appear in stores. The first products to be evaluated will be light bulbs, household cleaners, house paint, and tissue and toilet papers.

Fortunately, neither of these non-profits are operating in the complete absence of prior experience. Green Cross is a division of Scientific Certification Systems (SCS), a for-profit corporation that has been testing and awarding its NutriClean stamp to pesticide-free produce since 1984. (Patrons of Andronico's, Raley's, and Petrini's supermarkets are likely to recognize the NutriClean seal.)

And both groups can look for experience to green labeling programs in other countries, some of them well-established.

The first national green labeling program was set up by West Germany in 1977 after the United Nations Environmental Program suggested such programs as a means of educating consumers. The Blue Angel, as the logo is known, has been awarded to more than 3300 products in 58 categories. A survey taken in 1987 found that about 70 percent of the population was familiar with the Blue Angel, and the West German government has adopted Blue Angel specifications for all levels of government procurement.

The German program has aspirations to encompass full life cycle considerations, but in practice good performance in a single criteria has often been enough to get the mark. A much-criticized example is the gas-powered mower that got the seal based solely on the low noise level it produced.

Nonetheless, the Blue Angel has had some significant effects both on manufacturing techniques and on consumer habits. Paper manufacturers particularly have invested in new plants and products in an effort to meet the requirements of the Institute for Quality Assurance, which oversees the program. Blue Angel managers figure the program has been responsible for a 30 percent reduction in average emissions from oil and gas heaters, an annual reduction of 40,000 tons of organic solvents entering the waste stream from household paints, and an increase of market share for Blue Angel-bearing paints from five to 30 percent.

Canada and Japan have both started green labeling programs in the last few years, and several other countries are in the planning stages. To varying degrees, the governments in these countries are involved in the process of setting standards and running programs, although in most it is an independent or semi-private body that oversees the program.

A debate over what involvement the U.S. government should have in setting environmental standards for products is currently gearing up. Some members of Congress and officials of the Federal Trade Commission and the Environmental Protection Agency have expressed concern that two or more competing private green labeling systems will leave the public confused, and that the government should therefore be involved and at least standardize some terms — "degradable" and "recycled," for instance.

But some environmentalists, including representatives of both the new green labeling enterprises, are wary of government involvement.

"Our standards are very strict — much stricter than the government would ever set," said Linda Brown of Green Cross. "The government doesn't like to set standards nobody can meet. Our standard for pesticides in food, for instance, is zero. No pesticides. The government limits are so high that we very rarely encounter any produce that has more than 10 percent of their tolerance levels."

There is also a reasonable fear of running into conflicts with powerful government agencies. Simply by daring to set standards in an area of government jurisdiction, Brown said, SCS took on a lot of battles. "They tried to sue and legislate us out of business," she said.

In his proposal for Green Seal, Denis Hayes recounts the similar experience of the American Heart Association, which recently cancelled its proposed "Heart Guide" food labeling program after the Food and Drug Administration announced its intention to create its own health labeling program and threatened to seize products bearing the AHA label.

"At least for the time being," Hayes wrote, "Green Seal will not evaluate food products, pharmaceutical products, or other products that might put it into conflict with labeling programs conducted by the federal government."

At present there is plenty of room for the setting of environmental product standards far distant from likely territorial conflicts with the government. Green Seal intends to focus its first efforts to evaluating light bulbs, house paint, laundry products, toilet paper and facial tissues. So far, no government agency is concerning itself with such products or their environmental effects.

The only territory battles looming at present are between the two new private standard-setters, although they may also be able to peacefully co-exist or even complement each other. At any rate, interviews with each revealed it is unlikely that they will be cooperating or merging with each other in the near future. Their approaches are too different. Green Cross founder Dr. Stanley Rhodes is a scientist; Green Seal's Hayes is an activist and organizer. Green Cross is interested in strictly scientific investigation of products' claims about themselves; Hayes plans to put together an Environmental Standards Council to develop product criteria, which will then be submitted for public and peer review, before being returned to the council for final analysis and approval — in short, to take a broader view.

It will be a pleasant change to hear such authoritative voices speaking on subjects so confusing, even if they should on occasion quarrel. And once these green seals of approval make their presence known in the marketplace, environmentally conscious consumerism will be capable at last of being a powerful force indeed.

Green-Ribbon Panel Helps Small Businesses

The conservation ethic is being given a boost in San Francisco by the city's Small Business Advisory Commission, which gives voice to the needs of small businesses.

The commission has formed a "green-ribbon panel" that both gives advice to existing businesses on how to improve their environmental practices and is on the look-out for new niches for environmentally-minded enterprises.

The panel, composed of small business owners, can give advice on such divergent topics as how to recycle photo chemicals and where to take scrap materials from your apparel manufacturing shop. Businesses that subscribe to the commission's recommendations may also be able to obtain a logo and sticker to proudly display in the front window.

The panel was the idea of Gwen Kaplan, owner of Ace Mailing & Printing and president of the commission. Kaplan hopes that, as businesses bring their questions and needs to the panel, possibilities for green services will be identified, creating new entrepreneurial enterprises for the eco-minded, such as recycling unusual materials or selling one companies waste materials as another's raw materials.

For more information about the green ribbon panel and its work, call Exec. Director Sue Lee, at 554-8941.

The Stockbroker's Smile

By Paul Rauber

When you work on Wall Street, every cloud has its silver lining. "Worldwide shock at the consequences of the disaster in Alaska could increase efforts to protect the environment," Merrill Lynch told its brokers in the wake of the Exxon Valdez disaster. "Pollution is a problem, but for investors with a long-term outlook—it could be an opportunity." "Daily environmental news was a sales booster for many brokers," reports Stephen Parker, chair of the John Hancock subsidiary Freedom Family of Funds. "We are confident that environmental action will be a major growth industry throughout the 1990s."

The environmental services sector in the US is growing at a phenomenal 15 to 20 percent a year, with specialized sectors like hazardous waste removal growing even faster. At the same time, record numbers of Americans now consider themselves "environmentalists." In order to take advantage of this situation, Merrill Lynch, John Hancock and Fidelity Investments have all come out with new environmental "sector funds": stockfunds featuring companies expected to profit from the anticipated multi-billion dollar environmental cleanup of the '90s. Brokers are targeting environmentalists with "cold calls," trying to sell them social responsibility.

The pitch appears to be working. Hancock's Freedom Environmental fund sold $46 million in its initial offering; Merrill Lynch sold out its initial $50 million offering in three days, and $83 million more came in before brokers got the order to stop selling.

There is, of course, a catch. The portfolios of all three of these major "environmental funds" include some of the worst polluters in the country, polluters whose expansion is now being financed by the investment dollars of American environmentalists.

All three funds include, for instance, the giant waste disposal firms of Waste Management, Inc. and Browning-Ferris Industries, who together over the last decade have been forced to pay over $50 million in fines and penalties for violating environmental and anti-trust laws, and who have received more than a thousand citations for violations at their dump sites. Among their environmental crimes are the following:

- In 1979, Browning-Ferris admitted to getting rid of toxic nitrobenzene by mixing it with the oil used to surface dirt roads in East Texas. The practice came to light when a former employee claimed he was fired for refusing to go along with the illegal disposal (which allowed the company to clear an extra $800 to $1,500 per truckload of waste). The result: dead livestock, and local residents complaining of headaches, nausea, and respiratory problems.
- In 1984, Waste Management mixed PCB-contaminated oil with uncontaminated oil, and sold the resulting product to some fifty different customers in the Midwest—the names of which it has refused to reveal. This and other violations at its Vickery, Ohio dump site (including allowing 45 million gallons of waste to leak out) resulted in a $10 million fine.
- Employees at a dump site owned by the Browning-Ferris subsidiary Cecos International in 1984 intentionally pumped 27,000 gallons of phenol-contaminated water into Pleasant Run Creek, the source of drinking water for Williamsburg, Ohio.
- Waste Management, through its Recycle America subsidiary,has recently entered into the recycling field in a big way—charging its customers extra for the pleasure. Although it denies the practice, community recycling groups have observed WMI personnel unceremoniously dumping recyclables in landfills in Carsonville, Michigan, in Fayetteville, Arkansas, and in Arlington, Virginia.
- Charges of price fixing and bribery are commonplace for both companies. In 1984, Browning-Ferris paid Texas State Senator Jack Ogg $25,000 to scuttle a rival's application to open a landfill. In 1987, Waste Management admitted bribing a Chicago alderman with $6,500 and Super Bowl tickets. When Vermont trash company owner Joe Kelley refused to sell out, Browning-Ferris sought to "squish him like a bug." Finding that BFI had violated anti-trust laws, a jury awarded Kelley $6 million in damages.
- Waste Management's subsidiary, Chemical Waste Management—also a part of the Freedom Environmental Fund portfolio—is the only American company engaged in the ocean incineration of toxic wastes. Among Freedom's other nominally environmental investments is Combustion Engineering, which makes boilers for nuclear power plants. Fidelity's "Select Environmental Fund" includes Perkins Elmer, a nuclear weapons contractor, and Westinghouse, a manufacturer of nuclear power plants.

Based on its green prospectus, Peter Camejo, president of Progressive Asset Management (a genuinely socially responsible brokerage firm in Oakland), initially recommended Hancock's Freedom Environmental Fund to his clients. He has since changed his mind.

"People are investing in these funds, believing they're helping the environment, when they're not necessarily helping the environment at all," he complains. "Their money may actually be going to help promote companies that are major polluters in the worst sense of the word." Camejo's client Ruth Dyke, an active environmentalist who bought into the Freedom Environmental Fund, has filed charges against the fund with the Securities and Exchange Commission, charging false advertising and use of a misleading prospectus. "I am left with the impression," she wrote to David Beckwith, the fund's manager, "that you want to appear as though you are concerned about our environment so that people like myself will invest in your fund. But in reality you will invest in any company, in certain industrial sectors, regardless of their environmental record."

Beckwith believes that it does. In a recent article in the New Hampshire Business Review, he states that companies like Waste Management "are socially positive by definition. They contribute to a clean environment by what they do."

Since neither the SEC nor the National Association of Securities Dealers regulates what may be characterized as "socially responsible," the guileless environmental investor is left to the mercy of the Wall Street sharks.

The environmental investor, however, need not despair. There is an alternative. Several genuinely socially responsible funds now available both screen for polluters and go out of their way to find companies which are making a positive contribution to the environment. Three of these are "sector funds," investing exclusively in environmental companies: the New Alternatives Fund of Great Neck, NY, which specializes in alternative energy firms; Progressive Asset Management's Progressive Environmental Fund,and Merrill Lynch's new Eco-Logic Trust '90, which promises a portion of its profits to the environmental movement. Other ethical funds which invest more widely but also use environmental screens include Calvert, Parnassus, Pax World, Dreyfus and Working Assets.

The fact that a fund has an environmental screen, however, does not remove ultimate responsibility from the investor. The criteria employed by these screens tend to be subjective, and the degree of rigor varies widely. Dreyfus Third Century Fund, for example, calls merely for examination of a company's "protections and improvement of the environment," a screen large enough to allow Lockheed, Amoco and Atlantic Richfield. At Parnassus, all decisions are made by founder Jerome Dodson. "I actually can't give you what our criteria are," he says. "Each one is different."

Other funds have more formal mechanisms. Calvert's Social Investment Fund's advisory board, which includes Sierra Club Executive Director Michael Fischer, makes general policy recommendations, and tackles questions in the "grey areas" of corporate responsibility. One recent dilemma involved the case of a pharmaceutical company which used illegally obtained endangered species to test AIDS drugs; the board deadlocked on the issue, and Calvert did not make the investment.

Although Calvert's screen involves consultation with local environmental groups, that did not stop it (as well as Working Assets and New Alternatives) from investing in Hawaii Electric, the company seeking to develop a 500 megawatt geothermal powerplant in the Wao Kele O Puna rainforest on Hawaii's Big Island. The project is hotly opposed by the Rainforest Action Network andthe Sierra Club, among others. Calvert and Working Assets no longer hold Hawaii Electric stock, but the decision to drop the company was made for economic reasons, not environmental ones.

They don't call it filthy lucre for nothing. If you're looking for the most socially responsible place to invest your money, leave it in your local credit union. But if you want the higher returns of a stock fund, it is up to you to find out exactly what social and environmental criteria your fund uses,and not to rely on crocodile smiles. Ultimately, social responsibility is yours alone.

Paul Rauber is a writer living in Berkeley. This article was originally printed in Sierra Magazine, *and is reprinted here with permission of the author.*

Green funds and their screens

Calvert Social Investing Fund
Calvert Group
1700 Pennsylvania Avenue NW
Washington, D.C., 20006
1-800-368-2748
Seeks advice from local and national environmental groups; screens out major polluters and the nuclear industry.

Dreyfus Third Century Fund
666 Old Country Road
Garden City, NY 11530
1-800-645-6561
Minimal environmental screen, includes major oil companies.

Eco.Logic Trust '90
Merrill Lynch
(available from any major broker)
Mutual fund with a strict environmental screen, applied by Progressive Asset Management. Donates a portion of profits to the Environmental Federation of America.

New Alternatives Fund
295 Northern Blvd
Great Neck, NY 11021
516-466-0808
Sector fund specializing in alternative energy. Excludes nuclear power.

Parnassus Fund
244 California Street
San Francisco, California, 94111
1-800-999-3505
Vague environmental criteria, but screens out most egregious violators.

Pax World Fund
224 State Street
Portsmouth, NH 03801
1-603-431-8022
Invests in companies with "sound environmental policies and practices," among other criteria. Advised by Clean Yield *newsletter. Consults with local environmentalists, and screens out major polluters.*

Progressive Environmental Fund
Progressive Asset Management
1814 Franklin St., 7th Floor
Oakland, CA 94612
1-800-527-8627
Sector fund investing in environmental services, with strict screen to exclude environmental services which also pollute.

Working Assets Money Fund
230 California Street
San Francisco, CA 94111
1-800-533-3863
Fairly strict environmental criteria screen s out EPA violators; fund looks for, but is not limited to, alternative energy and energy conservation firms.

Can Green Consumerism Save The Earth?

By Debra Lynn Dadd

A 1990 survey, co-sponsored by a leading New York advertising agency and an advertising trade journal, revealed that 96 percent of consumers now say that environmental factors play a "very important" role in their purchase decisions. This is up from 1989, when the Michael Peters Group, the world's largest independent design and new products consultancy, found that 89% of Americans are concerned about the impact on the environment of the products they purchase, more than half of all Americans say they decline to buy certain products out of concern for their effects on the environment, and 78% of all Americans would pay more for a product packaged with recyclable or biodegradable materials.

In response to this interest, many manufacturers and retailers are putting new products on the market that claim to be good for the environment, or repositioning old products with new marketing campaigns. The underlying message here is "Save the Earth by buying green products!" But just what is a "green" or "earth-friendly" product anyway?

The most widely used definition apparently is that anything is green or earth-friendly as long as one aspect of the product or packaging is somehow better for the environment. With this definition, we get products such as:

- Toxic cleaning solutions packaged in plastic bottles that purportedly can be recycled.

- Nonbiodegrable polystyrene plastic food service containers that are not blown with chlorofluorocarbons (CFCs, which destroy the ozone layer).

- Sugared, pesticide-laden cereal packaged in 100% recycled paperboard boxes.

- Candy full of butter and sugar that also contains nuts from the rainforest.

While candy with nuts from the rain forest certainly will help keep the forests intact and cereal boxes made from recycled paper likewise will help reduce waste in landfills, this kind of green consumerism does little to help over-all environmental conditions. The recycled paperboard cereal box does not change the fact that the cereal is made from wheat and sugar grown with persticides and artificial fertilizers, which destroy topsoil and pollute groundwater, or that the artificial colors used to make it more attractive are synthesized from non-renewable crude petroleum products, or that the cereal uses a tremendous amount of energy being shipped from the factory in Michigan to your store in California. A preferable choice would be a fruit-sweetened cereal made from organically grown grains in a recycled paperboard box. Better still would be to eat those grains in the form of bread baked at home or purchased from a local baker in a paper wrapper.

In many respects, green consumerism is a flashy and (hopefully) passing fad. But it also marks a turning point, for it is serving to bring the slower-moving and more honest trend towards living in harmony with the Earth into the mass market limelight. If we are serious about "saving the environment," we are in for some major changes in our lifestyles. Manufacturers must alter their methods, and the government must shift its support to environmentally sound practices.

To truly make a difference, the first thing we can do as consumers is to simply buy less. Green or not, every time a product is manufactured, it takes a piece of the Earth. We suck oil from deep beneath the planet's crust, rip open mountains to mine metals and minerals, destroy primeval forests, kill animals, extinguish species, exhaust aquifers, erode topsoil, deplete the ozone layer — all for a few moments of pleasure with a product that ends up in a landfill, or an incinerator, or on a garbage barge circling the globe, or floating in the ocean.

All living things must take from their surrounding environment to survive, but we can choose whether to do so in a wasteful, harmful way, or in a thoughful and sustainable way. As members of ecosystem Earth, we are each allowed to take what we need to maintain a simple, yet high, standard of living. Though we may consume quantitatively less, what we do consume can be of high quality and value and satisfy our practical and emotional needs.

One of the main problems that allows misleading green marketing activities to exist is that there are no general, accepted standards for evaluating products. So far, the perceived needs of the consumer and the economy far outweigh what is needed by the Earth in order for it to continue to sustain us. The ideal product that would serve both consumers and the Earth would be:

- Practical, reusable, durable and well-made with a timeless design.

- Non-toxic to humans and the environment.

- Made with the prudent use of natural, renewable resources taken in a sustainable way, or be made of recycled materials.

- Biodegradable or recyclable.

- Compassionate to animals.
- Energy-efficient in its manufacture and use, and preferably use renewable energy.
- Sold without packaging (if appropriate) or minimal packaging that is reusable, recyclable, or biodegradable.
- Provided by businesses with socially responsible business practices.

Much of what is sold as green, or labeled otherwise as good for the environment, falls far below this description. For years, environmentalists have been accepting so-called "natural" products as the next-better step. True, these products are reasonable alternatives to using toxic and polluting petrochemical products, but the manufacture of many natural products produce their own environmental problems. While natural foods are additive-free and less processed, many still come packaged in plastic and are not organically-grown. Natural cosmetics may not contain mineral oil, but many are colored with carcinogenic coal-tar dyes. Cotton crops use toxic fertilizers and are repeatedly sprayed with toxic pesticides. The molecules of petroleum-based biodegradable detergents contain a benzene unit which, when broken down in a lake or stream, can be converted into toxic phenol, which can kill fish.

Each of us has our own priorities, and until more products are developed, manufactured and distributed with a true environmental ethic in mind, we all will have to make trade-offs. But we know now that we have tremendous power as consumers to affect the marketplace when it comes to environmental issues. And as each of us buy the best products available, and let manufacturers know how they can improve their products, we will increase their availability to all.

Consumer advocate Debra Lynn Dadd is the editor of The Earthwise Consumer *newsletter, and author of* Nontoxic, Natural & Earthwise *(Jeremy P. Tarcher, 1990) and* The Nontoxic Home *(Jeremy P. Tarcher 1986). See the Reviews section and the Books heacing in the Consumer Guide for more information.*

The Green Triangle

By Ernest Callenbach

Living a sane and ecologically responsible life doesn't mean self-sacrifice and austerity. On the contrary, it should mean a richer, more interesting, fuller, longer and healthier life.

But so far nobody has been able to dramatize this on a national level in the folksy, convincing way in which Ronald Reagan and Ivan Boesky made greed respectable. Jimmy Carter may have been our only recent president to understand that an equation has two sides, but his wan demeanor on TV in a sweater, urging us to save energy, did not exactly inspire the American people.

Is it possible to talk attractive sense about a new life style for Americans? It had better be, or we can start preparing a suitable tombstone for our nation — and the rest of the globe, which follows our cultural lead. And what we say needs to have both human verve and internal logical coherence to be memorable — more than a cafeteria menu of 50 or 750 ecological good practices we ought to choose among.

One way I've devised of talking about some critical everyday regularities is what I call the Green Triangle. It's a handy means of generating for ourselves ideas for personal and community and national change. (Matters ecological are also inevitably social; we can't do it all ourselves.)

The environment, health, and money are the three points of the triangle. The principle that relates these three points is: Anytime you do something beneficial for one of them, you will almost inevitably also do something beneficial for the other two—whether you're aiming to or not.

For example, let's suppose you decide to take a step to improve your health, like eating less fatty meat and dairy products. This will of course decrease your chance of circulatory disease and probably prolong your life; it may even make you stronger and give you greater endurance. But, since meat and dairy products are relatively expensive, you will also save quite a bit of money; moreover, you will also help the environment — since meat production is a very land-intensive and damaging use of our farm resources.

But the interesting thing is that you can start at any point of the triangle. Let's assume you do something beneficial for the environment, like walking or bicycling instead of driving your car. (Better yet, you push a bike-lane system through your local government.) You cut down pollution emissions, you reduce smog and lung damage, you decrease acid rain, and you may postpone the greenhouse effect. You'll also help your health because you get more regular exercise, and you'll save money on gas, oil, and car depreciation.

Some people are skeptical about good things stemming from thrift, which is an American virtue that has gone out of style temporarily in the well-to-do layers of our society. But the third point of the triangle is actually just as potent.

Anytime you do something beneficial for your pocketbook, like not buying an expensive gizmo whose manufacturing expends a lot of energy and uses a lot of raw materials, or taking an expensive trip that turns a lot of petroleum into atmospheric pollution and noise, you're also helping the Earth.

You're probably also do-ing your health a favor since you're less stressed out to earn the money to pay off the gizmo or trip; and not pouring a lot of emotional energy into interacting with the gizmo leaves time and attention for other human beings and the kind of spontaneous improvisation and fooling around that our species evolved to be good at.

Don't keep this good news to yourself, or show reinforcing enthusiasm (or envy) towards friends boasting of their latest acquisition.

If you apply the Green Triangle to your everyday life, examples of delightful synergistic effects can be found everywhere. You come out with many delightful new perceptions.

Some cases: Low- or no-cost fun with other people is almost always more ecologically and financially benign than hard work and heavy consumption; evidently evolution did not commit an ecological error in making us playful.

Exchanges outside the cash economy — trading massages, for instance, or passing on extra eggs knowing your neighbor will probably someday help you with a plumbing problem — don't have monetary ramifications you have to worry about, whereas if you pay for a massage, the money may go into a bank, and you know what they do with it.

Growing or making your own is usually cheaper and healthier, as well as more ecologically benign. Fun, isn't it? So go triangulate!

Environment

Money

Health

One last word: even using the Green Triangle, we must still remember that there is no such thing as innocent purchasing even in countries with eco-labelling programs that guide consumers to "less damaging" products.

Of course it's good to buy things that do less damage, and we ought to have an eco-labelling program in this country as soon as possible (see the story at the beginning of this chapter).

To keep a sense of proportion, however, the really ecologically damaging things we do are to use cars, eat meat, have more than one child per parent, and live in dispersed single-family dwellings. Apartment living is something like five times as energy- and materials- efficient.

Even the most devoted recycling and conserving will not outweigh the enormous effects of these basic factors. You may not be able to consider change in all of them at once, but how about trying just one?

The other difficult-to-accept principle is this: buy less in general. Jesus was perfectly correct in saying "Blessed are the poor" — because they do less ecological damage. There are a few things that we, the rich peoples of the North, can buy that really do positive good for the Earth: Photovoltaic cells and solar hot-water heaters, for instance, which move us toward a solar economy. But learning to live contentedly with less income and less consumption of goods vastly outranks all the other things we might choose to do to save the Earth.

That's why, in Ecotopia, people are appreciated for what they produce — in human relationships, in art, in community life, in work, in science, in politics — and to call somebody a "consumer," even a green one, is an insult.

Earnest Callenbach is the author of Ecotopia, Ecotopia Emerging, *and* The Ecotopian Encyclopedia, *which incoporates parts of his first and out-of-print book,* Living Poor with Style. *A slightly different version of this article will appear in In Context magazine.*

Chapter Five

Recycling

Somewhere in the minds' eye the world still stretches out forever, a bottomless treasure trove that manifest destiny has handed over for our enjoyment and endless exploitation. On this vision we have structured our lives, society, and economy. As a result, everything goes one way, from mine to dust bin, in the light headed assurance that we will never reach the bottom of the barrel.

It's been a good dream. But the vapors are beginning to dispell, and struggle to keep the illusion as we might, we are awakening to the dirty realities a century of this sort of dreaming has led us to: dwindling resources and mountains of waste, with fewer and fewer places to dump it.

Most of the landfills in the heavily populated parts of the Bay Area shut down in the last decade. In the next decade, half of the remaining dumps will close. If we don't find an alternative, we will bury ourselves in our own garbage.

This chapter is dedicated to the only solution that makes sense. When we close the loop, turning garbage into resource, we find that bottomless barrel of which we dreamed.

The Bay Area's Prospects for Total Recycling

by Daniel Knapp

Thanks to more than a hundred healthy and growing recycling enterprises, Bay Area cities are replacing unpopular landfills with a comprehensive recycling system. Many elements of this system are already in place. How long it takes to supplant the garbage manufacturing industry will depend on complex factors such as how high and how fast garbage user fees rise; how aggressively recycling entrepreneurs compete for scarce land and financing; and how successfully current attempts reform the overall legal and regulatory structures.

Garbage is not one of nature's own products. A well-capitalized industry manufactures it by routinely mixing all discards together, thus degrading them. Recyclers are developing an increasingly competitive disposal technology that keeps discarded materials separate by category, then classifies, processes, and upgrades them. The garbage system organizes its efforts according to the lowest common denominator, while the recycling system organizes by the highest and best use.

Recycling works because, underlying the chaos, mystery, and even taboo of garbage, there is an order. The Bay Area can take some of the credit for discovering its true profile. Part of the process has required looking at everything that arrives at the landfill or transfer station. The City of Berkeley, for example, commissioned five increasingly sophisticated garbage composition studies over the last three decades.

But a study can report only what its categories let it see. The early ones left out a lot of recyclable categories. That omission propped up the myth that we throw away a lot of undefinable "miscellaneous" stuff, and led to the conclusion that the curse of garbage would always be with us. In fact, this incompleteness was an underlying factor in the incinerator battles that raged in Berkeley and other Bay Area cities during the early 1980s. Burner backers reasoned that studies showed recycling couldn't handle everything, so the city needed a dump-equivalent, or we would have garbage in the streets. But Berkeley voters rejected the incinerator and called for intensive recycling.

Recycling the Twelve Master Categories

In 1988-89, Berkeley commissioned a year-long composition study using a set of categories that covered everything. When the results were published, the garbage fraction simply disappeared. It was replaced by twelve clearly-defined categories of material that are wholly or partially recyclable right now. This development has already allowed Berkeley planners to shift their attention away from creating a higher-tech garbage system and to focus on designing its replacement.

What are these categories? What proportion of what we waste does each represent? The answer for any particular locale can be given only after careful study of local discards. But in 1988 I proposed a generic picture that has proven very useful and versatile in various applications, including the Berkeley study mentioned above.

Only two of the twelve master categories — plastics and chemicals — pose any great technical difficulties for recycling, and they amount to less than ten percent of the total volume now wasted. All the rest are eminently recyclable given relatively modest investments of time, energy, land, and money.

Yard debris and putrescibles (such as food) are recycling soul-mates, completely recyclable if composted. Together they make up about one-third of what currently goes into our landfills.

Composting is the largest single disposal technology, because in addition to handling yard debris and putrescibles, it can take parts of the ceramics fraction (mineralized dust and sand); soils; paper recycling residues; and most otherwise unrecyclable organics, as long as they are nontoxic. Capital requirements are low. The limiting factor is land, although here, too, there is a solution if we have the political will and courage to use it.

Metals recycling has probably the best-developed industry of all the categories, and virtually all of the metals fraction is recyclable. About 60% of the steel produced in the US is made from scrap, and the metal scrap industry has been around for centuries.

Paper recycling is increasing rapidly as manufacturers invest in more fiber reclamation plants, responding to increased supplies of postconsumer paper and demand for recycled paper.

Glass manufacturers claim their product is totally recyclable as long as the batches contain no ceramic contamination. But even contaminated glass can be treated as lower-grade ceramic and made into sand and gravel for use in cement or asphalt.

Ceramics, which by molecular structure include rock, brick, crockery, cement, and even asphalt, are 100% recyclable using conventional quarry equipment. The crushed products are suitable for various uses, including road beds and landscaping components.

The ten-year experience of my company, Urban Ore, demonstrates that reusable goods are virtually 100percent recyclable, since the amount we spend on discarding our wares as unsalable is a miniscule half-percent of our annual gross income. We have also learned that virtually any piece of wood that is whole and denailed can be sold. What can't be sold for reuse can be ground up and composted.

Textiles have a substantial reusable component, but those that can't be sold for reuse can be fed into fiber reclamation systems. As a last resort, natural-fiber textiles can be composted.

Even the hard-to-handle categories of plastics and chemicals are currently the subject of intense research and development. The key players here are the manufacturers, who want to keep producing, and progressive politicians, who are banning products and increasing regulatory pressure on industries that willfully go on producing things that are troublesome or impossible to recycle.

A comprehensive system that recycles everything and wastes nothing — a total recycling system — will handle all (or nearly all) the discarded materials from every category.

Key Elements in Place

No place to my knowledge contains a fully developed system. But in the Bay Area, we have key elements already in place. Many businesses exist whose purpose is to recover discards and convert them into resources. These recyclers have formed a trade association, the Northern California Recycler's Association, which takes an active role in lobbying for improved recycling systems and programs.

Recycling businesses take many forms. There are sole proprietorships, partnerships, and nonprofit and for-profit corporations. There are municipal enterprises and privately-owned ones. There are businesses that collect and haul recyclables; that process and upgrade them; that manufacture new products using recycled feedstocks; and that sell the finished products. Some businesses combine two, three, or even all of these functions. There are garbage companies that are converting themselves as rapidly as they can to become recycling companies.

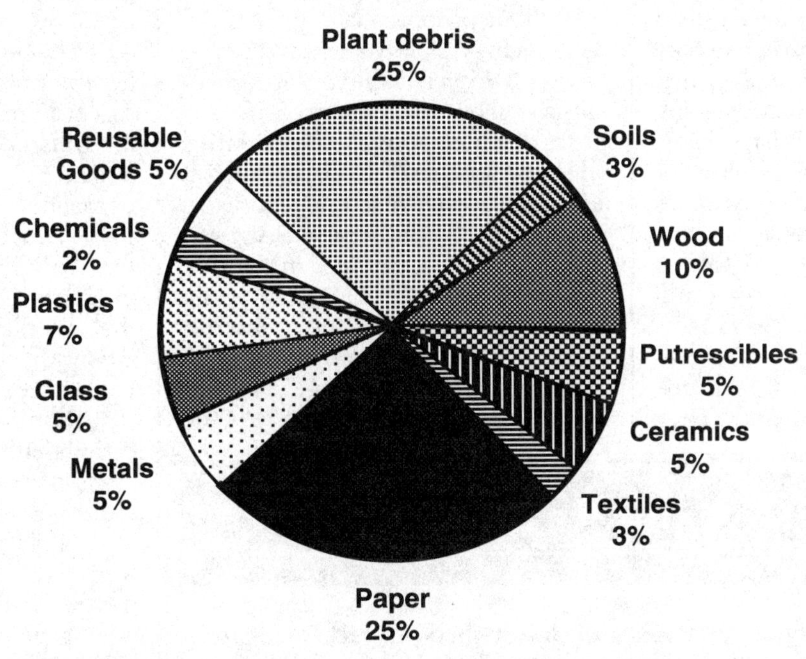

Pie Chart ©1989 by Daniel Knapp and Mary Lou Van Deventer. Excerpted from *Total Recycling: Realistic Ways to Approach the Ideal*, in progress, to be published by U. C. Press

The Golden Gate Disposal Company, the Sunset Scavenger Corporation, and Marin Sanitary Service are three large garbage companies that are innovators in the recycling field. They operate materials recovery facilities (MRFs), which handle feedstocks of mixed materials. The cities of El Cerrito, Berkeley, San Francisco, Palo Alto, Mountain View, and San Jose are noted for their strong commitment to recycling. Berkeley and El Cerrito even operate recycling businesses themselves.

Berkeley's Ecology Center and Community Conservation Center are nonprofit curbside and buyback programs that helped pioneer the field in cans, bottles, and paper, and are notable for outstanding promotion of recycling and educating the public on how and where to do it. Other nonprofits handling similar materials include San Francisco's Richmond Environmental Action and Haight-Ashbury Recycling Centers and the San Francisco Community Recyclers; Tri-City Ecology Center of Fremont; Contra Costa Community Recyclers; and Sonoma Community Recyclers. There are too many more to include here.

Urban Ore, Encore!, West Coast Salvage and Recycling, and Pacific Rim Recycling are some of the better-known for-profit corporations, but there are many others. Standard Metals and Levin Metals Corporation buy metals and upgrade them. American Soil Products blends and sells bulk soils, compost, and soil amendments. American Rock and Asphalt operates several quarries that accept cement, asphalt, and other ceramics that the company turns into various sand and gravel products. In 1989 they handled 350,000 tons of these materials, a large contribution to a growing recycling rate.

The list could go on, but the point is clear. All of the twelve categories are being recycled in some proportion right now in our area. The prospects for total recycling are excellent, as far as underlying business structure goes.

Legal change needed

Legislative underpinnings are another matter. Recycling is not structured into disposal rates, for example, so recyclers are expected to finance their entire businesses by selling materials. They are also frequently given short-term contracts with cities.

Meanwhile, garbage companies performing inferior disposal services and creating long-term liabilities are entirely financed by long-term exclusive franchises that provide them with comfortable profits. Recycling could use a little leveling of the playing field.

To outline the resources available to recyclers, more cities need to do composition studies that measure all twelve master categories. Too much of our local information about garbage and recycling is simply imported from national studies. Those studies extrapolate from manufacturing data instead of measuring actual discards, and they use limited or poorly defined categories. There is no substitute for actually going out and measuring what is happening. Lacking good empirical work, we must guess at quantities. Serious mistakes are inevitable — and expensive.

Recycling needs room, too. A compost operation takes acres, but it produces valuable humus. The surfaces of closed landfills are now slated to become parks, while recycling industries cry out for low-cost land. They should be re-opened for use as comprehensive recycling centers. An environmentally responsible concept of park development would let some parts of these lands be recycled into a solution to the garbage crisis. Visitors to such an environmental park would be told that garbage created the land, but using it for land-intensive recycling systems can slow or stop the march of landfills into unspoiled rural areas.

These ideas involve changes in legislative and regulatory structures. That means new laws.

One proposed law, the Alameda Recycling Initiative, was written by Bay Area recyclers and environmentalists to accomplish many of the purposes outlined above. The initiative will be on the November 1990 ballot. It adopts the twelve master categories as its conceptual base. It proposes a surcharge on garbage dumped in landfills, and the proceeds would fund recycling systems for any or all of the twelve categories. By driving up the cost of dumping mixed garbage, it makes landfilling less competitive. At the same time, it recommends that cities provide different-sized containers and charge garbage-user fees based on actual volumes wasted, so people who waste the most pay the most.

But although this homegrown measure is innovative, California as a whole has fallen far behind other states in legislation. Illinois, for example, banned plant debris from its landfills starting in July, 1990, and for three years has been collecting a garbage surcharge earmarked for recycling. Wisconsin passed a law last year that is even more ambitious. By 1995, yard debris, metal, glass, paper, and several other materials amounting to over 60% of the total discard supply will be banned from landfills. The same law appropriated almost $20 million for recycling development.

Obviously, good laws can stimulate recycling and help build a thriving industry. Other states have found that their people's interests lie in bold legislation. California passed AB 939, which has been highly touted, but it fails to provide the fundamental economic restructuring that recycling deserves. Its concepts of what constitutes recyclable materials are also seriously flawed.

But legislation is the map, not the territory. It is not the same as business development. California has always been a place where innovative businesses take root and prosper, and the overall climate for recycling businesses has never been better.

Even if the state's legislative development remains sluggish, recycling businesses have an enormous competitive advantage over garbage businesses. Whether they are helped or hindered by laws, they will find ways to go after the ever-more-attractive supply of discards. From small and large niches, they will whittle away at the supply. As they do, they will generate jobs, produce income, and pay taxes. Recycling provides a superb industrial frontier for people who want to make a living in an environmentally sound way.

Dan Knapp is a board member of the Northern California Recycler's Association, and the author of Total Recycling, soon to be released by University of California Press.

Sources and Resources:

For an extensive list of Bay Area recycling businesses and products made from recycled materials, please see the recycling section of the Consumer Guide.

READING

Mining Urban Wastes: The Potential for Recycling, Cynthia Pollock, Worldwatch Institute, 1987
A discussion of the potentials for recycliong of various materials with examples of successful recycling programs from around the world.

Biocycle - Journal of Waste Recycling, JG Press Inc, Box 351, 18 South Seventh St, Emmaus, PA, 18049
Phone (215) 967-4135
A professional/technical journal with lots of how-to information about setting up a recycling business.

Resource Recycling - North America's Recycling Journal, Resource Recycling, Inc, P.O. Box 10540, 1206 NW 21st Ave., Portland, OR 97210
Phone (800) 227-1424
Primarily designed for policy makers, this monthly national pulication keeps on the cutting edge of waste meanagement issues

ORGANIZATIONS

Northern California Recycling Assoc.
(415) 548-6659
A trade association. NCRA membership is open to anyone for $35 a year. People who want to join can mail a check for $35 to NCRA Treasurer, PO Box 5581, Berkeley, CA 94705. Anyone can also come to the monthly meetings to see what the group is like before joining.

Ecology Center
2530 San Pablo Ave, Berkeley, CA 94702
(415) 548-2220
Operates the oldest on-going recycling operation in the Bay Area, and provides a comprehensive information service about recycling and many other environmental topics.

The Recycling Cycle

By Alison Marquiss

To most of us, recycling means the somewhat space-consumptive chore of separating all of our garbage into its component parts and then, when all the containers are positively overflowing, hauling it down to the collection center or (if we're lucky enough to have curbside collection) out to the curb. As far as the majority of us are concerned, that's the end of it.

But the recycling process goes much further than just to the curb or the local collection center, which is actually only the first step in a multi-level cycle that should, if the system works, bring that same stuff back through our front doors as new goods.

Most collection centers do just what their name says— collect glass and plastic bottles, aluminum cans, paper, and, in some cases, tin cans. Very few collection centers process these materials. Usually they simply sell their collection to a larger recycler or processor, who then compresses and prepares the materials for sale to a "middleman." The preparation is different for the different types of stuff, and results are sold to different middlemen.

Aluminum is typically first crushed into bundles weighing 2500 pounds, which are then pressed into 120,000 pound blocks. These ingots are placed on railcars to be sold to nationwide aluminum dealers such as Alcoa, Brockway Owens, and Kaiser. The middlemen melt the ingots down into sheets of "canshield". This "second-class" aluminum (which is first-rate, despite its moniker) is then sold to manufacturers to create flexible aluminum products such as pie pans, or at a reduced price to beverage manufacturers. Although this aluminum is of equal quality to new aluminum, its recycled status makes it cheaper.

Many beverage manufacturers have taken advantage of this "discount," and some have reduced costs even further by setting up their own recycling systems. Both Anheuser Busch and Coors have separate recyling plants, created primarily for aluminum recycling, but which process plastics and glass as well, as mandated by state recycling laws. Aluminum reaps by far the most profits for recyclers, but glass, plastic, and paper are also moneymakers.

Glass processing is more difficult, as it requires removing all contaminants such as labels on bottles and jars. Once the glass is cleaned, the processor crushes and melts it in order to sell it as "cutlet" to, again, nationwide middlemen. The middlemen in turn sell to manufacturers of glass products and products requiring glass packaging, where the recycled glass is regarded as just as good as new.

Plastics, *if* they can be sorted into lots of the same type of plastic, are baled and sold to middlemen to be turned into non-food packaging plastic products (for a detailed account of the problems with plastics recycling, see the following story). Paper is also baled, then shipped both overseas and across the nation to paper mills where it is rehydrated into a pulp form. This pulp can be used to create the same products as the original wood pulp.

"Tin" cans, actually part tin and part iron, present a difficult problem. Although the cans are abundant in food packaging, many recyclers refuse to accept them because of the low price they bring — only $50 per ton. Recyclers actually spend more money processing and transporting the tin than they make in its sale. The few recyclers that accept tin are able to do so because all of their recycled materials are donated — the money-making recyclables such as aluminum cans finance the processing of the more costly materials.

Post-consumer waste composes only ten percent of the material used by tin re-processors. The other ninety percent comes from the trimmings left over from making the original cans. Only some of this tin is used to create new tin cans, which requires removing the tin coating from the base metal through a chemical wash, resulting in a tin "mud". The mud is then formed into bricks of iron which can be used to mold new cans for food manufacturers.

Once all of these materials have been processed and distributed to the appropriate middlemen, they face much the same markets as virgin materials. As more manufacturers become aware of the alternative, recycled materials are gradually capturing a larger share of their markets. The competition and growth of all components of the recycling industry are good signs for its continued development and additional progress toward the ultimate goal, conservation of limited resources. Still, the industry is young and has many problems to overcome, especially in the area of materials that can be recycled, but not competitively.

Consumers can help the recycling industry by demanding availability of, and then being sure to buy, products made from or packaged in recycled materials of all possible types. This is the too-often forgotten final link in the "cycle" of recycling, which will ultimately determine the success or failure of the industry.

Remember - if you're not buying recycled goods, you haven't really started recycling.

Alison Marquiss is a writer based in Berkeley.

The Plastic Perplexity

By Jo Ann Gutin

Plastic will soon be to modern Americans what the walrus was to the Aleut or the buffalo was to the Sioux: nothing less than the basis of our entire material culture. In some form or other we wear it, eat with it, write with it, cover our floors with it, insulate our houses with it — the list is practically endless.

Being less spiritually evolved than the Aleuts or the Sioux, though, we loathe plastic instead of respecting it; we're addicted to it, but we make fun of it. It's not just a substance, but a code word for a way of life that all right-thinking people despise. Plastic has gone in thirty years from a symbol for high technology and inventiveness to a symbol of rampant consumerism.

It is this schizophrenic attitude toward plastic that makes the current struggle over what to do with plastic trash so interesting to watch. We seem to have blundered through the looking glass: advertising that once touted the convenience and disposability of plastic are now trying to soft-pedal convenience (in deference to the fever of guilt and self-abnegation that is sweeping the land) and backtrack on disposability. Anyone who has a TV has been exposed to a barrage of television commercials announcing that plastic has been rehabilitated and is now a recyclable material. We've all been wrong to be so contemptuous of it and so guilty about using it, they imply.

Those commercials have done their job well: most people are now thoroughly confused. Are we supposed to hate it, or love it, or what?

Most environmentalists disagree passionately with the contention that plastic is — or even that it ought to be — recyclable. They cite chemical residues from plastics manufacturing, hint darkly about cancer-causing chemicals in styrofoam, outline the logistical problems in its collection, but mostly you can tell they just hate the stuff. They hate it because it's based on fossil fuel, they hate it because people throw it away on the sidewalk; they hate, in short, the culture that it epitomizes. Talking about recyclable plastic or biodegradable plastic to an environmentalist is like talking to Nancy Reagan about legalizing drugs. In their view, anything that encourages the use of the forbidden substance is absolutely wrongheaded. Abstinence is the only solution.

Which leaves the poor consumer caught in a philosophical crossfire. Given that most people would rather not spend their remaining years doing environmental cost/benefit analyses of every purchase, it's no wonder that we've snuggled right up to the idea of recyclable plastics. But really, the process bears as much resemblance to traditional recycling as R.V. camping bears to backpacking. In fact, it might be a good idea if there were some sort of linguistic flag for the difference, the way "kamping" has come to mean R.V. travel. "RecyKling," maybe.

Managers of city and private recycling programs all over report that people — at least the kind of people who worry about recycling generally — are desperate to recycle their plastic. For example, Susan Katchee, head of the El Cerrito recycling program, reports dozens of calls a week from concerned consumers asking what they can do. (El Cerrito alone accumulates a ton of plastic a month from curbside and dropoff programs.)

Piles of plastic bottles sprout overnight like mushrooms around locked drop-off igloos in supermarket parking lots. Clearly, the word that plastic is recyclable has gotten around.

But is it? Most people assume, given the traditional meaning of "recycling," that old two-liter soda bottles become new two-liter bottles. Unfortunately, that's not what happens in RecyKling. Although the resin produced by melting old soft drink bottles is the most high quality plastic resin, if we and the bottle are lucky, it ends up as fiberfill for parkas, carpet backing, or tennis ball fuzz. Although it would be comforting to think that old plastic milk jugs become new plastic milk jugs, they don't; mud flaps for trucks, maybe, or plastic park benches, but not milk jugs. One of the best prospects to which a styrofoam cup can currently aspire is reincarnation as styrofoam packing pellets. This is metamorphosis, but it's not recycling. Or, in the unintentional oxymoron coined by the plastics industry, it's "linear recycling."

When glass is recycled, bottles and jars are pulverized, combined with some virgin raw materials, and made into more bottles and jars. Ditto for aluminum — aluminum beverage cans into aluminum beverage cans. Theoretically, if all glass were recycled, and if population size and demand for bottled products stayed the same, few new bottles would have to be manufactured. The loop, as recyclers say, would be closed.

But in recycling terms, plastic is more like paper than it is like glass or aluminum. Paper that begins life as a wedding invitation will, after several cycles be on the bottom of a birdcage: because of ink on the surface and wear and tear in the washing process, it loses a little of its dignity on each go-around. The same holds true for plastic, though for different reasons; and since it starts life with fewer pretensions than paper, it reaches the nadir sooner.

Why can't the most ubiquitous consumer plastics be recycled into new incarnations of themselves? A number of factors make the prospect a recycling scenario from hell.

For one thing, the federal Food and Drug Administration requires that any food or beverage container to be refilled with food be heated to a specific temperature, and plastic won't tolerate that temperature; thus, it can't be adequately sterilized.

For another, some plastic has the tendency — unlike glass and aluminum — to absorb minute quantities of what it contains. If some tidy householder had stored motor oil in a plastic milk jug, and that jug were later recycled into a new milk jug, traces of the motor oil might conceivably leach back into the milk.

But the biggest catch is that for any kind of plastic recyling to work, plastic trash must be sorted. That's the reason for the separate containers at the plastic drop-offs: one undetected high-density polyethylene (HDPE) milk jug in a load of 2-liter polyethylene terephthalate (PET) soda bottles contaminates the resin. In what seems like a cosmic joke, the versatility that has enabled plastic to fill so many packaging niches is what makes recycling it so difficult. Glass is glass, more or less, and aluminum is aluminum, but plastic has a thousand faces.

There are two classes of plastics, the thermosets and the thermoplastics. Thermosets are rigid plastics that don't flow when they're reheated, so for all practical purposes they're not recyclable: once a tail light, always a tail light. Because thermosets constitute only 13% of plastic production, they're fairly low on the list of things to worry about. Thermoplastics are infinitely malleable, practically, because their component molecules polymerize; that is, small molecules combine easily to form large chains, and those long chains, tangled when cold, slip easily past one another whenever they're heated.

There are somewhere between 500 and 1,000 different thermoplastics, with more coming out of laboratories every day. However, almost all of the packaging materials we've come to know and love/hate are made from one of six groups. The catch is, to recycle anything made of any one of these resins, the source material must be pure.

So mingled plastics need to be sorted after collection. At this stage sorting requires humans, who need to be paid. If plastic were an intrinsically valuable commodity, that wouldn't be a problem. However, it emphatically is not, which is why so much plastic is shipped to the Far East, where labor is cheap. (In 1989, between 70 and 80% of the post-consumer plastic that was collected for recycling went to buyers in Hong Kong. Some of it was melted down into resin and reshaped; some of it was burned for energy.)

In this country, post-consumer plastic sorting is most often done either by volunteers (the Jaycees or the Boy Scouts, for example) or by people who are unpaid or paid very little. For example, a large-scale plastics recycling program in Trenton, New Jersey, employs prisoners as plastic sorters, in a kind of polymer chain gang. Closer to home, in Pittsburg, CA, a recycling program administered by the non-profit organization Many Hands employs unpaid mentally and physically disadvantaged adults to do the plastic sorting.

Given the current, if temporary, abundance of fossil fuel, consumer plastics are cheap, cheap, cheap to manufacture. It is, in fact, cheaper to make plastic, both from the standpoint of raw material and the energy consumed in manufacturing, than glass. This is because the resins that are the raw material of modern plastics are a by-product of petroleum refining.

Aluminum and glass recycling have been relatively successful because recycling those materials makes economic sense. According to industry sources, there is a 95% saving to the manufacturer who makes an aluminum can out of recycled aluminum. There's a similar, though less stringent economic imperative operating in the recycling of glass. The only plastics recycling that has been justified by the bottom line has been recycling of scrap from large manufacturing operations, which has been going on since the dawn of the plastics age. Since industrial-scale manufacturing inevitably generates big piles of stuff, recycling of industrial plastic has always made sense. Somewhere between 3.5 and 5 billion pounds of industrial plastics were recycled this way in 1988.

But the problem is post-consumer plastics, not post-industrial plastics. They're cheap to make; amassing a pile of them big enough to be worth something costs more than making them in the first place.

There are some domestic markets for crushed milk jugs (high density polyethelene or HDPE) and 2 liter bottles (polyethylene terephalate or PET). Wellman, Inc., in Johnsonville South Carolina, processes 100 million pounds a year of PET into fiberfill. Plastic Recycling, Inc., of Iowa Falls buys 15,000 pounds of low grade plastic waste a day, and makes it into car stops for parking lots, plastic lumber, and the like. But facilities like that are few and far between, and they have been plagued by technical difficulties; the upshot is that a broker who buys plastic waste may find himself with no place to sell it.

In the Bay Area, even as the number of concerned consumers looking feverishly for someplace to take their used plactic increases, the number of recyclers willing to take it and brokers willing to buy it from recyclers diminishes.

And what happens to plastic that isn't RecyKled? If it isn't littering a beach or a vacant lot, it's on the way to a landfill or an incinerator.

According to environmental sources plastic waste hovers around thirty percent of the wastestream by volume, and shows no sign of slowing down. It doubled between 1976 and 1984, and is projected to reach 75 billion pounds by the year 2000.

And that plastic will be there for a long, long time. The average, garden-variety squeeze bottle will last well over 100 years; like the erosion rates of sandstone and granite, the decomposition of plastic depends on atmospheric conditions and the type and density of the plastic. Exposed to rain, wind, and variations in temperature, a ketchup container might disintegrate in a century and a plastic grocery bag in a year. In a nice sanitary landfill, particularly one in the arid West, both can reasonably aspire to synthetic immortality.

If you can't bury plastic, can you burn it? That would really be a recycling coup: take an intrinsically worthless by-product of petroleum, then turn it into something putatively useful; when the useful object is discarded, incinerate it and create energy. It's an appealing vision, like turning straw into gold, and high-tech waste to energy plants are widespread in Europe and Asia. Incineration was the disposal method initially promoted by plastics manufacturers, who observe with satisfaction that plastic burns twice as hot as Wyoming coal.

More than a hundred so-called waste-to-energy plants are in use in the U.S., mostly in the East, where the landfill crunch is most acute. They've run into trouble in California, where a handful are in operation but more have succumbed to local opposition.

Incineration is fraught with problems, both those inherent in developing technologies and those inherent in burning anything. As everybody in the developed world must be heartily sick of hearing by now, burning fossil fuel results in emission of CO_2, the chemically innocuous but atmospherically insidious greenhouse gas. Plastic is fossil fuel once removed, so on that basis alone burning it is a bad idea.

Worse, many WTE plants do what's known as "mass burns" of unsorted trash, including batteries, leftover paint, old refrigerators, and plastic waste. The ash that results may be full of heavy metals, and the fumes, despite expensive scrubbers, contain minute amounts of dioxins and furans.

In other words, it begins to look as though the least environmentally costly strategy, despite the appeal of plastics recycling as a concept, is source reduction. In other words, boring old abstinence.

Jo Ann Gutin is a freelance writer living in Berkeley, California. A version of this article was originally published in the Express.

Snubbing Styrofoam

Berkeley ban sets sterling example

You used to see them everywhere in Berkeley. Not hippies, not yippies, not yuppies, not huppies (or rather, homeless urban professionals).

No, styrofoam packages.

Back before the Waste War, you saw styrofoam cups gripped tightly by the cappaccino-dependent and scattered across UCB's lawn. You saw creaky squeaky foodtrays wedged in gutters outside McDonald's and jammed in over-crowded trash cans. Sometimes, you saw them tumbling menacingly down empty alleys.

Before 1987, Berkeley, like most American cities, didn't blink in the face of styrofoam. But that year, when it became clear that the chloroflurocarbons (CFCs) used to make styrofoam packaging were eating a gaping hole in the ozone layer, people decided to act.

The time was right. Berkeley was already struggling with a mounting garbage problem and attempting to achieve 50% recycling. Glass, metal, cardboard, newspapers — all were taken care of. But it was clear that the fastest growing portion of the rest of the waste stream was plastics, particularly styrofoam.

In 1988, the city's Solid Waste Management Commission created a Styrofoam Taskforce to examine the issue. On that taskforce's recommendation, the City Council decided to ban all styrofoam produced with ozone-depleting CFCs. A second law, which was passed in 1989, banned styrofoam completely ("Who could tell which was produced with what?" explains councilwoman Nancy Skinner), mandated that all food packaging have 50 percent biodegradable or recyclable materials, and that all restaurants provide recycling containers for their customers to use.

Businesses were given a year and a half to comply voluntarily; the law became mandate in January of 1990. While several businesses complained bitterly when the law was first passed, none have suffered financial hardships as a result of the ban.

Berkeley was the first city in the country to pass such a complete ban. It has served as a shining example to other cities across the country. Portland, Minneapolis-St. Paul, Carmel, West Hollywood— even Newark, New Jersey— have since followed suit and banned styrofoam.

Of course, paper packaging has its problems as well, in terms of trees felled and waste produced. But at least it isn't chewing a hole in the ozone layer.

"I don't want to be put in the position of saying that paper is good," Skinner says. "But at least it's not a petroleum-based, ozone-depleting, ugly packaging material."

Recycling Carbon

A View from the Compost Heap

By Daniel Knapp

Carbon recycling? We've all heard about recycling bottles and cans, but recycling carbon? Why carbon?

Because as much as 80% of all the garbage that we bury in landfills every day is organic. Not organic in the sense that it's clean, natural, chemical free, and good for us if we eat it — but organic in the chemical sense, meaning that 80% of the various substances in the garbage have molecular skeletons made of carbon.

Carbon-based discards include all the paper, plant debris, putrescibles, plastics, wood, and textiles. That's roughly 75 percent of everything we put in landfills. Other categories that contain appreciable amounts of carbon include reusable goods, soils, and chemicals. Only ceramics, glass, and metals are mostly inorganic, and current estimates peg them at about 15 percent of what is landfilled.

All known disposal methods, including reusing, recycling, composting, burying, and burning either store this carbon in its current form, recycle it into another solid form, or transform it from a solid into a gas.

When reclaimed wood is used in a building project, the carbon in the wood is stored in exactly its current arrangement. Paper, on the other hand, can't be reused without being turned back into pulp. After pulping, the reusable fibers are extracted, and then the reclaimed pulp is rolled out into new paper. With paper and the other conventional recyclables, recycling involves a more drastic change in physical form than reuse, but in general, recycled solids stay solid.

Composting is the last recycling form before we enter the realm of wasting by burying or burning. Like reuse, and like all the other organic recycling methods, composting retains most of the carbon in a solid form. Composting treats organic discards as food for air-breathing organisms — mostly bacteria and fungi. After they have dined, and exhaled as carbon dioxide some of the carbon they have consumed, they leave behind a black rich organic humus, the best soil conditioner on the planet.

Burying and burning have very different effects on the carbon. Burying garbage stores carbon in a solid form, but it is wasted because it is first blended into an unwholesome mess, then locked away underground. There it decomposes slowly and very inefficiently due to the lack of moisture and air. The organisms that eat garbage in an airless environment exhale flammable gases like methane. Like carbon dioxide, methane is a gas that can trap heat in the earth's atmosphere. But methane is twenty-five times as effective as carbon dioxide at trapping heat, and after carbon dioxide it is the second largest contributor to greenhouse warming. It can also blow up if trapped in an enclosed space, or just ignite suddenly into a sheet of flame. Most closed landfills now must control methane by installing collection systems for the gas.

Burning garbage generates carbon dioxide more quickly than any other method of disposal. It also creates lots of other things from the carbon. The consuming fires break complicated carbon-chain molecules willy-nilly into smaller units, all within a swirling bath of chemically reactive elements like chlorine. Things combine and recombine in complex and unpredictable ways. Exotic molecules are created that weren't in the original fuel. Some are very poisonous. Some are very long-lasting. Some — perhaps most — are unknown.

Burning a highly refined fuel, like chips made from wood or shredded plant debris, produces fewer exotic chemicals, but it still transforms the solid carbon into carbon dioxide — again, contributing to global warming.

If we want to do something positive about global warming, we have to do more than plant trees. We also have to reduce, with the goal of eliminating, all discard disposal by landfilling and burning. For our carbon-based discards, we have to reuse, recycle, and compost everything possible.

That's a tall order. Here in the Bay Area, estimates vary as to how much garbage we produce. Some people (I am one of them) believe our annual production is somewhere between three and four million tons. The Association of Bay Area Governments thinks seven and a half million tons is the correct figure. Either way, it's a very large amount.

So far our citizens — in Berkeley, Brisbane, Cupertino, Fremont-Newark, San Jose, Redwood City, and Richmond — have rejected at least seven major incinerator projects, so we are burying just about all this garbage.

Reusing, recycling, and composting have different capacities for disposing of the carbon available in the supply of discards.

Reusing has the smallest capacity. Lumber, plywood, and other wood-based building materials can be cleaned up and sold as-is. Unspoiled food can be fed to animals. We can set up reusable goods depots so things like clothing and furniture are redistributed from those who don't need them to those who do. Some chemicals — oil, paint, some solvents — can be cleaned up for reuse.

Recycling can handle more than reuse. Most paper can be recycled. Fibers can be reclaimed from textiles. Some wood that can't be re-used can be made into particleboard.

But the big player in the organics recycling system is going to be composting. Right off it can handle all of the plant debris; all of the putrescibles that aren't fed to animals (including sewage sludge); much of the wood; all of the soils; and all of the leftovers from paper recycling. With further research and development it may even be able to dispose of some of the chemicals, plastics, textiles, and the finest particles from ceramics recycling. Composting is recycling's power hitter.

After getting off to a rocky start in the late 70s and early 80s, commercial composting shows signs of developing fairly rapidly in the 90s. Four big forces are pushing us toward large-scale composting: legislative initiatives, disposal fee increases, entrepreneurial vision, and landfill limitations.

Legislative Initiatives

Although California has not yet followed the lead of states like Illinois and Wisconsin in banning burial of yard debris, it has passed a law that requires cities to achieve 25 percent and then 50 percent recycling levels, or face fines of up to $10,000 per day.

There is widespread consensus developing among bureaucrats and politicians that these recycling goals cannot be met without developing composting for organics. San Jose and San Francisco are now working to implement major composting programs. San Francisco is studying the feasibility of composting not only yard debris, but also paper and food waste. Berkeley's compost plans were ready to be implemented a few months ago, but were put on temporary hold due to budgetary constraints. The city of Pleasanton is working on plans for composting yard debris and sewage sludge.

Disposal fee increases

The cost of dumping garbage is rising precipitously all over the Bay Area, and this is compelling people to save money by composting at home.

Disposal fees were down around $2.00 per cubic yard in 1980, but by 1990 the cheapest fee was twice that, and the highest was more than six times higher. Waste Management's Davis Street Transfer Station, which feeds the biggest dump in the area at Altamont Pass, is still a relative bargain at $4.00 per yard, $4.00 minimum. The Sunnyvale landfill charges $5.17, but restricts incoming traffic to Sunnyvale residents. The San Carlos and Oyster Point Transfer Stations feed the Ox Mountain landfill, and their fees are $6.50 and $7.00 per cubic yard, respectively. The Richmond and Zanker Road landfills are at $7.50 and $8.50. The highest dump fees I found were Berkeley's refuse transfer station and the Nubi Island landfill, both at $12.50 per yard.

Entrepreneurial vision

High and rising garbage disposal fees are not only an incentive to recycle, they also serve as an incentive to develop compost businesses that compete with the garbage dumps.

The cheapest tip fee I found for dumping brush is at Recycled Wood Products in Berkeley, which composts about two-thirds of everything it takes in, and charges $3.00 per cubic yard at the gate. Even though Zanker Road landfill charges $8.50 per yard for dumping brush or wood, that fee is still less than the nearby Nubi Island landfill. Zanker's composting manager says they are currently gearing up to produce over 200,000 yards of compost each year. Zanker has a sales agreement with the City of San Jose, and currently can't produce as much as the city wants to buy.

Landfill limitations

An epic ten-year struggle to keep the landfilling option open for the San Mateo County area came to an end in September, 1990 with denial of a permit to fill Apanolio Canyon with garbage. Browning-Ferris Industries spent an estimated $10 million seeking the 100-year landfill permit, only to be turned down after a long fight.

Contra Costa County has been trying to find a new landfill site that voters will approve for years. One of their last two dumps closed this year and the Richmond landfill will close in 1993. Alameda County has the most remaining landfill capacity of any Bay Area County; not surprisingly, its garbage tipping fees offer the least incentive to switch to recycling.

Big garbage companies like Norcal and Marin Sanitary Service, who do not own landfills and cannot or will not stay in the competition with industry giants over landfill siting, have moved aggressively into recycling. Norcal bought out Zanker Road landfill last year, but has continued financing Zanker's moves toward large-scale composting and ceramics recycling with the idea of using that facility as a prototype for its other operations in San Francisco and statewide.

Marin Sanitary has been an industry leader for years in recycling transfer station design, and recently joined with American Soil Products to develop a composting program.

What if we develop composting all over the Bay Area? Will there be markets for the products? Expert opinion varies. Dr. Louis Truesdell of American Soil Products is pessimistic. He is worried that municipalities will simply give compost away, which will undermine the ability of businesses like his to stay solvent.

Barton Blum, a composting consultant and former employee of American Soil, is an optimist. In his nearly-completed Master's Thesis at the University of California, he calculates that there is currently a demand for 200,000 to 230,000 tons of compost in the Bay Area. He is confident that legislation requiring public agencies to establish purchase preferences for locally-produced compost products will soon pass. This will further boost demand.

Perhaps we will one day see a revival of an old dream that those of us working at the Berkeley landfill used to talk about. We envisioned compost facilities operating on the surfaces of all the closed bayfill dumps, paying their costs by charging tipping fees, making high-quality compost by the millions of tons, and loading it onto barges for transport up to the Sacramento delta farms, where it would be used to raise the level and the fertility of farmlands there. It could still happen if we just decide to make it so.

Creating Markets for Recycled Paper

By Alan Davis

Most people assume that paper has always been made from trees. But only since the 1850s have trees been cut and pulped to make paper. For most of its history, paper has been a recycled product. The first paper, invented in China in 105 A.D., was made from reclaimed material-rags and discarded fishing nets- as well as hemp and China grass. The Arabs were probably the first to make paper from used linen. Even North America's first paper mill, built in 1690 in Philidelphia, recycled rags to manufacture paper.

By the mid-18th century, as the Industrial Revolution gained momentum and demand grew for more paper, European nations scrambled for a dwindling supply of cotton rags and discarded linen. As the "rag wars" intensified , Europeans spent the next hundred years searching for a new fiber source for paper. Inventors experimented with fibers ranging from swamp grass to marshmallow to asbestos, until testing in wood-pulping techniques yielded successful commercial processes. Over the final decades of the 19th century, paper mills converted to the new wood-pulping technology.

Today, many of us are trying to make environmentally sound decisions in our everyday lives. One choice is to use recycled paper, a high-quality product that cuts waste, pollution, and energy consumption. And it doesn't deplete as much of our forests as do conventional wood-pulping practices.

What is recycled paper?

Today's recycled paper is made from wastepaper generated by paper mills, envelope makers, print shops, homes, and businesses. If the wastepaper has ink on it, it must be "de-inked" to separate contaminants from the paper's fibers. The prepared fibers can then be made into new paper.

Only the process for preparing the fiber is different for recycled paper. Otherwise, papermaking is essentially the same for both recycled and non-recycled papers.

Why should I use a different kind of paper?

Recycled paper is not a " different kind" of paper. (High quality paper is also made from cotton, straw, linen, bamboo, and other fibers.) Recycling simply uses discarded paper as a fiber source.

A ton of paper made from 100 percent wastepaper, rather than from virgin fiber, saves 17 trees, 4100 kwh energy (enough to power the average home for six months), 7000 gallons of water, 60 pounds of air-polluting effluents, three cubic yards of landfill space, and taxpayers' dollars that would have been used for waste-disposal costs. (In many municipalities, waste-disposal costs in landfills and incinerators total nearly $100 per ton.)

Some recent studies show that paper takes up as much as 50 percent of our landfills. Much of our wastepaper, if properly sorted, can be reused to make high-grade printing and office papers. But recycling programs are little more than collecting programs as long as there are few buyers for recycled products.

Is recycling really more environmentally sound? What about the bleach and the sludge that are left over?

Not only does recycled paper protect natural resources, but it's also a less toxic paper-making process than conventional practices. The manufacture of recycled paper employs fewer chemicals and far less bleaching. It's also the most ecologically safe way to handle potential toxic materials in wastepaper. Here's the process:

Pulping: When fine paper (office and printing paper) is made directly from trees, the pulp is "cooked" in a soup of chemicals, many of them caustic and sulphur-based, to separate lignin from the fibers. (Lignin is what gives trees their structure and about half their weight. It also causes paper to yellow and become brittle. Lignin must be removed from pulp that's being processed into most office and printing paper.)

In a de-inking pulp mill, wastepaper is dumped into a pulper full of water. The pulper looks and acts like a huge blender. A combination of hot water, centrifugal force, caustics, detergents, and fine screening and washing separates the ink, clay, and fillers from the paper.

Recycling wastepaper does not require the heavy chemicals and chlorine that the wood-pulping process uses, since the lignin was removed when the wood fibers were first made into paper. The caustics used in de-inking are not on the federal Environmental Protection Agency's list of toxic chemicals.

Bleaching: Wood fibers are heavily bleached to erase their color and brighten the whites that you see in copier and printing paper. It's not really necessary for the paper to be that white, but paper mills keep producing bright, white paper.

Chlorines whiten printing paper more than any other bleaching chemicals. However, chlorine interacts with the lignins in the pulping process to form dioxins, furans, and other organic compounds. Recycling mills are pulping wastepaper from which virtually all the lignins have been removed, so there is far less opportunity for dioxin formation.

Because wastepaper fibers are already whitened, recycling mills use about 75 percent less bleach than non-recycling mills. Most of the recycling mills primarily use bleaching chemicals that don't produce dioxins for residues left after de-inking.

Sludge: De-inking in a recycling mill is the most environmentally sound way to handle any potentially toxic materials in wastepaper. In the past, inks on the paper contained high concentrations of lead, chromium, cadmium, and other heavy metals that produce color. Recently, toxic agents in inks have been reduced by the EPA's standards for ink manufacturing, air emissions and water effluents.

When wastepaper is de-inked, the ink residues end up in the leftover sludge. About one-fifth of the wastepaper material is drawn off as sludge, which contains not only ink residues, but also fillers, clays, fiber fragments, and other materials.

The question now is: How do we best handle potentially toxic ink residues? If de-inking sludge meets certain toxic levels, it's buried in a controlled landfill that is specially designed for hazardous materials. However, the sludge from many de-inking mills has tested as non-toxic even under strict state regulations, and is prized by farmers as clay-heavy soil conditioners.

Won't recycled paper end up as garbage anyway?

Recycling doesn't eliminate the solid-waste problem, but it greatly reduces it. We need to exceed even our most ambitious recycling goals. The American Paper Institute's recently announced target of 40 percent paper recycling by 1995 - laudable by today's standards - will barely keep pace with the expected increase in paper use by then! The current recycling goal is unlikely to cut even a single ton from the 60 million tons of paper presently dumped or incinerated.

Research indicates that paper fibers can be used up to a dozen times. If we bury the paper, we waste landfill space as well as fibers that could have been used repeatedly, while causing more trees to be cut for new fiber. If we burn the paper, we use energy for combustion, but we waste the opportunity to reuse the fibers.

Don't all papers have recycled content?

Papermakers used to argue that all paper included recycled material because it contained "mill broke" - the scraps produced in the papermaking process. However, while it's important that these scraps get reused, a mill would go out of business if it didn't use them. The wastepaper that needs to get used is "post-consumer waste"-discarded home or office paper that would have been burned or buried if not recycled.

You may think that the term "recycled paper" means that the paper contains post-consumer waste. But there is no standard definition for recycled paper. Several states and the federal government have issued their own widely varying definitions. The EPA minimum-content guideline allows some re-used mill wastepaper to be counted as recycled paper content. Many new "recycled" sheets take advantage of this guideline by containing only allowable mill and converter waste, which does not require de-inking.

When mill waste comprises all the recycled content in a paper, there is no real advance for recycling. Mills may argue that if they didn't use these scraps they would landfill them, but the economic reality is that they are always used, and will continue to be used regardless of how recycling fares.

Post-consumer waste is not being adequately collected and recycled. The EPA's guidelines are so loosely defined that all recycled paper purchased by the government could meet the standard and still not reduce the nation's solid waste problem by even one truckload.

Conservatree has developed a ranking system for recycled papers so that consumers can determine whether the paper they're buying truly meets their environmental goals. These rankings show on all of our paper, and on that sold by the paper merchants who carry our paper nationwide.

Our lowest ranking, C4, goes to the paper that meets the federal-government guidelines. This is the paper to choose if the only alternative is non-recycled paper.

Several states have used stricter recycled-paper definitions for years. They've been successful not only at buying recycled paper, but also in stimulating the development of new papers. A C3 ranking, essentially using Oregon's definition, counts as recycled content only those materials that have actually left the paper mill.

Higher rankings such as C2 (New York's) and C1 (California's) include de-inked and post-consumer fibers. Paper ranked C1+ exceeds all definitions, proving that high-quality, high-content recycled paper not only can be made, but is available even now.

Who wants to use recycled paper when it's all brown with dirt specks in it?

You're thinking of the early 1970s, when growing environmental concerns prompted paper mills to rush-produce recycled paper. Nowadays, recycled paper is virtually indistinguishable from virgin paper. You may be using recycled paper and not even know it. For years, it wasn't identified because mills were afraid that people wouldn't buy it.

Today's recycled paper comes in almost every grade imaginable — including bond, offset, copier paper, computer paper, coated paper, and beautiful text and cover stocks for designer brochures. It comes in bright whites and a wide range of colors, in sheets and rolls for printers' presses. It's strong and versatile, and it meets the same technical specifications as non-recycled paper.

My printer says recycled paper's quality is poor, and it'll cost more to print on it..

Most likely, the printer tried recycled paper in the 1970s, when it was not up to quality standards. However, because printers questioned the paper's quality, recycling mills worked hard to improve their standards. A recent *Paper Sales* magazine study found that more than 80 percent of the commercial printers polled reported that recycled paper's performance was equal to or better than that of non-recycled paper. Here are some of the complaints you might hear from printers unfamiliar with high-quality recycled paper:

"Recycled paper 'lints' in my press." All papers, including non recycled, sometimes lint- deposit fine particles. When this happens, the printer knows that the paper was not made properly, and arranges to replace it.

However, when recycled paper lints, printers often blame the recycled content. Often, they're mistaken. In one complaint we investigated, it was the recycled fibers that were holding the paper together, while the non-recycled fibers were flaking off. When printers have linting problems with recycled paper, they should make the same assumptions that they do with non-recycled paper, and return the paper to the mill for an exchange.

"Recycled paper absorbs more ink and takes longer to dry, so it'll cost more to use it." Ink absorption is related to the finish on the paper. Recycled paper uses the same finishing processes as non-recycled sheets. A properly made sheet will not differ in ink absorption or drying time.

"Recycled papers will jam my machines and just cause problems." Today's recycled papers are made to meet the same technical standards as non-recycled papers. Many printers are using recycled paper, and many report that they find no difference between running recycled and non-recycled paper. However, printers like to stick with the papers that they run the most, because often they must adjust their machines when they switch to any other kind of paper.

" Recycled paper is duller than non-recycled." The "brightness" in recycled paper is, on average, slightly less than that of non-recycled papers. Nevertheless, many recycled sheets do meet high- brightness specifications. When non-technical people speak of recycled paper as being "duller," they may be comparing apples with oranges, or coated with uncoated. For example, *GARBAGE* magazine is printed on recycled paper that has no clay coating, so it isn't "shiny" as the usual coated magazine stock, It is, however, comparable to other uncoated sheets. (Coated recycled papers are available only in limited grades, currently, but more will follow shortly.)

Isn't recycled paper weaker?

Paper needs a mix of fibers to meet printer expectations. Softwood fibers are the longest and give the paper its strength. However, if the sheet were compromised entirely of long fibers, copier and writing paper would have the consistency of a paper bag. Paper needs short, hardwood fibers to add flexibility. Some recycled fibers will be longer than virgin fibers, and some will be shorter. All of the fibers are part of the mix needed for high-quality papers.

There are some 100-percent recycled-content papers that perform well on printing presses. Most recycled sheets, however, are a blend of (typically) 50 percent recycled, and 50-percent non-recycled fibers. This combination maximizes the advantages of both types of fibers. Technical quality doesn't suffer.

Why does recycled paper cost more?

The answer, simply, is economies of scale. Recycled paper-making is actually a less expensive process than non-recycled papermaking. Compare the prices of recycled and non-recycled specialty papers like rag bonds and designer papers: The recycled sheets are equal in price or less expensive than the non-recycled. This is because both recycled and non-recycled sheets are made on similar-size paper machines.

But recycling mills are small by today's standards. Machines producing about 200 tons of recycled paper per day were competitive twenty years ago. Today, paper companies are building virgin supermills that daily produce 600 to 800 tons of paper on a single machine. These mills primarily produce office and offset papers.

The supermill's production costs average about 13 percent lower than the smaller recycling mills.' Virgin paper prices are also down right now because of a "soft" paper market, while recycled paper demand is strong. All of these factors can create wide differences in prices. When a recycling mill- sure of long-term consumer demand- builds a paper machine as big as those in virgin paper mills, you'll see the price of recycled paper drop below non-recycled, as common sense would indicate.

We also have only recently begun to collect enough post-consumer waste to support recycling supermills. Now that businesses are beginning office-paper collection programs, cities can collect and sell their post-consumer waste to guarantee a recycled-fiber supply.

Technology is also just catching up with recycling. Mills used to have trouble cleaning laser printed and xerographic-printed papers. In the past few years, new technological and chemical processes are resolving that problem, although these advances haven't yet found there way into most printing- and writing-paper mills.

Why should we collect more papers when there's already a glut?

East Coast recycling markets are flooded with newspapers, but not office papers. Actually, what the East Coast has experienced is not so much of a glut as a dearth — of consumers of recycled newsprint that would encourage mills to buy what's being collected. If you're not buying recycled products, you're not recycling.

Recent legislation requiring publishers in Connecticut and California to start buying recycled newsprint, as well as new legislative proposals in several other states, has spurred recycling markets. In the past few months, several newsprint mills in the U.S. and Canada announced that they will add de-inking capacity to make recycled newsprint for the consumer market. As it takes up to three years to build a de-inking mill, increased de-inking capacity won't affect the market for a while.

Meanwhile, we throw away about twice as much high-grade paper as newsprint. Fine-paper markets are expanding faster and using lower amounts of recycled content than any other paper category: We need to develop the market for recycled printing and office papers.

I've heard that coated papers, like those used in magazines, can't be recycled.

When paper is called "recycled," it means that it's made with recycled content. "Recyclable" means that it simply has the potential to be recycled. Some foreign elements that combine with paper — plastic from mail pouches, aluminum foil from juice boxes, latex used in graphic designing processes — make paper non-recyclable.

Coated papers can be recycled, but they're not in high demand. (It's true that some recycling mills are recycling them even now.) The real reason coated papers are often excluded from recycling programs is economic. The cost of wastepaper is determined by its weight, and the heavy clay coating reduces the amount of fiber in the paper by weight. That means the mill does not get as much fiber per ton from coated paper as from uncoated. Currently, of course, the recycling mills have their pick of wastepaper. When more recycling mills come on line, the demand for coated paper will grow.

I'm having trouble finding recycled paper. Why isn't more available?

In the 1970s, few mills were identifying recycled paper as "recycled," even though a fair amount of recycled fiber appeared in many papers. State governments were the first to make a major commitment to recycled paper. Even so, purchases of recycled papers amount to "small change" for most paper mills. All the paper that the federal government buys accounts for only two percent of the paper purchased nation-wide. When the major corporations and printers start buying recycled paper, the paper industry will respond.

In our opinion, five steps will lead to a sound recycling economy in the printing and office paper industry:

1. *A standard definition for recycled paper*

Over the years, the proliferation of different state definitions for recycled -paper content fractured the market into small segments, and paper mills had trouble responding to them all.

In 1988, it seemed that new EPA guidelines for minimum recycled content in paper purchased by federal agencies would finally provide a consistent, national definition for recycled paper. However, the guidelines are so flexible that they permit the manufacture of "phony" recycled papers that include only material that never left the mill or converter. They do nothing to stimulate the use of post-consumer wastepaper. Many environmental advocates are encouraging government officials to ensure that "recycled-paper content" includes de-inked and post-consumer fibers.

2. *Government buying of recycled paper*

Several states have led the way with effective procurement programs. These programs have expanded the market, stimulated the development of new papers, and provided leadership. The federal government is still in the process of implementing its program. Meanwhile, Congress needs to tighten EPA guidelines.

The Resource Conservation and Recovery Act (RCRA), passed in 1976 to protect human health and the environment but never really implemented, is now up for reauthorization in Congress. RCRA required the EPA to write guidelines for government procurement of recycled paper. As a result, all government agencies using federal funds to buy more than $10,000 in paper products are required to buy recycled paper.

In our opinion, the federal government needs to send a clear message that true recycling is the way to go. The paper industry is reluctant to invest in new equipment for what they perceive to be a "fad." Therefore, federal guidelines need to be tightened to eliminate counting mill waste as recycled content. They should also provide a 10-percent preference for de-inked and post-consumer fiber content. It's imperative that the reauthorization legislation incorporate effective definitions of recycled paper.

3. *Commercial procurement*

Printers and corporations purchase most of the paper in the U.S. When these major players start buying recycled paper on a large scale, the paper industry will move quickly to respond.

4. *Retrofit mills with de-inking equipment*

Even now, a few mills are locating de-inked fiber from recycling pulp mills that don't make paper. Buying pulp is a creative interim step, but it's far less expensive for a mill to have its own de-inking capacity. A clear message that major consumers want paper containing de-inked and post-consumer fiber will convince mills that the technology investment is worth it.

5. *Build a world-class recycling mill*

The paper industry is currently building two million tons of additional, high-grade papermaking capacity over the next three years, but none of this includes recycling capacity. Several years ago there were two dozen high-grade de-inking mills making fine paper. Now there are only four.

When the day comes that a recycling supermill is built to compete with the huge machines making today's non-recycled white papers, then the industry will quickly change. When recycled paper is finally on a level playing field, it will be obvious that its's a less expensive papermaking process than non-recycled papermaking, and virgin mills will quickly adapt to preserve their market share.

Alan Davis is president of Conservatree, a San Francisco paper company specializing in recycled paper. A version of this article originally appeared in Garbage *magazine, and is reprinted here with permission of the author.*

Chapter Six

Energy

We've heard the bad news again and again. In the short term, our dependence on fossil fuels has dirtied our air, destroyed our landscapes, and is now threatening our economy and our national security. In the long-term, rising temperatures from the greenhouse gases they generate could cause drought, famine, and catastrophic sea level rises across the planet.

The technology to reduce our dependence on fossil fuel — and its detrimental side-effects — already exists. Two of the most practical strategies for that reduction are switching to renewable energy sources — like solar and wind — and increasing the energy-efficiency of currently existing fossil fuel technologies.

Sixty years ago half the homes in Florida had a solar water-heater. Today, California's 17,000 wind turbines create enough electricity to fuel all of Washington D.C.'s residences. Even the conservative Department of Energy has said that 75 percent of the country's projected energy needs could be economically extracted from renewable sources.

Clearly, the technology exists. The question is whether we have the will to use it.

The Hidden Costs of Energy

By Richard Perez

One of the nice things about life in America these days is the easy access to cheap energy. Even as the price for oil skyrockets, we still pay a fraction of what the rest of the world pays for fuel. That's because when we pay our utility bill, or fill our car's gas tank, we pay only for the energy we directly consume. What we don't pay, or rather what we don't think we pay, are the hidden costs of that energy. Altogether, the American Solar Energy Society (ASES) estimates this country pays between 109 billion and 260 billion dollars yearly in hidden energy costs. That's over $740 each paid for health damages, crop losses, and military defenses.

"Our free market economy operates best when both the buyer and the seller have complete knowledge of which choice will benefit them the most," writes Michael Nicklas in the 1989 *ASES Roundtable: Societal Costs of Energy*. "With energy this is obviously not the case." In other words, we will continue sending troops to defend our oil supply until we sharply reduce our demand.

Let's take a look at what we're getting when we buy fossil fuels.

Health costs

The Environmental Protection Agency (EPA) estimates that more than 100 million Americans live in regions where the air is unhealthy at least several days a year. The American Lung Association (ALA) estimates that 30,000 people die prematurely of lung disease each year.If the new EPA standards regarding sulfur dioxide emissions for power plants are adopted, the ALA claims, Americans will save 82 billion dollars in health cost yearly. The health costs of vehicle emissions, however, including nitrogen oxides, carbon monoxide, and hydrocarbons will remain at $93 billion.

Military costs

The U.S. is actually more dependent on imported oil today than we were during the 1973 embargo. Fifty percent of our oil comes from foreign supplies— which means we pay our military a lot of money to keep the lines open. Before Saddam Hussein invaded Kuwait in early August, the U.S. spent between 14.6 and 54 billion dollars yearly just defending the oil supplies coming from the Persian Gulf. Our military presence there costs in excess of an additional million dollars a day.

When we look at the hidden military costs of oil, we must also consider the cost to national security, and the effect of diplomatic and foreign policy decisions made on the basis of oil. These are not easily quantifiable costs, but they have considerable effects on our lives. And, of course, there is the rather high cost of human lives lost defending our oil supply.

Crop Loss

The EPA reported in 1988 that ozone pollution alone is reducing crop yields by up to 12% yearly, at a cost of 3 billion dollars. Boyce Thompson of the Institute for Plant Research at Cornell University has revised this estimate to a 30% crop loss yearly with an annual price tag of 7 billion dollars. These estimates do not include crop losses due to global warming, acid rain, and other energy related forms of air/water pollution. The National Crop Loss Assessment Program estimates that auto emissions cause annual yield losses of $1.9 to $4.5 billion.

Corrosion

Besides hurting our lungs and our crops, auto emissions, especially sulfur dioxide, contribute to acid rain

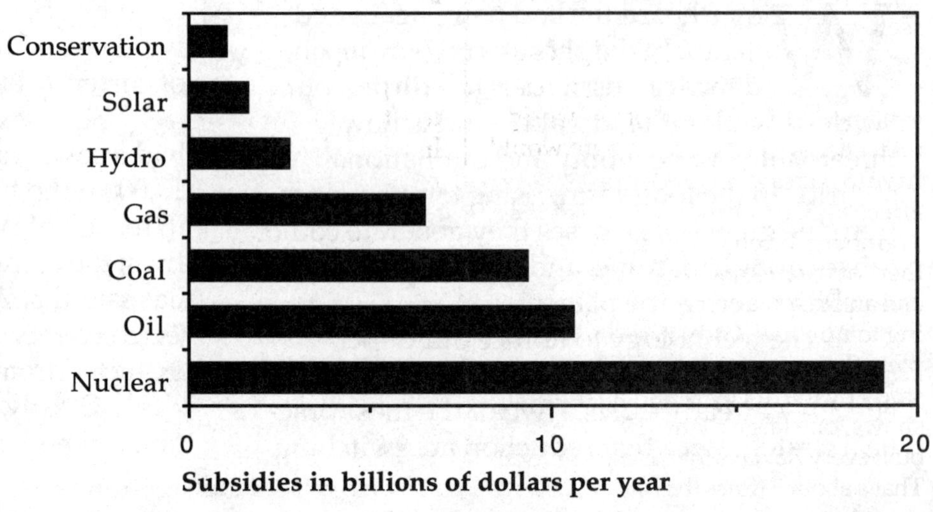

which kills vegetation, pollutes waterways, and corrodes metal. It is estimated that the damage to metal buildings alone is 2 billion dollars per year. That does not include the damage to sculptures, outdoor furniture, and other metal products.

Global Warming

The leading energy sources in the U.S. today are: 43 percent oil, 24 percent coal, 23 percent natural gas, 6 percent nuclear, and 4 percent hydropower. Our heavy reliance on fossil fuels means we emit more carbon dioxide than any other country in the world, which means we are one of the primary contributors to the greenhouse effect— which currently threatens ecosystems and economies the world over.

Radioactive Waste

Lots of people these days are saying that nuclear power will cure us both of the Hussein blues and the greenhouse blues. Secretary of Energy James Watkins is one of those people. His office is currently compiling the first comprehensive national energy strategy for this country. When it lands on President Bush's desk in December, it is likely to include recommendations for more nuclear power.

Advocates say nuclear power is safe, clean, and cheap, and that it will reduce our need for foreign oil. But 70 percent of the oil we import goes to transportation— not to homes and offices. Somehow the idea of nuclear-powered cars, each with its own plutonium supply, zipping along this country's over-crowded freeways, does not seem very viable.

In 1988, the International Atomic Energy Agency reports, nuclear energy supplied 12 percent of the world's total energy needs. That figure would have to increase six times to even begin affecting $C0^2$ emissions, says Alvin Weinberg, often called the "grandfather" of the pressurized water reactors, and an outspoken proponent of switching to nuclear power plants. To complete that expansion by 2025, a recent study by the Rocky Mountain Institute shows, one large plant will have to be built every 7.5 days for the next 37 years. That's about 3 times the rate of building between 1970 and 1985. That same study claimed that energy-efficiency would be 7 times as effective in reducing CO2 emissions as nuclear power.

Once the plant is built, the next question is what to do with the wastes. To date, no one has a viable disposal plan for the millions of pounds of radioactive waste generated by nuclear power plants. These wastes— which the World Watch Institute estimates cost between 1.44 and 8.61 billion dollars per year— won't just disappear. Plutonium (Pu239), for instance, which is the primary element used in fission reactors, has a radioactive half-life of 24,400 years and is environmentally dangerous for over 100,000 years.

Besides all this, of course, there is the danger of a nuclear accident. Three Mile Island and Chernobyl have been just hints of what could come if we increased the number of nuclear plants.

"Nuclear power is not an option— it's a Madison Avenue-type propaganda hype," says Brady Bancroft, a researcher at the Rocky Mountain Institute.

Energy Subsidies

In 1980, when oil prices were as high as $40 per barrel, the Carter Administration was spending as much as $160 million per year on solar energy research. In 1988, when oil prices were $12 barrel, the Reagan administration allotted only $35 million. Altogether, between1981 and 1989, the federal support for research on renewables dropped from $629.9 million to $108 million.

Despite the many and well-publicized problems of fossil-fuel combustion, the current federal government has continued to heavily subsidize the energy technologies with the worst health and environmental impacts. Indeed, nuclear and combustion technologies receive 90 percent of government funds allocated to energy. Renewable energy technologies, with few if any detrimental side-effects, receive the least government support. Photovoltaic and thermal solar technologies together receive only three percent of the government energy subsidies. Conservation, which could rapidly reduce our dependence on oil and its toxic emissions, receives only 2 percent of the subsidy dollars. Something is radically wrong.

What to do?

Write your elected representatives and let them know how you feel about energy issues. Our government agencies have access to the information presented here— they know what's going on. What they don't know is how YOU feel about it. Let your elected officials know that you consider energy a political issue. Ask them what they are doing to help solve our energy problems. Let them know that when you cast your vote, you will have energy on your mind.

Conservation, which can be practiced by everyone, will offer relief until we implement non-polluting, renewable energy sources on a broad scale. Whether you make your own power or buy it from the grid, conservation saves energy. Implement conservation techniques in your home. Install efficient lighting. Turn off unused appliances. Find and isolate those "phantom loads." When you buy an appliance, make efficiency your primary criteria.

Then switch to renewable energy sources. Every time any one of us puts up a PV panel, a hydro-turbine, or a wind generator, we directly mitigate America's energy problems, by reducing both the amount of pollutants released and our dependence on foreign oil supplies.

Richard Perez is editor of Home Power Magazine, *where a version of this article was originally published.* Home Power Magazine *is available at: P.O. Box 130, Hornbrook, CA*

Resources

Copies of the entire "1989 ASES Roundtable: Societal Costs of Energy" can be purchased for $20 from the American Solar Energy Society, 2400 Central Ave., B-1, Boulder, CO 80301. I want to thank the American Solar Energy Society for letting me use the data they have collected. The opinions expressed here are mine and do not necessarily represent the views of ASES.

The Energy Efficient Town of Osage, Iowa

By Craig Canine

North Americans are inclined to look west when searching for new ideas — especially new ideas in combating environmental woes. California recently pioneered a tough toxic disclosures law, for instance, and San Diego, Edmonton, and Portland are showing other cities how light rail transit can lessen air pollution and traffic problems.

But there are lessons to be learned elsewhere, too. In the area of energy conservation — an old idea whose time has now returned, thanks to newly rising gas and oil prices, continuing fears about nuclear energy, and coal-fired power plants' massive contribution to acid rain and the greenhouse effect — we should look to the small town of Osage, Iowa. Like many other small towns across the farm belt, Osage is worried about the future — including economic concerns as well as environmental ones. The 1988 drought on top of the ongoing depression in agriculture means that many small towns like Osage must struggle to stay alive.

To ensure its survival, Osage, population 3,800, has devised its own economic development program, the cornerstone of which is energy efficiency. Since 1974, Osage Municipal Utilities has led the state in setting up programs that help its electricity and natural-gas customers save energy, thereby reducing the amount of electricity and natural-gas the town must buy from out-of-state suppliers. Wes Birdsall, general manager of the Osage utility, estimates that his company's various conservation programs save local businesses and residents at least $1.2 million a year. "It's no different from having a new business in town," Birdsall says. "I don't see any difference between a dollar brought in by a new business and a dollar that's saved due to energy conservation."

Steele's Super Value grocery is a good example of the Osage approach to the goals of energy efficiency and economic development. Everett Steele, the supermarket's owner, put all his refrigerator compressors together in an insulated compartment, then installed two fans and a duct that carries waste heat from the compressors into the main part of the store. Steele now saves up to $600 a month in heating costs during the winter — enough, he thinks, to keep his prices low enough so Osage residents won't be tempted to take their business to the bigger supermarkets in Mason City, 15 miles away.

Steele's grocery store is also the site of an experiment in energy-efficient lighting. With the help of state funding, the Osage utility fitted 22 of the store's 44 fluorescent light fixtures with high-efficiency ballasts and tubes. The remaining 22 fixtures were left as they were, and each set of light fixtures was placed on a separate meter to measure its electrical consumption. "The project is still in its early stages," Birdsall says, "but the first month, the improved fixtures used 22 percent less energy than the unimproved ones."

The Spahn and Rose lumberyard is a short walk from Steele's Super Value. Rick Miller, manager of the lumberyard, takes a few minutes to talk about his town's record of energy conservation. "The general awareness of what is efficient and what isn't is pretty high around here," he says.

"Wes Birdsall over at the utility started doing infrared scans of all the houses in town," Miller recalls. "People could go into the utility office and see pictures of their houses that showed places where heat was escaping. Whenever the utility does those scans, we notice a big increase in insulation sales. We've had people come in and say they cut their fuel bills in half, just with insulation."

Ken Swenson's insurance office is a few blocks east of the lumberyard, on Main Street. Swenson is a member of the board of directors of Osage Municipal Utilities. Both as a board member and a local homeowner, he is enthusiastic about the utility's conservation programs. "After Wes initiated the infrared scans," he says, " the local Jaycees volunteered to go to all the houses of low-income residents and install weather stripping and water-heater jackets, which the utility supplied free of charge. Then we started our load-management program. Anybody who owns an air conditioner can call up the utility and have a free controller hooked up. On really hot days, a computer at the utility plant sends out a radio signal that causes the controllers to turn off the air conditioners' condensers, but not their fans, for up to 7.5 minutes every half hour. The customers can't even notice a difference. This has kept our peak load down.

Osage has six diesel generators of its own, which are capable of meeting the town's electricity needs. But it can buy power cheaper than it can generate its own, so most of the town's electricity comes from a coal-fired power plant in Wisconsin, 125 miles away. "Because of our load-management and conservation programs, we haven't had to add to our existing generating capacity in 15 years," Swenson explains. "The utility has been able to pass along the savings from all this to its customers. Osage Municipal has cut its rates five times in the last decade."

Besides giving away a free water-heater jacket to any Osage resident who wants one, the utility also bought a tree spade several years ago that it makes available to townspeople who want to plant shade trees around the houses. The utility even donates trees from its own nursery. In time, Birdsall expects that the investment in placing trees will keep down the town's air-conditioning load.

"The overall beneficiary of all these programs is the community," says Swenson. "When you consider that our dollars for natural gas go to Texas and our dollars for electricity go to Wisconsin, then the more we can conserve in town, the better off everybody is. Energy conservation has given this town a tremendous shot in the arm."

If the whole country followed the example of Osage, Iowa — or the example of Germany and Japan, where people use energy twice as efficiently as Americans do — the benefits to the U.S. economy, environment, and national security would be enormous. The United States now spends about 10 percent of its gross national product on energy; Japan spends 5 percent. This means that American products cost about 5 percent more on the world market than comparable Japanese products do — and the gap is expected to widen over the next 10 years as Japan continues to improve its energy efficiency more rapidly than the United States does.

If the United States were to pursue energy efficiency as vigorously as Japan does, it could save $1.3 trillion to $2.2 trillion (in 1987 dollars) over the next 20 years, cutting its oil imports by up to 3.5 million barrels per day and reducing its trade deficit by $20 billion to $40 billion per year. By avoiding the need to build hundreds of costly power plants over the next two decades, investments in energy efficiency would free up more than $100 billion each year for capital investments in other industries, including much-needed environmental projects. This would create more jobs and provide U.S. industry with a better competitive position in world markets.

Improved energy efficiency has already benefited the U.S. economy. Because of higher energy prices following the Arab oil embargo of 1973, the country's rate of energy consumption stopped growing at its former rate. The resulting savings have added up to $150 billion a year, a sum nearly equal to the U.S. trade deficit. Energy saved through conservation now contributes as large a share to the nation's energy supply each year as petroleum does, and far more than coal, nuclear reactors, or hydroelectric dams.

This slowdown in the growth of energy consumption has been accomplished with relatively little sacrifice for most people. Even so, Americans have become apathetic because of relaxed vehicle-efficiency standards and the dramatic dip in oil prices during the 1980s. But with increasing concern over global warming trends, acid rain, and the nation's ability to compete in the world market, improving America's energy efficiency is now more crucial than ever.

If the benefits of energy conservation are so great, why aren't we realizing them? In other words, why doesn't every town in the country follow the lead of Osage, Iowa?

Osage is fortunate in that it is served by a municipally owned utility, which supplies residents with both electricity and natural gas. The customers and owners of a municipal utility are the same people, so whatever benefits the utility's owners also benefits its customers, and vice versa. But this is not true of most utility companies in the United States, which are private businesses owned by investors, or stockholders. The stockholders and the customers of an investor-owned utility are two separate groups of people, and their interests are not necessarily the same. On some issues, in fact, they are at odds.

Energy conservation is one such issue. When a utility company's customers use less energy, those customers benefit because their bills go down. But the company and its owners lose out, because the utility's profits go down. Encouraging energy conservation is, quite simply, against the financial interests of investor-owned utility companies. You may as well ask a grocery store to encourage its customers to eat less.

Investor-owned utility companies are a curiosity of American capitalism. Utilities are generally publicly owned in other industrial nations. In most business enterprises, the desire to make a profit is balanced by the forces of competition. But the nature of the utility business is such that competition is impractical. Instead of competitors, utilities have government regulators. The regulators tell the electric companies how much they may charge the public for electricity, as well as how much they may profit from its sale. This can lead to some topsy-turvy economics. As David Moskovitz, a public-utilities commissioner from Maine, puts it, "They make money whenever they sell their product — kilowatt-hours — no matter how much it costs to generate those kilowatt-hours or how much they charge customers for them. Existing regulation always provides the incentive to sell more kilowatt-hours. The flip side is that existing regulation always provides a disincentive for utilities to encourage conservation."

Supporters of an energy-efficient America see this as something of a scandal. Once upon a time, our current regulatory system made sense. The postwar American economy desperately needed more electricity, and utility companies were expected to provide it. The costs of generating electricity went down as utilities built bigger and better power plants. The companies were financially rewarded for investing in those new sources of supply, since the nation needed and wanted electricity. Costs were low, demand grew at a steady and predictable rate, and utilities prospered. It was the industry's golden age.

Then came the Arab oil embargo. The price of energy of all kinds shot up, and the nation responded by using less. The slowdown in energy use startled most utility executives. As energy demand failed to grow at the rate they had predicted, they found themselves with billions of dollars tied up in new coal and nuclear power plants that were suddenly unneeded.

The utility business has since undergone many changes, but the regulatory system has remained the same — it continues to reward electric companies for investing in new energy supplies. But a new fact of life has emerged in the field: It costs more to build new supplies of energy than it does to "create" the same amount of energy by investing in efficient consumption.

For example, some utilities (PG&E is one) offer rebate programs that pay customers to trade in their old refrigerators for new, more efficient ones. In this manner, the utility might spend $200 to $300 for each kilowatt of electricity subtracted from the power load it must deliver to customers. Building a new coal-fired power plant, on the other hand, would cost the utility $1,000 to $2,000 per kilowatt — five times more than the conserved kilowatt of capacity.

Existing regulatory policies allow utilities to make a fixed rate of return (15 percent profits, for example) on their capital investments — such as new plants. But refrigerator rebates don't involve capital investments; they are considered only an operating expense. Utilities don't normally make a rate of return on such expenses. So the rebate program, while a more cost-effective source of kilowatts than the coal plant, would not make the utility a profit.

Amory Lovins, the well-known energy expert and co-founder of Rocky Mountain Institute (see "Soft path in the Rockies," Utne Reader, Aug/Sept 1986), says, "There is a perverse coupling, which serves no socially profitable goal, between sales of electricity and utility profits."

One way to get around this perverse coupling is to require utilities to sponsor conservation programs, despite their negative effect on the utilities' profits. Many state commissions have adopted this approach. Most homeowners regularly receive mailings from their electric companies offering anything from free energy audits and appliance rebates to low-interest loans for insulation purchases.

The utilities may give the impression that they are offering these programs out of the goodness of their hearts. And indeed, to some extent, that may be true. Yet, the incentives for utilities to promote efficiency pale beside the incentive to sell more electricity. But this may be changing.

In 1988, the Connecticut Department of Public Utility Control (DPUC) ruled that the $10.1 million that Connecticut Power and Light proposed to spend on conservation programs was not enough, and ordered the utility to raise its conservation budget to $18 million. "Conservation will reduce Connecticut's total energy bill while it reduces the risk of dependence on unreliable fossil-energy supplies," the DPUC reasoned in its decision.

As a result of this decision, Connecticut Power and Light — a subsidiary of Northeast Utilities, New England's largest electric company — collaborated with environmentalists, regulators, and consumer advocates to put together a comprehensive package of energy-saving measures for all classes of its customers.

The urge to save money by paying lower utility bills is contagious. Once a utility company gets the ball rolling and customers learn the benefits of conservation firsthand, there's no stopping it. No three people have proven this more conclusively than LeRoy Huegli, Richard Woodruff, and Collin Myers — high school teachers in Osage. For years, they have had a running contest. The winner each month is the one who brings the lowest utility bill to the teacher's lounge.

"Woodruff was always the guy to beat," Myers says. "His house is smaller than LeRoy's or mine, and it was more tightly constructed in the first place. He would come into the lounge bragging about how low his bills were. That kind of got to some of the rest of us, so we started trying to beat him.

"When my wife and I first moved into our house in 1982," Myers continues, "we spent about $2,100 a year for utilities. Around that time, the utility company did one of its infrared scans of all the houses in town. I knew our house was pretty leaky, and the scan made it extremely obvious. Our family-room addition was especially bad, so I put sheets of rigid-foam insulation on the inside walls, took off the trim around the windows, and caulked and taped the insulation to the window frames before putting drywall and trim back on.

"Later, we took advantage of another one of the utility's programs. They have several electric meters that you can borrow and hook up to any appliance to see how much energy it uses. I plugged our old refrigerator into one of the meters and learned that it was eating up about 30 kilowatt-hours a day, costing us about $50 a month to run it. We sold the old refrigerator for $200 and bought a new one for a net investment of $180. We operate our present refrigerator on less than 10 kilowatt-hours a day. It paid for itself in about a year.

"My electric bill is now about 40 percent of what it was," Myers says. "The gas bill is 50 to 60 percent as big, even though we added a gas stove. When you're saving more than $1,000 a year, the payback comes pretty quickly. Our total yearly bill went from twice as big as Woodruff's to less than his."

Woodruff grins. "Yeah," he says, "but you guys come over to my house and stand there with the door open in the middle of winter."

Spurred on by the others' results, LeRoy Huegli made a list of energy improvements in his house similar to Myers'. "When you see the benefits of conservation demonstrated so graphically," he said, "it's hard to resist. Osage has learned that conservation is good business for everyone."

Craig Canine is the former editor of Harrowsmith *magazine, where this story was originally published. He is currently writing a history of the environmental movement in America. He lives in Norwalk, Iowa.*

Here Comes the Sun

The potential for solar power

By Karen Kho

On Sun Day, May 3, 1978, over 20 million people in 30 countries attended conferences and workshops on solar power. Within months, governments all over the world had developed laws and tax credits to usher in the Solar Age. In this country, President Carter installed solar panels on the roof of the White House. His administration devoted $500 million to solar energy development. In the Bay Area, solar industry flourished. Progressive consumers leapt at the opportunity to retrofit their homes, and dozens of businesses benefited from tax credits provided to those consumers.

In 1986, the Reagan Administration slashed federal funding for renewable energy. The solar industry bottomed out. In the Bay Area, 80 percent of the solar businesses went bankrupt. Consumers, who lost both warranties and subsidies, also lost faith in the potential of solar.

With increased evidence of the dangers of fossil fuels, including air pollution, global warming, and threatened national security, it may be solar's time to shine again. The Inter-governmental Panel on Climate Change predicted this year that solar applications would not be commercially viable until 2025.

Other experts, however, note that the cost of an installed solar system dropped 50 percent in the last decade, while thermal output has increased by 35 percent, and that a solar water-heater is still one of the largest energy savings available. Indeed, combining active solar and passive solar systems can reduce energy consumption in buildings from 30 to 80 percent. Homeowners can save 2300 kilowatt-hours of electricity a year, savings of $300 to $800 per year for a family of four.

On a national scale, if 20 percent of the houses in the U.S. had solar water-heaters, we could reduce energy consumption by 500 trillion BTU per year, equal to 85 million barrels of crude oil or 235,000 barrels of crude oil per day for one year, according to Carlo LaPorta, an energy-consultant with R&C Enterprises in Washington, D.C.

Switching from fossil fuels to solar will also drastically reduce greenhouse gas emissions and other air pollutants. An individual homeowner can avoid emitting 3894 pounds of greenhouse-causing C02, LaPorta claims in *Global Warming: The Greenhouse Report*. Converting just 6.7 million houses to solar— a mere seven percent of the national housing stock— could eliminate 67.5 million short tons of carbon dioxide, 284,000 tons of sulfur dioxide, 165,000 tons of nitrogen oxides, and 10,000 tons of particulate matter.

On a utility-wide basis, a solar-generated power plant would emit 268 times less carbon dioxide than a coal-fired one, and 211 times less than oil-fired ones, if you count fuel extraction, plant construction, and power production, the Department of Energy has estimated. That's nothing to sneeze at.

Can we get that far from here? In 1985, buildings consumed 37 percent of the primary energy supplies in the U.S. Heating, cooling and ventilation, and lighting used 54 percent of the energy used for residences and 84 percent of that used for commercial buildings. Solar power can replace fossil fuels (like coal and oil) for any of these tasks. Combining solar technology with other alternative energy systems, including conservation, wind, and geo-thermal can give some consumers total energy independence.

In the 1930s, one out of every two new houses built had solar water heating systems installed. Did builders know then something the IPCC doesn't know now?

Three Kinds of Systems

You can choose from three kinds of solar technology: passive, active, and photovoltaic. If you develop a "passive system," you essentially turn your whole house into a solar collector. A passive solar house uses strategically-placed windows, sunspaces, greenhouses, and rooms to make use of the sun's radiant energy. Generally speaking, such attributes need to be installed as the house is built— but many people enlarge windows and add sunrooms on the south sides of their houses to reduce their heating energy needs.

Passive water heaters merely absorb the sun's rays to heat water in collecting pipes, and then store it either in water or masonry. There are no mechanical parts to break down, and they don't need external power to operate. Passive systems last as long as the house does.

Active systems actually absorb and distribute the sun's heat with the help of fans and pumps powered by electricity. Active systems cost more to install and generally only last 15 to 20 years, as do most heating and plumbing systems. In addition, the moving parts make the systems susceptible to corrosion and break-down.

Photovoltaic systems convert sunlight to usable electricity— without moving parts, noise, or harmful emissions. Here's how it works: PV cells are placed in a solar panel, which is mounted to your roof, transforms solar energy into direct current (DC) electricity. The energy runs through a charge controller, 12 volt battery and fuse box. From there it can directly power DC appliances such as fluorescent lights. Otherwise, energy needs to pass through an inverter, which will allow you to use standard alternating current (AC) appliances.

What can solar power do for me?

The Bay Area, where there's plenty of sunshine and where heating needs are generally quite low, is an ideal place for solar power.

You can use your solar system to do anything a conventional system does: heat water, heat space, and run appliances. The cheapest function is heating water, but as photovoltaic technology advances, solar for all energy needs will be competitively priced.

Solar Electric Systems

In general, photovoltaics is the most efficient and economical means of "wiring your home" through solar energy. Photovoltaics have been used for generating power for both on and offshore traffic control systems, crop irrigation systems, radio relay stations, and public lighting systems. Increasingly, they are also being used in homes to produce electricity. A large PV system can be linked up to the local utility company so that electricity will be supplied when needed or sold to the company when overproduced. This also avoids the need for a storage system.

PV electricity currently costs about 25-35 cents per kilowatt-hour to produce and is expected to drop to 6-12 cents per kw/hr by the year 2000. When that happens, the price of an installed PV system large enough to power a residence will drop significantly.

Conventional methods, on the other hand, produce energy at 12 cents per kilowatt-hour, and that cost is rising. California utility companies are starting to use peak metering, a system in which electricity is higher priced during high use periods, such as a hot summer day. (Those days, of course, are the best-suited for solar power).

With the phase-out of nuclear power, and the advent of such metering techniques, conventional costs to the consumer will increase as well. Solar power systems will then be directly competitive.

In remote areas, such as mountaintops, photovoltaics are already economical, because of the high cost of extending utility lines. PV panels in those areas can produce all of a home's electricity.

Space Heating Systems

Solar systems can generally provide 20-50 percent of your heating needs. The type of heating system you already have determines the complexity and expense of any solar heating installation. Lowering the amount of building heat loss through energy conservation measures improves a heater's success.

Both passive and active heating systems can be installed. Active systems need more monitoring and maintenance, but they can provide 50 percent of your heating energy. Sunspaces and solar greenhouses are a very popular type of passive retrofit. They reduce energy consumption and add living space. They can provide 20-40 percent of heating, and are a good option for those interested in adding more to their home than a lowered energy bill.

Solar Water Heaters

Hot water systems are far more common in the Bay Area than space heating systems. A solar water heater uses the sun's energy to heat water, thereby reducing your monthly utility bill. Over a long time period, solar water heaters are far more economical than electrically-heated ones.

In an all-electric home, 25 percent of the utility bill goes to water heating, at a cost of about $25 per month. A solar water heater can save 50 percent to 80 percent of that bill, or $10 to $20, if you keep a back-up element turned on. If you turn the back-up off, and rely only on sunshine, you will save even more.

Of course, as the prices of both fossil fuels and electricity rise, switching to solar will become even more profitable.

You can choose between a number of different system types, including: pumped systems, which circulate water from the water storage tank through the collectors and then back to the tank; thermosiphon systems, which use gravity to distribute heat throughout the tank; and integral systems, which combine the tank and collector in one unit.

Generally speaking, residences using a dishwasher and a clothes-washer need 10 to 15 square feet of single-cover, flat-black collector area and 20 gallons of water storage per person. (For a family of four then, 40 to 60 square feet of collector and an 82-gallon storage tank are recommended).

How much will it cost?

The biggest question for people considering solar energy as an alternative is - does it pay off? In general, the more a system costs, the longer it takes to pay off and the more energy it must provide to justify those costs.

The real price of solar collectors today is about one half of what it was a decade ago, due to mass production, and the tough market created by low fossil fuels. Generally speaking, the installed cost of solar water-heating system in the U.S. is $2000 to $2500. With minimum savings of $300 per year, consumers can expect a payback within five to eight years.

If you build and install your own system, it will be considerably cheaper. Factory-built solar collectors cost $15 to $25 per square foot. Pumps and controls cost $150 to $300. Tanks cost $300 to $500.

Photovoltaic systems can be quite a bit more expensive. A small system, which will power essential appliances and a water pump, may cost as little as $1,000. A more complex system that can meet all of a home's electrical energy needs has an installation cost of about $10,000.

Beyond the initial investment in hardware (solar collectors, PV panels, etc.), the cost of installing equipment is determined by the orientation of your building. The south side is generally the best collection area. The position of your building compared to solar south, which is the position of the sun at midday and the amount of shading the wall receives must be determined. Maximum solar efficiency demands that the collector surface be 30 degrees east or west of solar south. Roof and wall obstructions can complicate or even rule out solar installations.

The slope of your roof also determines your ability to use solar collectors inexpensively, as some angles collect a greater percentage of the sun's rays. If indoor installations, such as plumbing, wiring, and mounting have to be worked around, expenses also increase. Likewise, the outside landscaping and utility lines must be considered.

This is all just hard cash talk, of course. By converting to solar you dump less C02 and other pollutants into the air, you free yourself of the whims of OPEC countries and utilities, and you contribute to world peace by reducing this country's monstrous demand for oil. All of those are costs which are hidden in the seemingly cheap price of oil in this country. (See *The Hidden Costs of Energy* at the beginning of this chapter.)

A few subsidies are still available for solar power. At present, a 10 percent tax credit for commercially used systems exists in California. Loans of up to $150,000 at 5 percent interest are available for small businesses for energy conservation equipment. Homeowners also may be able to add a $3500 value to their home when applying for an FHA-VA loan if they plan to install a solar system. Contact your local bank for the most recent information.

Solar greenhouses and sunspaces may sometimes qualify for federal tax credits, but there are special guidelines that the IRS sets. Contact the Conservation & Renewable Energy Inquiry & Referral Service at the US Department of Energy for further information.

Where to Buy

Lists of contractors, dealers, and installers are available through this guide, through the Berkeley Energy Office, the California Solar Industry Association, and through some utilities.

A rating system currently exists for hot water heaters. Look for certification by the Solar Rating and Certification Corporation (SRCC) which lists the amount of energy in British thermal units (BTUs) that the collector can be expected to produce. These ratings apply only to the collector— not other parts of the solar system.

Practical Tips

• Before you actually choose a system, you need to do a home energy audit. This gives you your energy consumption needs and the solar potential of your home. Check under Energy Consultants in the Consumer Guide to find an energy consultant who can help you. You also can call your utility company to have an auditor come out and check your home free of charge. If you are dissatisfied with this audit, call the California Energy Commission and ask for a recommendation.

• You also can do your own energy audit via a computerized system offered by PG&E. They send the forms, you fill them out, and they run it through their data-base. This is also free of charge.

• Improve your home's energy efficiency before you buy. This will let you buy a smaller system. Properly insulating your home will save you far more money than merely turning off appliances that are not in use. The initial expense of adding insulation will be recovered through consistently lower utility bills as fuel prices rise.

• Before you invest in a system there may be legal considerations based on your local zoning restrictions. Roof height, yard requirements and limitations on the amount of glass allowed may rule out the use of certain types of solar systems. Contact your local city planning office about solar access rights. This can help avoid future problems, such as obstructive building or plant-

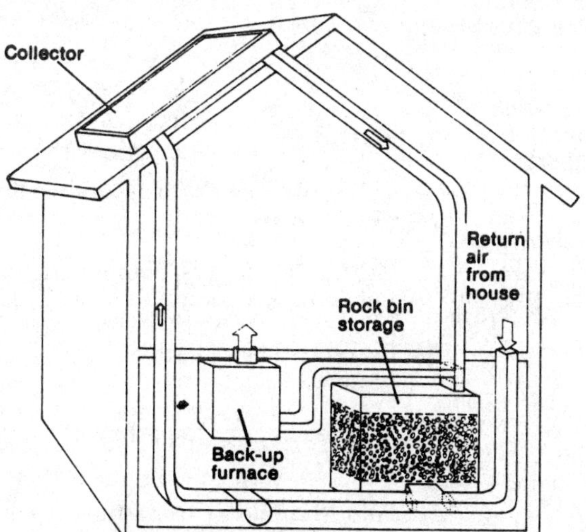

Figure 1: Active Space Heating System (Air Collector)

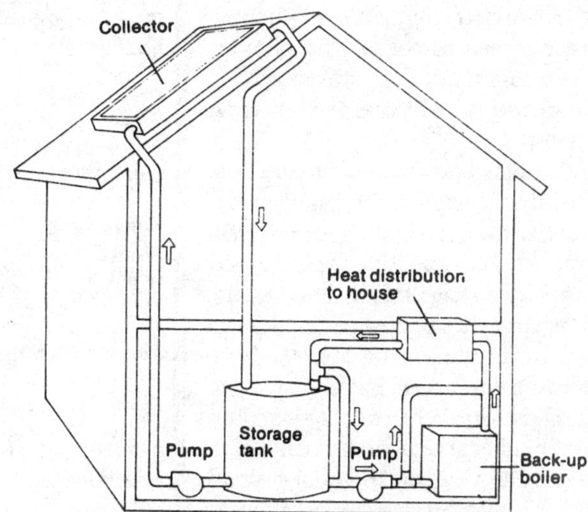

Figure 2: Active Space Heating System (Liquid Collector)

ing near your home. You may be able to prevent such projects from being implemented around your home if you plan ahead.

- Shop around to find the best system for your needs. Watch for suspicious claims. Small heating systems with no storage component cannot be expected to provide a large portion of your heating needs. Avoid buying high-priced, complicated systems that claim to have multiple applications. The more complicated the system, the less likely you will be able to understand why it breaks down. Be sure you also understand what the warranty covers.

A Sunny Future?

How far can we go with solar? Only a decade ago, at the height of the solar credits craze, some analysts predicted that solar energy would only penetrate into 10 percent of the market in this country. Today, with 1 million units installed, it's at about 1 percent. (About half of those units are in California, according to California Energy Commission data.) LaPorta claims that if all houses sold or resold in the future (nearly 3 million per year) were required under air pollution legislation to have solar water-heaters installed, we could have 20 percent of American homes solar retrofitted within 20 years.

Seem like a lot? It's nothing. Barbados already is 35 percent solar powered. In Israel, 65 percent of the homes have solar water heaters. All new homes and apartments of four stories or less are required to heat their water with solar power.

Photovoltaics offer the greatest potential for large-scale solar power. Most experts agree that PV power will become truly competitive in developed areas towards the end of this decade. But it is already the best alternative in remote and developing areas, where power line extension is expensive. As demand rises in such areas, sales of photovoltaic cells have been increasing 25 percent each year for the past several years. As sales grow, new economies of scale will be realized by manufacturers, and the cells will gradually become less expensive.

New technology may also soon lead to higher efficiency and/or price reductions. Experimental cells made of gallium-arsenide and gallium-antimonide have proven 35 percent efficient in the laboratory. Cells using thin films of semiconductor materials in layers are less efficient, but may cost as little as one-tenth as much as cells currently available.

As fossil fuels become scarcer and more expensive, photovoltaics will take their place as the world's premier source of energy. The U.S., unfortunately, will be left behind as that PV star rises. While Japan, West Germany, Italy, and Spain have been pumping money and resources into research and development, the U.S. has thrown away its lead in the field. Ten years ago, this country produced 95 percent of the world's photovoltaic cells. With the recent sale of Arco's solar division, the largest manufacturer of solar cells in the world, to the West German Siemens Company, U.S. market share has dropped below a third.

One way to get the U.S. back on track is to increase subsidies for solar research and development. Pressure your representatives to get more money allotted to this valuable, oil-saving, pollution-free technology. Currently we spend over $38 billion per year for crude oil and petroleum product imports, whereas appropriations for research development on all solar energy technology total less than $120 million/year.

Karen Kho is a staff researcher and writer for the Bay Area Green Pages.

Sources and Resources

In order to satisfy all the human energy needs in the world, we would need enough solar panels to cover 83,000,000 acres. That is only .2 percent of the world's land space.

FURTHER READING

Alternative Energy Sourcebook, Real Goods, 1990. 966 Mazzoni St., Ukiah, CA 800/762-7325

Berkeley Energy Resource Directory, City of Berkeley, Office of Economic Development - Energy Office, 2180 Milvia St., Berkeley, CA 94704.

Consumer Handbook of Solar Energy, J. Keyes; Morgan & Morgan: Dobbs Ferry, NY 10522, 1979. Guide for consumers purchasing equipment.

The Integral Urban House: Self Reliant Living in the City, Helga Olkowski, Tom Javits and the Farallones Institute staff, Sierra Club Books: San Francisco, 1979.

The Most Energy Efficient Appliances, American Council for an Energy Efficient Economy, 1001 Connecticut Ave., NW #535, Washington, DC 20036 202/429-8873. Yearly booklet rating major appliances.

Northern California Sun, Northern California Solar Energy Industries Assoc., PO Box 3008, Berkeley, CA 94703.

Solar Industry Journal and Directory of SRCC Certified Solar Collector & Systems Ratings, Solar Energy Industries Association, 777 N. Capitol St., NW, Suite 805, Washington, DC 20002 202/408-0660.

US Department of Energy, Conservation & Renewable Energy Inquiry & Referral Service, PO Box 8900, Silver Spring, MD 20907
800/523-2929 or 800/233-3071.
Free informational pamphlets on alternative technology and insulation.

SOURCES

California Solar Industry Assocation
916-782-4809

California Energy Commission
Public Information Line 916-324-3298

Berkeley Energy Office
644-6309

San Jose Energy Department
408-277-5538

Palo Alto Energy Department
(415) 329-2268

Florida Energy Center
407-783-0300

Chapter Seven

Bay Habitats

Birds flocked in such numbers they clouded the sun. Bears fished and rested at the shore, while herds of deer browsed the bunch grasses. When white men first laid eyes on the Bay, they found the abundance of wildlife here awe-inspiring

But as is happening almost everywhere, the swelling human population around the Bay has squeezed out most of the other animals that once lived here. Several species unique to the Bay Area hover on the brink of extinction, their habitats reduced to tiny fractions of their former range. What habitat remains is seriously degraded by the huge volume and variety of pollutants with which we soil our nests on a daily basis. This is despite some of the most stringent environmental laws in the world.

Sometimes we tell ourselves that an urban area such as ours can expect no better; that the wilderness is the appropriate place for nature; and that, besides, the Bay Area is still a beautiful place with many natural attractions.

But those attractions are fast vanishing, and what we are doing to nature here we are doing everywhere. If we can't solve these problems at home, we are unlikely to be able to solve them at all.

Going . . . Going . . . Gone .

The Bay Area's Vanishing Wildlife

By Laura Hagar

It is hard to remember that we live in a land of ghosts, that we walk not in the footsteps of giants, but in fading tracks of all the creatures that came before us. Where you walk your dog, a wolf once brought a pronghorn down, shaking it until its blood lit up the grass. In the morning as you wait in traffic, you sit where a tern once hunkered down, warming her eggs in the soft moist sand. A grizzly slapped a salmon from a stream near where you take your morning coffee.

The diaries of early explorers paint the San Francisco Bay Area as a paradise for wildlife. The Bay and Delta teemed with salmon and steelhead trout. Otters swam off Alcatraz and harbor seals gathered in such number off San Francisco that the rocks seemed to writhe with their bodies. Bellowing herds of elk browsed in the marshy meadowlands around the Bay and vast herds of pronghorn antelope moved across the oak-spotted hill sides. Grizzly bear, wolves, and mountain lions stalked their prey in the long grass.

And everywhere, there were birds: seagulls, terns, rails, ducks, geese, hawks, eagles, and even condors. Bay Area wetlands were home to innumerable local species, as well as hordes of migrating birds on their way south down the Pacific Flyway. French explorer La Perouse said the flocks of seabirds and water fowl were so huge that a single shot sent them rising "in a dense cloud with a noise like that of a hurricane."

For thousands of years, the Bay Area's human population lived in balance with the wildlife around them. The Ohlone and Miwok peoples gathered acorns and shellfish and hunted game. Sometimes the hunter became the hunted — early Spanish missionaries marvelled at the terrible scars that many Indians carried from their encounters with grizzly bears. To most Americans, the chance of becoming some other creature's hot lunch seems unbearably barbaric, but to the native people who lived here it was simply the price to be paid for living in a world of plenty.

The relentless waves of people who came later thought differently. When the Spanish arrived in the Bay Area in 1776, they came as missionaries, intent on building a Christian utopia peopled by newly baptized savages (i.e. Native Americans.). The savages disobliged them — at first by resisting Christianization and then by dying in vast numbers from smallpox, measles and syphillis.

The ranchers that followed in the wake of the mission system had an equally devastating effect on local wildlife. They slaughtered grizzlies and wolves to protect the newly arrived herds of cattle and sheep which were already replacing the native antelope and elk. The introduction of European grazing animals brought about a startling change in the landscape: as short European annuals replaced native bunch grasses, California's hills began growing golden in summer and fall. Before they had stayed green almost all year long.

The Anglos who arrived in Northern California in the middle of the nineteenth century merely stepped up the assault on the environment that the Spanish had started a hundred years before. They logged the old-growth redwood forests in Marin county and the East Bay, using the wood to build the opera houses, mansions, and Barbary Coast saloons of a booming San Francisco. The East Bay's redwood forests were completely gone by 1860, and so were the animals that depended on them for survival. Hunters finished off the elk and antelope populations and trappers wiped out the Bay's sea and river otters. Hydraulic mining in the Sierra foot hills sent mountains of silt down the Sacramento River Delta, smothering the rich beds of oysters, mussels, and clams which once carpeted the bottom of the Bay.

At the edges of the Bay, things were changing as well. Although we now think of the Bay Area as a fairly arid place, the new settlers found it downright soggy and set about "reclaiming" the seemingly useless marsh land for nobler purposes like agriculture, salt production, and urban development. They channeled the creeks in box culverts, buried them underground and began diking, draining and filling the Bay's wetlands. By the end of World War II, the war on the wetlands was in full swing. Real estate developers filled thousands of acres of tidal and seasonal wetlands to build homes, airports, freeways, and industrial parks.

Over the last 150 years, tidal marshes around the Bay have shrunk from 200,000 to 35,000 acres. Sewage and toxics have poisoned much of what remains. A 1989 report sponsored by the Sierra Club, Audubon, and Save San Francisco Bay, estimated that seasonal wetlands in the South Bay alone shrunk by 61 percent between 1956 and 1988. The number of waterfowl in the Bay has plummeted. The Bay Area's vanishing wetlands are also home to several endangered species, such as the salt marsh harvest mouse, the clapper rail and the fresh water shrimp.

According to Barry Nelson, director of Save San Francisco Bay, the plight of the clapper rail and the salt marsh harvest mouse symbolize the continuing ecological decline of Bay Area wetlands.

A native of cordgrass and pickleweed salt marsh, the clapper rail is a plump brown tear-drop of a bird with short legs and a long pointed beak. At the turn of the century, clapper rails were so numerous that hunters boasted shooting 200 per day. Restaurants hung strings of rail carcasses in their windows for decoration. By 1987 only 700 rails remained, living in eight marshes in the Bay Area. By 1990, there were only 500.

Clapper rails have lost 90 percent of their salt marsh habitat to diking and filling around the Bay. Fresh water flows from a San Jose sewage plant has rendered an additional 300 acres of salt marsh unusable. Non-native red foxes are raiding their nests, and now there's evidence that toxics in the Bay are damaging the rail's eggs.

"More than any other species," Nelson says, "the clapper rail is really on the brink of extinction. They're being assaulted from all sides. The rail's only hope is preserving what remains of the Bay's salt marshes and cleaning up the Bay."

Like the clapper rail, the salt marsh harvest mouse lives in pickleweed salt marsh. Subsisting on a diet of insects and grass seeds, the mouse is the only mammal that drinks salt water.

In the south Bay, the mouse has lost 95 percent of its original habitat. South Bay developers are still scrapping with the state Fish and Game department over the remaining five percent. In an attempt to get around wetlands regulations, landowners have taken to illegally plowing up their wetlands. According to Barry Nelson, this illegal degradation is now the biggest threat to the survival of the mouse and the other creatures which depend on seasonal and salt marsh wetlands for their survival.

Wetlands aren't the only endangered habitat. The Bay Area has been growing outward even faster than it's been growing in. The megalopolis is spreading north, east and south, gobbling up open space as fast as you can say "subdivision." That's bad news for the hundreds of natives species that depend on the Bay Area's fragile greenbelt. The Bay Area's grasslands, oak-covered hills, and redwood forests still support a host of large animals like deer, grey foxes, bobcats, and mountain lions, as well as smaller species. But the advance of civilization has pushed some animals to the limit. The Bay Area greenbelt supports several endangered and threatened species such as the San Joaquin kit fox, the San Francisco garter snake, the Alameda whipsnake, and golden and bald eagles.

Some of the greenbelt's endangered species never had much habitat. The Alameda whipsnake, for instance, lives only in very specialized environments in Alameda and Contra Costa. Endangered butterflies like the mission blue and the San Bruno elfin live only on the San Francisco Peninsula.

Species with a limited range are more susceptible to disease and less likely to recover from habitat disturbance. The building of San Francisco, for instance, wiped out the primary habitat for the brightly colored San Francisco garter snake, which now exists only in isolated pockets on the peninsula.

According to Seth Adams, a Contra Costa open space activist and program director for Save Mt. Diablo, the biggest threat to wildlife in the greenbelt, besides the overall loss of open space, is the fragmentation of habitat — the advertent or inadvertent creation of islands of wild land surrounded by city.

"Large species like mountain lion and eagles range over a huge area, sometimes a hundred miles. Animals that need large amounts of space both for foraging and breeding purposes are easily damaged by habitat fragmentation. What's happening in places like Contra Costa County is that these animals are finding their traditional movement corridors cut off by fingers of development."

One such animal is the cat-sized San Joaquin kit fox. The fox's large territorial requirements — one square mile per animal — make it particularly susceptible to habitat fragmentation. The San Joaquin kit fox once roamed the western edge of the Central Valley, from Kern County to Contra Costa, but has since lost 93% of its habitat to agricultural and urban development.

Although most of the remaining 7000 kit foxes live in San Luis Obispo and Kern counties, a small isolated population of kit foxes still lives in Contra Costa, where its habitat is threatened by suburban development and the proposed Los Vaqueros reservoir.

"This animal doesn't have anywhere else to run to," Adams says. "We've pushed it as far it can go."

So what hope is there for the Bay Area's endangered species and other wildlife? The answer isn't in yet. Some species are making slow progress back from the brink of extinction. The bald eagle, the peregrine falcon, and the brown pelican, all endangered species in California, have staged remarkable comebacks since DDT was banned in 1972. Humpbacked whales, virtually wiped out along the California coast by whaling companies in the middle of this century, are back in force.

But these are the exceptions. For most Bay Area wildlife, the outlook is grim unless we can stop the relentless destruction of habitat.

Acquiring private land for parks and wildlife refuges is the surest way to protect open space, but it's very expensive. In these days of tight budgets, buying habitat is often low on the political agenda.

Creating preserves for wildlife used to be the province of government agencies like the California Department of Fish and Game and the federal Fish and Wildlife Service, but in the last few years private organizations have gotten into the act. The Nature Conservancy operates a native grassland preserve near Fairfield called Jepson Prairie, and Ring Mountain, a native plant preserve in Tiburon.

More than half the habitat for endangered species in the Bay Area and a large proportion of the remaining open space is still in private hands. Most landowners don't want to sell. According to Greenbelt Alliance's Jim Sayers, the best way to protect wildlife on these lands is through zoning law. Throughout the Bay Area, environmental groups have sponsored citizens' initiatives to zone land as permanent open space.

"A lot of people are beginning to look around at what they have in terms of open space and wildlife and what they've lost," Sayers says.

"They're deciding to save what they've got left. The protection that zoning offers isn't as complete as park status, but it's a lot cheaper and it's better than nothing."

When zoning fails, environmentalists are reduced to waging a war of nerves with landowners who want to develop their property. Environmentalists demand endless studies and challenge every permit and Environmental Impact Reports, in hopes of preserving the land and ultimately wearing down the owner's resistance.

In the area of wetlands preservation, this scheme seems to be working.

"In the last ten years, we have really been able to stop cold almost all the developers who have applied for permits to fill their wetland." says Barry Nelson of Save San Francisco Bay. "We haven't been able to stop the illegal degradation of wetlands, but time is on our side, even if its not on the side of the clapper rail and other species. Ultimately some of those land owners are going to start selling."

Some already have. In the North Bay, the Fish and Wildife Service is negotiating to buy Cullinan Ranch, 1500 acres of diked wetlands west of Vallejo that are currently being used for pastureland. Fish and Wildlife plans to restore the area to salt marsh.

Cullinan Ranch will be the largest restoration project on the west coast. In the south Bay, the San Francisco Bay National Wildlife Refuge is considering buying 20,000 additional acres of wetlands habitat.

Habitat restoration is the new by-word of the environmental movement in the Bay Area — a more positive and proactive alternative to the "Don't do this. Don't do that." mentality that has characterized the environmental movement in the past.

Besides the large-scale restoration projects sponsored by the government and large environmental groups, small local groups are springing up all over the Bay Area dedicated to restoring creeks, hillsides and native plant communities.

Habitat restoration is not necessarily the answer, however, because once damaged, land is very slow to recover.

"It takes a long time for a restored marsh to develop to the point where it can support the kind of wildlife that a naturally-occurring wetlands can," says Carl Wilcox, a wildlife biologist at the California Department of Fish and Game. "Most of the restoration projects around here just aren't old enough yet. Some animals, like the clapper rail, need a special kind of marsh, with cordgrass and well-developed slough channels. White Slough (a marsh near Vallejo that restored itself when a levee broke and flooded some unused pasture land) is fifteen years old and we are only just now beginning to see clapper rails there."

Wilcox is also suspicious of the "technological fix" that restoration offers.

"I think the biggest mistake we've made in doing restorations is over-engineering things instead of adapting to the natural conditions of an area. In the past, we've tried to do too much in too small an area and we've tried to do it too quickly — dredging channels and building islands, trying to create an instant habitat. We've learned that that approach just doesn't work."

Several groups are working on re-introducing native species that were wiped out in the Bay Area over the last 150 years.Fish and Game has successfully reintroduced tule elk to Point Reyes, Grizzly Island in Suisin Marsh, and the Concord Naval Weapons Station. Unfortunately, since the state was unable to reintroduce predators as well, the elk are quickly overpopulating. Fish and Game now holds elk hunts to keep down the population.

The Predatory Bird Reseach Group from the University of California at Santa Cruz has successfully reintroduced the endangered peregrine falcon to Marin County and Contra Costa. The fastest bird alive, peregrines were once found throughout the United States, but their population began to decline in the 1940's after DDT was introduced. By the time DDT was outlawed in 1972, there were only two breeding pairs left in California. Thanks to the work of the Predatory Bird Research Group, peregrines are on their way back from the brink. Adaptable birds, peregrines have been seen nesting on top of the Bay Bridge. Several birds live and hunt in downtown San Francisco where they feed on the ubiquitous pigeons. (One incensed pigeon lover beat a falcon to death with an umbrellaafter it ate the smaller bird.)

Despite their progress, peregrines are still not free of the legacy of DDT. Their eggshells are still soft. Researchers are often forced to remove eggs from the nest so that the mother's weight won't crush them.

No one is quite sure how the birds are being exposed to DDT, but experts guess that the birds are picking it up south of the border. Both peregrines and much of their prey are migratory. Mexico still uses DDT, exported by American companies that can no longer peddle it here.ì

"It just goes to show the global connections," says Mark Palmer, a former Sierra Club wildlife activist, now head of the Mountain Lion Preservation Fund. "Raising animals in captivity and then letting them go is not a panacea. You've got to clean up the environment. You've got to give them a clean place to go home to."

According to Palmer, preserving and restoring habitat and reintroducing native species is only part of the solution to the problem of the Bay Area's vanishing wildlife. It may be the easiest part. The harder, long-term solution lies in changing our own attitudes.

"In the long run, we need to alter our concept of how we as human beings fit into the natural world." Palmer says. "We have to learn how we can share the land with wildlife. We're going to have to be much more mindful of our impact on other species. This will undoubtedly mean some inconvenience. It will mean changing our lifestyle. And that's going to be very tough to do."

Laura Hagar is a environmental writer whose work has appeared in Sierra, California Magazine *and the* East Bay Express. *She is editor of the* Sierra Club Yodeler, *a monthly environmental news paper for the San Francisco Bay Area.*

Threats to the Green Belt

By Jim Sayer

Look closely at a map of the San Francisco Bay Area and, like many people, you'll start picking out some favorite or familiar places: Point Reyes National Seashore, Mount Diablo, Napa Valley, an orchard on the hillsides of Santa Clara county, perhaps a pumpkin farm on the coastlands of San Mateo County.

But step back from the map and notice how these places start to connect with one another, forming a broad swath of open space around the Bay Area's urban settlements. That swath is called a Greenbelt, and the Bay Area is unique among major American metropolitan areas in having one at all. Consider this:

- Its productive two million acres of farmland yield more than a billion dollars' worth of food and fiber each year;

- Its scenic beauty is made up of some of the finest parks and other public lands in the entire United States: nearly 800,000 acres in all;

- Its protective qualities filter the pollution out of our air and water, minimize the impact of floods and other natural hazards, and shelter more than 60 rare and endangered species of plants and animals, and;

- Its special contribution to the Bay Area's quality of life has led thousands of local residents to fight for the permanent preservation of this remarkable landscape since the turn of the century.

Bay Area citizens are fighting harder than ever before, because an unprecedented surge of development threatens to shatter the Greenbelt into small fragments. Every two years, an area of farmland or rolling hillside the size of San Francisco — nearly 50 square miles — is being idled or plowed under for suburban tracts or rural subdivisions. The latest wave of this growth is spreading to previously untouched parts of the Greenbelt — to the vineyards, dairylands, orchards and fields at the edge of Solano, Sonoma, Santa Clara and Contra Costa Counties. It's getting harder to climb a hill — anywhere in the Bay Area — without suburban sprawl spoiling your view.

That fact is underscored by the findings of a special computerized study done by the Greenbelt Alliance and the University of California at Berkeley.

The study, entitled *At Risk*, revealed that while nearly 735,000 acres of land are currently urbanized in the nine-county Bay Area, another 780,000 acres are at serious risk of being developed in the next 20 to 30 years.

That is an area almost the size of Santa Clara County.

What would it mean to develop all of that acreage? The Bay Area's urbanized lands would more than double. Virtually all of the Bay Area's remaining buildable flatlands (with the exception of parts of Napa County and unserviced areas in Solano County), as well as large parts of the region's buildable hillsides, would be developed.

The Bay Area's productive croplands (except for some parts of Solano County's plains and some wine grape growing areas) would be wiped out. Good pasture for ranching operations would be seriously diminished.

The Bay Area's metropolis would extend in unbroken swaths to every part of the nine county region — a marked change from the Bay Area's traditional urban focus around San Francisco Bay.

In short, the development of these 780,000 acres — or even a large portion of these lands — would radically change the basic character of the Bay Area metropolis and destroy a large chunk of the region's Greenbelt, the basis for the Bay Area's uniquely attractive quality of life.

No longer would we be confronting sprawl in isolated areas. The Bay Area would not be too different from the megalopolis people like to point fingers at in Southern California. In fact, the similarities are already becoming discomforting.

That's the prospect that has regional and local citizens groups seeing red — and mobilizing to keep the Bay Area green.

The tragedy of the Bay Area's outward sprawl is that it does not need to be this way. According to a recent report by the Greenbelt Alliance, we have more than enough land in existing urban areas to provide for jobs and housing development.

The report, titled *Room Enough*, inventoried vacant lands available for residential development in the Bay Area's towns and cities. One finding was that by slightly increasing the building densities on these lands — from luxury half-acre and quarter-acre lots to the regional average of eight units to the acre (still single-family home densities) — we could dramatically increase the number of units available to Bay Area homebuyers. And if we add several strategies for "infill" housing (such as combining housing and commercial development downtown or recycling unused industrial lands for housing), we could more than double the housing units projected as needed by the Association of Bay Area Governments by the year 2000.

Room Enough's message is simple and heartening: if we want more (and more affordable) housing, we can have it without chewing up our farmlands and ridges.

The research looks good on paper — but how to make it a reality? First, we need a vision. Greenbelt Alliance has put together a comprehensive strategy for the Bay Area's metropolitan re-development in a document called *Reviving the Sustainable Metropolis*. The paper ticks off nine pragmatic policies to protect the Greenbelt and encourage more compact, sustainable development in our cities — to improve community life and keep development pressures off the Greenbelt.

Those policies include: establishing urban growth boundaries around cities and towns; reconcentrating jobs in major urban centers like San Jose, Oakland, and San Francisco, rather than spreading them among low density, auto-dependent business parks; linking job centers with more compact neighborhoods by public transit; and coordinating land use policies at the regional level to ensure that cities and counties are working together to balance conservation and development.

Second, we need action. Happily, local communities, citizens groups, and public agencies are starting to respond to the vision.

At the government level, the city of San Jose is actively pursuing a "sustainable city" program, in which it is concentrating new jobs downtown, serving public transportation needs with a new light rail line, and launching a housing initiative to provide as many as 20,000 new housing units in the city's central area. In Sonoma County, the city councils of Cotati and Santa Rosa have initiated measures to establish urban growth boundaries. And in Marin County, the Board of Supervisors continues to hold the line on sprawling development by concentrating new development along the Highway 101 corridor, while protecting productive dairylands through a strong general plan and zoning in West Marin.

At the activist level, conservation groups like San Francisco Tomorrow, Sierra Club and Greenbelt Alliance are working with planners to encourage more compact development and good public transit — and they're going so far as to endorse specific new housing and mixed use (commercial and housing) projects.

These groups are also forging new ties with housing advocates, like the Non-Profit Housing Corporation of Northern California, to promote good housing within urban growth boundaries and farming and other "Greenbelt-compatible" uses outside the boundaries.

Even at the regional level, the Bay Area is starting to take some steps towards controlling growth. The Bay Vision 2020 Commission, a blue ribbon panel composed of citizens from all parts of the Bay Area and different ethnic and professional communities, is evaluating new and better plans for the region's future land use. In early 1991, the Commission will release its own blueprint for the region's future, to better coordinate how local communities balance land conservation and development.

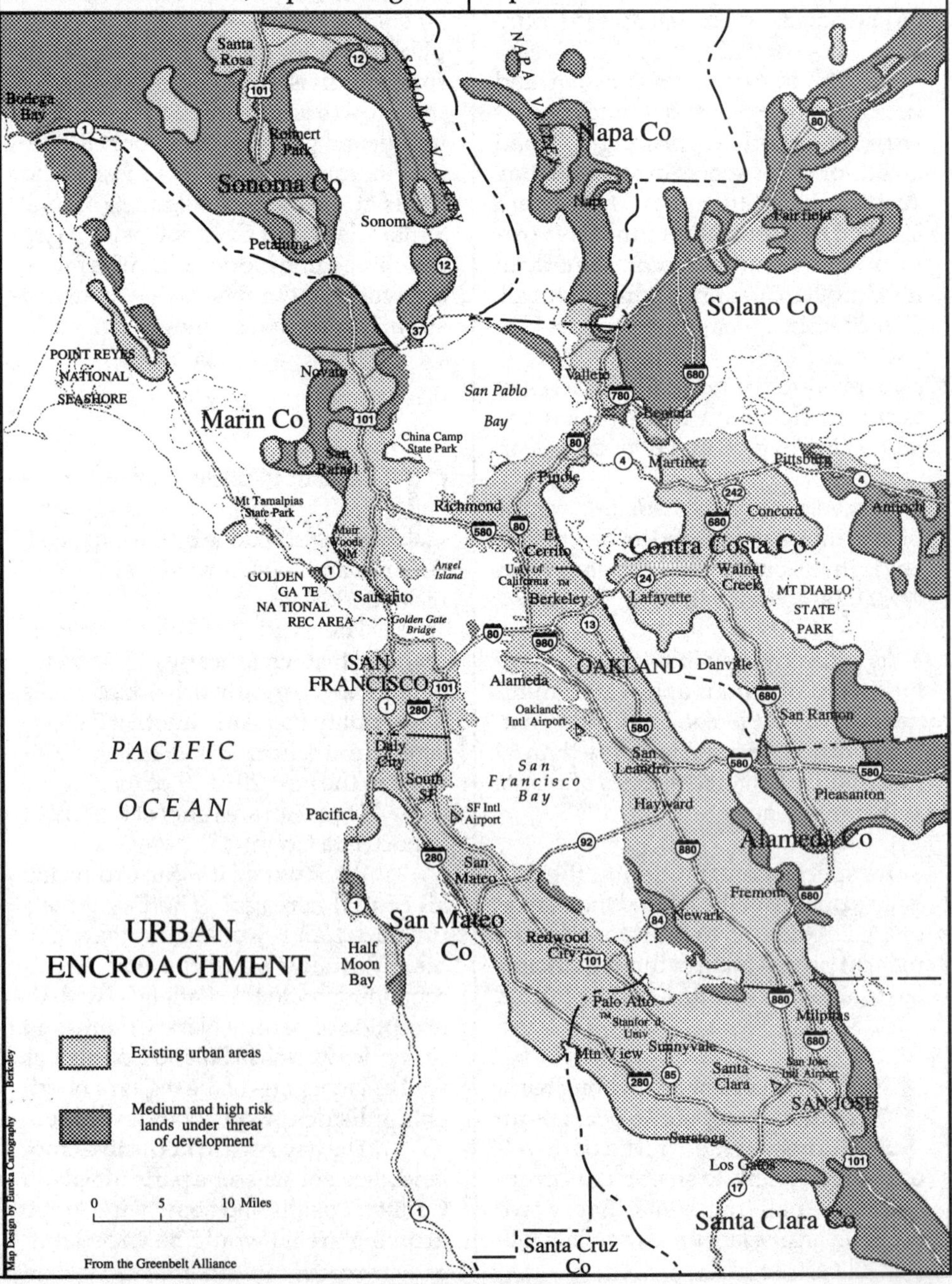

Despite all these efforts, the Greenbelt is under enormous development pressure. To keep these pressures at bay while more sustainable development practices take root, citizen groups all across the region have launched an unprecedented campaign to keep the Bay Area green.

In 1990 alone, citizens have put major open space and growth management measures on the ballot in eight of the region's nine counties. They include: farmland and wildlife habitat protection measures in Napa and Contra Costa County; new open space authorities to buy land in Santa Clara and Sonoma Counties; long-term growth management initiatives that will help protect the San Mateo coastlands around Half Moon Bay and the farmlands in Santa Clara Valley around Morgan Hill, and; measures to protect farmlands, valleys and ridgelands in Alameda, Marin and Solano Counties.

It is truly a regional open space revolution — a Greenbelt campaign — unprecedented in the history of the Bay Area. And now more than ever before, the Greenbelt needs people power. Use the guide on the right to see how you can get involved.

Jim Sayer is Director of Communication for the Greenbelt Alliance

Joining the Greenbelt Campaign

The Bay Area is fortunate to have an extraordinary array of land conservation groups working at the regional and local level. Following is a sample of some of the leading groups.

REGIONAL GROUPS

Greenbelt Alliance
Founded in 1958, Greenbelt Alliance is the Bay Area's regionwide citizen land conservation organization, with offices in San Francisco and San Jose. The organization works to protect the Greenbelt, promote better development, and educate the public about the benefits of the region's landscapes. Greenbelt Alliance has prepared a number of technical guides and strategy papers on land use and hosts a number of trips into the Greenbelt, including a one week, 440 mile bicycle tour in April. For information, contact (415) 543-4291 or (408) 983-0539.

Sierra Club
The Sierra Club has several chapters active in land use issues at the regional and local level. If you live in San Francisco, Marin, Alameda, or Contra Costa Counties, contact the club's Bay Chapter at (415) 653-6127. (The chapter also has an excellent bookstore at its office in Oakland and is planning another one in Contra Costa County.) In San Mateo or Santa Clara Counties, call the Loma Prieta Chapter at (415) 494-9901, and in the North Bay, the Redwood Chapter (for Sonoma, Napa and Solano) at (707) 544-7651.

Audubon Society
The Audubon Society has several active chapters in farmland and wildlife habitat protection. Bay Area chapters include Golden Gate Audubon (415/843-2222), Madrone Audubon (707/996-0601), Marin Audubon (415/383-1770), Mount Diablo Audubon (focusing on Contra Costa County at 415/376-8732), Santa Clara Audubon (415/329-1811) and Sequoia Audubon (for San Mateo County at 415/593-7368).

LOCAL GROUPS

North Bay (Marin, Sonoma, Napa and Solano Counties)
One of the most effective conservation groups in the Bay Area is the Marin Conservation League, which works on open space protection and urban planning throughout Marin County. Call (415) 472-6170. In Sonoma County, contact the Sonoma Conservation Council (707/578-0595) for information on active land conservation groups. One of the best there is the Sonoma County Farmlands Group, which is involved in many facets of environmental protection in Sonoma. Call (707) 576-0162. In Napa County, there are a number of active farm preservation groups, led by the Napa County Farm Bureau (707/224-5403). The Orderly Growth Committee is Solano County's leading land conservation group — and an effective one, at that. Call (707) 448-3892.

East Bay (Alameda and Contra Costa Counties)
The East Bay has a number of savvy groups working on land conservation and better urban planning. In Contra Costa County, a leading group is Save Mount Diablo at 415/549-2821. In Alameda County, there are many groups working to protect the Greenbelt: the Hayward Area Planning Association (415/538-3692), Save Pleasanton Ridgelands Committee (415/846-8148), Preserve Area Ridgelands Committee (415/447-0115), and the Tri-City Ecology Center serving the Fremont and Union City area (415/793-6222). An active group in Berkeley that works on remolding our cities into more environmentally sound places to live and work is Urban Ecology (415/548-7801).

Peninsula (San Francisco and San Mateo Counties)
Besides the Sierra Club, there are a number of groups in San Francisco interested in environmentally sustainable urban planning. Prominent city-wide groups are San Francisco Tomorrow (415/566-7050) and the Planet Drum Foundation (415/285-6556).

On the the San Mateo Peninsula, Committee for Green Foothills is the leading group working on open space protection. The committee also works on land use issues in Santa Clara County. Call (415) 494-7158.

South Bay
The Santa Clara Greenbelt Coalition is an active and effective combination of land use, neighborhood and outdoors groups (including bicyclists and equestrians) committed to protecting the county's hillsides and cropland. For information, call (408) 983-0539.

Rising Waters

The Bay in the Greenhouse Era

By Susan E. Davis

Looking into the shimmering blue Bay sky, or even into a cover of low-lying clouds scudding across the heavens, it's hard to imagine the myriad, invisible patterns of wind swirling with moisture, reacting to heat combined with pressure, the billions of molecular collisions and syntheses that create what we call "weather."

Even more difficult to imagine is that those skies contain a seal of gases which prohibit the earth's heat from escaping into the atmosphere, and which could create devastating effects for the Bay Area in the next century—or earlier.

Climatologists call this seal the "greenhouse effect." One of its most troublesome aspects is that people can't imagine how it will affect their communities. Since the crisis is framed as "global," it no longer seems local. It will happen everywhere, it seems, or nowhere. The Not in My Backyard credo affects our imaginations as well as our policy.

Computer models don't help, because the modelling grids are too large to render local effects. (The Bay Area grid, for instance, treats Yosemite, Santa Barbara, Death Valley, and San Francisco as one climate). Nor do predictions of "variable" conditions causing "probable" changes in "parts per million" and degrees Celsius bring the greenhouse effect home for us. The threat remains far-off, unreal, too damned numerical. It's not something we can imagine happening here.

A number of local experts, however, have begun formulating scenarios for the future of the Bay Area in the age of global warming. They're finding that rising temperatures and expanding sea levels could do considerable damage here. Indeed, say these researchers, climate change of the kind predicted now could ruin our water supplies, drown our wetlands, submerge our transportation system— and possibly destroy the Bay Area landscape as we know it today.

Rising temperatures

Researchers around the world have estimated that if the amount of atmospheric carbon dioxide doubles between the years 2030 and 2050, average global temperatures will increase 4 to 8°F. The California Energy Commission reported in 1989 that average temperatures in this state would increase 5.4°F. The Bay Area may experience slightly less warming, because it is by the coast.

That may not sound like much. In fact, if you like hot weather, a few degrees may sound like a boon for the chilly Bay Area. But even a minor increase in temperatures can have drastic effects on the quality and quantity of water supplies in a region already close to needing more water than it has.

The ironic twist of the greenhouse effect in the Bay Area is that we will suffer both from increased drought and increased flooding. During summer and fall months, we will be arguing over the fate of Northern California's freshwater supplies, and pumping groundwater to make up for the shortages. During late winter and early spring, however, we will be sandbagging levees and riverbanks to prevent the inundation of our cities and towns.

The drought will come as higher temperatures reduce the amount of water available for irrigation and urban use. Peter Gleick, director of the global environment program at the Pacific Institute for Studies in Development, Environment, and Security, in Berkeley, estimates that annual runoff could drop by 10 percent, and summer supplies could drop by as much as 62 percent. This could have severe effects on agriculture, urban water use, and hydro-electric power in this state.

It also could have severe effects on the health of the Delta and the Bay. During drought periods, there is not enough freshwater coursing through the delta to push back the salt water from the Bay. Disrupting the delicate balance between salt water and fresh water can ruin the nursery grounds for fish, plants, and wildlife, and can affect everything from phytoplankton to pelicans. Indeed, this year, which is the fourth year of drought in the Bay Area, and the most protracted period of drought since 1930, the striped bass index is the lowest it has been since monitoring began in 1959.

Salt water intrusion also threatens the primary source of the state's water. In cases of severe, or extended, drought, salt water can go as far east as the pumping stations for the federal Central Valley Project and the State Water Project, which currently divert eighty-five percent of the delta's water to Southern California.

Won't increased rains make up for the short fall? No one knows how climate change will affect the amount of precipitation in the Bay Area. The EPA has estimated, however, that if the amount of precipitation doesn't change with increased temperatures, the amount of unimpeded run-off from April to July will decrease 40-45 percent in the Sacramento Valley River, and 25 percent in the San Joaquin Valley Rivers.

If, on the other hand, precipitation increases, we won't be able to use the surplus water, because our reservoirs aren't large enough to contain it.

So far, no one has mentioned the golden words "greenhouse effect" in the Regional Water Quality Control Board's hearings for a new Bay-Delta Plan, due in 1993. Yet, in the case of global warming, demand for irrigation may increase 25 percent, while the supplies of the state water project alone may drop 7 to 16 percent, the EPA predicts. That water will have to come from someplace. Under current state law, fresh water quality in the Delta has to be protected before water can be exported. But with the Central Valley 8 percent to 44 percent drier during future summers, and California's population as a whole doubling by 2050, Bay and Delta needs may become even less of a priority.

The floods will come from changes in the snowpack and run-off cycles in the Sierras, from which most of our water comes. Increased temperatures mean that more rain and less snow will fall each winter, and the Sierra snowpack will shrink considerably. The state Department of Water Resources has estimated that the snowpack in the Northern Sierras' Sacramento River Drainage Basin would decrease 75 percent, while the snowpack in the southern San Joaquin-Tulare Drainage Basin would decrease 33 percent.

Under current conditions, the Sierra snowpack melts in the late spring and early summer, and creates the riverflow which fills our reservoirs. With less snow and more rain, however, and with snow melting faster and earlier each spring, we'll have earlier run-offs, with greater flows. If we capture that run-off in our reservoirs in the winter, we risk the danger of flooding. If we don't capture it, and the rainy season ends early, we run the risk of falling short in the summer and fall.

In other words, we'll have more flooding— in a state in which 75 percent of the communities already have land vulnerable to flooding. And we'll have less water available for use in a

Causes and Effects of Greenhouse Warming

The atmospheric "greenhouse effect" works much the way a real greenhouse does. As the sun's short wavelengths of energy pass through the earth's atmosphere, the earth re-radiates longer wavelengths of infrared energy, which are then absorbed by "greenhouse gases." The gases create an insulating blanket that prevents the heat from dissipating into the atmosphere, thereby warming the planet.

Greenhouse gases include carbon dioxide, methane, and nitrous oxides, and their effect is nothing new. Small amounts of each have existed in the atmosphere for billions of years. Without them, the planet would be 60°F colder than it is. In the last century, however, levels of these gases have been rising rapidly. Until the Industrial Revolution, carbon dioxide levels had steadied at about 270 parts per million. Since the last century, however, C02 concentrations have risen to 353 parts per million, according to the most recent reports by the Inter-governmental Panel on Climate Change.

At the same time, scientists have observed, the earth is warming at a rate 10 to 100 times faster than it has since the last ice age, 18,000 years ago. James E. Hansen, director of NASA's Institute for Space Studies in New York, has found that the earth's average temperature has risen more than 1°F in the last 100 years. Seventy percent of that rise occurred in the last 30 years. The five hottest years on record occurred in the last decade.

Whether greenhouse gases actually cause that warming is not yet firmly established. Some skeptics think that temperature records are skewed, because most recording stations are in cities, which tend to be hotter than the surrounding areas. Others have noted that the Northern Hemisphere actually cooled at mid-century, even as greenhouse gases accumulated. Such discrepancies lead some researchers to suggest alternative causes for warming, including sunspot activity, volcanic activity, and natural fluctuations in the earth's ever-changing, little-understood climate.

Still, scientists from agencies all over the world predict that when C02 concentrations double— around the year 2030— the average annual temperature will increase by four to eight degrees Fahrenheit, depending on the region, and the type of counter-measures taken. The earth hasn't experienced temperatures this warm in the last 100,000 years.

The effects would be devastating. In 1989, the EPA predicted a two to five foot rise in sea level by 2100, due to thermal expansion of ocean waters and the melting of glaciers and icecaps. Such a rise would inundate 30 to 80 percent of the country's coastal wetlands, 5,000 to 10,000 square miles of dry land if shores weren't protected, and 4,000 to 9,000 square miles of dry land if only developed areas are protected. Many islands would be submerged. Coastal countries could lose between 100 and 400 feet of beachfront from flooding and erosion, and billions of dollars in damage to structures including buildings, roads, and shoreline protectors. In California, between thirty-five and one hundred percent of the remaining wetlands could be lost. San Francisco, Santa Cruz, Santa Barbara, Ventura, and Los Angeles all would take a physical and economic beating, according to a recent Coastal Commission report.

It could get worse. If the Antarctic icebergs melt, sea levels could rise 16 to 19 feet, submerging an area greater than the U.S. or China, 70 percent of which would be in the Northern Hemisphere. New York, London, Seoul, and Beijing would flood. The Ganges, Amazon, and Mekong deltas would sink. Los Angeles, and San Diego could be flooded; the Farrallons, the Channel Islands, and the Delta could all but disappear.

state which is already threatened by water shortages. That's not a pretty picture. And it gets worse.

Rising waters

The Environmental Protection Agency estimates if average global temperatures increase 4 to 8°F, ocean waters will expand and ice caps may melt enough to raise average sea levels by three to three and one-half feet by the year 2100. (see sidebar) In California, where 60% of the population lives by the coast, communities from San Diego to Eureka will be inundated. Cliffed stretches, like Mendicino and Big Sur, will suffer the least. Low-lying areas with significant urban infrastructure— like Montery Bay, Los Angeles, and the San Francisco Bay— will suffer the most.

A sea-level rise of 3 feet within the Bay threatens $48 billion worth of commercial, residential, and industrial structures, Gleick estimated in a fall 1990 report entitled "Assessing the Costs of Adapting to Sea-Level Rise: A Case Study of San Francisco Bay." That property includes all the major freeways, the Oakland International Airport, the San Francisco International Airport, Southern Pacific Railraod tracks around the Bay's perimeter, Southern Pacific staging yards in San Mateo County, and major landfills and sewage treatment and disposal plants in each county. As each of these sites floods, of course, there is a high risk of toxics sweeping into the Bay's already polluted waters.

The harbors and ports— with their yacht clubs, restaurants, and port facilities— will also take a beating as the waters rise and waves get increasingly choppy. Large storm waves can do considerable damage to port facilities, pier supports, seawalls, and even ships. Storms that used to occur only once in 100 years will now occur once in ten years— even with a sea-level rise of only 5.9 inches, Gleick estimates. Harbor operations also may become more difficult as the sea level raises boats above the docks and loading areas.

Finally, sea level rise could devastate the Bay's already dwindling wetlands. California already has lost 90 percent of its wetlands to fill and dredging for harbor and port development. Although many of the remaining areas have been degraded due to reduced or polluted inflows and sedimentation, most continue to function as habitat for migratory birds and endangered species. With a sea level rise of three feet, wetlands all up and down the California coast will be submerged, resulting in a loss of vegetation, breakwaters, pollution filters, spawning and feeding grounds for estuarine and anadramous fish and resting places for endangered waterfowl and wildlife, including the salt marsh harvest mouse, the light-footed clapper rail, least tern, Belding's savannah sparrow and the California clapper rail.

In the Bay Area, where 95 percent of the wetlands have been destroyed, the 80,000 acre Suisun Marsh, which is the largest contiguous wetland in the state, the Estero Americano, Estero de San Antonia, and marshlands of the Petaluma River, Tomales Bay, Bolinas Lagoon, the San Francisco Bay, and Elkhorn Sough could all be affected. The North Bay wetlands are especially vulnerable.

Even the minimal predicted changes— a 4°F warming and one foot sea level rise in 30 years— could bring floods so high that violent waters from the San Joaquin and Sacramento rivers could wipe out the delta, deluge the bay, and leave behind a saltwater sea stretching from Suisun Bay to Sacramento, estimates Phil Williams, a San Francisco-based hydrologist who specializes in local sea level rise issues.

The Sacramento-San Joaquin delta is an 1100 square mile flood plain for the Sacramento and San Joaquin Rivers, which stretches from Antioch to Stockton, and from Sacramento to Tracy. Before 1860, all of it was uninhabited wetlands. High tides covered the marshes with one to two feet of fresh water twice daily; the flooding rivers submerged them.

Today, most of the delta is prime farm land, protected by 1100 miles of levees. Those levees also protect the quality of much of the state's water supply, which flows through the delta. Unfortunately, those levees are made of peat— a low-density, structurally-weak soil which can't stand much pressure from rising waters.

If temperatures rise as predicted, however, rising waters will come from two sources: the Sacramento and San Joaquin Rivers, and the Bay's tides.

The rivers will flood because changes in the snowpack will increase winter and spring run-offs. Gleick estimates that floods that normally occur every 100 years will come every ten years.

The tides, which sweep into the delta from the bay each day, will rise as the bay waters rise. The EPA has suggested planning for a two-foot rise in sea level in the next 50 years, and a four-foot rise in the next century. In the delta, a four-foot rise will mean a three to four-foot increase in the 100-year high tide elevation— from 6 to 8 feet, to 9 to 12 feet, Williams says.

That may not seem like much of a difference. But perimeter levees in the delta now are only 6 to 8 feet high. In some older developments, they are only 2 feet high. Most of them are in "fair" to "very poor" condition, according to a 1980 DWR inspection. That means high tide waters can either roll over the levees, or smash right through them.

It only takes one levee break to flood an island, and it only takes one flooded island to flood the whole delta. If one levee fails, others will also fail as the waters gain wind and wave force. And as those waters gain momentum, they will knock out more and more levees. In the case of a major flood, which is not far-fetched, the roiling waters could take with them the 300 miles of interstate, state and local roads, 13 miles of railroad, 35 natural gas fields, 4 miles of the EBMUD Mokelumne Aqueduct— which supplies the Bay Area's water, and all the farmlands, parks, homes, businesses, sewage treatment plants, power lines, stores, and industries which comprise the human face of this once wild and wide-ranging wetland.

If that happens, the Bay and Delta together will contain twice as much water as they do now, and spread across three times their current area. The San Francisco Bay may need to be re-named the San Francisco Sea. "If the level rises that one meter," notes Maurice Roos, chief hydrologist for the Department of Water Resources, "we'll have a calamity."

Just how much water pressure will it take to set this tragic game of dominos into effect? "It's impossible to

THE SEA-LEVEL RISE THREAT TO SAN FRANCISCO BAY

Area Threatened by a 1-meter sea-level rise

0 5 10 Miles

Map Design by Eureka Cartography, Berkeley

Santa Rosa
Bodega Bay
SONOMA VALLEY
NAPA VALLEY
Rohnert Park
Napa Co
Sonoma Co
Napa
Sonoma
Petaluma
Fairfield
Sears Point Raceway
Solano Co
Marine World Africa USA
POINT REYES NATIONAL SEASHORE
Novato
Vallejo
San Pablo Bay
Benicia
Marin Co
China Camp State Park
San Rafael
Martinez
Pittsburg
Pinole
John Muir NHS
Antioch
Concord
Mt Tamalpias State Park
Richmond
Muir Woods NM
El Cerrito
Contra Costa Co
Univ of California
Walnut Creek
GOLDEN GATE NATIONAL REC·AREA
Angel Island
Sausalito
Berkeley
Lafayette
MT DIABLO STATE PARK
Golden Gate Bridge
SAN FRANCISCO
OAKLAND
Danville
Alameda
Oakland Coliseum
Oakland Intl Airport
San Ramon
PACIFIC OCEAN
Candlestick Park
Daly City
San Leandro
San Francisco Bay
South SF
Pleasanton
Pacifica
SF Intl Airport
Hayward
Alameda Co
San Mateo
Fremont
San Mateo Co
Newark
Mission San Jose
Redwood City
Half Moon Bay
Palo Alto
Stanford Univ
Milpitas
Great America
Mtn View
Sunnyvale
Santa Clara
San Jose Intl Airport
SAN JOSE
Saratoga
Los Gatos
Santa Clara Co
Santa Cruz Co

Bay Habitats

predict just how many inches of sea level rise it will take to break the levees," says Robert Buddemier, formerly an associate division leader for environmental projects with Lawrence Livermore National Laboratory. "The impact of the combination of flood, high seasonal tides, and high winds is hard to predict. But any increase in sea level will increase the chances of failure." Seven record high tides already have been recorded in the delta since 1979.

If the levees fail, the state's primary source of water again will be threatened. If water fills in even one island, a vacuum is created in the western delta which sucks salt water up from the Bay. That water can move far enough east to reach the pumping stations of both water projects.

In past floods, project directors have used freshwater releases from the Folsom, Oroville and Shasta Dams to push back the encroaching salt water. To protect pumping stations during a three-foot sea level rise, freshwater releases will have to be doubled, the EPA estimates. That option is not likely to be available to future water-managers, however. DWR already predicts an end to surplus water after 1990 and statewide water shortages of 290,000 acre-fet per year by the year 2000, due to increased demand and decreased supply That's without taking global warming into account.

Rising costs

Is there anything we can do about this? The three basic alternatives given in most analyses of sea-level rise are to reduce the threat, protect the shorelines, or retreat. Reducing sea level rise would entail immediately stopping global warming by curtailing carbon dioxide emissions. Unfortunately, this does not appear to be a practical alternative at the moment.

Protection includes adding sand to beaches, building dikes, as in the Netherlands, or building breakwaters.

This is no small task. Gleick suggests it would cost more than $940 million to build new defenses or modifying existing protection around the Bay. That figure includes neither the costs of protecting wetlands (through artificial barriers or restoring wetlands on higher ground) nor the cost of pumps, drainage systems, and navigation locks, nor the cost of maintaining those protections.

Protection also includes prohibiting further development in the new flood plains— a policy taken by Contra Costa Planning Department. BCDC also has upped levee height and floor elevation requirements for new structures along the Bay. And all permit applicants are required to consider how sea level rise will affect their shoreline structures.

A number of researchers have suggested building master levees, called "polders," to hold the delta. Such a project would cost billions of dollars. Even rehabilitating the levees to minimum flood-resistant standards would cost $1 billion, the Army Corps of Engineers estimated in 1980.

The Peripheral Canal also counts as protection— but one which makes citizens and environmentalists around the Bay nervous. This 43-mile canal would divert Sacramento River water around the Delta, thereby avoiding the risk, supposedly, of salt water contamination. Opponents of the plan have long argued that it would deprive the delta of the fresh water needed to maintain its water quality and its wildlife. In 1982, voters rejected the plan. But in a 1989 report on the greenhouse effect, entitled "Possible Climate Change and Its Impact on Water Supply in California," DWR hydrologist Maurice Roos suggested that the canal could be reconsidered.

Unfortunately, far too many agencies have chosen a fourth action, called "ignore." In dozens of interviews conducted in 1989, state legislators called the threat of sea level rise everything from "not concrete enough" to "a marshmallow subject." Neither global warming nor sea level rise have been mentioned in the State Water Resource Control Board's three-year hearings on water diversions and the quality of the Bay and Delta. Only a few municipalities have begun re-zoning for estimated flood plains. DWR personnel have been saying both that there is no evidence for greenhouse effect, and that we have a long time to wait before a one meter sea level rise. As Gleick's study shows, however, we don't have to wait for the exact day of a 3 foot rise to see destruction— even 6 inches more of water could do serious damage.

A few agencies have undertaken studies— but without sufficient follow-up.

The California Energy Commission wrote a draft of a report on how global warming will affect the state's energy supply and demand, economy, agriculture, and water supplies. BCDC issued a report summarizing sea level effects on wetlands and planning along the Bay. A draft of this report was released in 1989. The California Coastal Commission also did a draft report of how global warming will affect the coastline. The Coastal Conservancy devoted one whole issue of its *Waterfront Age* (now called *Coast*) to the global warming and sea-level rise in California. But none of these studies are very helpful in suggesting solutions.

Many people have asked "Is there any way to take local action against a global problem?" After all, we are told, the problem is global in nature. No one country is responsible and no one country will escape its effects. But just as framing a crisis as "global" makes it difficult to imagine its effect on our communities, so too does it make legislation difficult. We can wait for years for countries to agree on carbon dioxide, CFC and deforestation policy that will satisfy developed and developing countries alike.

By taking a local perspective we can honestly confront the way global warming will affect our lives over the next 10, 20, or 50 years— and the way it will affect the lives of our children in the next century. It also allows us to take more responsibility for the ways in which our region is contributing to the problem, and the ways in which we ourselves can take responsibility for alleviating it. Only by understanding how the acts of each one of us— in using fossil fuels, in ignoring the warnings, in failing to lobby our policymakers— can we begin to make a change.

Susan Davis is a Berkeley-based journalist whose work has appeared in the Washington Post, Newsweek, *and* California Magazine, *among other publications.*

From Beneath Our Feet

Restoring Creeks and Streams in the Bay Area

By George Spies

In the hot summer grasslands above the bay, beyond the marshes and below the wooded hills, only birds and insects move amidst the grasses and scattered shrubs and occasional bare rocks on a hot hillside. The deer and the squirrel, the fox and the bobcat, all seek the cool refuge of the serpentine meanders of the small creek arched over with oak and bay trees. Now, in late summer, the creek is a mere trickle, but the small, cold pools harbor steelhead hatchlings and stickleback, and the stream banks support currant and snowberry in the shade of the trees.

When the winter rains come and swell the small creek one thousand fold, the steelhead and salmon run up from the bay to spawn in the gravel beds sheltered by fallen logs and tree roots. Then the bears are busy in the rushing water, too. This is truly an oasis for both animal and human, not only for comfort but for survival as well.

When the Spaniards came up the east side of San Francisco Bay, they were struck with homesickness because the tree-lined creeks reminded them of *las alamedas*, the tree-lined streets of their native Spain. But the demands of civilization came before sentimentality. First the Spaniards, then other Europeans, brought ranching and large-scale agriculture to the Bay Area and the decline of the native creeks and streams began.

Today, these creeks and streams are, for the most part, unrecognizable or invisible. Pause a moment and look out the window. Is it raining? If so, could you describe the path the rain on your window takes in reaching the ocean? Unless you live very near the Bay or ocean, that rain will most likely spend some time in your local creek — possibly, right under your house. In our urban areas, more than 50 percent of the creek mileage runs in buried culverts. Almost all of the rest have been modified or contained in some way.

For the most part the creeks have been buried or diverted to make way for building on what is considered valuable real estate. The problem is that creeks meander — that is, they shift their banks and spread and re-direct themselves. Nobody wants them to re-direct themselves through their basement or living room. So many of our creeks have been channelized, their banks armored with concrete to prevent shifting. Or metal pipe has been laid in the creekbed and the entire thing covered over with soil.

In addition to these ruinous modifications, local agriculture diverts water out of local streams for crop irrigation. Cattle ranchers water their herds in the creeks, which erodes the banks and poisons the water with feces.Water returns to the stream system tainted with fertilizers and pesticides. Otherstreams have been dammed by the water districts, and flows out of the dams are usually not allowed. The frenetic and seemingly endless march of construction in the Bay Area deposits large amounts of loose sediments into dry creekbeds in the summer, which then choke the streams when the rain comes.

Add to this pollution both from industry and private citizens. Look out of your window again. Is your neighbor changing the oil in his/her car? All too often that oil will end up being poured into the storm drain. The storm drain leads to the local creek, culverted or not, and that creek ultimately empties into the Bay.

These factors all add up to a lot of damage: ruined habitat for fish and other wildlife, a grossly simplified environment, and a lost opportunity to interact with a natural setting that once was so soothing and beautiful.

Restorationists, bioregionalists, fishermen, neighborhood organizers, scientists and even some urban planners agree: our planning and development paradigms need to change. The basic fact that water is life and it flows downhill must take a central place in our thinking. When a raindrop strikes a ridgetop, it rolls down to the bottom of the ridge where it joins a first-order watercourse. This watercourse, or creek, flows downhill until it meets another stream. This combined second-order stream continues to flow downhill until meeting another, and so on until all the ridges, valleys and plains are drained into the sea. All of the land area drained by a given stream or river is its "watershed," from the peak of one ridge to the next and all of the valley in between. To understand the watershed as the basic unit of planning is very important.

A watershed is united in that everything and everyone in it relies on the same water for health and survival. The health of the watershed is reflected in the health of its stream or river, which drains the whole unit, and contains a set of habitats unique to itself. If the stream or creek can be kept clean and healthy, then we will have come a long way in understanding and aiding the watershed as a whole.

Unfortunately, up until now, very little attention has been paid to planning with watershed health in mind. That's why our watersheds are afflicted with all of the problems enumerated above.

How to restore a creek

Approximately 100 creeks and streams flow directly into the San Francisco Bay System, and most of them are in trouble. Over the last decade, a few organizations have sprung up around the Bay with the goal of restoring local streams to something like their original natural state. What follows is a kind of step-by-step guide, based on the experience of these groups, for those readers who may be interested in trying to restore a stream near home.

Bringing buried streams back to the surface in urban areas is, for obvious reasons, a very tricky, expensive, and intrinsically political operation. Convincing politicians and landowners that such a project is worthwhile is a major hurdle. Finding the funds with which to do it are another; the Urban Streams Restoration Grants Program of the California Department of Water Resources is a good place to start .

Once you have political suppor and funding, the technical work begins..

First, analyze your stream. The creek's course from the top of the watershed to its connection to the Bay or the ocean must be traced. The local water district should be contacted regarding the location of culverted sections.

Then make a list of the negative impacts upon the creek. If the water in the creek is muddy, erosion is occurring upstream. Get a water sample tested for pollutants, and then try to find the source of any that you find. Are the banks of the stream littered with trash? Is there sufficient tree cover over the open sections of the stream? Whatever your creek needs to return to health will become apparent with this kind of assessment. Some problems can be addressed within the neighborhood, like trash clean-up and getting people not to pour their motor oil into the storm drain. Others, like re-vegetation and erosion control, require consultation with informed and experienced restorationists or scientists. The work itself can often still be done by neighborhood people.

Local streams used to be much shallower (2-3 ft.) than they are today (10-20 ft.), according to activist John Steere of East Bay Citizens for Creek Restoration (EBCCR). When an area is paved and urbanized, less water is absorbed and more consequently runs off, which then causes the channel to deepen and the water to move faster. Channelizing and culvertizing also speed up the water.

A restoration project seeking to help an eco-system recover its equilibrium first must consider energy dynamics. Check dams and other small structures are often used to slow the water down. This reduces bank erosion and improves fish habitat. "Pools and riffles are what you want," Steere says.

Eroding banks are then lined with local rock to stabilize them. Streamside trees such as alder, redwood, and elderberry are planted both to shade the stream and to limit erosion.

"Always try to go with native species when you can. Exotics sometimes bring unknown variables. For understory plants, fern, snowberry, and currant were here before. As for animals, we can only set the stage, provide good habitat, and wait," says Steere.

Jim Hamilton, of California Trout Inc., agrees: "Good habitat is essential. Minimum flows are necessary and the fish need to be able to pass from the spawning grounds to the ocean and back again." CalTrout is helping that to happen in one peninsula creek by removing barriers and installing a fish ladder to help spawning steelhead. Although EBCCR and CalTrout are membership-supported groups, both acknowledge that government support is important, especially from such agencies as the state Department of Fish and Game and the Wildlife Conservation Board.

Some of the problems involved with creek restoration are quite complex and involve the whole community. Creek advocates have found that there is rarely a voice heard in the community in behalf of the health of the creek. They also have discovered that a creek can be a great focal point for a community. The ever-growing demand for urban greenways, for example, can be nicely satisfied by creek restoration designed with both nature and people in mind. Their efforts to bring Strawberry Creek back to the surface in downtown Berkeley has encouraged public attention to urban waterways.

These efforts are a great inspiration to people who hope to see city living more aligned with the natural world in the future. As John Steere points out, creeks are the key to understanding where we live because they reveal the health of the eco-system they serve. We would be wise to attend to his admonition for humility in our dealings with our waterways, for water truly is life. 

George Spies makes films on environmental subjects, and lives in San Francisco.

Sources and Resources

ORGANIZATIONS & AGENCIES

Urban Creeks Council
2530 San Pablo Blvd., Berk., CA 94702
(415) 540-6669
An umbrella organization that supplies support services to local creek restoration efforts.

California Trout, Inc
870 Market St, Ste 859, S.F., CA 94102
(415) 392-8887

Cal. Dept. of Water Resources
Division of Local Assistance
Urban Streams Restoration Grants Program
Earle Cummings, Director
P.O. Box 94836, Sacramento, CA 94326
(916) 323-9544

TECHNICAL INFORMATION:

Wildland Resources Centers
Division of Agriculture and Natural Resources
145 Mulford Hall
Univ. of California Berkeley, CA 94720
Contact: Robert Z. Callaham

Environmental and Water Quality Operational Studies
Army Corps of Engineers
Waterways Experiment Station
P.O. Box 631, Vicksburg, Miss. 39180
Studies E82-4, E82-7, E84-11, E85-3, E85-7contain useful details and illustrations.

Ecological Restoration in the Bay Area
a publication of Restoring the Earth
1713C Martin Luther King, Jr. Way
Berkeley, CA 94709
(415) 843-2645

Sold to the Highest Bidder

Habitat Destruction by Restoration

By Monica J. Fletcher

In 1982 developers won a great victory over a battered enemy — the na-tion's endangered species. Congress, under pressure from a major lobbying effort, weakened the Endangered Species Act by allowing developers to "take" endangered species or their habitat as long as a larger area elsewhere was "restored" to enhance the species. The act allowed for "Habitat Conservation Plans," —elaborate long-term documents that set guidelines for mitigation to "protect" and "enhance" an endangered species' chance for survival in an area slated for development.

This major weakening of the nation's key environmental law gave strength to an already burgeoning practise in the Bay Area of off-site restoration to make up for destruction of the Bay's ever-shrinking wildlife habitat. The Bay Area's first Habitat Conservation Plan (HCP) was drawn up in the same year in response to a developer's plans for the lower reaches of San Bruno Mountain in San Mateo County. The two species in jeopardy are the endangered mission blue butterfly and the threatened San Francisco silverspot butterfly. The plan aimed to stop further encroachment of non-native plants on the mountain and to revegetate a site, away from the development, with the butterflies' host native plants (crucial as a larvae's food source).

The thirty-year development permit and plan immediately drew criticism from entomologists, biologists and environmentalists. They charged that the plan was based on inadequate data, overly-optimistic population surveys, and experimental theory. The U.S District Court upheld the HCP, however. Condos and expensive homes, with more development planned, have now taken the place of a good portion of these butterflies' very small world. Unfortunately (and as the opponents predicted) the effort to re-introduce crucial plants for the butterflies survival hasn't gone so well. Restoration crews are having trouble establishing the golden violet, the host plant for the silverspot. Non-native plant control has been unsucessful in some areas, and no one knows what effect the development's own landscaping may have on the mountain or the butterfly if it spreads.

The San Bruno Mountain case is just one example of the tragic loss of an endangered species' natural habitat in return for a pale man-made imitation. It is typical in the controversy over its science and imperfect execution. More disturbing than the losses to the mission blue or the silverspot is the trend this case represents: the increasing use of restoration, a technique still essentially experimental, as a tool for the sanctioned destruction of natural habitats already severely diminished. Developers are allowed to destroy these scarce lands in return for a promise of providing a copy of an ecosystem.

California's environmental laws are considered among the strongest in the nation. We have the California Environmental Quality Act, our own endangered species list, and a mitigation maintenance law. We have enforcement and protection through the California Department of Fish and Game, the Bay Conservation and Development Commission, the Army Corps of Engineers, and the U. S. Fish and Wildlife Service (they even have marshals). We have a vast array of experts to carry out restoration - hazardous waste management specialists, soil remediation experts (to either seal a site, or haul off toxic material), erosion control specialists, surveyors, seed and seedling suppliers, wildlife biologists, academic specialists in ecology, botany, marine biology or endangered species, landscape architects, foresters, range managers, and hydrologists. The Bay Area is also home to a large number of conservationists and environmentalists. It would seem that here, more than anywhere else, we should be able to stop the destruction of habitat that supports our endangered wildlife. Why can't all of these experts stop the destruction of the minute parcels of surviving natural habitat around the Bay?

The answer is simple. Development generates a great deal of profit for the developers, restoration experts get a piece of the pie, and the governmental agencies charged with regulating the whole process are often split by jurisdictional in-fighting and are far from immune to the influence of the perceived economic benefits of development.

The Law

The foundation of environmental protection — our state and national law — is not very strong. In 1969, the first national legislative response to environmental deterioration was passed. The National Environmental Policy Act (NEPA) forced public agencies to report on, and evaluate the environmental effects of, any federal actions, projects or developments.

One year later (1970) California passed a similar but more ambitious law, the California Environmental Quality Act (CEQA - pronounced see-qwa) which was intended to address loss of habitat, noise, water, air pollution, development's effect on other species, and the health and well-being of Californians. Unlike the federal law, CEQA would not be merely a procedural outline but would have the definite goal of minimizing or reducing deleterious effects of development on the environment.

CEQA procedures require documentation, notices of preparation, and environmental impact reports to insure that no project slips through without reasoned attention given to it. It was hoped that the process itself would slow development and lessen impacts on the environment.

SALT MARSH HARVEST MOUSE

CEQA has led to a whole new industry of planners, consultants and experts who are employed to protect and provide for its implementation. It has created specialty fields of lawyers to fight development in courts on minute details of procedural error.

Unfortunately, CEQA has fallen short on its central purpose of protecting key habitats. This is because its framers left one huge loophole.

When significant impacts like loss of wildlife habitat or agricultural land can not be significantly mitigated, the approval agency may nonetheless approve the project by filing a "statement of overriding considerations." This statement allows for the development of projects such as housing, stadiums, racetracks, hotels or other large projects solely on the basis of the project's ability to employ people, increase the tax base, or diversify the economy. It is precisely these sorts of large developments that wreak the most havoc on sensitive environments.

Since the passage of CEQA real losses have continued to occur. Pete Sorensen of the U. S. Fish and Wildlife Service's Endangered Species Office estimates that, since CEQA became law, 1500 acres of non-tidal wetlands — critical habitat for the endangered southern salt marsh harvest mouse — have been lost to development or environmental degradation. This reduces the remaining non-tidal salt marsh acreage in the Bay to between 700 and 800 acres

Very few official figures have been compiled for the loss of other, less dramatic habitats such as chapparal or oak woodland. We are not even in a position to quantify the loss currently. We may lose it all before anyone takes official notice.

Weak enforcement

While there are significant weak spots in the law, there are also provisions for enforcement and governmental oversight of sensitive species and areas. Unfortunately, those agencies are susceptible to the whims of political winds.

The California Department of Fish and Game's (DFG) Region 3 headquarters is the first line of defense for an area that stretches from Mendocino county to San Luis Obispo. The department is charged with maintaining California's species of fish and wildlife for their natural and ecological value, as well as for their direct benefit to people. One of the most important duties of the DFG's wildlife biologists is the review of Environmental Impact Reports. They monitor proposed development's effects on wildlife, make site inspections to confirm data presented, and advise national and regional agencies on the particular needs of wildlife in the area.

That's a lot of work for a staff of four wildlife biologists — some say far too much work. But in 1990, the even this small staff was threatened with lay-offs when the Department of Fish and Game was confronted with a $12.6 million budget shortfall.

In the end the lay-offs were prevented by last minute legislation attaching fees to commercial fishing and development that covered the biologists' salaries. But the episode pointed up the low priority accorded to wildlife study and protection by government. What to cut in such situations is basically a political decision; with few wealthy backers, wildlife supporters have traditionally had very l;ittle political pull. The threatened cuts show the overall vulnerablity of enforcement agencies to the political process. Any cuts would directly affect the number of field biologists available to inspect development sites in the Bay Area. Without regional biologists, thousands of environmental documents would be checked in Sacramento by relatively few biologists with little knowledge of the areas described.

Another environmental protection agency, the Bay Conservation and Development Commission (BCDC) is a regional planning and regulatory agency that oversees and protects the Bay, its sloughs, tidal marshes, creeks and wetlands. BCDC monitors filling and dredging activities and issues per-

CLAPPER RAIL

mits for bayside development. It is the main agency responsible for the institution of off-site mitigation and restoration in exchange for development permits in wetland areas. The restoration work is paid for by the developer, and the compromise is deemed beneficial because the areas to be restored were judged by BCDC to be degraded and in need of restoration.

There are two serious flaws with BCDC's permitting process. The first is the manner in which restoration is carried out. The second is that a restoration trade-off is allowed at all.

The traditional bias held by BCDC is that restoration is best carried out by returning areas to tidal influence. Non-tidal wetland areas (a preferred habitat for salt marsh harvest mice) have been sacrificed in order to renew tidal flow. Permits that require the return of tidal action to an area are not just the result of a scientific bias held by the commission. BCDC has limited jurisdiction over non-tidal and seasonal wetlands, and shares control of these areas to a lesser degree with the Army Corps of Engineers. By returning wetlands to tidal influence, and thus increasing the acreage of the bay, the BCDC increases its jurisdiction.

Perhaps the worst example of this bias occurred at the Warm Springs Marsh restoration in Fremont. Beginning in 1981, tidal action was restored to the area in the hopes of creating what is considered by BCDC the best form of marsh - the brackish salt water marsh. This was done regardless of the fact that the non-tidal marsh supported a large number of salt marsh harvest mice. The project ended up not 'restoring' the area but rather converting one kind of marsh to another. Two-hundred and fifty acres of prime habitat was opened up to the tides, and in a last-minute effort, a small 'preserve' of twenty-three acres was designated to shelter the mouse. This difficult-to-manage donut in the middle of the marsh currently holds only approximately five acres of suitable mouse habitat. Trapping in the area has produced only two mice in two separate studies.

BCDC has begun to listen to the biologists and scientists who have been proposing a less dramatic restoration goal known as "muted tidal influence." BCDC seems to have become more sensitive to the many criticisms that have come its way. We may yet see a decrease in off-site mitigation permits with less drastic alteration inflicted on those sites. Habitat Conservation Plans have also fortunately not yet become commonplace in the Bay Area. Of three HCP's proposed in the Bay Area only the San Bruno butterfly plan has been approved.

But BCDC's use of restoration as mitigation for destruction of wetland habitat, and the possible use of Habitat Conservation Plans wherever an endangered species is threatened by a project, means that more loss of habitat is possible.. The methods of legal destruction of habitat rely on the use of restoration. Without restoration the wildlife and habitat loss is too great on paper. With restoration developers can make a promise to increase needed habitat. But a promise is only a promise, and an experimental restoration is not the same as a functioning habitat.

Restored to what?

Bay Area habitats, or biomes, have never been static. Prior to the appearance of humans on the scene, California's plants and animals had been dealing with fire, volcanism, drought, plate tectonics, and climate change for eons. Humans, first Native Americans and later European settlers, played their part in habitat alteration and destruction through fire, cattle grazing, agriculture and mining. Estimates of the percentage loss of wetlands, so frequently cited in calls for preservation, often ignore the fact that what we now consider pristine wetlands were actually the by-product of Sierran placer mining that brought tons of silt into the Delta. There is in fact, no natural state to be restored to. There are only older, more complete ecosystems, ones that supported more species, and that have been around for longer periods of time.

This reality of habitat change, and variable ecosystem viability, has left the science of restoration at a loss to truly know what makes-up a habitat. What is chapparal? What factors allow a coastal scrub community to flourish? The hundreds of thousands of organisms that make up a habitat can barely be studied, let alone duplicated. In a forest or a wetland there are thousands of fungi, bacteria, symbiotic microorganisms, insects, plants, and animals with unknown and perhaps essential inter-relationships. There are soil and chemical factors, water regimes, climatic cycles — in short, ecosystems are very complex.

The fact is, we just don't know enough to allow the destruction of a functioning ecosystem in trade for a habitat look-alike containing a few selected key species.

How we define a functioning ecosystem, and therefore a successful restoration, is another problem. If only the base vegetation is considered, many restorations have been successful. A number of projects also monitor benthic fauna (animals living in a mudflat) in an attempt to assess the restored area's ability to support higher forms of life such as fish or birds. Our ability to recreate is only one of a very limited degree. Successful restorations, such as those defined by the Bay Conservation and Development Commission's 1988 report, may contain the required benthic, tidal and vegetative components for a marsh, but still lack the ability to support key species.

One oft-cited 'success' is the Hayward Marsh Restoration Project near the San Mateo Bridge. In the Hayward case viable habitat that supported both the salt marsh harvest mouse and the California clapper rail was sacrificed for the construction of the Dumbarton Bridge further south. The restored environment that was created in exchange currently supports no clapper rails. This situation leaves only a hope, not a guarantee, that someday the clapper rail will return. In other restored areas it has taken at least fifteen years for the clapper rail to return. There is no way of knowing what the Hayward site will support fifteen years hence. We exchanged something that definitely worked for an unknown.

We can 'restore' a few plant and animal species to an area, allow natural systems like fire or tides to act upon an area once again, but we cannot be assured of our ability to create a habitat. The most intelligent course is to leave intact those areas that are already supporting endangered species, marsh vegetation and functioning ecosystems.

A Growth Industry

The destruction of wild areas usually benefits only a very few. The profits for developing land in the Bay Aea are enormous, and the motivation to develop is correspondingly high. Developers do not make money by allowing land to just be.They build business parks, condominiums, marinas, housing, factories, or race tracks Land is real estate, not substrate. When developers destroy habitat for profit they are easily labeled as "bad guys."

It is not a simple extrapolation to apply this description to the restorationists and public servants that are also involved with the process These are regular people making a living. They are 'doing something for the environment.' Unfortunately, the present growth industry in restoration expertise, and the many people who are supported by it, survive because of a general tolerance of destructive business practises. What appears to be an ecologically-virtuous profession is in reality a support system for the continued destruction of land for development interests.

It is not enough to keep cleaning up after oil spills, replanting after clearcuts, or restoring areas as an excuse for further destruction of unspoiled ones. It is time to stop the destruction. The people and agencies involved must refuse to participate or permit the practise, and stop the loss in any way they can.

How to get involved

The Bay Area is replete with environmental activists and organizations that excel in the local fight for the environment. The best form of support that can be given to these groups is a personal one. Our remaining wild land needs monitors. A monitor is someone with an interest in getting to know an area, or the local government's minute machinations, or the local developer's next project, A monitor could use that information to stop further destruction. Many organizations need and depend on volunteers to help their efforts to save pieces of the planet; they need ears and eyes to keep track of what is going on. Our wild areas are not just backdrop landscapes to an urban reality. They need to be known, watched-over and protected.

A number of local groups are acting as protectors of the Bay's natural wealth. All of them need information and help. The groups can differ in outlook and intent but ultimately what they do is watchdog the players in a destructive process. Some specialize, others cover a wide range of concerns. Groups with various specialties include anti-nuclear groups, agricultural preservation groups, groups that specialize in wetlands or forests or rivers, or groups that fight to change the law. They all need help. So peruse the organizations section of this book to see how you might get involved.

Monica Fletcher is the research director for the Bay Area Green Pages .

Sources and Resources

GOVERNMENT AGENCIES

U. S. Fish and Wildlife Service
Interior Building
C Street, N. W.
Washington DC 20240
Direct your commentary on the Endangered Species Act to this office and your representatives in Congress.

U. S. Fish and Wildlife Service
Endangered Species Office
(916)978-4866
This office oversees enforcement of the Endangered Species Act and has a number of wildlife specialists knowledgeable in our area's wildlife.

San Francisco Bay National Wildlife Refuge
USFWS
P.O. Box 524
Newark CA 94560
(415)792-0222.
SF Refuge's biologists monitor the status of the South Bay's endangered species.

ORGANIZATIONS

Bay Area Mountainwatch
P.O. Box AO
Brisbane CA 94005
(415) 467-6631
Dedicated to the preservation of San Bruno Mountain, its landforms and unique ecosystems.

Brisbane-Northeast Ridge Legal Defense Fund
933 Humboldt Road
Brisbane CA 94005
Fighting to preserve San Bruno's northeast ridge where Brisbane has approved 578 units of residential housing, a school, and commercial space.

Save San Francisco Bay Association
P.O. Box 925
Berkeley CA 94701
(415)452-9261
Monitors all types of development along the bay, currently doing follow-up studies of some restoration sites.

Save Our South Bay Wetlands
784 Danforth Terrace
Sunnyvale CA 94087-1225
(408) 720-1955

READING

Restoration? San Bruno Mountain. From Ecology Center Newsletter April 1990. This month's newsletter is devoted to the discussion of restoration's advantages and controversies. (415) 548-2220.

Citizen's Report on the Diked Historic Baylands of San Francisco Bay. - prepared by the Bay Institute of San Francisco (415) 331-2303.

Endangered Habitat: A report on the status of seasonal wetlands in San Francisco Bay and recommended plan for their protection. Available through the National Audubon Society's Richardson Bay Sanctuary in Tiburon (415) 388-2524 (Sat. or Sun.)

Ecological Restoration in the San Francisco Bay Area: A descriptive Directory and Sourcebook. Available from Restoring the Earth. (415) 843-2645. This directory lists and describes most of the restoration projects in the San Francisco Bay Area.

Helping Nature Heal, Whole Earth Review, Spring 1990. This entire edition of the Whole Earth Review was dedicated to a discussion of restoration.
Call (415) 332-1716 for availability.

Chapter Eight

Transportation

When the car was first invented 90 years ago, nobody thought there would be one in nearly every garage in the country. Only the elite would buy these vehicles, inventors said. Everyone else would still use horse power.

It wasn't until Henry Ford introduced mass production that car sales really took off. Soon, automobiles stood for far more than easy transportation. They symbolized affluence, modernity, and innovation. The whole country, it seemed, was moving into in a new age of speed, power, invincibility.

Today's cars don't really fit our society. Sure, we're all still addicted to speed and style. But our highways are crowding out farmlands. Our cars' emissions are crowding out fresh-air. And our freeways are so crowded we can't get anywhere without fuming in traffic jams first.

Are there any alternatives? Absolutely. Alternative fuels, public transit, and better urban design will all help. The Bay Area, now one of the most congested areas in the country, could be a model of change— if we just take those first steps down off the road.

Bay Area Traffic Shake-Down

By Susan E. Davis and John Holzclaw

Ask most people what the benefits of living in the Bay Area are, and you'll get any number of responses, ranging from beautiful landscapes to diverse populations, from nice architecture to wacky philosophies, from "laid-back attitudes" to educational and job opportunities. Ask most people what the liabilities are, and you'll get a stock response: "Earthquakes. And traffic."

Last October, the Loma Prieta earthquake highlighted both risks. The jolt hit us right at rush hour, and it hit us right where it counts: the freeways. The Bay Bridge broke. The Cypress freeway crumbled, killing 41 people. The Embarcadero suffered enough damage to render it useless and to fuel one of the hottest environmental debates in The City this year. Had the quake lasted any longer, the Embarcadero might have suffered the same fate as the Cypress. Hundreds more could have been killed.

For just a few months, it looked as though the earthquake might be a much-needed shot in the arm. BART ridership rose by 100,000. Bridge crossings dropped. Five ferries, servicing 15,000 stranded commuters, went into service. There was a sense that the Bay Area would pull together and pull itself from a transit snarl that has been fraying people's nerves, dirtying the atmosphere, and wreaking havoc with municipal and state politics for the last four decades.

No such luck. One year later, the road to freedom is blocked again. Bay Bridge traffic is normal, the ferry programs have sunk, and BART lost 75 percent of its new riders. The Bay Area remains the most congested region in the West, with portions of every major highway stalled every day. Yet the California Air Resources Board projects car travel rising 70 to 75 percent by the year 2005. Is there anything we can do?

Those Wonderful, Polluting Machines

The primary culprit in Bay Area transportation woes, as in most American cities today, is the automobile. Some people say that we have too many autos for the number of highway miles, and that the way to end this shimmering gridlock we call "freeways" is to widen and lengthen them. In truth, the real problem is the form and function of the auto itself.

Let's look at form first. Most cars use between one and two tons of steel, alumimun, rubber and plastic. Ninety-five percent of them are dependent on petroleum, they consume 65 percent of the country's oil (and 76 percent of the California's oil), produce 20 to 30 percent of the country's C02 emissions, and 70% of our other air pollutants. The National Crop Loss Assessment Program estimates auto emissions cause annual crop yield losses of $1.9 to $4.5 billion. As Berkeley Mayor Loni Hancock said at a recent conference, "Driving a single occupancy vehicle today is the equivalent of smoking a cigarrette in a crowded room."

If motorists had to pay directly for free parking, defending the petroleum supply line, smog-related health, crop and structural damage costs, or time lost to traffic congestion — they would end up paying about $4 dollars per gallon of gas.

Meanwhile, highway construction for autos displaces thousands of acres of wetlands and 3 million acres of farmland per year, according to the American Farmland Trust. Runoff from those highways sends salt, oil, gas, and other chemicals into our waterways.

Now let's look at the function. Supposedly, cars let us go farther, faster. This is still true in some parts of the Bay Area, and in many rural parts of California and the rest of the country. But in most urban and many suburban areas, cars don't move us faster at all. When time spent parking, servicing, and paying for cars is included with travel time, American drivers average only five miles per hour — one mile faster than walking, says John Holtzclaw, a sociologist and transportation expert with the Sierra Club.

It's still thrilling to cross the Bay or Richmond Bridge, at speeds slightly faster than legal, when the sun is setting behind Mount Tam and the fog is just rolling in. But increasingly, that crossing is mere memory. Nearly 6 million people, with 4.2 million registered cars, live in the Bay Area. Daily traffic on the Bay Bridge and the Golden Gate Bridge has increased 30% in the last 15 years, has doubled on the San Mateo, Carquinez and Richmond Bridges and tripled on the Dumbarton Bridge.

Congestion has increased in the Bay Area by 25 percent in the last three years, the Bay Area Council reports. Today, Bay Area commuters sit in stop and go traffic for 20 minutes each day — at a cost of $1194 in wasted gas and time, according to a recent study by the The Road Information Program (TRIP). The morning and evening rush hours have risen from one hour each to three.

By 2005, the Bay Area's population will have increased 20 percent. In general, California's auto mileage outpaces population growth by 2.5 times. In ten years, TRIP predicts, commuters will waste 60 minutes per day in traffic. Every hour will be "rush hour," meaning no one will rush anywhere. People won't be pulling out shotguns on the Bay Bridge; they'll be blowing up each others' cars to get them out of the way. Unless something changes.

Is there life beyond freeways?

The cure for the Bay Area transportation headache is two-fold. First, we need more public transit. And second, we need to re-design our cities.

Public transit seems like an obvious choice. The American Public Transit Association estimates that every full van removes 13 cars from traffic, every full bus removes 40 cars, and every full rail car removes 75 to 125 cars. Commuters who ride on mass transit for a year cut their personal hydrocarbon emissions by 90% and carbon monoxide and nitrogen oxide emissions by 75 percent.

A recent report by the Natural Resources Defense Council showed that mass transit serving dense pedestrian-oriented neighborhoods can be more effective in saving energy and improving air quality than has been thought in the past. A Danville-San Ramon house, the study found, consumes 31 times as much land — former wetlands, farmlands or hillsides — as a northeast San Francisco house. Residents of the latter neighborhood also have nearly 200 times as many restaurants and markets within walking distance, as well as vastly superior transit. Consequently, Danville-San Ramon residents own three times as many cars and drive four times as much as the northeastern San Franciscan resident.

Providing good public transit, and building three and four story apartment buildings and condos, the study claimed, with markets and restaurants, near the transit stations, would let residents walk both for the nearly 80 percent of trips that are not job related and to transit to get to work.

But the Bay Area's path to more public transit has been blocked a number of times. The ferry system was put out by bridge openings in the 1920s and 1930s. The electric rail system died when General Motors, plus major oil, steel, and tire companies bought out more than a hundred street car systems across the country, and dismantled them to pave over for buses using — you guessed it — GM motors and Firestone tires. San Mateo County refused to pay for their end of BART when the system was being built, and only now are considering an airport extension. The Caltrain still terminates over a mile from downtown San Francisco, which forces commuters to walk or transfer to Muni.

Autos and freeways also rule transit policy. Eight-ninths of the federal transportation tax dollars, including the gas tax, which goes into the Highway Trust Fund, is disbursed to State Highway Departments. The remaining fraction goes to public transit. Only last March, President Bush revealed his transportation policy in a

Save gas. Drive a Car

Even a cursory look at transportation in the Bay Area reveals two things. First, cars are bad. And second, we have to use them. Until we get reliable and extensive public transit in this area, we will continue to drive our cars. Or rather, we will continue to sit in our cars in traffic. Is there any way to cut down on oil consumption and toxic emissions while we're doing so?

You bet. And here's how.

• Don't push the pedal to the metal. Starting fast burns more gas than starting slow. Don't "floor" your car as you accelerate from a stop.

• Carry less freight. Each excess pound uses fuel. Check for extra tools, forgotten bags of soil amendment, and long overdue library books. You'll be cleaning up your life while you clean up the environment.

• Don't idle. Folk wisdom holds that turning a car on and off uses more gas than idling. Folks are wrong. Starting a warm engine uses less gas than idling in traffic for 60 seconds. If you're jammed, turn the car off. Similarly, 30 seconds is plenty of warm-up time for a cold engine — if you remember to drive slowly at first.

• Be mechanically minded. Any inefficiency in your car means it needs more fuel to work. So keep your spark plugs and air filter clean. Have regular tune-ups. Keep your front wheels aligned, and all tires properly inflated.

• Keep steady. Maintaining a constant speed, as opposed to pumping the accelerator, uses less fuel.

If you have the opportunity, of course, don't use your car. Get into the habit of walking and biking to close destinations. Half of the trips made to work in this country are less than five miles— a nice ride length. (Already, bicyclists save about 50,000 barrels of oil a day, according to the New York City based Transportation Alternatives, a bicycle advocacy group). Contact your local ride-sharing organization, or find neighbors who work in the same area you do. Combine errands to save one-stop trips.

And remember — keep lobbying your representatives for better mass transit systems.

report named "Moving America." That report called for expanding the highway systems and building more airports.

Since World War II, we have sought mobility by building freeways and starving transit. Freeways bulldozed out from San Francisco, Oakland, and San Jose spawned leapfrog sprawl, requiring residents to drive long distances to jobs, markets and restaruants and spreading congestion ever further. The response to this congestion has been to build more highways. Experience has been a poor teacher. Currently, over two dozen highway projects are being constructed, or planned for, in the Bay Area, including a 74-mile, six-lane toll road between I-680 in Sunol and I-80 near Vacaville, which will cost between $6.80 and $10.20 to traverse. Critics are already worried that the new road will encourage further development in Alameda, Contra Costa and Solano Counties.

Taking it from the streets

Annual polls show that residents still consider congestion a more serious problem than drugs, crime, shabby schools, homelessness, or any other urban problems in the Bay Area. Even the Metropolitan Transportation Commission (MTC - see sidebar) admits that if Bay Area commuters took BART or carpooled once a week, much of the congestion here would cease. But public transit in this area is in strikingly poor shape.

The 71-mile long BART system, built between 1972 and 1975 to sweep commuters between the East Bay and the West Bay, is in a dismal state of disrepair. The 18 year old rail cars have electrical viruses that cause breakdowns but cannot be traced. Each of the 34 stations has to be upgraded. Parts for fare collection machines are scarce. The company's general manager, Frank Wilson, has said he thinks it will cost $500 million to rehabilitate the system.

Ferries have been one of the most neglected aspects of public transportation in the Bay Area. Before the construction of Bay Area bridges, we had one of the finest ferry fleets in the world, with five lines that transported 6 million vehicles and 60 million passengers each year. To this day, there has been almost total lack of legislation allocating public transit funds for waterborne transit. Of the 17 agencies responsible for public transportation, none have been willing to take responsibility for the development or regulation of ferry services.

Five lines were opened after the earthquake — all but three have been discontinued. Only Alameda and Oakland are still offering their service. Their contract is up for review in March, 1991.

Taking it to the courts

Two years ago, it seemed as though the 1988 California Clean Air Act would help the Bay Area's traffic problem. That Act, which required that the Metropolitan Transit Commission and the Bay Area Air Quality Management District cut auto emissions by 35% by 1997, empowered the BAAQMD to adopt Transportation Control Measures (TCMs) to limit auto use.

Today, Citizens for a Better Environment and the Sierra Club are suing MTC and other agencies for approving new highways without controlling auto emissions to meet requirements of that Act. MTC has agreed to improve their evaluation of the environmental impacts of new projects. Now the plaintiffs are requesting that all projects, including those already approved by MTC, be delayed and re-evaluated as well. In June of 1990, MTC recommended a number of TCMs, including improving transit and High Occupancy Vehicle (carpool and bus) lanes, increasing the cost of driving, and cleaning up auto exhausts. Most of these cannot be implemented without new state legislation. In addition, MTC is considering levying higher bridge tolls, more gas taxes, odd-even driving days, mandatory employer-based carpool programs and parking restrictions, some of which also require new state legislation. However, MTC avoided proposing a moratorium on highway construction, for which they have authority.

Also in June, California voters approved Proposition 111, which would provide $15.5 billion for transportation projects. It sounded good at first. Since then, the California Transit Commission has targeted $15 billion of that money for highways. Another $3 billion, slated for "congestion management relief" and promised to public transit, will also go to more highways. Propositions 108, which provided $1 billion for transit construction, and Proposition 116, which provided $2 billion for specific rail projects, including Caltrain, a San Jose to Sacramento train, and BART Muni and AC Transit rail extensions, may be more helpful. These funds, however, fall far short of what is needed to build the high-speed trains between major California cities needed to reduce airport loads and expensive expansions.

It's not all bad, of course. BART plans to open 25 miles of new extensions in the 1990s, including lines to West Pittsburg, Dublin, Warm Springs, and the San Francisco Airport. AC Transit has revamped 16 of its lines in the East Bay to provide better connections, has added bus service in western Contra Costa and is considering light rail construction; Caltrain is considering extensions to downtown SF and Gilroy. The Golden Gate Bridge and Highway Transportation District is planning to replace its ferries and buy the right-of-way from Northwest Pacific to build light rail. Muni has proposed building light rail extensions to Mission Bay, out Geary Boulevard, and down 3rd Street, adding cable cars to Taylor Street, and adding historic streetcars on Market and Embarcadero. And the Santa Clara County Transit District is planning light rail extensions in the Guadalupe and Tasmans corridor.

But is that enough? Probably not. We also need a massive extension of rail services in the Bay Area, including: extension of the Caltrain to downtown San Francisco, improved BART service, frequent interurban rail passenger service between San Jose, Oakland, and Sacramento; light rail (streetcar) extensions all over the region; and expanded ferry and bus service. We could pay for this by charging motorists the full cost of their driving, and by diverting more of our transportation funding to public transit.

Planning cities and growth

The existence of automobiles has ruled urban design and development for over 40 years.

In the nineteenth century, cities grew up around transportation centers. Houses and shops were clustered around harbors and railroad and trolley lines. People walked or rode horses. After the automobile was mass-produced, city-dwellers began sprawling out away from cities. Suburbs, designed to take middle-class folks away from inner-city squalor and deposit them safely in "natural" surroundings, depend on automobiles and freeways to get commuters from home to work and shops each day.

In the last few decades, this decentralization has been exacerbated by a parallel decentralization among businesses. Indeed, since 1980, most of the Bay Area's new jobs and office space have been created outside San Francisco, Oakland, and San Jose. The 1980 Census reported that more than a third of Bay Area workers cross a county line to reach their jobs, and less than half live in the same region where they work. But only 11% of Bay Area residents commuted via transit at the beginning of this decade.. Nationwide, half as many people walk to work today as they did in 1960, due to reduced rail lines and the difficulty of servicing more remote areas.

"Society is structurally addicted to the automobile," says Richard Register, founder of Berkeley's Urban Ecology. "For the vast majority of us, walking, bicycling, and taking transit to work, to socialize, to shop, to entertainment, and out into the country is simply not feasible. We're hooked."

Unhooking entails re-designing our cities so that they are centers of commerce and community, not just office buildings that stand empty on nights and weekends. To do this, we need to stop sprawling outwards and start filling inwards, by bringing small businesses and housing into the city centers. Cars will automatically disappear if networks of intercity rail, local transit and pedestrian routes are provided. NRDC found that when residential density doubles, per capita auto driving decreases by 25 to 30 percent, if neighborhood commerce grows as well.

The rallying cry of such "urban ecology" is "access by proximity" — not by cars. To make that cry mean anything, we'll have to turn away from freeways, turn towards public transit, and transform the face of our cities. This time, let's not wait for an earthquake to do the work for us.

Susan Davis is a writer living in Berkeley. John Holzclaw is an urban sociologist and regional planner who works on transportation issues for the Sierra Club.

How activists can influence transportation projects

Basic highway and transit "improvements" are planned and approved by the Metropolitan Transportation Commission (MTC). MTC models future trip scenarios based on the Assocation of Bay Area Governments' projections of housing and job locations. The model divides trips between existing and planned roads and public transit. Then it evaluates traffic levels on highways.

The MTC's modeling serves as the technical justification for the commission's decisions. Generally, the highway projects included in the model have strong advocates among local city and county officials and local developers. But except for representatives of CalTrans, the U.S. Department of Transportation, and the Bay Conservation and Development Commission, MTC's commissioners are elected county supervisors and city officials, and are subject to local political pressure.

Pressuring the commissioners to support more public transit and less highway transit could have considerable influence.

If they get approved by MTC, highway expansion and extension projects go to the State Department of Transportation (CalTrans) to be segmented into short projects for planning (including environmental evaluation), design, land acquisition and construction contracting. Activists can have some impact on projects which cross wetlands or navigable waterways because they require Army Corps of Engineers approval. Similarly, projects which cross the bay or are on bayshores need BCDC approval. And projects which cross a state or national park need the park service's approval. Lobbying these departments can have some effect.

In addition, you can join or contact any number of activist groups concerned with transportation in the Bay Area, including:

California League of Conservation Voters: 415-896-5550

California Rail Foundation: 415-592-6580

California Transit League: 916-447-9639

FORCE Gilroy Extension: 408-778-6198

San Francisco Tomorrow: 415-929-8520

Sierra Club
San Francisco Bay Council: 415-653-6127
Loma Prieta Chapter: 415-494-9901

The Alternative Fuel Options

By Lisa Levy

There is new activity in the alternative fuels field, an area of research which nearly ran out of gas when oil prices dropped in the 1980's. Congress recently passed a law which encouraged car makers to develop and manufacture vehicles powered by alternative fuels. Of course, in an ideal future, mass transit, bicycles, and walking would serve the majority's transportation needs. Until then, however, alternative fuels offer alleviation to some of the ills caused by cars. Here are the pros and cons of five alternative fuels.

Compressed Natural Gas (CNG)

Compressed natural gas burns more efficiently and cleanly than gasoline, producing fewer hydrocarbons and greenhouse gases. It is not, however, a completely "clean" fuel. CNG is also plentiful in North America.

According to Pacific Gas & Electric, CNG is the fuel most likely to be in your car by the year 2000: "dual fuel" cars, which run alternately on CNG and gas, will be available by 1994. CNG-exclusive cars will be on the market two years later. There are already kits that enable drivers to "retrofit" their car to be a dual fuel vehicle for $2,500, and six CNG public access stations will be open in California by 1991, including ones in Concord, San Rafael, San Jose, and Hayward. Filling a car with CNG takes around two to five minutes, the same as gasoline.

Methanol

There are more methanol vehicles in California than in the rest of the world combined. Methanol is popular with both state and federal officials for public transit vehicles because it does not require expensive retrofitting like CNG; in fact, the California Energy Council wants to convert 30% of California's automobiles to methanol within the next thirty years. Methanol is cheaply produced from natural gas, although it can also be made more expensively from coal, wood and agricultural wastes, and is most often used in a 15% gas blend called M85.

The problems with methanol, however, are sizable. Methanol corrodes pipes, engines, and storage tanks, and is only half as powerful as gasoline, so larger amounts are required to power a vehicle. There is also evidence that methanol emissions may be carcinogenic.

Ethanol

Ethanol is a distant cousin of methanol, and is commonly used in a 10% gasoline blend called gasahol. Currently one-tenth of the gasoline used in the U.S. contains ethanol. It is produced by fermenting agricultural goods, such as corn in the United States, or sugar cane in Brazil, where ethanol is often used in its pure form.

Because we lack the raw materials needed for fermentation, ethanol is not very economically feasible in California. It would require two acres of corn to run a car for one year using ethanol, and since food is already scarce in parts of the world, corn could probably be put to better uses. There is a possibility, however, that ethanol could be derived from wood cellulose, which would make it a more feasible energy option in the future; because growing trees would absorb carbon dioxide emissions, the level of carbon dioxide could theoretically remain the same with widespread use of methanol or ethanol.

Electricity

Electric vehicles run on batteries charged in an ordinary socket. They emit no pollutants at all, except for those resulting from power plants, and are very inexpensive to run. Currently there are about 37,000 electric cars in California, according to the DMV and most can run 80 miles between charges, although some "super-lightweight" cars can go 250 miles. These vehicles have no problems reaching freeway speeds.

A solar electric car is already in production by Solar Electric Engineering in Rohnert Park, (707-586-1987) and conversion kits and manuals are available from Electro Automotives in Santa Cruz (408-429-1989). Chrysler and GM are both developing electric cars.

Hydrogen

Hydrogen is touted as the fuel of the distant future by many alternative fuel proponents. It is pollution free, the only-emissions being water vapor and nitrogen oxides. It can be made from water, an abundant resource, using electric current to separate the hydrogen from the oxygen molecules.

Unfortunately, this is a fairly energy-intensive process, making hydrogen a relatively expensive fuel at the present time — about $3.50 per gallon. Eventually, solar power could do this job, which would make hydrogen a much better option.

There are also problems with storage. In pressurized tanks hydrogen holds the potential for explosion. A solution to this problem is promised by technology that combines the hydrogen with a metal alloy in a weak chemical bond, but this storage method also poses cost problems.

Lisa Levy is a staff researcher and writer for the Bay Area Green Pages.

Chapter Nine

Pollution

It seems like a pretty nice place, as urban areas go: fresh air, fresh water, cool sea breezes, plenty of nature left. The smog usually isn't anywhere near as thick as it is in Los Angeles, and the water isn't as bad as in Boston Harbor. The Bay has never caught fire the way the Ohio River once did. It seems we must be doing all right.

Don't be fooled. Nearly six million human beings inhabit what was once a wild, pristine, and bountiful estuarine area, and we're having some nasty effects. Our cars and factories belch tons of pollutants into the air each day. Industry, sewage plants, and plain old storm drains feed a steady stream of toxics into our water. The Bay itself, while fairly clean in some spots, is terribly polluted in others. There are a large number of toxic waste dumps and nuclear hot spots here.

Dozens of federal, state, local, and non-profit organizations are trying to clean up the Bay Area. The picture they paint of current conditions is pretty grim. But a healthy Bay is not an impractical vision, if each of us does our part to reduce our own impacts, and to be vigilant against the harm done by others.

How Polluted is the Bay?

By Susan E. Davis

Picture the Bay Area, and the first thing that comes to mind, of course, is the Bay itself — 460 square miles of choppy blue waters, spotted with islands, spanned by four bridges, rimmed with marshes and wetlands, backed by Mt. Tamalpais to the west and the low-slung Berkeley Hills to the east, shimmering beneath the sun sometimes, and lying flat and grey beneath the fog at others. Something about the hills, the wind and the water, the light and the fog, intoxicates us. So incredible to have so much nature here, we say. We're so lucky to have fish, birds and wildlife close to home.

But the Bay is more than just mountains, water and dazzling sunshine. In 1981, the U.S. Fish and Wildlife Service announced that no other major estuary in the United States had been modified by human activities as much as the San Francisco Bay. Today, the Bay is home to more than six million people and two major ports. Its 300 miles of shoreline house thousands of industries, including six major oil refineries, 40 hazardous waste sites, and 30 municipal and 40 industrial waste treatment plants. It hosts houseboats discharging wastes from humans and companies discharging wastes from processed fish, animals and chemicals. Shipyards spill oil, and restaurants spill cleaning fluid. Sand mining operations stir up silt, and dredging operations stir up toxics. Some fish have ulcers and burns. Some shellfish are riddled with heavy metals. Some birds cannot reproduce.

In other words, the Bay may look lovely from Russian Hill, or Mount Tam, or the top of Berkeley's Marin Avenue. But up close, you see it's not in great shape.

It's better than it was. Thirty years ago, bacterial and other organic wastes filled the Bay and their odor filled the air. With the passage of the Clean Water Act, over $3 billion was spent on Bay Area waste treatment modernizations. These sharply reduced the amount of oxygen-consuming organic matter and ammonia in the Bay.

Still, the Bay is plagued by a number of problems. Reduced freshwater flows, industrial and domestic wastes, urban run-off, and incessant dredging all take their toll on what once was a pristine and bountiful estuary.

Water, Water, Nowhere

Before 1850, the Sacramento and San Joaquin rivers discharged about 34 cubic kilometers of fresh water into San Francisco Bay annually. As the population grew, however, and as the need for irrigation increased, the state built a monumental water system to transport billions of gallons of water south. Today, eighty-five percent of the water in the Delta is sent to southern California via the State Water Project and the federal Central Valley Project. The flow of fresh water into San Francisco Bay is less than 40 percent of its historic levels.

This is having serious effects on the health of the Delta and the Bay, both of which depend on a delicate balance of fresh and salt water to maintain their habitats. Without sufficient levels of freshwater, too much saltwater creeps into the Delta. Phytoplankton — the basis of the oceanic food web — die off. Migratory fish that depend on miles of cool freshwater cannot thrive. The Bay cannot flush out contaminants, especially the South Bay, which has no major tributaries. Add to this reduced in-flow from four years of drought, and you have an estuary that's in serious trouble.

One of the most important signals of the bay's health is the "striped bass index," which measures the survival ability of striped bass during their first 6 weeks of life. That index is now down 80 percent— its lowest since monitoring began in 1959, the Department of Fish and Game announced in August. Many juvenile bass die as they get sucked into the water-project pumping stations in the south-east Delta. Many others die simply from lack of habitat.

Chuck Armor, an associate fishery biologist with the Department of Fish and Game, notes that many other populations in the Bay have been reduced at "frightening rates" as well. Long-fin smelt, English smelt, crabs, starry flounder, and other species have "disappeared."

"What's scary is that those fish were too small to emigrate back to the ocean," he says. "They're coming in and not surviving. And we don't know if it's from pollution — or lack of food."

Armor is particularly concerned about the Northern Anchovey population. "Usually when we take samples for Northern Anchovies, we're up to our arms in them. This year we haven't seen them at all. It's kind of spooky. Anchovies support the entire food chain for the bay and the gulf. It really bothers me to see a big part of the food web disappearing."

The State Water Resources Control Board is in its fifth year of a court-ordered re-drafting of a Bay-Delta Plan to balance urban, agricultural, and estuary water needs. So far, no such balance has been achieved. This summer, agricultural and urban interests in the south — faced with diminishing water supplies from the Colorado River, Mono Lake, and Owens Valley — called for a Peripheral Canal-type structure to divert Sacramento River water around the Delta, and send it directly south. Environmentalists, farmers, and citizens in the north are claiming the bay and Delta needs that freshwater to survive.

The first draft of the plan, released in the fall of 1988, suggested establishing minimum freshwater flows. That draft was so hotly criticized by water developers, however, that the State Board withdrew the plan. The second draft, released in the summer of 1990, replaced minimum freshwater standards with those based on salinity levels and temperature. This could have drastic implications for the Bay.

"Freshwater flows are important for more than just salinity levels," notes John Krautkramer of the Environmental Defense Fund. "We need adequate flows for circulation in the Bay and for the survival of fish in the San Joaquin River. Minimum flows are fundamental to the health of the estuary. It is the crux of the issue, and they are side-stepping it."

Dishonorable discharges

Unfortunately, while we've reduced the amount of freshwater coming into the Bay, growth around the Bay has contributed more pollutants, thereby creating a fine broth of toxins.

Industrial discharges began when gold miners in the Sierra Nevada used hydraulic pumps to wash out hillsides during the mid-19th century. This sent massive amounts of copper, zinc, silver, gold, and mercury into the Bay. Today, many of those toxic sediments are still here. Many human-caused trace elements also come from the built-up areas which line the Bay's shores.

In 1987, in an effort to understand the composition and volume of Bay pollutants, the State Water Quality Control Board and the EPA commissioned the Aquatic Habitat Institute to study both the volume and composi-

Toxic Hot Spots

In 1987, Citizens for A Better Environment identified toxic hot spots in every part of the Bay and on the shores of almost every County with Bay shoreline.

Sediments in Coyote Creek, Channel Creek/China Basin, Hamilton Air Force Base, Concord Naval Weapons Station, Redwood Creek, Oakland Outer Harbor, Richmond Harbor, the area around Candlestick Cove and Brisbane Lagoon, San Leandro Bay and Oakland Inner Harbor all have between 5 and 9 contaminants which exceed toxicity thresholds. Islais Creek, Alameda Naval Air Station, Channel Creek, Mare Island Strait, and Hunter's Point Naval Shipyard all have ten or more contaminants which exceed toxicity thresholds in sediments.

Shellfish at ASARCO-Selby Slag Pile, San Leandro Bay, Point Isabel, Redwood Creek, Richmond Harbor, Candlestick Cove, Brisbane Lagoon, Islais Creek, Mare Island Strait, Oakland Inner Harbor, Oyster Point, and Palo Alto Outfall had toxicity levels which exceed standards for human consumption.

Ducks at Dumbarton Bridge, Coyote Creek, Guadalupe Slough, Newark Slough and Plummer Creek, Redwood Creek, Richmond Harbor, Castro Cove, and Point Davis all had selenium and mercury concentrations in their livers which exceeded toxicity thresholds for reproduction.

Some areas had contaminants exceeding toxicity levels for all three categories, most notably Redwood Creek and Richmond Harbor. In addition, a number of places have pollutants for which no toxicity levels are available.

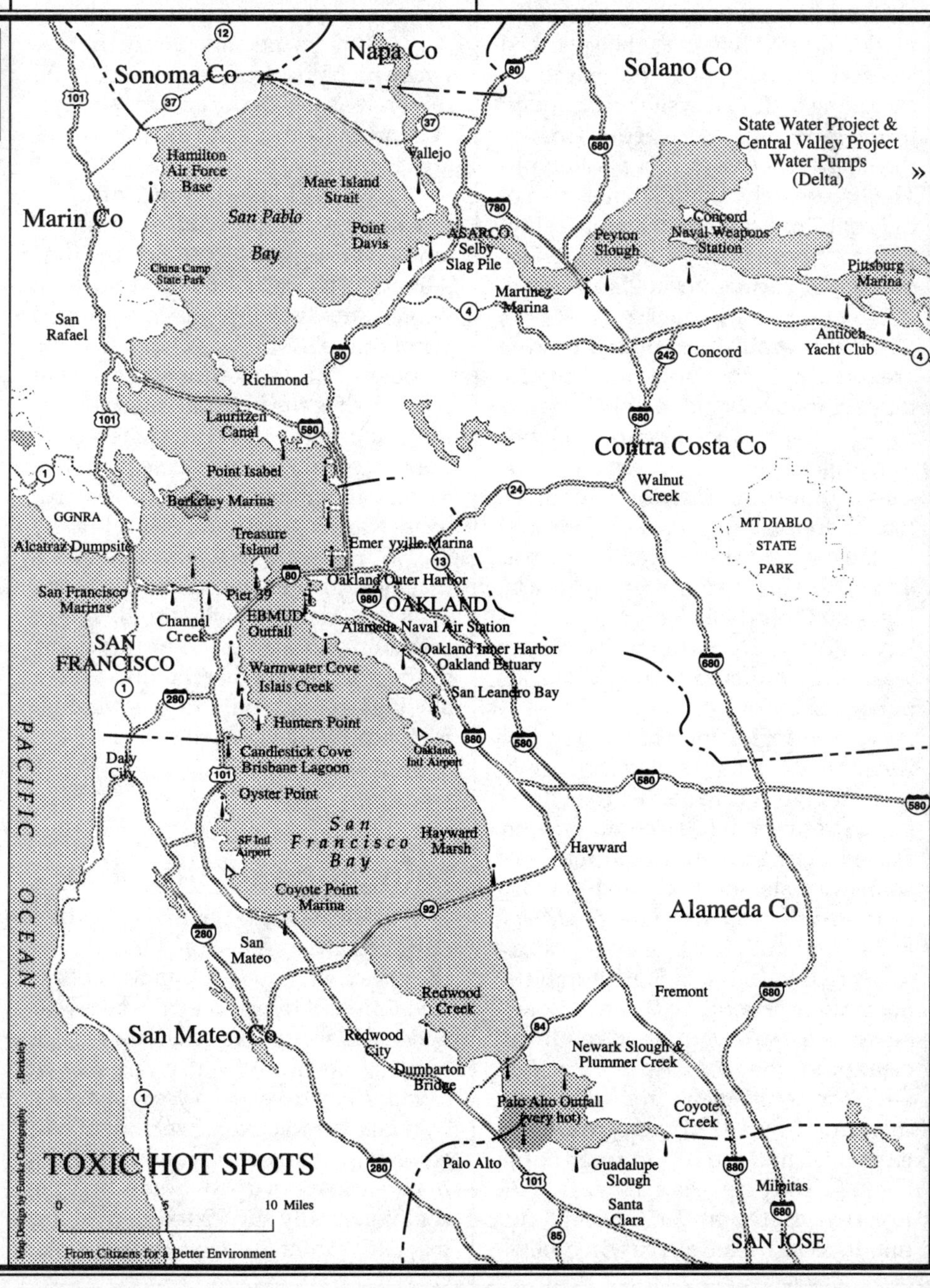

tion of Bay pollutants. AHI found that it is difficult to trace pollution to specific sources in the Bay Area, because so much comes from the rivers and from city streets. But AHI also found a number of major point sources around the bay, including sewer plants, oil refineries, and industries along the Bay's shorelines.

A report done by Citizens for a Better Environment (CBE) that same year more specifically named 39 "toxic hot spots" where chemicals had reached threatening levels in the water, sediments, shellfish, or ducks. The most abundant contaminants were 10 heavy metals, the pesticides DDT and TBT, oil, grease and PCBs. Some of their levels are the highest of any estuary in the world. Twenty-nine of the hotspots had at least one pollutant at concentrations exceeding toxicity thresholds for aquatic life. Fifteen spots exceeded toxicity thresholds for five or more pollutants. The tissue of the shellfish in eleven of the spots exceeded international standards for human consumption. Eight exceeded toxicity thresholds for selenium and mercury in ducks.

Generally, harbor and marina areas are the most polluted, because they are the site of industrial discharge, run-off from sewage treatment plants, and urban drains. Currently, over 100 sites of industrial discharge exist in the Bay. Unfortunately, those areas are used for fishing, often by Asians, Hispanics, and other ethnic groups trying to supplement their diets. Few public piers have posted Department of Health warnings for mercury and PCB levels in Bay fish.

The region most affected by toxics is the South Bay, below the Dumbarton Bridge, where neither river flows nor currents help flush contaminants. Indeed, during the dry season, the major source of water for the South Bay is discharge from waste water treatment systems. The South Bay is one of eight Bay spots that both the U.S. EPA and the State Water Resource Control Board agree is a "toxic hot spot"requiring clean-up under the Clean Water Act. Despite improvements in sewage treatment, more than 1700 tons of toxic trace elements enter the Bay through Santa Clara County sewers each year, CBE reported last December. Levels of cadmium, copper, lead, mercury, nickel, silver, and zinc exceed EPA water-quality standards for safety by 2 to 140 times on 75 percent of all sample days.

During dry seasons, the San Jose-Santa Clara, Palo Alto, and Sunnyvale plants are responsible for 75 percent of the silver, copper and nickel in the South Bay, AHI reported in 1987. In addition, metals found in sewer outfall have been traced to more than 500 Silicon Valley industries. As a result of waivers on water quality standards, Silicon Valley industries have been allowed to discharge ten times more toxic metals into South Bay sewers than these companies could do elsewhere. Indeed, more than 2300 facilities in Santa Clara County have no sewer discharging permits at all.

"This is the only part of the Bay where such practices are allowed," CBE research associate Greg Karras says. "Unfortunately, it's the most sensitive area in the Bay."

The released contaminants threaten aquatic life, wildlife, and wetlands throughout the South Bay. Already, changes in salinity due to wastewater discharges have destroyed significant parts of the salt marsh at the National Wildlife Refuge there. That salt marsh is the last major habitat for the California clapper rail in the world. Populations there have decreased "five to ten fold" in the last decade, according to Karras. An unreleased report by the U.S. Fish and Wildlife Service shows that the remaining population is laying eggs with dangerously high levels of selenium and mercury. "At this point, the question is whether this area will be a wildlife refuge or a toxic waste dump for Silicon Valley," Karras says.

Mitigation by litigation

Fortunately, the South Bay has also become a "hot spot" for toxics litigation. In 1988, Clean South Bay, a coalition of over 40 local, state, and national groups began one of the biggest battles over pollution in the Bay in decades. Fed up with 15 years of delays by state and local governments in enforcing the original Clean Water Act's mandate to clean up the South Bay, the coalition finally succeeded in pushing the state's Water Resource Board to take action in October, 1990. The board's order tells South Bay cities to set limits on industrial toxic discharges by April of 1991, with a complete clean-up due in three years. In effect, the order tells the cities that they've had long enough to study the problem, and that it is now time for action.

"I think we're over the hump," Karras says. "This overturns 16 years of a 'study the Bay to death' attitude."

Cities have been extremely reluctant to take on the industrial polluters that also form a significant portion of their tax base, Karras says. He credits citizen activism with being the influence that ultimately led to the decision to enforce the law. "The big story is that we here in the Bay Area can and are having an effect," he says. "It's unfortunate that government isn't standing out in front of the effort to clean up the Bay. But they aren't."

Meanwhile, the quality of effluent pouring into the South Bay has improved slightly. The Palo Alto Sewage Treatment Plant, for instance, is dumping far less copper into the Bay than it used to, mostly as a result of scientific studies and press attention. That plant "has really cleaned up its act," says Sam Luoma, a biologist with the United States Geological Survey who studies the South Bay. "You can see a positive reduction of copper traces in the mud-flats now." Similarly, 40 metal platers and electronics manufacturers have eliminated measureable copper discharges, CBE reports.

Perhaps the "hottest" industrial waste spot in the Bay is the Lauritzen Canal. This 13-acre site, located in the Richmond Inner Harbor, has been the site of a half-dozen plants processing a number of chemical products, including napalm and DDT. Today it is a metal recycling plant.

Since 1951, the state Department of Fish and Game has recorded several instances of chemicals being discharged into the Lauritzen Canal. In 1960, an illegal DDT disharge killed at least 180 striped bass in the area. The State Department of Health Services found elevated levels of DDT, lindane, BHC, endrin, and other pesticides in soil and sediment samples at the site. In 1981, when the canal was dredged, between 50,000 and 60,000 cubic yards of spoils were dumped off Alcatraz Island, researchers realized that the canal was heavily contaminated with DDT.

Clean-up is difficult, because as soon as the sediments are disturbed, there is a risk of having the DDT circulate out into the Bay and beyond. "This canal is a toxic nightmare," says Barry Nelson, director of the Save The San Francisco Bay Association. "It's probably the most highly-contaminated place in the Bay."

In addition, a number of the military bases around the Bay, including Alameda, Concord, Treasure Island, Hunter's Point, Fort Cronkhite, the Presidio, and Mare Island have been designated as Superfund sites by the EPA. Substances discharged into groundwater, or directly into the Bay, over the last decades include lead-based paints, waste solvents, PCBs, asbestos, cyanide, benzene, acids, oil, gasoline, and a whole host of chemicals whose individual and combined effects have yet to be reckoned with. The government currently is documenting and making plans to clean up those sites. (See the following article, "Marching on Toxics.")

Crude contributions

The six refineries around the Bay, which together process nearly 700,000 barrels of crude oil into fuels, asphalts, lubricants, and other products each day, are also real trouble-spots. In addition to discharging arsenic, copper, nickel, and zinc into the Bay, the refineries together discharge as much selenium as the San Joaquin and Sacramento Rivers combined, according to CBE reports.

They also contribute large amounts of petrochemical-related toxics. Concentrations of petroleum hydrocarbons as high as those in LA and San Diego have been found in mussels here — even though we have only a fraction of the spills the southern ports do. Mussels placed by DFG scientists in Castro Cove near Richmond in February were found to contain some of the highest levels of toxic petrochemical compounds ever found in the U.S.

Skin lesions, tumors, and increased parasites in striped bass have been linked to exposure to synthetic organic compounds, and concentrations of petroleum-derived hydrocarbons have been found in their eggs.

Agricultural wastes have been a serious problem in the Bay since farmers in the Central Valley began using fertilizers, soil amendments, herbicides, and pesticides in the 1920s. Agricultural waste water now makes up 20 percent of the total flow of the San Joaquin River. Since 1950, the maximum annual concentrations of sulfate and nitrate in that river have increased three and five-fold respectively, USGS research scientists reported in a 1986 *Science* article. In addition, high concentrations of selenium — which occurs naturally in arid alluvial soils, but is concentrated when it is leached by irrigation — have been found in ducks in Carquinez Strait and the South Bay. Selenium causes reproductive failures and deformities in embryos.

Residues from DDT and other pesticides are also flowing into the Bay as the soils in the Cental Valley erode, and carry with them pesticide residues from years ago. The danger here is that no one really understands the patterns of sediment transport in the Bay, so we don't know where all those pesticides are going.

Organic wastes in the Bay have been sharply reduced, but are still troublesome in some spots. The Clean Water Act forced local cities to clean up their effluent so successfully that today, while the amount of wastes being taken into plants has doubled, the amount reaching the Bay has dropped sharply. For instance, both the biochemical oxygen demand of sewage entering treatment plants and suspended solids doubled between 1955 and 1985. The amount reaching the Bay, however, dropped by 75 percent and 60 percent respectively, according to Regional Water Quality Control Board figures.

The Regional Board is charged with enforcing the federal Clean Water Act. But until this past year, no one had monitored bacteria levels in the Bay itself since 1981. The BayKeeper, a volunteer "police force" on the Bay, offered to collect samples in 50 sites, if the Regional Water Quality Control Board would analyze it. So far, the two have found, 75 percent of those 50 locations have levels of contamination higher than RWQCB's objectives for freshwater contact or shellfish harvesting. In several areas, including houseboats and anchor-outs in the Richardson Bay, the bacterial counts exceed safety thresholds.

Spoiling the Bay

If Lauritzen Canal is the hottest toxic spot, the Alcatraz Island dump site may be the largest. The Army Corps of Engineers has been dumping dredge spoils from around the Bay there for over fifty years. The problem will be exacerbated if the Oakland Port goes through with its plans to deepen its channel to allow larger container ships into the harbor. The Port has proposed deepening an inner harbor channel first from 35 to 38 feet, and then to 42 feet. The second phase will include the creation of an 1100 -foot wide turning circle for the biggest ships. Together, the two dredging phases will create 7 million cubic feet of spoils. Annual maintenance will produce millions of tons more.

The problem with dredging is two-fold. First, sediments on the Bay's bottom are often toxic. Urban sewer outfalls, industrial discharges, Central Valley drainage, city run-off, and human wastes have all settled to the bottom of the Bay, and with them traces of mercury, lead, zinc, DDT selenium, and pathogens. Digging them up sends them swirling into the Bay's waters. At Alcatraz, at least half of what is dumped is taken by currents, the Bay Conservation and Development Commission (BCDC) and CBE have found.

Second, we're running out of sites to put the spoils. Traditionally, we have dumped 6.5 million cubic yards of spoils annually on five sites around the Bay. That's a lot of mud. Alcatraz has been nearly filled in. Options for dredge placement include the outer continental shelf, uplands areas like the Delta, and new Bay sites. So far, none have been ideal. A site off the coast of Half-Moon Bay was shot-down several years ago, after it was discovered that the sediments threatened Dungeness Crab, Dover, English and Rex sole, salmon, herring, albacore, and most of the sea mammals at the Farrallones. When the Port proposed leaving spoils on Delta levees last summer, the Contra Costa Water District sued, saying it risked contaminating their water supplies. Representatives Nancy Pelosi, George Miller, and Barbara Boxer have called for a GAO investigation of the impact of disposing dredge spoils in the Bay and off the coast.

Uncharted waters

Even if we get the necessary permitting processes implemented in the South Bay, get Superfund sites cleaned up, and stop the dredging, untreated urban runoff pouring into the Bay remains a severe problem. No one knows exactly what is in Bay Area run-off — or what it does in the Bay — because few studies have been done so far. Researchers do know, however, that the run-off, which enters the Bay through more than 50 small local streams, contains oil, grease, chemicals, debris, and numerous other toxins. The Association of Bay Area governments reported in 1982 that runoff sources of heavy metals represent about 32 percent of the total heavy metals found in the Bay.

Runoff is particularly a problem after storms, when the combined loads of sewage and urban run-off overwhelm plant capacities, and end up pouring into the Bay — untreated. This is happening now at Islais Creek, near Candlestick Point, and a number of other sites.

In addition to direct sources of pollution, dredging, and urban and agricultural run-off, there is a certain level of ambient pollution in the Bay. Until 1988, this hadn't been well studied. "Generally speaking, money goes into analyzing the stuff coming into the Bay, not what happens when it's in the Bay itself," Luoma says. But for the last year, the Regional Board and scientists at Lawrence Berkeley Laboratory have been studying these "ambient toxicity" levels, by exposing fish, invertebrate, and plant species to water taken from 12 samples around the estuary. In a report released this fall, the researchers concluded that much of the water in the Bay may be "moderately toxic," and that ambient toxicity definitely exists in the Contra Costa Canal, and several Bay marshes. Water from the Hayward Marsh, site of a reclamation/sewage treatment project was so toxic that all of the test minnows exposed to it died.

Cleaning up our act

One of the primary problems with monitoring and mitigating pollution in the Bay is that while many studies have been done, few are precipitating action. Indeed, one critic noted, "Studies are often undertaken to delay taking action. We have plenty of data to support more stringent regulations at this point."

The Regional Board is currently considering plans to monitor the waters through the Bay Protection and Toxic Cleanup Program. At present, much of the monitoring comes from industry reports mandated under the National Pollutant Discharge Elimination System (NPDES) established under the Clean Water Act. While some say the dischargers are fairly honest, and the record is probably complete, others, including Barry Nelson, say "this is like asking people to report when they break the speed limits." Those critics are calling for more regulation and enforcement around the Bay.

"The level of monitoring had been non-existent," Nelson says. "Now it's moving towards inadequate. No one is monitoring water quality in the Bay now, or watching the polluters."

The BayKeeper program, a non-profit organization founded in the summer of 1989, has been the most visible monitoring program on the Bay in decades. With a flotilla of 40 ships, several planes and 150 volunteers, the BayKeepers have filed over 175 complaints against illegal polluters in the last year. Some of these complaints have resulted in action.

Assemblyman Tom Bates introduced a bill calling for the Regional Board to do spot-checks on dischargers. It was vetoed in March of 1990.

But even before we begin catching the dischargers, many people note, we need to stop the discharging at its source. The most important stage of clean-up needs to occur before treatment. That is, citizens, industries, and municipalities have to stop producing the amount of toxics poured into the treatment plants in the first place.

To participate in cleaning up the Bay, you can act on a number of levels. In your home, properly dispose of household hazardous waste — like car oil, pesticides, and paint thinners (better yet, don't use them). Stop driving your car so much. These actions help reduce urban runoff and sewage treatment loads. Don't support companies that contribute to the Bay's pollution. Let your representatives know you're concerned about the Bay by attending Regional Board meetings and pertinent city meetings. Join the BayKeeper force, or volunteer at the many organizations devoted to the health of the estuary. And keep talking to people about the Bay; word of mouth is often the most effective educational tool.

Susan Davis is a Berkeley-based journalist whose work has appeared in the Washington Post, Newsweek, *and* California Magazine, *among other publications.*

How to get Involved

ORGANIZATIONS

BayKeeper (also known as the San Francisco Bay-Delta Preservation Association)
(415) 567-4401
Building A, Fort Mason
San Francisco, CA 94123

Citizens for a Better Environment
(415) 243-8393
501 Second St., Suite 305
San Francisco, CA 94107

Bay Institute of San Francisco
(415) 331-2303
10 Liberty Ship Way
Sausalito, CA 94965

Citizens to Complete the Refuge
(415) 493-5540
453 Tennessee Lane
Palo Alto, CA 94306

Save San Francisco Bay Association
(415) 452-9261
P.O. Box 925
Berkeley, CA 94701

Save Our South Bay Wetlands
(408) 720-1955
784 Danforth Terrace
Sunnyvale, CA 94087-1225

Committee for Water Policy Conscensus
(415) 682-6633, 1485 Enea Ct. #1330

Marching on Toxics

Military toxic build-up in the Bay Area

By Saul Bloom

Hidden among the rolling hills and crowded cities of the Bay Area, government investigators have identified 270 sites where the military may have dumped toxic waste. That's more military toxic hot spots than were found in 44 entire states.

More than a century ago, General Ulysses S. Grant observed that "an army marches on its stomach." The former president died too soon to learn that in the 20th century, a world-class military force also needs an ocean of hazardous chemicals and metals to extend its global reach. The larger the military force, the greater the environmental hazard.

As a consequence of fielding the most powerful military in the world, the U.S. Department of Defense has also become one of the world's largest handlers and disposers of toxic materials.

Everything within the military generates waste: the ships, planes, tanks, rocket launchers, barracks, maintenance yards, and storage areas generate solid and liquid hazardous waste — and sometimes radioactive waste. There are toxics unique to the military, such as propellant packs, explosive shells, other ordnance, obsolete chemical and biological weapons, and radioactive materials. But the bulk of the hazardous waste produced by the Pentagon —though poisonous — is rather mundane.

The military's waste is mostly industrial and health-related. Paints, petroleum products, and solvents are used mainly at air bases; electroplating acids and heavy metals are used primarily at shipyards. Throughout the military, most waste consists of PCBs from old transformers, asbestos in old buildings and ships, herbicides, pesticides, and infectious waste like syringes from hospitals.

Military installations produce large amounts of waste through their industrial activities. According to a 1986 report of the U.S. General Accounting Office, defense installations generate 500,000 tons of effluents from degreasers, heavy metals, pesticides, PCBs from old transformers, asbestos, and fuels every year — not including medical waste, inert ordnance, or radioactive materials.

The problem is particularly acute in the San Francisco Bay Area. Favored for its large, sheltered harbor, central location on the western coastline, and cosmospolitan culture, the Bay Area has for the past hundred years been a major center for U.S. military operations.

According to the 1989 Defense Environmental Restoration Program's annual report to Congress, the 25 military installations in the nine Bay Area counties have a combined total of 270 potential toxic hot spots. Of these, 171 have been cited for further study.

Records from the Bay Area's local military installations indicate that potent carcinogens like trichloroethylene and poisonous chemicals like mercury and lead were routinely dumped on-site at the installations themselves, particularly before the 1970s, when environmental regulation came into vogue.

In 1986 it became clear that the concentration of military activities in the Bay Area had caused serious environmental consequences, when the General Accounting Office released its report entitled "The Department of Defense's Efforts to Improve Hazardous Waste Management." In that report, the GAO specifically selected two Bay Area Navy facilities —Mare Island Navy Shipyard and the Naval Air Station in Alameda — as examples of installations severely out of compliance with the federal Resource Conservation and Recovery Act, the hazardous-waste-management law.

The chart on the next page is the most complete list of the Bay Area's contaminated military facilities to date.

Alameda's West Beach Landfill is a good example of the types of waste sites under investigation at military installations around the Bay. Operated from 1952 through the late 1970s, the site received municipal waste, as well as water from Oak Knoll Naval Hospital, Naval Supply Center Oakland, and Treasure Island. It is estimated that up to 500,000 tons of hazardous waste was dumped there. The contaminants include solvents, oily wastes and sludges, paint waste, strippers, thinners, PCBs, acids, plating wastes, industrial strippers and cleaners like trichloroethylene, mercury, low-level radiological wastes, inert ordnance, asbestos, pesticides, tear-gas agents, and infectious wastes.

Installations with an industrial or maintenance focus, such as air fields like Alameda, Moffett Field and Travis, munitions depots like Concord, Sharpe and Tracy, and shipyards like Hunters Point and Mare Island tend to pose the most serious hazards.

In the Defense Authorization Report for the fiscal year 1990, the Department of Defense offered a useful analogy when it likened the military's environmental impacts to those of cities due to the logistical complexity of handling the sheer quantity of wastes produced.

Cities are by far the largest single source of environmental contamination. Like cities, military installations frequently encompass a variety of industrial activities, employ a commuter workforce, and maintain residential housing. Also like cities, defense installations produce large amounts of sewage, garbage, and industrial pollution. Unfortunately, the report goes on to say, the environmental performance of the military — despite the reputed military ability to run a tight ship — is worse than the chaos of local government, more than half the time.

But the Pentagon's hazardous-waste problem persists, not simply because cleaning it up is an enormous task. The Arms Control Research Center has learned some very disturbing facts about the DOD's attitude toward its environmental responsibilities—an attitude important sections of Congress seem to share.

For example, the Pentagon has been investigating military contamination since 1980. Meanwhile, the number of toxic sites identified in the U.S. by DOD has grown from 3,526 in 1986 to 8,139 in 1988, a 230 percent increase in just two years. In fact, 60 military installations were added to the Superfund's National Priority List in 1989 alone.

Why have so many new hot spots been discovered in a program begun nine years ago? DOD says the increase is the result of the remedial investigations the Pentagon commissioned to follow up its initial finding. But the experience of the Arms Control Research Center and other organizations active in this field doesn't bear out this claim.

Consider the Presidio. In a 1980 letter to the Environmental Protection Agency, the Presidio's facilities engineer reported that "this installation does not transport, treat, store or dispose of hazardous waste." Despite this adamant denial, inspections of the Presidio by the California Department of Health Services in 1981 and 1982 found, among other things, PCB-laden transformers and drums of waste chemicals improperly stored on site. In 1982, the DHS inspectors cited the Presidio for 14 violations of state hazardous-waste control laws.

In 1983, the Army completed its initial investigation of toxic-waste contamination at the Presidio. Although that investigation identified four con-

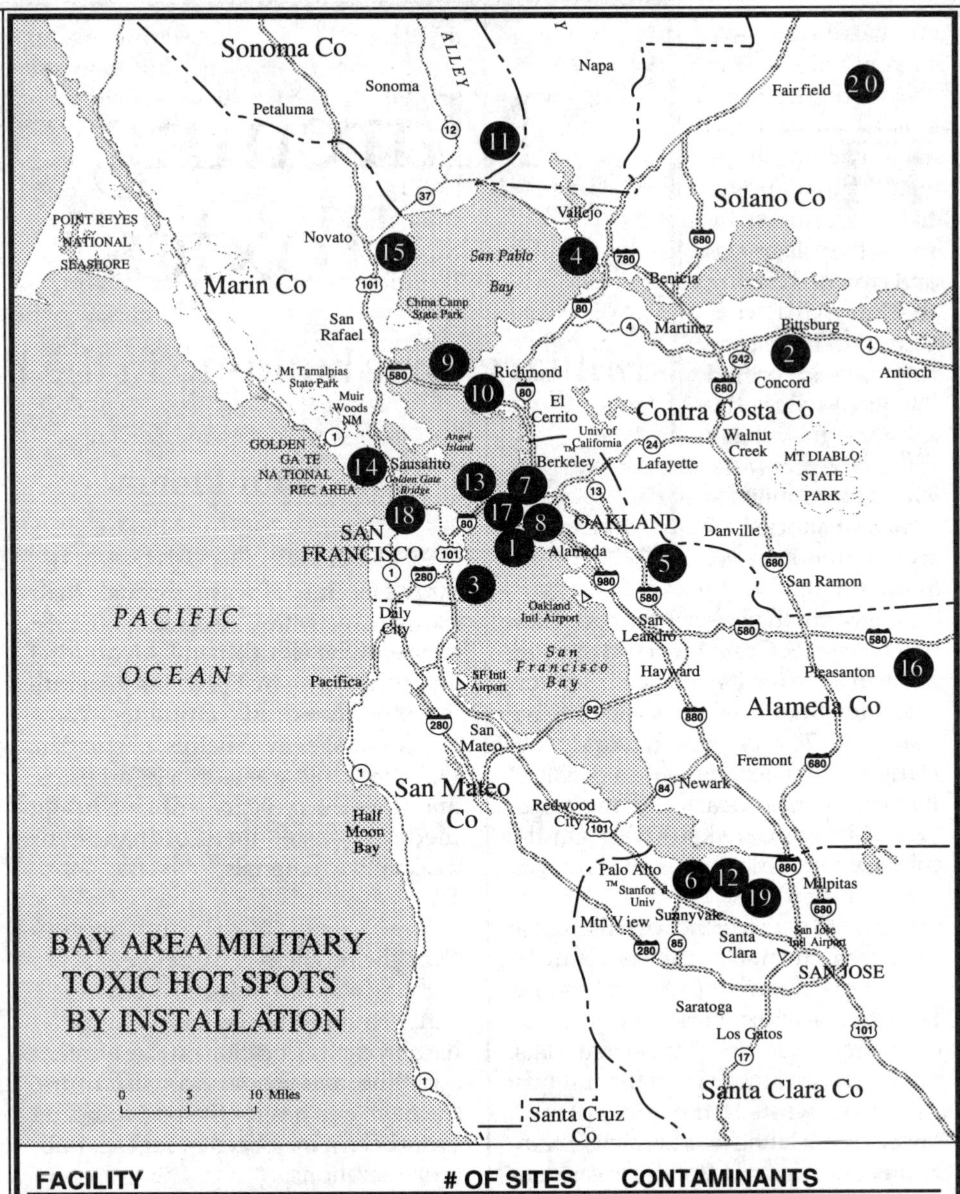

FACILITY	# OF SITES	CONTAMINANTS
1. Alameda Naval Air Station	22	PFOLS, Metals
2. Concord Naval Weapons Station	24	Metals, PFOLS, EW
3. Hunters Point Naval Stock Yard	17	PFOLS, Metals, Asb
4. Mare Island Naval Stock Yard	24	Metals, PCB, Solvents, EW
5. Oakland Naval Marine Center	1	
6. Moffett Field Naval Air Station	23	PFOLS, PCB, Toxic Chem.
7. Naval Public Works Center		
8. Oakland Naval Supply Center	6	PFOLS, PCB, Metals, Asb
9. Point Molate Field Depot		
10 Richmond	2	
11.Skaggs Island Naval Supply Center	1	
12. NIROP, Sunnyvale	3	
13. Treasure Island Naval Station	22	PFOLS, Paint, Asb, Org., Met.
14. Fort Cronkhite	1	Waste Explosives
15. Hamilton Air Field	7	PFOLS, Metals, Arsenic
16. Lawrence Livermore Laboratory	1	
17. Oakland Army Base	1	
18. Presidio Army Base	2	PFOLS, PCB, Asbestos
19. Sunnyvale Air Force Station	5	PFOLS, PCB
20. Travis Air Force Base	26	PFOLS, PCB

ABBREVIATIONS: PFOLS = Petroleum, Fuels, Oils, Solvents; PCB = Polychlorinated Bi-phenyl; EW = Explosive Wastes, Asb = Asbestos; Metals = Toxic metals, Org = Organics

taminated sites —including one major solvent spill — it took the Army six years to begin a remedial cleanup investigation for the Presidio. The Army is now compiling the research done in that investigation, and has identified at least eight toxic sites on the facility. The spill — 3,000 gallons of oily solvents that included the carcinogenic chemical benzene, at 2,200 times acceptable EPA levels — has now entered the Presidio's groundwater system and is headed toward the Bay. It has been awaiting a thorough cleanup since 1981.

A major factor in the Pentagon's attitude toward its toxic-waste problem is its obsession with secrecy and its perception that there is a conflict between environmental and national security. Because virtually all of a modern military force's equipment, ordnance and maintenance activities employ some form of toxic or hazardous material, it goes against the grain of the armed services even to coexist, let alone cooperate, with environmental regulation and the public's right-to-know that accompanies regulation.

In the past, the typical DOD response to criticism has been to argue that, while it strives to obey environmental law, in the final analysis, national security puts the military above the law. This attitude continues to be a problem.

Although congressional action can be credited with forcing the military into greater compliance with environmental law, the DOD's attitude problem has been reinforced by the Armed Services Committees of both houses of Congress, whose members are concerned that the costs of cleanup will take resources from other military priorities.

In the 1990-91 House Armed Services Committee report on defense authorizations, the Committee's staff compares DOD's environmental efforts favorably with those of industry. This argument sadly illustrates the ethical malaise afflicting Washington D.C. these days, and reveals a basic lack of understanding of the fundamental difference between government and private industry. Unlike industry, the military and Congress are public trusts whose first and only responsibility is the public good.

Solving the DOD's environmental problems won't come cheaply. The federal government plans to spend $500 million of its Defense Environmental Restoration Account and another $500 million from accumulated other DOD accounts (such as Operations and Maintenance, Construction and Procurements) this year to clean up these faciltiies. Between 1984 and 2008 the DOD will spend a total of nearly $34 billion for environmental restoration and protection.

If you think this is a lot of money, you're absolutely right—but the question is: Is it enough money to do the job? Unfortunately, it's not.

Dividing that $34 billion by the number of installations with toxic waste hot spots, the result comes to $38 million per installation. But the Navy estimates it will take $100 million to clean up the toxic contaminants at Hunters Point alone. Of course, Hunters Point is a Superfund site, with severe problems by definition. But even for the Presidio, which, despite some big problems, is not considered severely contaminated, the cleanup bill is estimated at about $80 million.

Defense advocates in Congress are not willing to take the necessary millions from unproven multi-billion dollar weapons systems. As a result, millions of Americans face a daily threat from groundwater contaminated by the military. The entire expected DOD expenditure for cleaning and protecting the environment amounts to 0.3 percent of the total Defense budget ($250 billion) for 1989.

Congress would have to spend a lot more money to protect the health of Americans from the military forces created to defend us. But Congress may be considering just the opposite. The professional staff of the House Armed Services Committee stated in its report on the National Defense Authorization Act of 1990-91: "...it might also be appropriate to take a hard look at environmental requirements to determine whether they should be modified to take into consideration the peculiar needs of the defense industrial base"

There is a myth of a sword so beautiful and so lethal that its mere possession was sufficient to poison its owner over time.

Over the past 27 years, our government has built a military force on the assumption that at any moment the world could be destroyed with the simple turning of two keys. In the madness of such a time, every living thing is at risk from moment to moment, and as a result most living things become acceptable losses if their deaths can be brokered against Armagedon.

The government enlisted the public in the Cold War. Some were asked to die in Korea, some in Vietnam. But the military's contamination of the environment made the Cold War a kind of ultimate war for the TV generation. Without dog tags or the inconvenience of basic training, the government has enabled the public to make the ultimate sacrifice without having to leave the living room.

Now that the Berlin Wall has opened and pundits have declared an end to the Cold War, Congress needs to reassess the effects of a large military on the public and the environment and:

- Undertake a series of public hearings in heavily militarized communities across the nation to determine the local environmental impact of the defense burden.

- Provide the necessary financing for installation cleanups and environmental protection.

- Direct the Department of Defense to undertake a comprehensive interdepartmental regional analysis of the military's environmental impact on the Bay Area. This Environmental Impact Statement should be in compliance with the National Environmental Policy Act, and provide adequate opportunity for public input.

- Once the Department of Defense's regional EIS has been released, hold local hearings to determine if any congressional action is needed to ensure adequate protection of the Bay Area and its residents.

Saul Bloom is a freelance writer living in San Francisco. This story was reprinted with permission from the San Francisco Bay Guardian.

Nukes Take to the Streets

By Saul Bloom

The collapse of the Cypress structure and the closure of a section of I-880 in the Loma Prieta quake of 1989 created a nightmare of traffic, rubble and construction noise for the West Oakland neighborhood surrounding the crumbled freeway.

But the neighborhood faces what could be a far more serious problem — one that has attracted almost no public attention and remains shrouded in official secrecy.

Interstate 880 was a major transportation route for nuclear weapons and hazardous military and civilian nuclear waste-and with a key section of the highway closed, that deadly material may be routed through the streets of West Oakland.

Since federal regulations allow nuclear transporters to change their routes by significant amounts without public notice or hearings, the residents of the neighborhood may never be told that some of the most dangerous cargo in the world is passing by their houses in the backs of non-descript commercial trucks.

West Oakland isn't the only area threatened by the possibility of a nuclear-transportation accident. Nuclear material is regularly shipped along several major highways in the Bay Area, and in many cases those roads are in bad repair, are susceptible to major earthquake damage, and run right through the middle of some of the region's most densely populated areas.

Evidence that has come to light in part through the federal court battle over Oakland's nuclear-free zone ordinance reveals that:

- More than 20 public and private facilities in the San Francisco Bay Area regularly handle and transport radioactive materials. The shipments range from radioactive medical supplies to fissionable weapons components and quite possibly nuclear bombs.
- The ports of Oakland and Richmond are major transfer points for radioactive materials shipments bound for the U.S. and abroad. The major Bay Area routes for radioactive materials transport are I-80 from Richmond, I-880 from Oakland, I-580 from Livermore, and I-5.
- At public hearings in 1979, 1981 and 1985, Bay Area residents and elected officials complained repeatedly that the highways used for transporting radioactive materials, including the downed Cypress overpass, might not be safe. The collapse of the Cypress structure confirmed those fears, but there is no indication that the federal government is making an effort to reconsider its nuclear transportation plan for the Bay Area.

Instead, the government went to court and prevented Oakland from banning nuclear materials transportation within city borders.

The Bay Area's link to the transportation of nuclear material goes all the way back to the beginning of the nuclear age. One of the first known shipments of an atomic weapon passed through the East Bay to the Hunters Point Navy Yard in 1945. The purpose was to load the atomic core of the Hiroshima bomb onto a ship bound for the Pacific.

From 1946 through the early 1970s, radioactive waste from decontaminated observer ships of U.S. nuclear tests in the Pacific and the research facilities at Livermore Labs was stored at the Hunters Point Navy Yard awaiting disposal at the Farallones Islands, just off-shore of San Francisco's Ocean Beach.

By the early 1970s, hazardous materials — explosives, solvents, chlorine gas and the like — were crisscrossing and crashing on America's highways at an alarming rate. Poorly maintained trucks carrying deadly cargos, unsupervised by any of a baffling array of federal and state agencies that theoretically had jurisdiciton over them, were spilling their contents into people's backyards.

Poorly trained police and emergency service crews frequently responded to these accidents without the slightest idea of the potential stew they were getting into.

In 1976, Congress responded to chemical and trucking industry pressure by passing the Hazardous Materials Transportation Act, which gave the federal government (specifically, the Department of Transportation) the authority to pre-empt all local ordinances regulating hazardous material transport. The act also created a unified code setting standards for moving the dangerous stuff.

Some transport safety activists at the time called it "The Lowest Common Denominator Act." It certainly didn't eliminate the danger. Hazardous materials continued to flow freely over our highways and through our neighborhoods.

Interestingly, however, the drafters of the hazardous materials law omitted any mention of radioactive materials transportation—and with this loophole began America's first nuclear-free zone movement.

From Solano to Santa Clara, a total of 22 public and private facilities handle and transport radioactive materials. Some of the material has low-level radioactivity or is the by-product of scientific experimentation. Some contain deadly plutonium in 16-foot-long rods of highly radioactive reactor fuel.

Some shipments contain highly radioactive, but relatively short-lived isotopes used, among other things, for medical research and treatment. (Nuclear medicine, by the way, has never been banned in any nuclear-free zone ordinance).

The routes are designed by the Department of Transportation, with input from the California Highway Patrol, local elected officials, industry, fire marshals and, occasionally, the public. The CHP administers the state process.

Radioactive materials shipments come in all shapes and sizes. Commercial shipments are posted with identifying diamond-shaped signs mandated by the Department of Transportation. Department of Energy shipments are occasionally marked in that manner, when the cargo is not military or classified. Department of Defense radioactive materials shipments are unmarked. The most dangerous commercial nuclear material—spent fuel from nuclear reactors—is shipped in 25-ton lead-and-steel containers. Because these shipments are thermally as well as radioactively hot, the containers are frequently refrigerated.

These container "casks" come in two forms. One looks a bit like a dumbbell, with two large impact rings on either side. The other resembles a large corrugated pipe.

Shipments of nuclear weapons, on the other hand, use large, unmarked tractor-trailers, accompanied by a military police escort. What has given these shipments away to anti-nuclear transport activists is the forest of antennae sprouting from the roofs of the truck cab and the accompanying escort vehicles.

According to Sergeant Munyer of the CHP, all routes currently in use in the Golden State follow the DOT procedure of using interstates as the primary conduit. The problem is, many of the Bay Area's interstates don't directly connect with the potential points of pick up and departure. So federal regulations allow the use of surface streets for moving between highways.

Advance notice of some kinds of shipments—fissionable material and extremely radioactive waste—is provided to the local fire marshal, if requested, and made public only after the shipment has taken place.

In the course of its lawsuit against Oakland, seeking to overturn the nuclear free zone, the federal government made some significant admissions about the Bay Area's radioactive transportation routes. The information is the most extensive available to date of the roads, highways, waterways and railroad tracks that carry nuclear material through the Bay Area.

For the most part, the nuke haulers stick to the major highways. The central routes include I-80 from Richmond, I-880 south to San Lorenzo, Route 238 east to Livermore and I-680 south to Sunol.

The government never acknowledges the routes used by military transport vehicles hauling nuclear weapons or weapon components. But based on the locations of the major local weapons stations and the material those bases almost certainly send to other specific local bases, it's possible to put together a reasonable picture of the unofficial, secret nuclear transport routes. They include highways 4 and 24, several surface streets in San Francisco and Oakland and quite possibly the Bay Bridge.

Since the collapse of I-880, there's a new element in the picture. The I-880 section that collapsed and that may now be closed for as much as two years was one of the key links in the Bay Area nuclear transport system. With the highway closed, the trucks have to detour through surface streets—almost certainly in West Oakland. Under federal law, deviations of as much as 20 percent of the total trip mileage are allowed without public notice or hearing. So in many cases, the nuclear shipments are taking place without the knowledge of neighborhood residents.

The federal government's refusal to give the public meaningful input into nuclear transportation decisions was among the key factors that led to the growth of the nuclear-free zone movement across the nation. Marin County, San Francisco and Oakland are now among the 168 cities, towns and counties that have passed local ordinances regulating or banning the use, manufacture or shipment of nuclear materials.

The heart of the lawsuit by the Department of Justice to overturn Oakland's nuclear-free zone law was the federal government's contention that it has the right to pre-empt local law and subject the citizens of any locality to the dangers of nuclear materials. The government won the case, and despite vociferous calls for the city to appeal the decision, Oakland's city council has so far demurred, arguing that an appeal would be too costly.

In essence, the ability of Americans to choose their risks and to reject those they find unacceptable was on trial, and the people lost. Even the dangers so eloquently demonstrated by the Loma Prieta earthquake ultimately held no weight.

The truth is simple: No roadway is really safe for the transportation through populated areas of material that is so deadly that even a teaspoonful could easily kill thousands.

Maybe some residents of the Bay Area believe the wonders of nuclear technology — and the supposed security provided by nuclear weapons—are important enough to be worth that risk. But we have repeatedly shown at the polls that the majority don't, and the fact remains that the federal government won't let the citizens make that call.

Saul Bloom is a freelance writer living in San Francisco. This article was reprinted with permission from the San Francisco Bay Guardian.

Resources

The following organizations work on nuclear issues in the Bay Area:

Alameda County Nuclear Free Zone
(415) 843-8143

SONOMore Atomics
(707) 526-7220

Abalone Alliance
(415) 861-0592

Center for Economic Conversion
(415) 968-1126

Greenpeace
(415) 512-9025

Silicon Valley Toxics Coalition
(408) 286-0315

Highest Disregard

Silicon Valley and the Ozone Hole

by R. Dennis Hayes

F. Sherwood Rowland co-authored one of the most significant scientific findings of this century. His finding was troubling. Very troubling. But it came early enough to allow us a chance to head off a disaster of global proportions. Rowland discovered, back in 1973, that chlorofluorocarbons (CFCs) shred stratospheric ozone, our planet's delicate shield against harmful solar radiation.

In the sixteen years since, Rowland's peers have honored his ozone work profusely; most recently he was awarded the prestigious Japan Prize for Environmental Science and Technology. But in that same time the threat to the ozone layer has grown bigger than ever. Over Antartica, a hole in the ozone the size of the United States is widening. If it keeps widening, if the world's ozone layer becomes too thin, the result could be more skin cancer and cataracts, weaker immune systems, and the obliteration of many animal species. CFCs also speed up global warming and its projected effects: melting ice caps, flooded shores, dwindling forests, more smog.

And so Rowland the atmospheric chemist has become a kind of ozone detective, gathering air samples from all over the world, searching for clues that can help measure the real damage, tracking down the worst ozone offenders where they live. One day in May of 1989, Rowland stood in his laboratory at the University of California at Irvine, sorting through air sample results. He inspected a graph showing the molecular footprints of CFCs in air from Samoa: a negligible bump. From Alaska's Point Barrow: a slightly bigger bump. From Amsterdam: a bigger bump again. Then Rowland came across a steep spike that dwarfed all the rest, on a graph bearing the label Santa Clara, California. Rowland held in his hand clear evidence that Silicon Valley — the design and development center for the Information Age — may be the world's leading hot spot for the ozone-shredding solvent called CFC-113.

"We weren't surprised," says Rowland. He knows that the electronics industry is the leading user of solvent CFCs. (The refrigeration and foam industries lead in nonsolvent CFC use.) He also knows that unlike older CFC users, U.S. electronics firms began using CFCs long after Rowland's 1973 discovery, after it was known that CFCs attacked the ozone, after CFCs had been outlawed in aerosol sprays. As a result, the "industry of the future" more than erased progress made by the aerosol ban in the 1970s. There is strong evidence, in fact, that the electronics industry had the knowledge and means to help lead us away from ozone destruction, but chose not to.

Today, the U.S. electronics industry acknowledges that CFCs harm the atmosphere. But behind the scenes it works to defeat and dilute local, state, and federal CFC-phaseout legislation. In doing so, it increases the odds that high technology's future will unfold under different, dangerous skies.

"It is quite common on the scientific side of industry to believe that there aren't any real environmental problems, that there are just public relations problems," observes Rowland over the creaking of his desk chair in his fifth-floor UC Irvine office. His brow furrows when he recollects the puzzling chain of events his 1973 discovery set in motion.

In 1975, a fourteen agency U.S. task force suggested it may be necessary to regulate CFCs in aerosols, then the largest use for CFCs. In 1976, a U.S. National Academy of Sciences report confirmed the need for regulation. By then, the debate had spilled out of the forums of science into the chambers of state and the boardrooms of industry.

In 1976, the EPA and FDA announced the aerosol CFC ban, effective 1978. In the interim, consumers boycotted aerosol spray products. Most of the U.S. aerosol industry, after predicting ruin, got out of CFCs by 1977, "swiftly and smoothly, " recalls Rowland. DuPont, then as now the largest CFC producer, soon reported that substitutes for other, non-aerosol CFC applications were a few years away. So concluded Phase 1 of the regulation. Phase 2, a closeout for remaining CFCs, was next on the agenda. But then the momentum to regulate CFCs dissolved under the corrosive influence of the Reagan EPA and lobbying by the chemical and refrigeration industries.

EPA administrator Anne Gorsuch discounted ozone depletion in 1981 as just another environmental scare issue; as late as 1985, Interior Secretary Donald Hodel proposed hats, tanning oil, and sunglasses as an alternative to CFC regulation. Scientists following up on Rowland's work made the policy reversal possible by injecting uncertainty into ozone depletion estimates. Their findings never challenged Rowland's hypothesis, but rather his estimate that CFCs eventually would destroy 7 to 13 percent of the ozone layer. As it turned out, those critics who had placed the figure lower had based their argument on computer models soon recognized as flawed. As early as 1976, the scientific consensus had re-formed in support of Rowland, and in 1979 the

National Academy of Sciences published a report that boosted Rowland's estimates to an eventual 15 to 18 percent ozone loss.

The most serious challenge to Rowland's scenario was short-lived, but it lasted long enough for corporate lobbyists and obliging government officials to redefine the decisive debates over the next step in CFC regulation. As the Phase 2 total ban on CFCs evaporated, Du Pont scaled back plans to bring CFC substitutes on-line, and an ecological crisis was reincarnated as a public relations problem.

Rowland reflects somberly on the logic that prevailed during the Phase 2 debates. "If there are any uncertainties, then the almost universal assumption is that those uncertainties will, of course, work in the direction of making [environmental] problems less important. The assumption is that the person who is raising the problem is exaggerating. In this case, it turns out we weren't exaggerating."

If the comprehensive CFC phaseout Rowland's discovery initiated had won the day, the United States, the world's largest producer and consumer of CFCs, might have been within spitting distance of zero dependency by 1984. Instead, a new stage of CFC production led by the United States, much of it for the new U.S. electronics market, had surpassed the rolled-back levels achieved by the 1976 aerosol ban.

The electronics industry had very little to do with derailing CFC regulation during the 1979-81 period. That's because U.S. electronics had only just begun to get into CFCs in a big way. Since 1979, CFCs have become "critical elements in the manufacture of all types of electronic equipment," Terry McManus, a manager at chip maker Intel told the U.S. House Oversight and Investigations Committee this year.

Representing the American Electronics Association (AEA), McManus explained how his industry "came to rely on CFCs as the chemical of choice in many operations and built whole new technologies based on the compounds' special properties. CFCs are now critical in one hundred to two hundred different electronics applications."

Most of those applications involve cleaning, where engineers have found use after use for a solvent so perfectly gaseous and compatible that it penetrates microsludge in the tiniest crevices of chips, disk heads, and circuit boards and then evaporates.

CFCs' ability to clean in small places helped make big fortunes. During the 1980s, microdevices shrunk in size every year. Miniaturization was the key to faster processing speeds, and a faster product could render the competition instantly obsolete. That's why the industry hustled, at great expense, to redesign its product lines and retool its processing technologies for use of CFC solvents throughout the 1980s. As it did so, it dismantled ozone-safe water-and-detergent-based solvent approaches widely used in the 1970s. It also chose to ignore an atmospherically clean "organic" solvent derived from citrus rinds and wood pulp that is now, ten years later, proving a more effective circuit cleaner than CFCs.

The CFCs in refrigerators and air conditioners are contained by plumbing until junked or refurbished. By contrast, in electronics, CFCs' primary mission is to clean and evaporate. Throughout the 1980s venting CFCs into the atmosphere was a standard electronics industry practice. Only recently, as a result of EPA regulation, are most electronics firms beginning to install emission traps for CFCs.

All along, the Pentagon has discouraged recycling the chemical by specifying that its contractors use "virgin CFC." Indeed, a co-dependent in the electronics industry's plunge into CFC addiction was the Pentagon, which still insists on using CFCs in an array of electronic product specs. This sent a message to high technology firms the world over: to simultaneously qualify your commercial products for U.S. military acquisition, a huge market for electronics, use CFCs. "As much as 40 to 50 percent of CFCs in electronics today is driven by military requirements," claims Joseph Felty, a process engineering manager with longtime Pentagon prime contractor Texas Instruments. In a recent trade magazine report, he explained that "what happened was, the [military] spec [for CFCs] became a de facto world standard. It's a badge of quality."

If anyone had been in a position to know about the characteristics of CFCs, it was the U.S. electronics industry, with its brain trust of top chemists and engineers. Yet it discounted Rowland's well-respected discovery that CFCs deplete the ozone. Then it ignored a popular political movement to ban CFCs.

"It's almost as if they existed on another planet," Rowland notes, recalling the questions he took from electronics industry people after addressing a Silicon Valley conference called by the EPA and AEA last February to discuss the industry's threat to the ozone. "It is clear that a large fraction of the problem we have with CFCs now is with companies that expanded their major uses enormously during the time period in which it was known to be a danger. It has to be either massive ignorance that they were involved with CFCs, " says Rowland, "or they didn't believe there were any real environmental problems."

Those at the top are still incorporating CFCs into their next-generation technologies. Sematech is the government-subsidized industry consortium charged with, as its press kit puts it, "reclaiming worldwide semiconductor manufacturing leadership." Sematech's Austin facilities this year vented hundreds of gallons of CFCs into the atmosphere. When asked to comment, Sematech spokeswoman Ann Marett said, "Well, isn't everybody using CFC?" Another spokesperson called back to say Sematec was planning to scale back its ventilation, if not its use, of CFCs in 1990.

"We are, without doubt, the biggest electronics user of CFCs in the world," said an IBM executive in the spring of 1989. IBM uses over forty-four thousand pounds of CFC-113 every workday. IBM's Endicott, New York, facility is the United States' single largest industrial source of ozone-depleting chemicals. IBM's San Jose plant is not only the biggest CFC-113 user in Silicon Valley, but in all of California.

Where "Big Blue" goes, its competitors are inclined to follow. IBM led its industry to CFC-113 and another solvent, methyl chloroform (TCA), in the late 1970s and early 1980s. (TCA shreds the ozone, although not as prolifically as CFC-113, and is a significant greenhouse gas.) The computer giant liked the economics of CFCs, but the chemical held other attractions as well. Evidence suggested trichlorethylene (TCE), a solvent used extensively by IBM and the industry during the 1970s, was carcinogenic. TCE's successor, TCA, was soon implicated as a reproductive health hazard. In 1982, TCA solvent leaked from IBM's storage tanks into a San Jose neighborhood's drinking water and a sharp rise in birth defects occurred locally. In 1986, IBM settled a class action suit brought by the neighborhood residents, keeping the terms secret. IBM still denies that its chemicals helped cause the defects, despite a corroborating follow-up study. And IBM continues to use TCA, though the company uses over twice as much CFC solvent.

The company says its big switch to CFCs was motivated by the fact that CFCs are less toxic to workers than TCE. "IBM's development of CFC uses in and before the early eighties was fueled by the desire to use the safest process for our employees," reads a company statement. (IBM would not grant a face-to-face interview with those who pilot its CFC policy, insisting instead on providing written answers to written questions.)

Didn't turning to CFCs invite danger of another sort? IBM writes that from 1973 to 1986, "calculated estimates of the seriousness of ozone depletion were constantly changing , and we could see no clear consensus that made it necessary to eliminate CFC use."

After Rowland's initial discovery, IBM explains, "estimates of total stratospheric ozone depletion due to CFCs fluctuated between 7 and 13 percent in the period between 1974 and 1978." Indeed, this was the range Rowland originally published in 1974.

"Between 1979 and 1983," IBM continues, "changes in model data input reduced estimation of total ozone loss to 5 percent. By the end of 1983, calculations ranged from 4.2 percent depletion down to even smaller lost under some scenarios."

IBM leaves out key context. The 4 to 5 percent rates that it cites were the product of a computer model developed at Lawrence Livermore Laboratory, that was soon proven incomplete by more accurate models, as well as by evidence showing more rapid ozone loss rates. As Rowland points out in the very paper IBM cites, the surest sign of the Livermore model's flaw was the wide range of estimates it produced; in 1984 other experiments using the Livermore model placed the steady ozone depletion at 24 percent and 31.7 percent.

In this debate, IBM chooses to rest on temporary, selective views for support, pointing to technical detours like the Livermore model as evidence that the ozone debate has never been free of scientific uncertainty—even though absolute certainty, or full consensus, is almost impossible to achieve in science.

But if there is one fact about the ozone that almost all scientists have agreed on since Rowland's first discovery, it is that ozone in the most sensitive, upper layer of the stratosphere is being seriously eroded by CFCs. Data indicating the eventual loss of up to half the ozone in the upper stratosphere has never been sigificantly called into question; the limited debate IBM points to has to do with estimates of "overall" ozone depletion rates from 1979 to 1984.

While the distinction might be a bit confusing to a reader new to ozone issues, it wouldn't be to any Silicon Valley scientist up to speed on the properties of CFCs and the atmosphere. And yet, IBM's official version makes it seem as if the company was the last to be informed. The spring 1989 issue of Visions, an IBM magazine for employees, reports that data presented in 1988 "gave the first convincing evidence that CFCs were depleting the world's protective ozone layer, which might allow more of the sun's ultraviolet rays to reach the earth's surface."

Why does IBM still equivocate that ozone depletion "might" allow more of the sun's ultraviolet rays to reach the earth's surface? Is there any question today? The company's answer: "Because of the role of clouds, atmospheric particulates, and ozone in the lower atmosphere, the magnitude of changes in UV radiation at the earth's surface due to stratospheric ozone depletion is not at all certain. We agree that the prudent course is to assume that UV radiation will likely increase if the stratospheric ozone is seriously depleted, but the future extent of the changes is not known."

The American Electronics Association, like IBM, prefers to write its own history of CFCs and the ozone, clinging just as tenaciously to the concept of "scientific uncertainty" as a defense. In testimony before a Senate Environmental Protection Subcommittee in May 1989, AEA Director of Environment and Occupational Health Cheryl Russell said that when U.S. electronics firms switched to CFC solvents, a date the AEA identifies as 1979, CFCs "were believed to be environmentally benign, neither contributing to air pollution nor to any hazardous waste problems."

It was in 1979 that the National Academy of Sciences published its 15 to 18 percent ozone loss rate estimate, a range well above Rowland's original 7 to 13 percent estimate that prompted the aerosol ban. As Rowland observes, "The now-widespread use of CFC-113 for cleaning electronic components has been developed almost entirely since the bans on CFC-11 and CFC-12 as aerosol propellants, even though it has been obvious all along that all three CFC molecules are roughly equivalent in their ability to deplete stratospheric ozone."

IBM now claims to be leading efforts to reduce CFC consumption. "We believe we will achieve reductions of between 40 and 50 percent by the end of 1989," the company writes. Much of IBM's effort involves trapping and controlling CFC emissions, a short-term approach since it prolongs dependence on CFCs and courts accidental emissions. Already, IBM has reported leaks in the ducts of its current CFC-solvent recovery system.

Solid progress would be the development of solvent processes that don't use CFCs. In 1988, IBM San Jose prominently announced that it was experimenting with a water-based cleaning process that could replace up to 30 percent of its CFC use. The "new" process involved a mild detergent called Triton X-100 and an ultrasound drying technique. "The entire project went from conception to operation in less than two years," and it could be operational by 1990, the company magazine reports. That would make a total of three years in development. However, a retired IBM production worker says that IBM used a very similar water-based Triton X-100 ultrasound process in 1979, before IBM switched to CFCs.

When pressed on this, IBM wrote, "You are correct, the 'new' way is a return to the 'old' in a sense, but that still means a substantial investment in time, engineering skill, equipment design, and effort to prove that the process can work for our specific application in a manufacturing environment." IBM was saying, in effect, that it requires three years to redevelop an already proven technology in widespread use ten years ago.

Why are IBM and the AEA at pains to revise the CFC saga? One answer is that they hope their version of history will yield credibility and goodwill in their politcal battles against proposed CFC legislation today.

By 1987 it was hard to find anyone, outside of the electronics industry, who would deny that the planet's stratosphere was in bad shape, that CFCs were a culprit, and that something drastic had to be done about it. That year officials from around the world gathered in Canada to sign the Montreal Protocol on Substances that Deplete the Ozone Layer, an international pledge to phase out said chemicals. Today, the U.S. electronics industry is in the awkward position of publicly supporting the Montreal Protocol process but, behind the scenes, lobbying against legislation that seeks to implement the current Montreal Protocol agreement — a total, global CFC phaseout.

The measure with the most teeth, Senate Bill 491, would phase out CFCs, along with TCA, by 1997, with exemptions for "national security" and "medical" use. IBM and the AEA have many objections to SB 491, just as they object to virtually every city, county, and state initiative seeking to infringe on CFC solvent use.

"Trying to reduce CFC emissions in the U.S. on a unilateral basis will have a negligible effect on ozone depletion," asserted the AEA's Cheryl Russell in her Senate testimony against provisions of SB 491.

Russell argues that, until the year 2000 and possibly longer, CFCs must remain available to U.S. electronics firms. This, the AEA insists, will keep the United States afloat in the high seas of global electronics, an imperfect but real world in which the U.S. electronics industry would be disadvantaged because of CFCs' presumed use by foreign competitors.

Rowland recalls hearing a similar argument in the 1970s. He calls it

Other Sources of Chlorofluorocarbons

Although the electronics industry is the biggest Bay Area culprit in venting CFCs into the atmosphere, the amount leaked from refrigerators and routinely released in the course of repairing auto or home air conditioners is considerable. These two sources alone contribute more than a quarter of the CFC releases in the United States, according to EPA estimates (see the accompanying graph).

The coolant used in virtually all cooling equipment is composed of CFCs. It is contained in sealed systems, and thus is not a threat to the ozone layer until the system either leaks or the coolant is released during repair.

CFC recycling units, popularly known as "vampires," that suck the used coolant out of the appliance's tubing, filter and clean it, and then re-inject it are currently available. At present, only large industrial refrigeration firms routinely use the vampires when overhauling equipment. Their expense — averaging about $3000 — has served as a deterrent to most garages and refrigeration repair companies. As one mechanic we interviewed put it, "I'm not going to spend that kind of money unless the law says I have to."

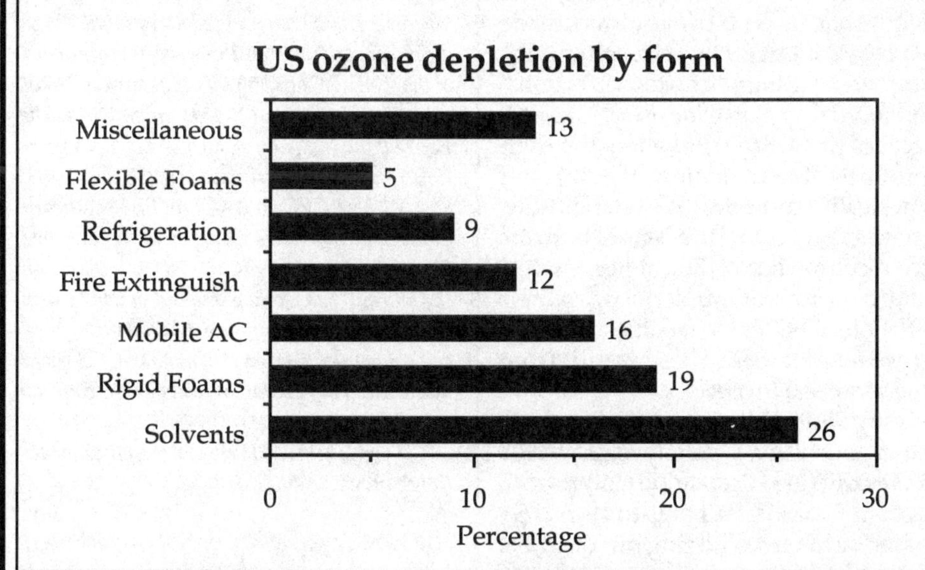

A bill currently before the state legislature proposes to do just that. But until such a law passes, when your auto air conditioner goes on the blink the chances are very high that the mechanic will simply disconnect the hose and let the CFCs bleed out. It is even more likely that nobody will reclaim the CFCs when a refrigerator is junked.

For this reason, we went to considerable trouble to ferret out repair shops that do use vampires. The next time your air conditioner needs repair, or if for some reason you must dispose of a refrigerator, please refer to the Consumer Guide to find someone who will do it with a vampire.

the Chicken Little episode. "Chicken Little" is what some called Rowland for suggesting "the sky was falling." But Rowland says the aerosol firms were the real Chicken Littles, predicting the collapse of their industry and U.S. jobs if CFC propellants were banned. The 1976 EPA-FDA regulation gave aerosol firms two years to get out of CFCs. Most were out in less, without the predicted losses. Couldn't the U.S. electronics industry do this today?

Current EPA regulations, framed in response to the 1987 Montreal Protocol, propose mere 50 percent reductions in CFCs by 1998. That timetable is fine with IBM and the AEA, but it has become unacceptable to the EPA and parties to the Montreal Protocol, as Rowland and his colleagues have weighed in with new evidence.

The Ozone Trends Panel, which included Rowland and over one hundred other scientists, was formed in 1986 by the National Aeronautics and Space Administration in collaboration with the National Oceanic and Atmospheric Administration, the Federal Aviation Administration, the World Meteorological Organization, and the United Nations Environment Program. After a "critical reanalysis" of all available ozone date, the panel found that stratospheric ozone levels over North America and Europe fell by up to 3 percent from 1969 to 1986. And it confirmed that the ozone hole over Antarctica, which lost up to 60 percent of its ozone, was the result of CFCs. In response the EPA acknowledged that, even with full global compliance with the 1987 Montreal Protocol, the concentration of ozone-depleting chemicals will at least double over the next eighty-seven years. Citing an "even worse scenario than anticipated," the EPA urged a faster global phaseout. That was in September 1988.

Once again, when the electronics industry had a chance to take a leading role in saving the ozone by acting on the latest information, it preferred to cling to the most favorable, if obsolete, finding of the past. The EPA has for over a year now been urging Congress to legislate a speeded-up CFC phaseout; IBM today still maintains that "additional legislation is not needed. We, like all other industries, are already under EPA regulations that are limiting the nation's supply of CFCs."

While the electronics industry protests that a faster phaseout would inflict unmanageable hardships on the industry, not all insiders agree. Nick Pasch, manager of special projects at chip maker LSI Logic, says, "Any federal regulation on compounds galvanizes the industry like lightning to pay attention...the industry can move very quickly to stop using materials." Northern Telecom, an IBM competitor, bears out Pasch's point. Northern Telecom received commendation from the EPA for cutting back CFC use by 50 percent in 1988-89 and for targeting a complete CFC phaseout by 1991. Northern Telecom has a large facility in Silicon Valley. But it is a Canadian-held company affected by Canada's aggressive, 100 percent phaseout of "electronic cleaning" CFCs in five years — by 1994. Sweden has scheduled a complete CFC solvent phaseout by December 31, 1990. But U.S. electronics firms continue to insist they will need a decade or more to drop their CFC habit.

"A global problem needs a global solution," Intel's Terry McManus told Congress recently, defending the American Electronics Association's opposition to Senate Bill 491 and other domestic CFC legislation. In fact, the purpose of the Montreal Protocol process is to encourage countries to implement global agreements through domestic legislation. And SB 491 does attempt a global solution, by including a clause that would ban imports of electronics made with CFCs. But for that clause, the AEA reserves its "single greatest objection." A thoroughly global industry, U.S. electronics firms now make most of their products abroad, mainly in Asian and Pacific Rim countries. As it stands now, there is no law to prevent U.S. firms from cutting back CFCs at home while shifting more manufacturing to their offshore, CFC-using facilities, many of them sited in countries exempt from the Montreal Protocol.

The AEA admits that firms in other nations may "simply replace CFC-113 with TCA," despite TCA's own suspected hazards. What is to stop U.S.-based firms from getting out of CFCs, only to make another bad technology choice? In July the Irvine City Council passed extremely restrictive controls on the use of chlorofluorocarbons and TCA. According to Jim Jenal, acting environmental program administrator and an author of the CFC controls, a Western Digital corporate representative "met with practically everyone on the city council to lobby against the TCA clause." Western Digital, as part of an effort to transcend its use of CFC-113, is building a $100 million plant in Irvine that will make use of TCA.

The continued use of TCA is a reminder that CFC regulation could prompt the U.S. electronics industry to further jeopardize the health of its nonunion, largely minority, and mainly female work force. Contrary to the AEA's testimony before Congress that the U.S. industry is "one of the safest," its semiconductor workers suffer systemic poisoning rates in excess of three times that of the manufacturing average.

The U.S. electronics industry's recent adoption of CFCs was the product of considered opinion among corporate scientists, the experts who remain at the helm of choosing new technology for us all. This suggests that our chances for survival may depend on changing the way we choose, and who has a voice in the decision.

This story was reprinted with permission from the December, 1989 edition of Mother Jones Magazine. *Since that time, IBM has kept its commitment to reduce CFC use in manufacturing, and has promised to phase out all use of CFCs worldwide by 1993, seven years ahead of the schedule called for by the Montreal Protocol. The most recent figures submitted to the EPA by the company show a 54 percent reduction of CFC113 use between 1988 and 1989.*

The CFC control law SB491, meanwhile, was passed by Congress in a weakened form as part of the 1990 Clean Air Act. Although the law still calls for a phase-out of CFC use in the U.S. by 1997, language disallowing American companies from simply moving CFC-based manufacturing operations outside the country was eliminated.

R. Dennis Hayes is a freelance writer living in San Francisco.

-Ed.

Chapter Ten

Solutions

There is no simple formula for unravelling the tangled mess we've gotten ourselves into. Our environmental problems are intimately interwoven with the way we live and think, and changing habits as old as we are — or as old as our culture is — is not going to be easy. Nonetheless, it is time to take the plunge.

We need to approach the problem on a number of levels. On the personal level we must think about the effect each of our actions will have on the big picture, and abandon the false impression, held by many of us, that we are somehow separate from the problem. We are all part of the problem.

Then we must devise ways of living better as communities. Besides improving the quality of life, community living in which resources and responsibilities are shared lightens our impact on the planet.

On a political level there are many battles to be fought. In the Bay Area we must fight, among other things, for coherent regional planning by an agency with the power to make and enforce decisions on issues that affect all of us.

The Eco-City Concept

By Susan E. Davis

In 1925, French architect Le Corbusier exalted the idea of the town as "working tool." Cities are an "assault on nature," he wrote proudly, a "human action against nature." As such, this "human organism designed for shelter and work" was a major triumph of geometry over natural disorder.

Sixty-five years later, on a stage bedecked with ferns, orchids, flowering photinia and chrysanthemums in a small auditorium in bucolic Berkeley, organizers of the first international "Eco-City Conference" called for a new vision. More than 800 architects, urban planners, environmentalists, hippies, legislators, organic farmers — even an astronaut— attended the conference. Their dress ranged from high-tops and jeans to high heels and silk; their language ranged from New Age idealism to the nuts and bolts of transportation technology. Despite their differences, however, all attendees were dedicated to building cities that are ecologically sustainable and a movement that integrates environmental ethics with urban design.

To some, "urban ecology," may sound like an oxymoron. But proponents believe it is the single most important wave in the environmental movement today— and one we simply cannot afford to ignore. For unless we begin to change our current patterns of land use, transportation, and community living, cities everywhere will suffer severe political, economic and environmental crises.

"Since their origin, cities have exploited nature and expressed the notion that humans beings must conquer nature," Register said. "Such a philosophy has become folly." Now, he said, urban centers must switch from being centers of economic activity to centers of "ecological repair and rebuilding."

Urban ecology, a term first coined by Richard Register in 1975 when he founded a Berkeley-based organization by the same name, depends on a number of principles. An eco-city is one "that satisfies our needs without diminishing prospects of future generations," said systems theorist Fritjof Capra, author of the *Tao of Physics* and *The Turning Point*. The building blocks of that city are limited urban sprawl and high-density, mixed use neighborhoods linked by public transportation and served by renewable energy sources.

A number of planners and environmentalists— including Paolo Soleri, Peter Calthorpe, and Ernest Callenbach— have been working on these design problems for over a decade. But the Eco-City conference, held in late March, was the first time any group of people have really asked the question, "How can we sustain a city?" said Jean Gardner, author of *Urban Wilderness: Nature in New York City*, and a professor of architectural design and criticism at Parsons School of Design.

To answer that question, participants in the four day conference attended dozens of presentations on topics ranging from eco-city theory and visions to recycling, bicycling, transportation, "green" investments and livelihoods, creek restoration, minority contact, urban street trees and rooftop gardens, preservation of urban wildlife, and creating non-toxic homes.

"What was most important was that a number of different interests were brought together — environmentalists, architects, planners, legislators — who don't normally work together," Gardner said. "I've never seen anything like it. It really expressed the point that there is no separation between the built and natural environment, that the two are a systemic whole."

In addition, Register said, the "urban ecology" vision is one of a few new or profound ideas to emerge from the Earth Day hoopla. "Earth Day 1990 didn't add much to the events of Earth Day 1970," he said in a recent interview. "In general, the '50 Simple Things You Can Do' idea is irrelevant; it means we'll neglect the 50 Big Things that are really important. One of those Big Things is to look at basic healthy ways of living collectively."

The "healthy city" vision is emerging in an era plagued by worldwide urban crises and global environmental damage. Media coverage of recent events, including the greenhouse effect, the ozone hole, pollution in Eastern Europe, the green summit, and Earth Day 1990, have brought global environmental problems to the public's living rooms. Less familiar, however, are the massive problems afflicting cities the world over, including overcrowding, transportation nightmares, and lack of affordable housing and sanitation.

"By now we are all aware of the environmental crises we face," said architect and planner Carl Anthony. "But few connect them with the way we build."

Le Corbusier was not the first to exalt the urban dominion over the wild. For centuries, humans have built cities as bulwarks against the perceived disorderly and threatening elements of nature. Built as a spiritual, political or economic center, the city represented civilization at its highest, a place specifically for humans. Today that attitude is not only outmoded, but unsustainable. Our emphasis on city as haven has left a legacy of smogged air, clogged waterways, gridlock, paved landscapes, and relentless urban sprawl. In short, these apogees of "civilization" are the major contributors to planetary pollution.

Conditions aren't likely to get better. Urban populations worldwide are skyrocketing. Already, 75 percent of the population in North America lives in a city. By the year 2000, 50 percent of the world's population will be urban. By the year 2025, 100 of those cities will have more than 5 million residents. The human species, as author and bioregionalist Peter Berg said, "has become a city species."

In order to transform the habitat of this new species, urban ecologists say, a number of steps need to be taken. First, urban sprawl has to be stopped. The 1950s dream home — a single-family home, linked by freeways to distant offices and stores — has engendered an environmental nightmare. We no longer have the resources — especially fossil fuels — to support such auto-dependent lifestyles, and we cannot afford the amount of waste generated. Fully 40 percent of our energy use today is transportation-related. Sixty percent of the urban smog, and 80 to 90 percent of carbon monoxide is from cars.

Besides, noted architect and planner Peter Calthorpe said, sprawl developed for single families is no longer culturally appropriate. Of the approximately 17 million new households formed in the 1980s, he said, 51 percent were comprised of single and unrelated persons and 22 percent were single parent families. Only 27 percent were married couples.

To build for the single family is to use planning strategies that are "40 years old and relevant to a different culture," Calthorpe, also author of *Sustainable Communities*, said. "We are still building World War II suburbs as if families were large and had only one breadwinner, as if the jobs were all downtown, as if land and energy were endless, and as if another lane of the freeway would end traffic."

In place of urban sprawl, eco-planners suggest, we should create a series of high density, mixed-use neighborhoods, in which residents enjoy "access by proximity," rather than by freeway. Calthorpe's "pedestrian pockets," are a strong model of mixed-use. Such pockets are between 50 and 120 acres, include housing, offices, retail, recreation and open space, and are all located within walking distance of each other and within 1/4 mile of a light rail station.

"The impersonal and complex nature of management and organization in today's large cities should be broken down into more decentralized smaller communities of 20 to 40 thousand people, set in rural and forested surroundings, each having its own identity and substantially self-contained," agreed architect Albert Gillison, a professor at University of Adelaide, South Australia. Gillison envisions strings of such communities, spread along a linear transportation corridor that would form an urban ecological system of about half a million people and provide both the amenities of city living and proximity to rural areas.

Such planning could have dramatic effects on our energy consumption. A recent study by the Natural Resources Defense Council showed that residents of Bay Area suburbs drive four times as far per year as residents in San Francisco's Nob Hill. That same study showed that doubling the density of residential areas can reduce annual auto mileage by 25 to 30 percent.

It also could have positive effects on social life. "Diversity at close proximity means just about everything beneficial," Register exclaimed in a panel entitled "Eco-City Berkeley." In addition to land conservation, pollution reduction, and energy conservation, he said, residents would enjoy "greater social vitality, cultural potential and physical access of the city's benefits to a much wider range of people."

In addition to increased public transportation and mixed-use development, eco-cities would include elements such as replacing fossil fuel and nuclear power with solar power, increasing energy efficiency, mandatory recycling, inspiring more local self-reliance, increasing links between urban and rural areas, creek restoration and wildlife preservation, and urban rooftop gardening.

A point that several planners repeated was that an eco-city isn't merely one that has been blessed with a street tree planting program, or even community gardens. "We are not talking about 'pretty' cities," said Paul Downton, president of Australia's Greenhouse Association. "We're not talking about planting a few more trees just to hide the mess of our soulless streets and the banal buildings we call home. Eco-cities go much further than this. Eco-cities are about substance rather than style, about process rather than superficial aesthetics."

Is it possible to really implement all of these ideas? Planting trees, uncovering creeks, and recycling are certainly viable enough. Ideas like mixed use, pedestrian pockets, and new public transportation have massive implications for the topography of the city and the lives of its inhabitants. Judging by conference attendance and the wide range of sponsors, including the U.S. Forest Service, Waste Management Inc., and the German Marshall Fund, eco-cities are a hot topic. But can a small conference held in a notoriously progressive enclave on the West Coast really make a difference? Is the public ready to convert an auto-city, like Los Angeles, into an eco-city?

"Probably not," Register admitted in an interview. "Getting rid of the auto is not a popular topic in suburbia. Urban ecology is frightening to people. We're not protecting something 'out there.' We're asking people to change themselves. That takes courage.

"But history is on our side. The present infrastructure really is not sustainable. At some point, you have to say, 'this is ludicrous,' and look beyond band-aid approaches to the ways in which our cities our built."

In the meantime, Register said, supporters of the concept can immediately begin a number of eco-city strategies, including implementing hiring and renting practices that favor locals, facilitating store deliveries, increasing telecommuting opportunities, and developing satellite offices in suburbs, thereby converting bedroom communities into boardroom communities.

Only a few cities have been built for ecological sustainability. Paolo Soleri's 14 acre Arcosanti, established in Arizona in 1970, was to be a self-sustaining "organic city," but is now more tourist site than thriving community, more famous for its windbells than its contribution to urban design. In his presentation, however, Soleri called Arcosanti a "partially successful" example, because it is "an embryo for what a larger community could be. "

Auroville, a 20 year old community in southeastern India, has had somewhat more success. That community, begun with 30 members in 1970, now has 700 members, from 25 countries. Each home is built of natural materials, and runs on renewable resources. In addition, members have embarked on a massive reforestation project with surrounding villages. Over 2 million trees have been planted, transforming what had been barren red clay into a complex rainforest in all directions.

Closer to home, Peter Calthorpe's pedestrian pocket idea is coming to life in "Laguna West," a small new town being built south of Sacramento. That community will have 3000 units, including affordable housing, a lake and bridge, open park land and public transportation to Sacramento. Already, 2500 people have signed up for units. "People are just tired of being tied to the automobile," Calthorpe said.

Representatives from Norway, Denmark, France and West Germany also spoke of eco-city projects, most of which are in fledgling states, but which are beginning to incorporate mixed-use design, water recycling, and use of renewable resources.

Other cities in this country have incorporated eco-city aspects. Berkeley, a community of what Mayor Loni Hancock called "aggressive dreamers," has taken the greatest number of steps. It has banned styrofoam, blocked the expansion of a nearby freeway, built a "slow street" for bicycles, and restored 41 miles of urban creeks.

In addition, cities all over the country have initiated tree planting programs, curbside recycling, and limited growth policies. Still, noted Tom Cook, director of the Housing and Land Use Committee of the Bay Area Council, limiting expansion is not the same as replacing sprawl. "Simply stopping expansion is only half the battle. We've stopped moving hillsides and converting agriculture, so we've stopped the consumption of open space. But the construction of low energy housing and public transit isn't going well. The detached single-family home remains the favorite. Retail areas remain low-density."

Utopian visions more novel than pedestrian pockets and social equality also came out of the conference. Paul Downton suggested developing commuter buses that provided newspapers, telephones— and comfortable seats. He also coined what could be the green quote of the weekend: "If there's no planet, there's no profit."

More difficult, than the infrastructural challenges are the social challenges, especially those presented by race and class. Before environmental problems can be solved in an eco-city, social problems have to be faced, said a number of community and minority leaders attending the conference. "The vision we have is a promise, and it is also a threat to people of color.,"said architect and planner Carl Anthony in his remarks the opening night. "It is a promise because it is life-affirming. It is a threat because we have the possibility of having another round of a sort of gentrified vision which does not deal with issues of social justice.

"There are many ways of restructuring urban areas," Anthony remarked at a later panel on social justice. "But how do we do it and who pays? We need to ask the questions, 'what's more important, the presence of open space or making public transit? Open space or jobs? Protecting views and property values or protecting jobs and housing opportunities?'"

Minority and working class groups are most often the neighbors and primary victims of toxic waste sites, nuclear power plants, and other polluting industries and activities. But the environmental movement has been slow to realize, or respond to, this demographic reality. Indeed, in February, 1990, the NAACP accused eight of the big 10 environmental organizations of being racist in their hiring and campaign practices.

This neglect, in turn, has alienated minority and working class groups that would be interested in the movement. "It is not that 'people of color don't care about the environment, " said building consultant and author Victor Lewis, as he read from his article, "A Challenge to the Greens," in *Creation* magazine, "though we have been so accused.

"It's just that we find it hard to listen to white activists who find it easy to emotionally identify with the threats to the ozone layer, the mountain goat, or the California condor, but disengage themselves out of feelings of guilt, fear, grief, powerlessness, or whatever, when confronted with the 'environmental effects' of racism on Afro-Americans. If you cannot listen or speak to me of the lives of Afro-American people you cannot speak to me of condors."

The social justice session was one of the most crowded in the entire conference. The room was filled to overflowing with mostly white activists. Despite this good-faith effort, however, some attendees felt the issues remained marginalized. "The social justice issues were not as completely integrated as they could have been," Gardner remarked. "The truth is, in most cities, environmental atrocities are in poor neighborhoods. This could have been covered in more detail."

Throughout the conference, organizers also tried to place urban ecology into the context of the widespread social changes occurring in Eastern Europe and the Soviet Union. Indeed, they said, restructuring the economic and political system to provide for environmental sustainability should be called "eco-stroika."

"International development along the American model has resulted in disastrous ecological consequences," Anthony said. "This demands a rebuilding of cities and transportation systems worldwide." Announced Jan Lundberg, president of the Fossil Fuels Action Institute, "A Conservation Revolution is practically upon us, yet our political and economic systems seem incapable of action."

Susan Davis is a Berkeley-based journalist whose work has appeared in the Washington Post, Newsweek, *and* California Magazine, *among other publications.*

Re-inventing Environmentalism

Towards a Multi-cultural Movement

By Victor Lewis

It will come as news to very few people that the U.S. environmental movement has traditionally been an overwhelmingly "white" cultural phenomenon. The concerns of the movement have reflected the imagination and priorities of the largely white, middle-class constituency that has dominated its ranks. As a result, the movement has focused much of its energy on problems in the wilderness to the neglect of the inner city,while both demand our attention.

If environmentalists are to be successful in securing the future of ours or any species, however, we must take immediate steps to "reinvent" environmentalism within an explicitly multicultural context that sheds the conditioned limitations of its privileged birth, and acknowledges and embraces the environmental concerns of people of color and the poor. This next environmental movement must also recognize the leadership that disenfranchised communities can and do provide towards building a sustainable, global community. The seeds of this transformation have already been planted, but they still require our thoughtful cultivation and care in order to bear fruit.

The other environmentalists

There are at least three distinct movements involved in environmental activism in the U.S. The "traditional" environmental movement, which includes organizations like the Environmental Defense Fund, The Audubon Society, and the Sierra Club, make up one group. Organizations like the Greens, Earth First!, Greenpeace, and the bioregionalists compose a second group, which includes but is not limited to a strong "counterculture" membership. The third group is characterized by its roots in communities — often poor and minority communities. This movement is known as "grass-roots", community-based, or "populist" environmentalism.

A wound that must be healed if environmental activism is ever to succeed in bringing about a just and sustainable society is the lack of mutual recognition between the traditional and grass-roots movements. While both groups would learn and profit from mutual understanding and cooperation, the traditonal movement is likely to undergo greater transformations in the early stages of such an association.

Luke Cole is an environmental poverty lawyer in San Francisco, working for California Rural Legal Assistance (CRLA). He sees the rift between traditional and populist environmentalism as based on the fundamentally different motivations behind them. "The mainstream environmental movement, characterized by the 'Group of Ten" big environmental groups, has been slow to recognize that poor people and people of color face a disproportionately large impact from most environmental hazards, including air pollution, toxics disposal, hazards in the workplace, and pesticides. The relationship between the community-based activists who are often acting out of a survival instinct, and the traditional environmental groups which are often acting out of a aesthetic instinct, is tenuous at best."

Grass-roots environmental groups organize in response to threats to the survival and well-being of their communities and an awareness of being treated unfairly. They look at what environmental hazards are being put where, and why, and typically find that they are being subjected to environmental risks while others reap the benefits. The demands of these groups for fair, just, and respectful treatment has the potential to bring a new coherence and moral strength to the environmental movement as a whole.

Building coalitions

Dr. Robert Bullard, a leading exponent of the grass-roots movement, is a professor of Sociology at U.C. Riverside, and the author of *Dumping in Dixie: Race, Class, and Environmental Quality*, a pioneering work which "chronicles the efforts of five black communities, empowered by the civil rights movement, to link environmentalism with the issues of social justice."

Bullard believes strongly in the role that grass-roots environmental activism plays in shaping our common future. "I see this emergent, grass-roots, coalition-type organization that's springing up and criss-crossing this country as the heart and soul of the movement."

According to Bullard these grass-roots groups are not only "the conscience of the environmental movement," they can also provide a critical link between the front lines and the national groups. They can provide crucial, up to date, "on-the-ground" information about environmental problems, he says.

"As issues crop up, these organizations know what's going on, they can filter the information upward, and its not like a top down kind of thing," he says.

The relative nonparticipation of people of color and poor people in the traditional environmental movement has been misinterpreted. The neighborhoods and work-places of these people are often highly stressed physically, economically, and culturally, and they rightfully prioritize those immediate environments first. White, middle-class environmentalists are more likely to live and work in relatively clean, non-toxic communities. This gives them the freedom to narrowly define wilderness areas and wildlife as their "environments" of concern. In addition, they are more likely to have the discretionary income, mobility, physical access, leisure time, social sanction, and sense of ownership to support their enjoyment of wilderness areas.

Disenfranchised communities are profoundly concerned with their physical environments, but they have never had the luxury of separating "environmental" problems from the complex of other issues confronting them. They are bound by necessity to address the total webwork of threats to their communities in a wholistic fashion. Long years of experience in grappling with many interwoven problems simultaneoulsy are a potential gift from people at the grass-roots to other sections of the environmental movement.

In defining environmentalism too narrowly, mainstream environmentalists sometimes inadvertently alienate their natural grass-roots allies. Bullard points out that this process can be as simple as ignoring a group that doesn't have the right kind of name. "If you define groups that emerge and mobilize as non-environmental groups simply because they don't have "environment" in their name, or simply because they emerge out of civic groups, church groups, civil rights groups, etc., and deal with housing issues, with education issues, pollution issues, with public health issues, and not just focus in on one issue, that's a mistake." Such mistakes, he says, do a disservice to groups, often led by people of color, that are in many cases more comprehensive in scope than the traditional environmental groups.

We need to learn a deepened appreciation of the fact that human beings live in "environments" and ecosystems. Our cities, our rural communities, our industrial areas, as well as the wilderness areas, should all be understood as the biological and social communities that they are.

The urban dimension

Human beings are rapidly emerging as the "urban species." More than 70 percent of people living in the U.S. live in cities. Significantly, there are massive concentrations of people of color in our urban centers. Locally, 51 percent of Oakland's population is African-American, and 85.6 percent of San Francisco's school children are people of color. The ecology of metropolitan areas is inextricably involved with the ecologies of communities where people of color live. Urban ecology is a growing concern for the environmental movement, but to understand it fully we need to appreciate its cultural as well as physical dimensions. Specifically, we must come to terms with the institutionalized inequities in urban centers that are established and distributed along the lines of race, class and gender. This will give us a context for the transformation of the destructive consumption and waste patterns which now dominate our cities and their industrial machinery.

Carl Anthony is an Oakland-based African-American architect, urban planner, and activist. He is President of the Board of Directors at Earth Island Institute, the Director and co-founder of the Urban Habitat Project for Social and Environmental Justice, and consulting planner for the Center for Economic Conversion. One of Anthony's central concerns is deepening our awareness of metropolitan areas as ecosystems and charting a path to the future that will shape cities in keeping with social as well as ecological wisdom.

"Almost by definition the city is a multi-cultural phenomenon." he said in an interview. "In order to bring the city into balance with nature we have to respect the diversity of the human communities, and in fact, in order to set an urban ecological agenda we need to have multi-cultural leadership that brings forth the strength and diversity of the communities that inhabit a city."

Two Bay Area examples of this emerging grass-roots urban leadership are the District 7 Democratic Club in San Francisco, and the West County Toxics Coalition, in Richmond, California. Both organizations are composed primarily of people of color living in the shadow of massive production facilities and waste sites owned by powerful corporate and government interests. Both organizations have earned a reputation for getting things done.

Sam Murray, of the District 7 Democratic Club, lives and works in the Bayview-Hunter's Point area, one of the most economically depressed communities in San Francisco. He finds the daily hardships of his neighbors no impediment to his organizing them to respond to ecological issues. On April 18, 1990, Sam hosted a neighborhood forum on toxic wastes for which over 350 people turned out. "A lot of people in the public sector don't believe that minorities really care about toxic wastes, and the reason for that is they feel we are hung up on food, clothing, shelter, and jobs. I proved that to be untrue. I had well over 350 people at my toxic waste forum from just this community. City Hall failed to get one tenth of this kind of turnout with a fist full of money to fund them."

The West County Toxics Coalition (WCTC), based in Richmond, California is a jewel in the crown of the national grass-roots environmental community. Not content to just put up with the smelly air, and the unusually high rates of cancer, respiratory problems, and skin conditions in their community, this small organization, nestled in one of the most industry-intensive areas in Northern California, has attracted national attention for its war against the polluting practices of Chevron and other corporations operating there. In April they received supportive visits to their community by Earth Day 1990 director Denis Hayes, National Toxics Campaign director John O'Conner, and 1988 presidential candidate Jesse Jackson.

Henry Clark, the president of WCTC, says his organization is fighting for the health of the planet as well as for the lives of his community. He has important words for members of traditional groups that don't know what grass-roots activists do.

"We contribute tremendously to the total protection of the environment," he said. "We organize people who are living in toxic hotspots, who are on the front lines, facing the frontal toxic assault, living around these companies that are polluting the communities and the greater environment, contributing to the destruction of the ozone layer, the greenhouse effect and all that."

As members of the National Toxics Campaign, groups like WCTC work on all the same issues as the traditional groups—issues such as air pollution, global warming and ozone depletion, but they also go a step further: using the proven strategies of the civil rights movement they organize and empower people in unfairly-treated communities to represent their own interests in survival and justice. Clark says it is a question of environmental democracy. "We organize these people to empower them to protect their communities and their lives, and demand their environmental rights. This is about environmental democracy, as we call it. The right to know what type of chemicals we're exposed to, the right to inspect these facilities that are polluting our communities, and the right to negotiate with these companies around our health and safety concerns."

The commitment of groups like WCTC to reduce and stop the flow of pollutants coming into their communities, rather than attempt to control them makes them essential allies to the traditional organizations that concentrate on tactics such as lobbying and litigation.

The Rural Dimension

On September 30, 1962, in an abandoned theatre in Fresno, California, the charter convention of the group that would become the United Farm Workers AFL-CIO took place. *Silent Spring*, Rachel Carson's classic invective against the thoughtless use of agricultural pesticides, had been published only three days earlier. Carson's powerfully written and well-documented work made several references to the probable human costs of pesticide exposures, but lacked detailed information of the subject, primarily because farmworkers, who have always received the most direct and lethal pesticide exposures, were routinely denied adequate health care for their ailments and were quickly fired from their jobs if they complained to anyone about being poisoned. These conditions still prevail today.

I wish Carson could have attended this convention. She would have learned first-hand about the skin rashes, dizziness, respiratory ailments, miscarriages, birth defects, cancers, and deaths, faced by farmworkers every day due to pesticides. It would have confirmed her gravest suspicions, and perhaps resulted in a stronger social justice voice in the environmental movement that she was to inspire.

In their landmark 1970 contracts with growers, won through a successful grape boycott, the UFW negotiated a ban on the use of DDT and a whole family of related pesticides on fields worked by union members. The EPA eventually followed suit with a federal ban on DDT 2 years later. Since then, these poisons have been sold and used primarily in Third World countries. At the time of the ban, the eagle, sacred to many indigenous peoples—a bird chosen as the emblem of this country, and of the UFW—was on the very brink of extinction. While Rachel Carson alerted the world that DDT mimics and fatally disrupts the sex hormone metabolism of birds, Cesar Chavez, president of the UFW, taught us that DDT and other more lethal pesticides were killing farmworkers. While fighting for their lives, the UFW played an important role in preventing the disappearance from the Earth of this land's great birds of prey. Seldom, if ever, did this group of struggling Latino farmworkers think of themselves as "environmentalists."

Dr. Marion Moses, a nationally-recognized authority on pesticide-related illness, is the former director of the National Farmworkers Health Group, and currently directs the Pesticide Education Center in San Francisco. Here she recalls the day she was first awakened to the dangers of persticides. "The grape strike started in September of 1965. I worked in a trailer-clinic that they had for the strikers because the local doctors wouldn't treat the striking farmworkers. And it was at that time, Cesar asked me, I'll never forget it, he said 'Do you know anything about pesticides?' Well, I didn't know anything about pesticides. Frankly, I really didn't. And he says, I can still hear him say it, he's still saying it today, 'I know they're killing our people.'"

The farmworkers of this country are the backbone of the U.S. food production system, the most generative one in the world. Due to hazardous working and living conditions, the life expectancy of a farmworker is more than 20 years lower than the national average. Almost all of our farmworkers are people of color. Their wages, access to housing, healthcare and education are abominable. At the same time, they are the most organized, dedicated, and tireless proponents of healthy, non-toxic agricultural practices to be found anywhere in the country. If we are to be successful in creating a sustainable agricultural program in the U.S., it will be in partnership with the farmworkers. What they want for themselves is in the interests of all people, and the entire life community. For them, the fights for social and environmental justice are indistinguishable.

Conclusion

We are all in the same boat, although not necessarily on the same deck. We need to bear in mind that women, people of color, and the poor have often been unwillingly restricted to damaged and flooding compartments of this boat, while many of those on the upper decks have shut their ears and turned away, looked on helplessly, or focused their efforts on preventing any spreading of the damage. We are beginning to learn, I hope, that if we are to survive the dangers ahead and flourish undiminished in our humanity, we must all work together to turn back the waters of destruction where they affect those among us who are the most vulnerable

Victor Lewis is a contributing editor for CREATION *magazine, andCo-editor of the* Race, Poverty and the Environment *newletter, and is on the policy board for the Urban Habitat Program. He is a faculty member at the Institute In Culture and Creation Spirituality at Holy Names College in Oakland, California, and the founder of Environmentalists Against Racism.*

Failed Plans

The Elusive Dream of Cohesive Regional Planning

By Lisa Levy

Growth produces problems— physical, social, and economic. Today we find such problems of increasing importance throuhgout the Region in the form of cramped and insufficient highways, the need for Bay crossings, limited water supply, and demand for improved transit and transportation."

Sound familiar? The sad fact is that this passage was written in 1925, as part of one of the first regional plans proposed for the Bay Area. Since that time, despite repeated road widenings, the building of hundreds of miles of freeways and four bridges, numerous dams, and a new mass transit system, the infrastructure problems have only grown worse and worse. We still have too little affordable housing, and urban sprawl has placed huge distances between homes and work places. The highways, though longer and wider, are still cramped. Bay crossings are a traffic nightmare, we have no comprehensive mass transportation system and the water supply is still constantly in doubt. Decades of poor planning have taken their toll.

Fred Dohrmann's Regional Plan Association of the 1920's, and virtually every successive effort, have failed to accomplish the goal of establishing regional rule. This is not because of a lack of public interest. A recent poll conducted by KQED and the Bay Area Council revealed that three-fourths of Bay Area residents identify themselves as citizens of the "region", and feel that regional coordination, between city and county governments, should be a part of planning policy. Respondents also felt that regional agencies should control freeways and highways, air and water quality, public transportation, and economic growth.

Are there any regional agencies trying to solve these problems? Yes and no. There are several government-sponsored organizations that deal with specific regional problems, such as the Metropolitan Transportation Commission. The rub lies in the power vacuum in which most of these agencies operate. Without the power to implement their regional proposals, all agencies can do is complain about present conditions or make suggestions which no one need consider in making planning decisions.

Local officials concerned mainly with protecting their own turf are also an obstacle to creating regional government. The "nimbys" and the "nimeys" often seem to have the upper hand in local planning decisions: the nimbys are citizen groups who respond to development queries with "not in my backyard", and the nimeys, their elected officials, respond with "not in my election year." They are often criticized for advocating regionalism because their constituents feel that their particular problems are being neglected, or sacrificed at the regional altar.

A Historical Overview

Regional planning theory has a checkered history in the Bay Area, where geographic and lifestyle factors led to some of the earliest land use movements. The 1920's were the decade that built two single purpose, multi-community districts: the Golden Gate Bridge and Highway District and the East Bay Municipal Utility District. In 1927, the California Legislature passed the law which delineated city planning, and authorized a primitive form of regional planning; two years earlier, however, the Commonwealth Club had fought for the creation of Fred Dohrmann's Regional Plan Association, which studied issues like harbor and port development, highway building and maintenance, recreational and park lands, pollution, and regional zoning. Unfortunately, the association was disbanded before implementing their broad based development plan.

The Depression and the Second World War effectively stalled regional planning in the 30's and 40's, but there was a renaissance of regional planning from the fifties through the seventies, which could be called the "alphabet" period. The fifties were the era of the BAAPCD (Bay Area Air Pollution Control District), the Alameda-Contra Costa Transit District (AC Transit), and BART (Bay Area Rapid Transit).

The sixties were a crucial time for regional planning. In the late fifties, Governor Ed Brown's "Commission on Metropolitan Area Problems" proposed regional government through the state legislators; alternatively, a group of Bay Area officials, the Bay Area Metropolitan Council, wanted a regional agency independent of the state. A compromise was reached in 1961, the year ABAG (the Association of Bay Area Governments) was born.

One of the most influential and visionary advocates of regionalism was California Tomorrow, a group of concerned citizens, journalists, and government officials. The "regional planning failure" which they documented in the early 1960's is almost a forecast of present conditions: public health threats, such as air and water pollution, which cross town and county lines, and conservation efforts that affect whole ecosystems and cannot be circumscribed by human borders.

California Tomorrow proposed an antidote to these ills in the form of an area-wide government, a level of legislators between the local government and the state, which had just appointed regional planning commissions. California Tomorrow applauded this effort but warned that unless regional planning was permanently incorporated into the system, land would end up misused and abused. The growth which seemed out of control in the post WWII years was producing 'slurbs:' "chaotic mixtures" of parking lots, supermarkets, housing tracts, and clogged roads, where farmland and parks used to exist. California Tomorrow sought to control the slurbanization of California through regional planning, under a system where each local government would help create a regional plan, and adopt it as part of local planning measures.

It sounds very logical and persuasive, but California was not ready to plan for the kind of tomorrow this group envisioned. Instead, we got the Association of Bay Area Governments (ABAG), which had an equally compelling vision, but no way to make it reality.

ABAG: A Growth Parable

The formation of ABAG, created to discuss and study regional issues, was applauded by citizen groups like the League of Women Voters and People for Open Space (now the Greenbelt Alliance) as well as business organizations, like the Bay Area Council. The first plan ABAG published in the early sixties excited conservationists because of the large areas designated as open space, and lack of land-fill.

Local governments, however, were less than thrilled with ABAG's idea of "growth", and pointedly ignored their plan. Because ABAG was, and still is, only an advisory group, there was no way for it to stop city and county governments from developing as they saw fit. ABAG built a Bay Area on paper with graphs and sentences which bears little resemblance to the region we live in today. Furthermore, the state government which created ABAG dealt it a crushing blow in 1966, when a Home Rule Agency Bill was defeated in the legislature which would have given ABAG an elected board with considerable power in regional development.

ABAG published a new regional plan a few years later which covered the next twenty years. Once again, lots of open space was set aside and landfill was avoided, thanks in part to the work of the newly formed Bay Conservation and Development Commission (BCDC). This plan preserved agricultural land, predicted growth in existing cities, and provided for sewage and water systems to supplement the present services for cities. Local governments were pleased with this report because growth was seen as enhancing the community and meant additional revenue. The plan was unanimously approved and heralded as a symbol of future unity.

Unfortunately, ABAG's predictions missed the mark. The growth which has occurred in the Bay Area in the last twenty years has been decentralized. Deconcentration, the term that urban planners apply to the phenomenon in which commercial growth occurs in suburban areas rather than in cities, has surrounded the Bay Area like a medieval wall. Counties like Contra Costa, Santa Clara, and more recently, Solano and Sonoma, have experienced unprecedented growth. New industries have located in the suburbs, like the high tech firms of Silicon Valley, and old firms have fled from cities like Oakland and San Francisco, in search of cheaper rents and more land.

It is this uncontrolled, decentralized growth, along with the effects of Proposition 13 and the Gann Limit, which has led to three related problems: the shortage of affordable housing, increasing distances between residences and jobs, and the resultant transportation crisis which includes chronic freeway congestion and mass transit splintering.

New Visions

The most exciting aspect of regionalism now is that this spirit of regional cooperation may be leading us to California Tomorrow's vision of regional government. The next chapter of Bay Area regional history is being written by Bay Vision 2020, the commission which was formed last fall through the efforts of the Regional Issues Forum, devised by the Bay Area Council and Greenbelt Alliance, and a committee of local officials led by representatives from ABAG and the MTC. Working independently, both of these groups concluded that the best way to study regional problems and offer solutions would be the formation of a citizen commission, which would consider proposals for regional government in the Bay Area from a variety of perspectives.

Bay Vision 2020 is also advocating solutions to the problems listed above, such as an integration of mass transit systems and a call for convenient, affordable housing, but the centerpiece of their plan is that the "Bay Area should have a democratically chosen, regionally accountable" level of government which would work in tandem with county and city planners to manage growth in the region.

Although it is natural to associate planning with the future, it is the action taken now which will influence the region into the year 2000 and beyond. The respondents in the Bay Area Council Poll felt that regional government is an imperative, and Bay Vision 2020 is laying the necessary groundwork by gathering data from all the major industries, environmental groups, and local governments in the Bay Area. The poll also vocalizes a real fear among Bay Area residents of creating this new level of government, when it seems like bureaucracy, or its practices, is part of the problem.

At a smaller level, San Francisco 2000 has been created by Mayor Agnos to provide information to the Bay Vision commission. SF 2000 is spending around two years trying to "create a shared vision for San Francisco in the year 2000".

The state assembly is also trying to create a regional framework. Speaker Willie Brown has introduced AB4242, a bill which advocates the formation of seven regional "development and infrastructure" superagencies to deal with transportation and land problems, air and water quality, and utilities and sewage systems. So perhaps sometime soon the elusive dream of regional planning will come true.

Lisa Levy is a staff researcher and writer for the Bay Area Green Pages.

De-subsidizing Sprawl

The First Step Towards an Answer

By Sherman Lewis

More than any other country in the world, the United States has covered hundreds of square miles of good land with dispersed, car-dependent development. In the process we have achieved the world's highest standard of living and created a terrible mess. The San Francisco Bay region, for all our vaunted environmentalism, is no better than any other region. Indeed, we are now sprawling into the central valley and out I-80 to Sacramento, becoming Los Angeles as fast as developers find farmland and pliant local governments.

Suburbanization has been greatly accelerated through phenomenal subsidies, of which most people are completely unaware. The FHA and other mortgage and mortgage guarantee programs have played a role due to their sharp bias against multiple-unit dwellings. Farm water, price, labor, and research subsidies undermined the agricultural competitiveness of land close to cities. The gas tax does not cover the cost of local roads and, equally important, of services to road users—police, fire, medical, legal. Employers subsidize employee parking, reinforced by an income tax loophole. Lacking a means of charging more when too many people want to use highway capacity, we let all comers onto the road, resulting in congestion. We then believe there is a "demand" for roads, build more roads without charging the users, and thus encourage more traffic. We do not charge for pollution, not in a gas tax, not in a registration fee, not in a mileage fee. We force developers in dense areas to have parking, which then encourages traffic in places where transit could work. We tend not to allow people to rent their living space separately from their parking place.

These land, agricultural, road and road service, employer parking, congestion, air pollution, and property market subsidies add up to a huge amount of money. They are a powerful economic incentive provided by government for people to buy their own private escape from the problems of the city. Without subsidies, there would no doubt be much development of suburbs to accommodate population growth. However, suburbs would be far more compact, and the tragedy of our central cities might well have been avoided.

The environmental crisis created on the fringe by suburbia is directly linked to the social crisis created in the center by suburbia. We have turned our backs on the disadvantaged of cities, and denied their governments the resources to prevent the crowded jails and prisons, hospitals with top notch emergency care but third rate general care, poor school performance, violent and broken families, premature babies born with crack in their veins, a declining life expectancy for black males, and the victimization of the law-abiding poor. The capacity of the old city to absorb growth—and to provide a standard of living comparable or even superior to suburbia—is almost limitless. However, we must change the way we think about density. It is not people in neighborhoods that create congestion, it is vehicles. As densities rise, mass transit becomes more economically feasible, especially if reinforced by pricing reform to take cars off welfare.

Spacious, well-designed homes on peaceful streets can exist in a low-rise, high density neighborhood. No more than four stories are needed for densities of over 100 people per acre. At those densities, walking distances to local business and frequent transit become very short — no more than eight minutes, and typically five. Half of the ground area can be in open space and walkways. The problem is to distance the car from the dwelling, orienting routine trips inwardly to a mostly pedestrian main street, and cars outwardly to the highway system for special trips.

If we can succeed in redirecting growth, particularly to redevelopment within half a mile of BARTand other high quality transit service, we will have many benefits: open space saved, cleaner air, greater economic competitiveness, a more balanced, equitable, and safe transportation system, affordable housing, and a high sustainable standard of living.

Thus, the return to the city can fulfill environmental goals as well as our more selfish material goals. Unfortunately, if this is all that happens, the tragedy of the suburbs will be repeated in the city through "gentrification." The affluent can just as well ignore the poor as neighbors as they can from the distance of suburbia. Propinquity does not guarantee or even assure understanding and compassion. We need to commit ourselves and our resouces to the problems of the disadvantaged of all races and ethnicities. The problems are complex. We must avoid creating dependency on welfare and we cannot condone criminality. Nevertheless, the opportunities are there for a new urban class to act privately and through government to overcome the social crisis. We cannot be just environmentalist, or just business, or just any one thing. We should grasp the nature of our times as a whole and live accordingly.

Sherman Lewis is a professor of political science at California State University, Hayward.

Cohousing

Bringing Community Back into the American Dream

By Kathryn McCamant and Charles Durrett

Traditional housing no longer addresses the needs of many people. Dramatic demographic and economic changes are taking place in our society, and most of us feel the effects of these trends in our lives. Things that people once took for granted—family, community, a sense of belonging—must now be actively sought out. Many people are mis-housed, ill-housed, or unhoused because of the lack of appropriate options. This article introduces a new housing model that addresses such changes. Pioneered primarily in Denmark and now being adapted in other countries, the cohousing concept re-establishes many of the advantages of traditional villages within the context of late 20th century life.

Several years ago, as a young married couple, we began to think about where we were going to raise our children. Already our lives were hectic. Often we would come home from work exhausted and hungry, only to find the refrigerator empty. Between our jobs and housekeeping, when would we find the time to spend with our kids? Relatives lived in distant cities, and even our friends lived across town. Just to get together for coffee we had to make arrangements two weeks in advance. Most young parents we knew seemed to spend most of their time shuttling their children to and from day care and playmates' homes, leaving little opportunity for anything else.

So many people we knew seemed to be living in places that did not accommodate their most basic needs; they always had to drive somewhere to do anything sociable. We dreamed of a better solution: an affordable neighborhood where children would have playmates and we would have friends nearby; a place with people of all ages, young and old, where neighbors knew and helped each other.

As architects, we had both designed different types of housing. We had been amazed at the conservatism of most architects and housing professionals, and at the lack of consideration given to people's changing personal needs.

Single-family houses, apartments, and condominiums might change in price and occasionally in style, but otherwise they were designed to function pretty much as they had for the last 40 years. Perhaps our own frustrations were indicative of a larger problem: a diverse population attempting to fit into housing that is simply no longer appropriate for them.

Contemporary post-industrial societies such as the United States and Western Europe are undergoing a multitude of changes that affect our housing needs. The modern single-family detached home, which makes up 67 percent of American housing, was designed for a nuclear family consisting of a breadwinning father, a homemaking mother, and two to four children. Today, less than one-quarter of the United States population lives in such households. Rather, the family with two working parents predominates, while the single-parent household is the fastest growing family type. Almost one-quarter of the population lives alone, and this proportion is predicted to grow as the number of Americans over the age of 60 increases. At the same time, the surge in housing costs and the increasing mobility of the population combine to break down traditional community ties and place more demands on individual households. These factors call for a thorough re-examination of household and community needs, and the way we house ourselves.

As we searched for more desirable living situations, we kept thinking about the housing developments we had visited while studying architecture in Denmark several years earlier.

In Denmark, people frustrated by the available housing options developed a new kind of housing that redefines the concept of neighborhood to fit contemporary lifetstyles. Tired of the isolation and impracticalities of single-family houses and apartment units, they built or developed out of existing neighborhoods new housing that combines the autonomy of private dwellings with the advantages of community living. Each household has a private residence, but also shares extensive common facilities with the larger group, such as a kitchen and dining hall, children's playrooms, workshops, guest rooms, and a laundry. Although individual dwellings are designed to function independently and each has its own kitchen, all common facilities, and particularly common dining are an important aspect of community life.

As of 1988, 67 of these communities had been built in Denmark, and another 38 were planned. They range in size from six to 80 households, with the majority between 15 and 33 residences. These communities are called *bofoellesskaber* in Danish ("living communities"), for which we have coined the term "cohousing." First built in the early 1970s, cohousing developments have quadrupled in number in the last five years.

The Netherlands now features 30 cohousing communities and similar projects are being built in Sweden, France, Norway, and Germany.

Imagine....It's five o'clock in the evening, and Anne is glad the workday is over. As she pulls into her driveway, she begins to unwind at last. Some neighborhood kids dart through the trees, playing a mysterious game at the edge of the gravel parking lot. Her daughter yells, "Hi Mom!" as she runs by with three other children.

Instead of frantically trying to put together a nutritionus dinner, Anne can relax now, spend some time with her children, and then eat with her family in the common house. Walking through the common house on her way home, she stops to chat with the evening's cooks, two of her neighbors, who are busy preparing dinner—broiled chicken with mushroom sauce—in the kitchen. Several children are setting the tables. Outside on the patio, some neighbors share a pot of tea in the late afternoon sun. Anne waves hello and continues down the lane to her own house.

After dropping off her things at home, Anne walks through the birch trees behind the houses to the child-care center where she picks up her four-year-old son, Peter. She will have some time to read Peter a story before dinner, she thinks to herself.

Anne and her husband, Eric, live with their two children in a housing development they helped design. Not that either of them is an architect or builder: Anne works at the county administration office, and Eric is an engineer. Six years ago they joined a group of families who were looking for a realistic housing alternative. At that time, they owned their own home, had a three-year-old daughter, and were contemplating having another child—partly so that their daughter would have a playmate in their predominantly adult neighborhood.

Responding to a newspaper ad, Anne and Eric discovered a group of people who expressed similar frustrations about their existing housing situations. The group's goal was to build a housing development with a lively and positive social environment.

In the months that followed, the group further defined its goals and began the long, difficult process of turning its dream into reality. Some people dropped out, and others joined. Two and a half years later, Anne and Eric moved into their new home—a community of clustered houses that share a large common house. By working together, these people had created the kind of neighborhood they wanted to live in—a cohousing community.

Today Tina, Anne and Eric's eight-year-old daughter, never lacks for playmates. She walks home from school with the other kids in the community. Her mother is usually at work, so Tina goes up to the common house, where one of the adults makes tea and toast for the kids and any adults who are around. Tina liked her family's old house, but this place is much more interesting. There's so much to do. She can play outside all day, and as long as she doesn't leave the community, her mother doesn't worry about her.

John and Karen moved into the same community a few years after it was built. Their kids were grown and had left home. Now they enjoy the peacefulness of having a house to themselves; they have time to take classes in the evenings, visit art museums, and attend an occasional play in town. John teaches children with learning disabilities and plans to retire in a few years. Karen administers a senior citizens' housing complex and nursing home. They lead full and active lives, but worry about getting older. How long will their health hold out? Will one die leaving the other alone? Such considerations, combined with the desire to be a part of an active community while maintaining their independence, led John and Karen to buy a one-bedroom home in this community. Here they feel secure knowing their neighbors care about them.

Cohousing is a grass-roots movement that grew directly out of people's dissatisfacton with existing housing choices. Its initiators draw inspiration from the increasing popularity of shared households, in which several unrelated people share a house, and from the cooperative movement in general. Yet cohousing is distinctive in that each family or household has a separate dwelling and chooses how much they want to participate in community activities.

In many respects, cohousing is not a new idea. In the past, most people lived in villages or tightly knit urban neighborhoods. Even today, people in less industrialized regions typically live in small communities. Members of such communities know one another for many years; they are familiar with one another's families and histories, talents and weaknesses. This kind of relationship demands accountability, but in return provides security and a sense of belonging.

In previous centuries, households were made up of at least six people. In addition to having many children, families often shared their homes with farmhands, servants, boarders, and relatives. A typical household might include a family with four children, a grandmother or an uncle, and one or more boarders who might also work in the family business. Relatives usually lived nearby. These large households provided both children and adults with a diverse inter-generational network of relationships in the home environment. The idea that the nuclear family should live on its own without the support and assistance of the extended family or surrounding community is relatively new, even in the United States.

To expect that today's small households, as likely to be single parents or single adults as nuclear families, should be self-sufficient and without community support is not only unrealistic but absurd. Each household is expected to prepare its own meals, do its own shopping, and so far as finances permit, own a vacuum cleaner, washing machine, clothes dryer, and other household implements, regardless of whether the household consists of two people or six, and whether there is a full-time homemaker or not.

People need community at least as much as they need privacy. We must re-establish ways compatible with contemporary American lifestyles to accommodate this need. Cohousing offers a new model for re-creating a sense of place and neighborhood, while responding to today's needs for a less constraining environment.

This article is excerpted from Kathryn McCamant and Charles Durrett's book Cohousing: A Contemporary Approach to Housing Ourselves *with the permission of Ten-Speed Press, the book's publisher. To order the book, contact Ten Speed Press, P.O. Box 7123, Berkeley, CA, 94707. $19.95*

Working for Change

Jobs in the Non-profit Sector

By Dave Smith

You'd love to change the world—but by the time you get off work, you're too tired to change anything but the TV channel. You need to make a living and pay the rent, but you wish you could find a job that rewarded more than your bank account.

Relax: It's not hopeless. Aside from Washington D.C. and possibly New York City, the Bay Area offers more job opportunities in non-profit organizations than any place else in the country.

Those jobs, however, often aren't advertised and job hunters must aggressively use every available contact and resource to find—and land—an interesting position.

Non-profit employers and job placement counselors surveyed for this story agree that competition is tough for most jobs because plenty of highly qualified, well-connected people desire work in organizations with objectives higher than the bottom line.

"There are more jobs than people realize," says Michael O'Neil of the Institute for Non-profit Management at the University of San Francisco, "but it is difficult to find jobs when and where they are available.

There are ways to go about finding those jobs if you know a little bit about how the non-profit sector works here in the Bay Area. A San Francisco Bay Guardian survey of 33 Bay Area non-profit organizations identified several surprising trends in the local non-profit employment market. In particular, the survey found that:

- The Bay Area offers more job opportunities in the non-profit sector than almost any other city in the U.S. Several local nonprofits can be expected to grow as demand increases for services for people with AIDS, the homeless, children and the mentally ill.

- The profile of the non-profit sector that emerged from the survey is remarkably similar to the small business sector profile reported by a 1985 study of small businesses in San Francisco. That study, conducted by David Birch of MIT, concluded that almost all new net growth in San Francisco employment occurred in businesses with fewer than ten employees.

  Like the small business sector, the dynamic non-profit sector is experiencing high start-up and failure rates that translate into greater employment opportunities. Employee turnover is high, particularly in entry-level positions.

- Non-profit organizations and their employees are becoming more professional as they attempt to expand their services in the face of Reagan and Bush budget cuts. Non-profit employers are hiring more people with specific management, technical and especially fundraising skills that can help the agency survive lean times. Commitment to the cause is no longer sufficient to get a non-profit job. Nevertheless, the non-profit sector is still the best place for entry-level applicants to work their way into responsible positions and to acquire a wide variety of skills needed for more advanced positions with other organizations.

- Non-profit jobs are hard to find primarily because openings are poorly publicized, competition is intense for desirable jobs and job hunters do not use the many resources that are available to them. But if you are clearly committed to a cause and are willing to volunteer, or in some other way prove your usefulness to the organization, you can find out where the openings are and eventually land a paid position.

The non-profit sector is "healthy and thriving," reports Paul Harder, co-author of a 1985 study of Bay Area nonprofits by the Washington D.C.-based Urban Institute. Harder estimates that locally more than 3,500 non-profit organizations employ about 35,000 people — a total equal to the number of primary and secondary school teachers in the area. In fact, Rick Smith, executive director of Support Center/CTD, which supplies management consulting services to nonprofits, says the Bay Area has more nonprofits per capita than any other American city.

Nonprofits can be divided roughly according to their primary function—service providers such as the Family Services Agency of San Francisco and the San Francisco Symphony, member service organizations such as the Film Arts Foundation and Media Alliance and advocacy organizations like the Sierra Club and American Civil Liberties Union.

The Urban Institute study found that health care and social service providers comprise the largest segment of the non-profit sector, accounting for perhaps two-thirds of all expenditures. Nearly half of the Bay Area nonprofits devote at least part of their resources to promoting culture, the arts and recreation—a figure far above the national average.

The Public Interest Clearinghouse Directory lists more than 600 Bay Area nonprofits that focus on public interest advocacy work. In particular, the Bay Area has long been a center for environmental activism, and several national environmental groups are headquartered here, including the Sierra Club, Earth Island Institute, Rain Forest Action Network, and Global Exchange.

In addition, many service providers, such as homeless shelters, also do advocacy on behalf of their clients, says Ira Okun, executive director of the Family Services Agency of San Francisco. Employment is growing fastest, he says, in service and advocacy programs concerned with AIDS, the homeless, and childcare, due to the growth in program demand and funding. By comparison, the Urban Institute study found that organizations particularly hard hit by cuts in government assistance include community development, job training and legal aid programs.

California and, more specifically, the Bay Area have long been hotbeds of non-profit innovation. Groups here often take the national lead in developing both new approaches for working on issues and new ways for managing organizations.

In recent years, innovation has been necessary to adjust to steep cutbacks in funding to social programs. Changes in listings in the PIC directory indicate that as many as 30 percent of public interest nonprofits in the Bay Area failed between 1984 and 1987. Although some of the change is explained by the decline in public interest in some issues, much of the decline was caused by funding crises, according to David Goldsmith, who until 1989 was the PIC project coordinator.

However, new organizations have sprouted at roughly the same rate as older groups have folded, Goldsmith says, noting that the rise of the Reagan administration precipitated two seemingly contradictory trends: Nonprofits began to find it more difficult to survive funding cuts, but opposition to the Reagan agenda provided a rallying point for new groups.

The surprising result is that despite the funding setbacks suffered during the late 1970s and the 1980s "the non-profit sector has done pretty well during the Reagan years," Harder observes.

Surviving nonprofits have had to diversify their funding bases and learn to squeeze every penny out of the budget.

The Urban Institute found that nonprofits have made up for cuts in government assistance by collecting more fees for services, charging membership dues, marketing products and publications and soliciting private donations more aggressively.

As a result, many non-profit organizations are hiring staff with professional backgrounds. With all the cuts in funding, nonprofits have to pay more attention to running the business like a business.

Anyone looking for a job in the higher levels of a non-profit should keep in mind the change in skill requirements this trend has produced. Trina Ostrander, director of Communication Works, contrasts the early days of the non-profit community, when dedicated service providers and activists often moved quickly into management and leadership positions, with the new trend toward professional management. "Now non-profit organizations from small to large tend to have people who specialize in marketing, development, long-term financial planning and computers," she says.

But while competition for the top slots is getting tougher, there are many obvious incentives to try to break into the non-profit sector.

Non-profit organizations offer a relatively casual work environment, where status is often defined less by hierarchy and credentials than by the work produced. And they offer opportunities to work with people who share common goals and beliefs, on issues you care about.

In contrast to the private sector, where jobs increasingly require workers to specialize in one function, employees in smaller non-profits often do many jobs. For example, Stana Hearne, coordinator of the Citizens for an Eastshore State Park, was once a paralegal for a private attorney. Now she's involved in practically every aspect of the citizens group's work, including grant writing, supervising volunteers, lobbying and office management.

In some nonprofits, and particularly in grassroots advocacy groups, talented employees will have excellent opportunities for advancement. Ostrander, for example, started her non-profit career as a secretary and advanced to become an executive director in only three years.

Working at a non-profit, however, is not for everyone. Paul Harder observes that to work in the non-profit sector, "You give up material rewards for psychic benefits."

The potential for advancement, for instance, is partially offset by the potential for burnout. The combination of poor pay — about 75 percent of the pay in similar private sector jobs, according to Independent Sector research — and long hours working on society's toughest issues eventually drives many talented people out of non-profit jobs. Some larger and better established groups such as the Sierra Club strive to offer competitive wages, but most cannot compete with the private and public sector.

But Ostrander claims, "Burnout may not be any worse in the non-profit sector than in the private sector." And Sierra Club Staffing Specialist Lynn Lutz observes that once an employee has been in a job for two years, he or she tends to stay an additional five to eight years.

So, if increasing professionalism and the risk of burnout and economic insecurity haven't scared you away, how do you go about trying to land a job? First, you need to know where to look for an opening.

As Okun points out, "Monies do follow social problems." One way to determine where future job opportunities will emerge is to look at the "social pathology of the day." Today's problem will become tomorrow's non-profit program mission.

A few issue areas are currently attracting more attention—and resources—than others. The environment is one of these.

But Okun notes that most of the available jobs are unglamourous and demanding. For example, he says, an applicant who is willing to work the night shift in a homeless shelter would have more options than someone who wants to do marriage counseling in an attractive office.

Goldsmith warns, "More public interest groups are freezing levels of hiring than expanding." The Nature Conservancy, for example, which grew from five to 50 employees during the last ten years, expects to grow only about 5 percent over the next five years, according to Suki Molina, former director of administration.

Nonetheless, entry-level fundraising and community organizing jobs are available in several groups that work on environmental and consumer protection issues. Organizations such as Access To Justice that work to qualify and pass ballot initiatives often need canvassers to raise money and collect signatures. Other groups are performing more basic research to support advocacy efforts, but the Sierra Club's Lutz says few new researcher positions are available.

In most nonprofits, there is frequent turnover in entry-level jobs such as canvasser, organizer, administrative assistant and secretary. Employers fill these positions by hiring volunteers and people from outside the organization. But even in these entry-level positions, most employers look to hire applicants with a broad range of skills that might prove useful to the organization.

Kunofsky, for example, looks for employees who can speak and write well, coordinate group actions and motivate fellow workers —and raise money.

Goldsmith says, "Where you see people hiring is in development positions. They're not hiring political science majors or political advocates as much —they're going more for the MBAs."

Sharon Johnson, listings coordinator for Opportunity NOCs, a biweekly listing of non-profit job openings, says, "Development field opportunities are unlimited for people who can provide proof they can raise money."

Although Molina says "fundraising experience is usually gotten through nonprofit organizations" themselves. She adds that private sector marketing experience would be useful in finding a job doing direct mail fundraising.

The skills used in many private sector jobs transfer well to the non-profit sector. Hearne, for example, believes her paralegal experience helps her manage a non-profit citizens group write grants and fight for changes of land use law. Caroline Tower, executive director of Northern California Grantmakers, says many private sector lawyers and bankers have made successful career changes as executive directors and fundraisers in non-profit groups.

Turnover in upper-level management, fundraising and program positions, however, is much less frequent than in entry-level positions, according to Lutz. Nonprofits often promote from within if current employees have the skills needed to fill these jobs, but Lutz notes the trend toward filling upper-level openings with people who obtained their skills in other nonprofits, government agencies or the private sector.

So what do you do if you lack the breadth of experience or extensive training in particular skills needed to compete for these kinds of jobs? If you don't have ideal credentials, but are willing to work hard, carefully plan your approach and take advantage of available resources, you can find the "right job."

First, you need to figure out how you will fit into the non-profit world. Kunofsky says the key to happiness and success in a non-profit job is commitment to the issues you work on. It pays to learn as much as you can about the issues that concern you. Employers say you often won't get the job unless you are familiar with the organization's cause.

You also have to determine, in general, what kind of work you want to do and how your skills can be used in a non-profit. The best source of information about work is the workplace. You can begin to find out who is active in your field of interest by using the directories and resource centers listed below.

You can also obtain free or inexpensive vocational counseling at services such as the Jewish Vocational and Career Counseling Service, and students and graduates can turn to school placement offices for assistance. Counseling may help you decide what kind of new job would be right for you, says Ann Sparks, program director of Alumnae Resources, a career development service for women graduates of liberal arts schools.

Because employers stress the need for skills that are tailored for a particular position, job hunters should consider basing their searches on the skills they want to use as much as the issues they want to address.

Then comes the most important advice from employers: You have to get out and talk to people. Not only is this the best way to find out what is happening in the sector, but Ostrander says, "Most nonprofits hire by word of mouth." John Peterson agrees. "Networking is the structure of the non-profit business. Become part of the network," he says.

The best way to get into a nonprofit workplace, many experts say, is to volunteer or get an internship. As a volunteer or intern, you can gain indepth knowledge about an organization, develop skills you can use in a paid position and meet people who may be able to help you find a job. Sometimes you can create your own paid position by making yourself indispensable to the group through your good work. Once inside, you're also in a better position to find job openings that fit your interest and talents. This is the most difficult part of the search because "perhaps 80 percent of the jobs are never listed," according to O'Neil.

You can also learn a lot by arranging informational interviews, but Lutz says you shouldn't be surprised if the person you want to see is too busy to talk to you.

Ostrander says, "The sector is pretty close-knit," so you often only learn about jobs by networking.

Experts interviewed for this report mentioned three good ways to work in nonprofits without pay. Most people who are working in the sector began as volunteers in a non-profit, according to Kunofsky. The Volunteer Center of San Francisco is one place to learn about volunteer opportunities. It interviews and refers more than 3,000 people each year for volunteer positions in 530 organizations.

You can also find internships in many nonprofits. These usually unpaid positions often provide opportunities to work more on the issues side of a non-profit group. Some listing services list internship opportunities, and college placement offices often have information about internships students have held in the past.

Finally, if you are in school, you may be able to arrange to do academic work about or for a non-profit. Your research may lead to a non-profit job. For example, the Nature Conservancy recently hired an MBA graduate who had written his thesis on non-profit fundraising.

If you can't afford to take an unpaid position just to gather experience and contacts, training may be your next best bet. Many kinds of non-profit jobs require specific academic training and credentials. For example many social service positions require a degree in social work. A degree in ecology, wildlife management, or environmental studies can help with environmental organizations.

Non-profit managers need increasingly specialized skills to run growing organizations in an era of financial instability. Training in non-profit financial planning, fundraising, marketing and administration is available at many local colleges and in organizations that serve the non-profit community such as the Support Center/CTD. Your studies could range from a half-day seminar on fundraising at Marin Community College to a two-year Masters program in non-profit management at the University of San Francisco.

Vocational counseling organizations, including Alumnae Resources and the Jewish Vocational and Career Counseling Service, also offer seminars in career skills development. Ostrander says more intensive training is far more useful than short-term seminars, which are often "minimally helpful."

But training can be expensive and certainly doesn't pay the rent. If you need to have money coming in, two basic kinds of entry-level jobs are available for job hunters without non-profit experience—canvassing and office support work.

Chris Steel, canvas director for Greenpeace, estimates about 25 non-profit groups are operating canvasses in the Bay Area. The Sunday newpaper classified advertisements usually list several openings for activists. In most of these jobs, you begin as a door to door fundraiser and educator. At CALPIRG, for example, you start as a canvasser earning $170-$200 per week.

You also may be able to get an entry-level office job in a non-profit if you offer good clerical skills or knowledge of computers. Lynn Lutz says about 85 percent of the openings in the Sierra Club are for experienced administrative assistants and clerical people. These jobs are often listed in major newspapers but may not involve real issue work. However, they do give you the opportunity to see the organization in action and, perhaps, develop new skills for future non-profit work.

Whatever route you choose—volunteering, interning, training or an entry-level job—make the most out of the many resources designed to help people find their way through the non-profit-sector job maze. Some of the best places to find listing of non-profit job opportunities in the Bay Area are described below. Publications such as Opportunity NOCs and the Public Interest Employment Report focus almost exclusively on jobs in the non-profit sector, but nonprofits also advertise in the major dailies, ethnic papers, progressive newspapers and newsletters.

The Nature Conservancy and many other groups distribute job notices to other nonprofits for posting on bulletin boards. Kunofsky says the Sierra Club bulletin board is a particularly good place to look for notices.

Don't forget, however, that some organizations such as the Environmental Defense Fund often do not advertise openings at all. Your networking efforts may help identify these jobs.

Nonprofits rarely recruit on campuses, says Michelle Dethke, placement officer at the University of California at Berkeley's Graduate School of Public Policy, and college placement centers put much less emphasis on placing graduates in non-profit jobs than in accommodating private and public sector recruiters.

Once you've found a job listing that appeals to you, many of the same strategies that helped you identify the openings will help you land the job.

"Face to face is important," says Harder. "It's hard to judge people by what you get in the mail." Don Horn, human resources director of the San Francisco AIDS Foundation says you should "learn something about a potential employer and present yourself in that context," both on your resume and in interviews.

Okun says, "Random resumes are not effective job-seeking devices." So don't just send in a resume unless that is the only way the group will let you express your interest. It is easier for an employer to ignore your resume than your person when you are talking to them about a job.

Kunofsky advises people to go straight to the program office where you want to work instead of the personnel office (unless you want to work there). Whenever you can, talk to people with the power to hire you or influence the decision.

How should you present yourself to your prospective employer? Be honest about past experience, even if their background doesn't seem suited for the job. You may not realize how your past experiences prepare you for the job. For example, that volunteer experience organizing a college outdoor club could help a person get a membership coordinator position. Your enthusiasm about the group and the job can also make a big difference. Ostrander says nonprofits "want someone who is committed and will put out."

Now you have explored the non-profit sector, found issues you want to work on, identified ways to use your skills in a non-profit setting, developed contacts, improved your skills, found openings and convinced a group to hire you. Will you be happy?

If you have the persistence and creativity you need to land a job in the non-profit sector, you probably have what it takes to thrive in one.

Only you can know the answer, but following the rigorous process of finding a non-profit job will give you some clues. If you have the dedication, persistence, creativity and persuasiveness you need to land a good non-profit job, you probably have what it takes to thrive in a non-profit setting.

This story was reprinted with permission from the San Francisco Bay Guardian. Author Dave Smith lives in Oakland and works for the Environmental Protection Agency.

Where to find Jobs with Environmental Organizations

The following are organizations, both career and environmental, which carry listings for jobs related to the environment.

The first section consists of organizations whose purpose, or partial purpose, is to help launch environmental and other social change careers. They often print newsletters including recent job listings and are available to help you during their open hours.

The second section lists public interest and social change organizations, most of which are environmental. As a public service, these groups keep listings of socially-conscious job opportunities. However, career assistance is not their primary purpose. Since these organizations are usually extremely busy keeping up with their workload, they often will not have as much time to spend with you as they would like. Please keep this in mind and if you are particularly interested in working with the organization, ask to make an appointment to speak with someone who can help you. It is also a good idea to call in advance since it is possible these organizations will not have any jobs listings on their clipboards or folders that particular day. Job listings usually include opportunities within the organization as well as with other organizations.

ENVIRONMENTAL CAREER ASSISTANCE SERVICES

Alumnae Resources
660 Mission St.
San Francisco, CA 94104
(415) 546–0125
A non-profit organization which specializes in career development, Alumnae Resources has a extensive library including many job listings which is open to members only. Membership is $50.00/year. Workshops and one on one counseling is available for non–members. The workshops vary in price from $10.00 up and one on one counseling is $55.00 for non–members and $45.00 for members.

CEIP Fund Inc.
(Formerly the Center for Environmental Intern Programs)
512 2nd Street,
San Francisco, 94107
(415) 543–4400
Hours: 9–5, Mon.–Fri.
CEIP matches people with paid internship positions in all aspects of the environmental field and provides other environmental career counseling services. Call for an application. Application fee is $15.00 for one region and $10.00 for each additional region. The CEIP Fund also publishes *The Complete Guide to Environmental Careers*, which discusses not only careers working with non-profits, but possibilities ranging from solid waste management to parks and forestry.

EcoNet
Bill Leland, Director
3228 Sacramento St.
San Francisco, 94115
(415) 923–0900
Hours:9–5, Mon.–Fri.
EcoNet is a part of IGC (Inter Global Communications) which is an electronic mail networking service. EcoNet is a service for environmental organizations, but also has a folder entitled EN.JOBS of available environmental jobs on line in the EcoNet system. Call to inquire about the use of this service.

Jewish Vocational and Career Counseling Service
870 Market Street
San Francisco, 94162
(415) 931–3592
Hours: 9:00–4:45, Mon.–Fri.
Offers career counseling for various jobs. Cost is on a sliding scale. Call for an appointment. TheY also carries job listings when they come in and anyone may view them for $1.00.

Opportunity Nocs
The Management Center
944 Market Street, Suite 700
San Francisco, CA 94102
(415) 362–9735
Opportunity Nocs is semi–monthly publication of the Management Center which lists a wide range of public interest jobs, including many environmental jobs. A subscription is $20.00 for a year (24 issues), and $7.00 for three months (6 issues)

Options for Women Over 40
3543 18th Ave., San Francisco
(415) 431–6944
Hours: 10–5, Mon.–Thurs.
A non–profit resource center for women over 40. Offers counseling (free to members and $8–10 for non–members), Vocational testing ($20 members, $25 non–members) and support groups. There is also a job listing center which carries all types of jobs including public interest and is open to the public for $1.00.

Public Interest Clearing House
200 McAllister St.
San Francisco,94102 (mailing address)
110 McAllister St.
San Francisco,94102 (location)
(415) 565-4695
Hours: 9–5, Mon.–Fri.
PICH carries various public interest career newsletters as well as publishing an excellent one of their own. Job listings include government, non profit and legal jobs. Use of the Career Resource Center is free and a subscription to their newsletter is $15.00 for unemployed or low income and $30.00 for all other subscribers.

San Francisco Renaissance
1453 Mission St., fifth floor,
San Francisco, 94103
(415) 863–5337
Hours: 8:30 –5, Mon.–Fri.
The San Francisco Renaissance offers courses and seminars on how to begin one's own business. Course and sem-

Jobs, continued

inar fees vary from $40–$350. Scholarships are available .

Volunteer Center of San Francisco
1160 Battery, Suite 400,
San Francisco, 94111
(415) 982–8999
Hours: 8:30–4:30, Mon.–Fri.
Provides guidance and extensive listings for people seeking to volunteer. Consultations are free and by appointment only.

Volunteer Center of Alameda County
1212 Broadway, Suite 6622
Oakland, 94612
(415) 893–7147
Hours: 8:30–4:45

Hayward Volunteer Center
21455 Birch Street
Hayward, 94541
(415) 538–0554
Hours: 8:30–4:45
Both these volunteer services provide guidance and listings volunteer jobs in the non–profit arena. Call for a consultation.

Y–House
2600 Bancroft Way
Berkeley, 94704
(415) 538–0554
Hours: 9–5, Mon.–Fri.
The Y-House offers guidance for general job seeking in three general areas: resume and interviewing; personal and professional development; and a resource center which is free on the first visit and a $10.00 membership fee upon the second visit. The Y–House also offers drop in support groups which are $5.00 and a brown bag lunch series ($2.00/each) where you can learn about different jobs from a selected guest speaker in that profession. Other classes offered range from $20.00–100.00.

ORGANIZATIONS WITH JOB LISTINGS OPEN TO THE PUBLIC:

American Friends Service Committee
2160 Lake St.
San Francisco, 94121
(415) 752–7767
Hours: 9–5, Mon.–Fri.

California Lawyers for the Arts
Fort Mason Center, Building C.,
San Francisco, 94123
(415) 775–7200
Hours: Mon. –Fri. 10–5

Earth Island Institute
300 Broadway, Ste 28
San Francisco, 94113
(415) 788–3666
Hours: 9:30–5:00, Mon.–Fri.

Ecology Center
2350 San Pablo Ave., Berkeley
(415) 548–2220
Hours: 11–5, Tues.–Sat.

Foundation Center Library
312 Sutter Street, San Francisco, 94108
(415) 397–0902
Hours: Mon., Tues., Thurs., Fri., 10–4:45, Wed., 10–7:45

Friends of the River
Fort Mason Bldg. C, 3rd floor
San Francisco, 94123
(415) 771-0400
Hours: 9:30–5, Mon.–Fri.

Greenpeace Action
139 Townsend (between 2nd and 3rd)
San Francisco, 94107
(415) 512–9025
Hours: 9–5, Mon.–Fri.

Greenbelt Alliance
116 New Montgomery St., Suite 640
San Francisco, 94105
(415) 543–4291
Hours: 9–5, Mon–Fri.

Lindsay Museum
1901 First Ave, Larkey Park
Walnut Creek, CA 94596
(415) 935–1983

Natural Resource Defense Council
90 New Montgomery, Suite 620
San Francisco, 94105
(415) 777–0220
Hours: 9–5, Mon–Fri

Peninsula Conservation Center
2448 Watson Ct., Palo Alto
(415) 494–9301
Hours: 9–5, Mon.–Sat.

Pesticide Action Network
965 Mission St., Rm 514
San Francisco, 94103
(415) 541–9140
Hours: 9–5, Mon.,–Fri.

Ocean Alliance
(415) 441–5970
Bldg E, Fort Mason Center
San Francisco
Hours: 9–5, Mon.–Fri.

Sierra Club
730 Polk Street
San Francisco, 94109
(415) 776–2211
Hours: 9–5, Mon.–Fri.

Sierra Club Job Line
(Sierra Club job openings only)
Phone only. Recorded message of Sierra Club job opportunities.
(415) 978–9085
Hours: 24 hrs/day

The Support Center/CTD
70 Tenth Street, Suite 201
San Francisco, 94103
(415) 552–7584
Hours: 9–6, Mon.–Fri.

Trust for Public Land
116 New Montgomery St., 4th floor,
San Francisco, 94105
(415) 495–4014
Hours: 8–5, Mon.–Fri.

Green Guidelines

A short list of things you can do

By Eric Ingersoll

The planet may be poised on the brink of destruction, but there are many things we can all do to help pull it back from the edge. We can be responsible consumers. We can take a little time to get informed about the issues, so we can make responsible choices when voting. We can make a phone call or write a letter to a politician or an organization once a week.

It helps to get systematic. Do an environmental audit of your household. Put a sheet of paper on your refrigerator on which to note your "ecological infringements." At the end of the week decide which seem the worst to you and then do something about them.

Being environmentally responsible can often be enjoyable, like when you return junk mail unopened with instructions to strike your name from the mailing list, or cancel subscriptions to newspapers and magazines you never read anyway.

Here is a short list of things you can do. For more information on how to go about any of the ideas listed here, just refer to the index at the back of the book under the appropriate heading. There you'll findreferences to articles, organizations, and relevant Consumer Guide entries.

• **Recycle everything.**

Try not to buy disposable items that can't be recycled. Resist packaging which isn't made from recycled materials and can't be easily recycled. Tell the businesses you patronize that you care about this, and that if affects your decisions about where to shop. If the manager of your local supermarket or fast food restaurant has ten people a day complaining about the polystyrene containers at the deli counter. He or she will probably make some changes pretty quickly.

• **Buy recycled products.**

If you're not buying recycled products, you're not recycling. There must be consumer demand for recycled products if recycling is to become widespread. So if there is a recycled alternative, choose it. When you know about a recycled alternative, ask your local merchants to carry it.

• **Pressure others to use recycled paper.**

For instance, call your local newspaper and tell them you will cancel your subscription unless they switch to printing on recycled newsprint. Tell them that recycled newsprint is cheap, of equal or better quality, and is made from old newspapers, not trees. There is essentially no excuse not to be using recycled paper except in a few specialized applications. We need to let the paper companies know that we are serious about making paper from paper, not from trees.

• **Plant trees.**

The human quest for building materials, paper, and fuel is deforesting the planet at a frightening pace. So plant trees and encourage others to do the same. Volunteer with a tree planting group, or organize your neighborhood, church, business or community to do group plantings. If each American planted one tree per month there would be three billion new trees each year!

• **Save the forests.**

Temperate and tropical virgin forests are threatened as never before. If we cannot change the course of this destruction, we will be the last generation to live with these complex and irreplaceable ecosystems. Don't buy redwood, unless it is used, and don't buy tropical hardwoods. If they are selling these kinds of wood, let your local lumberyard know that you are concerned about their role in this destruction. Look for sustainably-farmed alternatives. Try to use recycled building supplies whenever possible. (The building supply section of the *Consumer Guide* lists several companies that practice sustainable forestry, as well as sources for recycled or reusable building materials.)

• **Buy Organic.**

Organically grown food is more than just health food. Organic farming is based on sustainable practices which are actually good for the land. They don't damage the ecosystem, and they don't contaminate water supplies with pesticide and fertilizer runoff. Farmworkers are not being poisoned in the process of growing the food. Buying organic food supports farmers who are doing the right thing, and it encourages other farmers to join in. So when you go to the grocery store, look for certified organics. (the Food section of the *Consumer Guide* is full of places to buy organic foods. See also the articles in the Food and Agriculturechapter.)

• **Drive Less.**

Find a job closer to home or move closer to work. (Who ever said everything listed here would be easy?) Ride with friends whenever possible. Pretend gas costs $10 a gallon. This may be close to its real price if you factor in hidden costs. Consider sharing a car rather than owning your own. Get to know public transportation. Become in-formed on transportation issues, and pressure officials to improve mass transit.

• **Ride your bicycle**.

Make your bicycle your preferred mode of transportation — it's healthier, cheaper, and much better for the environment. Find all the good streets to ride on. Lobby your city to make it easier to commute on a bicycle by creating bike lanes. When you're driving, be kind to bicyclists. The next time you go on vacation, consider making it a bicycle trip.

• **Defend open space.**

Once open space is built upon, it will most likely never be open again. We need to stop trading this diminishing resource for short-term benefits. Imagine what the Bay Area would be like with 1.5 million acres *less* open space That is the amount which is either already under development or threatened with development in the next 20 years. We can and should preserve it now. So find out which open spaces in your area are endangered and join the groups working to protect them. (See *Threats to the Greenbelt* on page 87.)

• **Support sensible housing policy.**

Increasing density in mixed-use urban settings has many advantages over growth by sprawl, from savings in transportation to improvement in community values. Suburbs aren't going to work in the long run. We need to support our cities and help turn them into safe, pleasant, and sustainable places to live, play and work.

• **Grow food at home**.

The average morsel of food on the American dinner plate travels 1400 miles by truck, rail, or plane to get there. That energy investment is added to the environmental costs of chemical-intensive agricultural techniques. So if you have a garden, growing your own food makes a lot of sense. When we grow our own food, we have more control over what goes into it, and consequently what goes into us. It can also bring a tremendous sense of satisfaction; it brings you into touch with natural systems, with the earth, and with the mystery of life.

• **Make compost, not garbage.**

Compost is valuable and garbage is a problem. The only difference between them is a bit of care. It takes only a little more effort to make and keep a compost heap than it does to take out the trash — and it's a lot more fun. America loses billions of tons of topsoil every year. Composting is the beginning of returning some of those lost nutrients back to the land. Support municipal composting operations, and lobby to get closed landfills converted to this use. Compostables now make up about 45 percent of municipal waste.

• **Use energy-efficient appliances.**

When it comes time to replace household appliances, be sure to take energy efficiency into account. Remember, you will save money with lower utility bills and, if enough people do likewise, we will need fewer power plants(and can retire nuclear ones). Less pollution will result and less carbon dioxide will be contributed to the atmosphere. Your choice of an energy efficient product also sends a signal to the manufacturer and to others. (Even better is to let the manufacturer know directly why you chose their product, and let others know why you didn't choose theirs.)

• **Support solar energy.**

Most of the Bay Area has an excellent climate for solar energy, and although system installations can be expensive, operation afterwards is essentially free. The development of solar energy creates more employment and less pollution than other forms of energy. It adds nothing to the greenhouse effect, and supplies are projected to last another two billion years or so. Since the federal government stopped supporting the solar industry during the REagan Administration, our country has fallen far behind in the development of solar technologies. But solar is the safest and cheapest kind of energy in the long run. So lobby for government support for solar energy. (No large oil companies are going to do it for us.)

• **Improve your workplace**

There are many more improvements businesses can make than we can list here. All officesshould recycle their paper. Setting up an office recycling systemmay seem like a hassle, but its not difficult. Once in place, it's easy to maintain and may even save your company money. No one needs to use styrofoam coffee cups. Most office paper needs can be met with recycled paper. Your purchasing decisions add up to a change in the way our economy uses resources. (See Chapter 3 for other ideas)

• **Get and stay informed**

Pick some issues you feel are important, and learn all you can about them. An effective public is an informed public; we cannot change direction if we don't know which way we're going now. There is no need to become an overnight expert in everything. Rather, choose a few things to concentrate on and become informed and effective in a certain area. You will find that eventually the issues all connect anyway.

• **Join an organization and participate.**

It is worth making the time in our busy lives to ensure that we and our children have a healthy planet to live on, clean air to breathe, good food to eat, and pure water to drink. We can no longer take these things for granted, nor can we delegate the responsibility for them to someone else. We've tried that and it hasn't worked. Find a level of involvement that will work for you. Your contribution is important. Most of the organizations listed in this book have openings for volunteers; fill one of them.

• **Let someone know what you think — once a week.**

If we all called a representative or someone on a board or commission once a week to let them know how we felt, our democratic system might perk up and begin to work a bit better. Call the League of Conservation Voters (415-896-5550) for a copy of their environmental voting guide, which contains all the names, addresses, phone numbers and voting records of your representatives. (It also includes an explanation of some important issues.) The forces that lead to much environmental degradation have the money to pay lobbyists; these are the voices our legislators and officials hear the most. Be sure they hear other voices — your voice — as well.

Eric Ingersoll is publisher of the Bay Area Green Pages.

Section Two

REVIEWS

There is a great deal more that could be said about any one of the topics covered in The Green Digest, as well as on quite a few topics that aren't covered there at all. That's why we've included this section.

The books, periodicals, and other publications reviewed here were selected for the value of the information they provide, their local focus, or the originality of their thought. They are not necessarily the most recent works on their respective subjects; just (in our estimation) the best.

Attentive readers will notice that the children's education portion has more entries than most of the others. This is not only because there is at present a plethora of excellent environmental books for children on the market, or because we firmly believe that it is important to inculcate our young with sound environmental values.. The other part of the reason is that these books are often quite educational for adults as well. We certainly enjoyed reading them.

There are undoubtedly titles omitted that deserve to be here. We hope our readers will inform us of such omissions, and we will make every attempt to correct them in our next edition.

The reviews are organized by subject:

Agriculture

BOOKS

Alternative Agriculture
National Academy Press
National Research Council
Washington, D.C., 1989

More than 200,000 farms folded in the 1980s. Pesticides, nitrates and manure have become the largest non-point sources of groundwater pollution in this country. Mono-cropping is wearing out our soils. Yet federal crop programs continue to mandate the use of chemicals and punish the practice of crop-rotation.

Alternative Agriculture, a report released by the National Research Council in response to crises facing U.S. agriculture, is an engaging and highly informative book. Somewhat technical in tone, it is nonetheless accessible to lay people interested in comparing chemical and alternative farming methods.

The authors begin with a detailed discussion of the damage resulting from heavy use of fertilizers, pesticides, and irrigation water. They also describe the ways in which government programs, including commodity price and income support programs, prevent alternatives to these practices. A description of alternative farming systems is accompanied by eleven case studies which the authors use to illustrate these unconventional— but very viable— solutions in action. Concluding in their *Executive Summary* that such techniques are often more productive and profitable than traditional chemical methods,. This book is vital to understanding how federal programs do and don't work, and what alternatives are available.

California Farmer to Consumer Directory 1990
California Department of Food and Agriculture

This is a directory of growers involved in direct farmer-to- consumer sales and is an excellent resource for those interested in buying produce fresh from the farm (See **Farmers Markets** in the *Consumer Guide*). Bulk supplies for canning and freezing and smaller quantities for immediate consumption are both available for purchase.

Information about local farms contained in this directory includes: products sold, growing seasons, organic status of crops, and whether produce is pre-picked or can be picked by consumers. Additional farm services, including tours, demonstration plots, refreshments, picnic areas, and petting zoos are listed as well. Receive directories free of charge by calling 1-800-952-5272 or writing to P.O. Box 9427-10001 Sacramento CA

Circle of Poison
David Weir and Mark Shapiro
Institute For Food And Development Policy
San Francisco, CA 1981

Although Circle of Poison was written ten years ago, the issue it addresses, global pesticide dumping, remains relevant today. This book traces the cycle of these toxic chemicals: pesticides, illegal or restricted for use in the U. S., are manufactured here, exported to underdeveloped countries and then returned as residues on imported food shipments.

Weir and Shapiro carefully follow the path of destruction and death these toxic chemicals leave across the globe: American workers are disabled, poor nations fall deeply in debt, improper or careless storage of pesticides leads to contamination of crucial water resources, and every minute someone in the third world is poisoned by pesticides.

In addition to providing detailed documentation of the pesticides, corporations and countries involved, the authors also suggest ways to help break this toxic circle.

Healthy Harvest III: A Directory of Sustainable Agriculture
Potomac Valley Press
Washington D.C.

In its third year of publication, this directory is a helpful tool for anyone interested in locating organizations working in the field of sustainable agriculture and horticulture. The guide provides addresses and brief descriptions for over 1000 organizations ranging from the Urban Farmer Store in San Francisco to the North American Fruit Explorers in Illinois to Approtech Asian the Philippines. Listings include: suppliers of beneficial organisms, organic farms and certification groups, extension services, botanical gardens, libraries, volunteer opportunities, publications, research centers, development groups, cooperatives, and universities with agroecology programs. If you cannot find *Healthy Harvest* in bookstores, it can be ordered through AgAccess, 603 4th Street, Davis CA, 95616

Farming on the Edge
John Hart
UC Press
Berkeley CA !990

Part celebration of farm community, part case study in sucessful land use planning, *Farming on the Edge* offers what could be a nationwide model for farmland preservation.

During the 1960's Marin county had plans to build suburbs from shore to shore. Freeways were about to cut across the farmland and land prices were rising.

In 1970 a group of politicians, farmers, and citizen activists forged an alliance to protect the county's agricultural land from suburban "conversion." What was an uneasy alliance at first, with ranchers and environmentalists used to battling each other over over zoning, has grown into a friendship based on mutual respect and trust.

At a time when the United States is in danger of becoming the world's first suburban country, this book profiles the sense of stewardship and rooted way of life that make our family farm inheritance vital to preserve.

1990 National Organic Wholesalers' Directory and Yearbook

California Action Network
Davis, CA, 1990

Unlike *Healthy Harvest* the listings in this directory are limited to individuals and organizations involved with organic food, including wholesalers, farmers, farm suppliers, certification groups, and supporting businesses. As the title suggests, this is primarily a resource for farmers wishing to learn how to certify and market their products, or for retailers looking for organic suppliers. The book may also interest consumers seeking items like beneficial insects or farmers to supply community cooperatives. Articles covering organic issues and definitions and summaries of organic laws nationwide are also helpful to lay readers.

PERIODICALS

Agrarian Advocate

California Action Network
P.O. Box 464 Davis, CA, 95617

A quarterly publication, the *Advocate* is a newsletter of California Action Network, the same organization which produced *The 1990 National Organic Wholesalers Directory*. Like the *Directory*, this resource is geared mostly to farmers and others involved in sustainable agriculture. The *Advocate* offerings include legislative updates and hearings on agricultural matters, articles on issues such as pesticides and water reclamation, and discussions of CAN activities and employment opportunities. A subscription to this publication is free with a membership in California Action Network.

CCOF Newsletter

P.O.Box 8136 Santa Cruz, CA, 95610.

This quarterly publication of the California Certified Organic Farmers covers certification standards, sustainable agriculture legislation and innovations in organic production. Each issue also includes book reviews, CCOF news and chapter reports, job announcements, and a calender of sustainable agriculture events. A one year subscription is available with a CCOF membership.

Organic Food Matters

Committee for Sustainable Agriculture
P.O. Box 1300, Colfax, CA.

This quarterly journal provides a wealth of information on sustainable agriculture issues. Innovations in alternative farming, European organic systems, and legislative updates are just a few of the topics covered.

Although agricultural issues are emphasized, farmers and consumers alike will enjoy articles on everything from pesticide contamination of water supplies to overflowing landfills and environmental degradation in the USSR. Subscriptions are available by becoming a supporting member of the Committee for Sustainable Agriculture.

RESOURCES

Agricultural Information and Publications

University of California, Davis, CA, 95616

This is a catalog of reasonably-priced publications produced by cooperative extension offices. Each publication is written by experts and subject to review by UC Davis. Topics include: gardening, landscaping, pest control, livestock management, and fruit and nut tree-planting tips.

At 50 cents this catalog is a real bargain. You can call the extension office at (916) 757-8930 Fax (916) 757-8940.

Atmospheric Change

BOOKS

Global Warming

Stephen H. Schneider
Sierra Club Books 1989

This thoroughly researched and well documented book, by one of this country's leading climatologists, starts off with a jolt. Schneider presents a prophetic look at a year in the "greenhouse century." It doesn't look good: California smothers under heat, smog, water shortages, and raging forest fires; the great lakes and Mississippi River run so low that commerce is threatened and toxic sediments are exposed: New York experiences summer heat waves so intense that hospital emergency rooms are jammed with victims: and the Great Plains "breadbasket" shifts north to Canada as the United States dries up. Meanwhile other parts of the globe experience devastating floods and hurricanes.

In subsequent chapters Schneider leaves speculation aside and examines the arguments and evidence for increasing carbon dioxide levels, increasing world average temperatures and the relationship between the two. What saves this book from being boring and technical is that Schneider is a good writer and he is able to transmit his own enthusiasm for the study of weather and climate. Sierra Club Books, 730 Polk St., San Francisco, CA 94109

Global Warming: The Greenpeace Report
Jeremy Leggett
Oxford University Press, 1990

Since 1988, when news of the greenhouse effect hit the press, at least a dozen lay books, hundreds of technical reports, and thousands of newspaper and magazine articles have been published. Climatologists, physicists, hydrologists, biologists, sociologists, economists, anthropologists, international forums, Congresspeople, senators, federal agencies, state agencies, local agencies, and private citizens have all jumped on the greenhouse bandwagon.

If you want a book that combines technical acuity with easily-read language, sharp analyses of greenhouse politics with practical suggestions of what to do instead, read *Global Warming: The Greenpeace Report*. Written in response to the Inter-governmental Panel on Climate Change's most recent report, the Greenpeace Report documents the science behind global warming and the politics behind the science. This is eye-opening in itself. But the Report goes on to present well-researched arguments for the viability of switching to renewable resources and energy-efficiency to power the world's future— recommendations that were glaringly lacking in the IPCC report.

It's hard to recommend science and policy as pleasure reading. But besides being technically complete and politically persuasive, the Greenpeace Report is a great book. Even if you're not directly involved in greenhouse action, it is a compelling book. It's the best one published on our earth's climatic future so far.

Greenpeace plans to distribute to each of its 24 member countries, and Oxford University Press will distribute to all English-speaking countries, including its former colonies. Summaries are already available in German, Japanese, and Spanish. Available in bookstores or through Oxford University Press, 200 Madison Ave., Section EC, New York, New York 10016. Add $1 for mail orders.

Ozone Crisis
Sharon L. Roan
John Wiley & Sons
New York, NY 1989

This book is a good journalistic account of the "ozone crisis". The story starts with the 1974 discovery by two U.C. Berkeley chemists that CFC's, thought to be one of the most benign families of chemicals, were destroying the planet's protection against the dangerous end of the UV spectrum. What amazes and dismays the reader is how long it took before anybody in government was willing to do anything. It is a classic case of industry saying that the evidence is not conclusive, and the government commissioning more studies which are then ignored by the policymakers because they are subjected to a new round of pressure from industry groups. What this book makes abundantly clear is that, with the way our society is set up, we just cannot respond fast enough to crises which demand broad–ranging and comprehensive action. It also shows that we need to change our licensing procedure for chemicals from the current "innocent until proven guilty" mode which allows corporations to go on producing and using dangerous chemicals as long as they can delay or confuse the hearings. It is astonishing that there are still people in government and industry who, in the face of an impressive amount of evidence, are willing to risk the future of the planet for a few million dollars.

Building Design

BOOKS

Bioshelters, Ocean Arks, City Farming
Nancy Jack Todd and John Todd
Sierra Club Books
San Francisco, CA 1984

Nancy Jack Todd and John Todd are the founders of the New Alchemy Institute on Cape Cod, where teams of biologists have developed complex self-supporting artificial ecosystems. These are excellent models for appropriate technological development land restoration. They use natural cycles to develop renewable energy and agriculture systems.

Bioshelters offers a number of environmental precepts "design should follow, not oppose, the laws of life; design must reflect bioregionality; projects should be based on renewable energy sources; design should be sustainable through the integration of living systems and coevolutionary with the natural world, and should help heal the planet." The redesigning of communities by working with existing structures, recycling, transportation, power, and employment is explored.

Current industrial culture is not sustainable. Our present systems are fueled by consuming, and ultimately destroying, our life support systems. The New Alchemists are exploring the real possibilities of living *with* nature.

More Other Homes and Garbage: Designs for Self Sufficient Living

Jim Leckie, Gil Masters, Harry Whitehouse, Lily Young
Sierra Club Books
San Francisco, 1981

This book is a comprehensive do-it-yourself guide to building and installing the ultimate energy-efficient and independent home. It explains how site-selection and climatic factors influence a building's design, and how solar lighting, heating, insulation, and wind-protection can be used to maximize your energy-efficiency and comfort. Solar, photovoltaic, wind and hydroelectric energy systems are also discussed.

Unlike other alternative home guides, this book includes a section on waste treatment, which includes information on methane digesters and greywater systems. The last section of the book describes gardening, composting, and aquaculture. This book is packed with information.

The Integral Urban House: Self Reliant Living in the City

Helga Olkowski, Tom Javits, and the Farallones Institute
Sierra Club Books
San Francisco, 1979.

The Integral Urban House was an experiment in sustainable city living. It proved that you can grow your own food, recycle wastes, and produce energy in a limited amount of space. This book tells the story of how the house was planned, designed, and operated, and how it served as a working example of the enormous potential for residential self-sufficiency in urban areas. Concise text and pleasing diagrams instruct you in building your own greenhouse, greywater system, composting bin, and solar systems, and on raising rabbits and chickens to round out your diet. The section on plants covers everything from the Ph of soil to regulating humidity.

The Smart Kitchen

David Goldbeck
Ceres Press
Woodstock, NY, 1989

Welcome to the ideal kitchen: thrifty, efficient, spacious, comfortable, safe, environmentally responsible, non-toxic, and recycling-friendly. If you are curious about appliances, building materials, compostable kitchen wastes, or safety devices, this book has the answers. David Goldbeck describes the safest place to locate electrical outlets and teaches you to measure the warm and cool air flows within your kitchen for efficient cabinet location. Simple designs for building a compost container into your countertop make it easy to recycle wastes. In addition, Goldbeck uses ingenious and practical designs to solve age-old kitchen problems like poor lighting or lack of sink and storage space.

PERIODICALS

Design Spirit

438 Third Street, Brooklyn, NY 11215

A beautifully designed and refreshing new magazine, *Design Spirit* reaches out to artists and designers who wish to explore artistic and architectural alternatives. It investigates environmentally-sensitive building methods, and profiles craftsmen and architects whose work harmonizes with the natural world. A "Resource Roundup" section lists various organizations which relate public information on non-toxic art supplies, appropriate technology, and other products. The calendar of events contains a comprehensive directory of national conferences for those who are interested in sacred sites, non-toxic building, other alternative architectural methods, artists' workshops, and exhibitions that emphasize the environment. Anyone interested in striking a harmony between the human-altered and natural environment will enjoy this magazine .

Home Energy

2124 Kittredge, Suite 95
Berkeley, CA 94704-9942
415-524-5405

This bi-monthly publication is aimed at energy industry professionals, including auditors, retrofitters, contractors, and utility representatives. But for consumers who need accurate, detailed, and easy-to-understand information, this is the best energy-efficiency magazine available. "Home Energy" covers everything from leaky ducts to lighting, and from radon to weatherization. Each issue includes feature articles on energy efficiency and conservation measures, plus boxes on how the newest technology works, how it's installed, and how cost-effective it is.

Children's Educational

BOOKS

50 Simple Things Kids Can Do To Save The Earth

Universal Press Syndicate
The Earthworks Group
Kansas City, 1990.

This is a scaled-down version of the adult neo-classic, *50 Simple Things You Can Do To Save The Earth.* It is divided into two main sections. "What's Happening" explains environmental problems like the greenhouse effect and air pollution, and "Kid's Can Save the Earth By...," is divided into specific topics and activities.

This second section is the key to the Earthworks' philosophy that even kids can help save our planet. The book is full of little quizzes, things to send away for, eco-experiments, and activities from tree-planting to buying non-toxic toys. It's a well-written, non-paternalistic guide for kids concerned with their world, and it may help them get over any feelings of helplessness or being overwhelmed.

Hug A Tree
Williams, Rockwell & Sherwood
Gryphon House
Mt. Ranier, MD,1986

This book is chock-full of fun activities to do with children aged two and up, though it may be more useful for educators than parents. It provides excellent tips for organizing outdoor education programs and an extensive bibliography of books for kids about science. Moreover, a section called "Expanding the Outdoor Experience" has many useful suggestions for helping kids narrate the activities in the living room or classroom.

The activities themselves are written in a simple, step-by-step fashion, with a section called "Want to do More?" providing further suggestions. All of the materials are listed in the margin, and most are easy to obtain. The delight of the book is in the wide range of activities the authors have devised, which range from "Bury the Sock," in which kids learn about organic materials to "Talking About Monsters" and a discussion of "icky" things in the wild which culminates with a celebration of nature's diversity.

Manure, Meadows, and Milkshakes
The Hidden Villa Environmental Project
Tioga Press
Palo Alto, 1989

"At Hidden Villa we made animals our brothers. Sometimes I think about people who don't care about beautiful forests and pollution, but if they went to Hidden Villa they would just go to work to clean it up," said a fourth grader in the Hidden Villa Environmental Program in Los Altos. This book isgreat because the teachers and students of Hidden Villa are people who care about nature. The activities in this book are designed in that spirit: to stimulate, educate, and nurture children and the land.

This book is for environmental educators, and each activity is described in terms of "insight," "preparation," "action(s)," and "follow up." Creativity is valued, and the authors recommend using singing, puppetry, writing, and dance to help children describe their experiences. Stories about John Muir and World Food Day are inspiring, and a great bibliography and discography are found at the end. This is a valuable book for educators interested in teaching children to respect nature and find their proper relationship with the outdoors; it is full of love and wisdom.

Sharing Nature With Children
Joseph Cornell
Dawn Publications
Nevada City, CA 1979

From Joseph Cornell's dedication "to those who experience nature's inspiring, transforming moments, and who desire to share with others their love for the natural world," to the end, this is an activity book brimming with compassion and soul. Cornell urges you to explore nature with your heart as well as your mind, and his writing is so full of awe and respect for nature and kids that he inspires with every sentence. He describes many activities for ages 3 and up, for groups of all sizes, which are defined by one of three "moods": the calm- reflective bear, the active- observational crow, or the energetic-playful otter. He sprinkles the activities with personal anecdotes, which are amusing and revealing. This book could turn any parent into a nature educator, and any child into a nature lover. His five "suggestions for good teaching": "teach less, and share more; be receptive; focus the child's attention; look and experience first, talk later; and a sense of joy should permeate the experience" are tenets for any work and life with children, not just environmental education.

The Sense of Wonder
Rachel Carson
Photographs by Charles Pratt and others
Harper and Row
New York, 1956

Rachel Carson muses, "if I had influence with the good fairy who is supposed to preside over the christening of all children I should ask that her gift to each child in the world would be a sense of wonder so indestructible that it would last throughout life, as an unfailing antidote against the boredom and disenchantments of later years, the sterile preoccupation with things that are artificial, the alienation from the sources of our strength."

That one paragraph shows that this book is not exclusively for children, but for adults who enjoy the twin "wonders" of children and nature. Carson's story of her adventures with her nephew over a number of years spent in an isolated Maine cabin, and the accompanying photographs of the setting and children enjoying it, inspire Carson's "sense of wonder" in her reader.

PERIODICALS

Conservation and the Water Cycle
U.S. Department of Agriculture
Soil Conservation Service
P.O. Box 2890, Washington, DC 20013.

This USDA pamphlet is brief, informative, and free. It illustrates and describes the water cycle in a straightforward fashion, though younger children might have some trouble with the vocabulary (e.g. evaporation is mentioned, but not defined). Additional text describes the relationship between soil conservation and the water cycle. Concepts like erosion, run-off, and mulching are discussed intelligently, but there is no excuse for the sexism implicit in sentences like, "It is man's obligation to return water to streams, lakes, and oceans as clean as possible and with the least waste."

Parents' Press

1454 Sixth St., Berkeley, CA 94710

This monthly newspaper is a great resource for families in Alameda, Contra Costa, Marin, and San Francisco Counties. It includes a community calendar arranged county-by-county, as well as a "Family Calendar" with everything from concerts to nature walks, regular departments like "Growth and Development" and "Projects," and feature articles with a local focus. There are a good number of "Media Reviews," which cover books, music and movies from a parent's perspective, paying special attention to violence levels. The advertising section lists many educational opportunities, grouped by region, for parents and children, places which buy and sell used children's clothes, furniture and toys. Free at supermarkets.

RESOURCES

How to Organize a Rainforest Awareness Week in Your School

Creating Our Future
398 N. Ferndale, Mill Valley, CA 94941
415/381-6744

Although designed for high school students, this is a very empowering resource for almost any type of organization. It has a double duty. First, it discusses the rainforest, by covering the geographical areas of Hawaii, Malaysia, Central America and Brazil, and environmental issues, including cattle grazing and the rights of indigenous peoples. Second, the booklet serves as a tool for organizing and executing activities, complete with a sample schedule and many ideas for teachers and students to make a political statement inside or outside the classroom.

This is an extremely thorough and impressive sixty-page booklet, presently being updated to include even more great ways to save the rainforest or promote your favorite cause.

EPA Materials

Office Of Public Affairs
Washington, DC 20460
(202)475-7751.

The Environmental Protection Agency provides a wealth of information to educators and parents. Teachers can order lesson plans and other materials about recycling, as well as a booklet called *Earth Trek*, which contains information and activities related to pollution, an index of EPA laws, and a dictionary of environmental terms. Information about "The President's Environmental Youth Awards" is available, as is an *Earth Day Educator's Sourcebook* for grades K-6 or 7-12, which contains classroom activities with environmental themes.

Books for Young People On Environmental Issues is an excellent bibliography for parents, divided into grade levels for younger children and by subjects for kids 12 and up. The EPA also has information for adults concerned with children's exposure to pesticides.

Community Design

BOOKS

A Green City Program for SF Bay Area Cities

Peter Berg, Beryl Magilavy & Seth Zuckerman
San Francisco, 1989

This book is the result of a series of *Green City* meetings held in San Francisco in 1986, and it contains an excellent blueprint for achieving urban sustainability. The book is divided into nine chapters on topics including *Urban Wild Habitat* and *Neighborhood Character and Empowerment*, which are extensively crossed referenced. Each chapter addresses the topic by describing current problems, such as lack of neighborhood unity, and suggesting ways to solve it, like encouraging local celebrations in order to strengthen connections among residents.

The most powerful aspect of this book, however, is its vision of a green future. Each chapter ends with a *Fable*, based partially on real experiences, but not too far out-of-reach. The fables, along with a description of conditions in the semi-mythical Green City, make this book both utilitarian and a mini work of art. A resource guide at the end of the book empowers the reader to design and build the Green City of tomorrow.

Appropriate Technology Sourcebook

Ken Darrow and Mike Saxenian
Volunteers in Asia, .
Stanford, CA, 1986

This 800-page bibliography of the best available literature on small-scale technology focuses on developing countries. Energy– and ecology–conscious consumers in the overly developed world will also find much here. The Introduction is an excellent summary of appropriate technology issues. The following chapters, provide both background readings on A.T. and specific readings on agriculture, aquaculture, water supply and sanitation, energy, transportation, health care, education, communications, disaster preparedness, and many other subjects. The numerous technical illustrations facilitate a better understanding of the technology. There may be no better collection of appropriate technology materials in existence.

Bolo'Bolo

P.M.Foreign Agent Series
Semiotext(e)522 Philosophy Hall, Columbia University, New York 1985

This is an inspiring and clear vision of full and fulfilling human expression and co-existence, in a society that lacks a monetary system or government. Communities are based on the "Bolo," a voluntary gathering of people in groups that have a common interest. A Bolo is a collection of communal houses around plots of land that are worked by its inhabitants. It is more than a cooperative. It is the personal expression of a common desire. There could be any groupings imaginable, including "Blue-bolo, Paleo-bolo, Dia-bolo, Punk Bolo, Krishna-Bolo, Taro-bolo, Jesu-bolo, Tao-bolo, Marl-bolo, Necro-bolo..."It all depends on the wishes of an intelligent and self-directed populace. There is also life outside of the Bolos, allowing for expression by anti-social elements among the people.

Exchange is key in Bolo'bolo (The Bolo of all the bolos), but it is not simply a barter system. People take pride in their identities, their community's wares, cuisine, and expertise. People express this pride through the exchange of their particular specialties. Travel among bolos is unlimited, and exposure and reputation are of great importance to each person.

The book adopts a new language that assists the reader in stripping away assumptions and biases about how a society must be run. A society does not have to have a currency, rules, or conformity. It can allow for violence, peace, joy, pettiness, rock and roll, hate, love, salsa or marmite. It can have all, some, or none of these aspects of human expression. Society should be the willful expression of all human traits unhindered. Anarchy is only relative to a norm, but self–respect, connection to the land, and creativity are real and essential.

Bolo'bolo gives an exciting glimpse of a possible human existence. Unfortunately, the path to that world is not as well envisioned by the author as the destination. P.M. has outlined an elaborate global process for the removal of this planet's resource-hoarding and eco-gutting systems. It is a somewhat unwieldy technique of establishing cross-cultural links of understanding around the globe. Though it may be possible for people of the First World to travel to link up with others, this is an unlikely scenario. Nevertheless, Bolo'bolo is an attractive place all of itself. There should be no trouble convincing others of this fact even close to home.

Cohousing: A Contemporary Approach to Housing Ourselves

Katherine McCamant &Charles Durret
Ten Speed Press
Berkeley, 1989

Cohousing is a term coined by the two Berkeley-based architect/authors to describe the phenomenon they observed first-hand in Denmark, where self–organized communities live in clustered housing and share meals, open space, child care, and maintenance. Everyone from toddlers to the elderly can benefit from cohousing. Children have lots of playmates and supervision, working adults have a meal and some company waiting for them in the evening, and senior citizens have neighbors who look out for them.

This book presents cohousing from two perspectives. The first part of the book is devoted to the authors' "home stay" program in a cohousing development in Denmark, while the second part contains more technical architectural information. The writing is intelligent and accessible, and the illustrations, including photos and plans, are beautifully presented. There is a good chapter on energy use, and the concluding chapter, *Translating Cohousing to the United States*, is interesting and inspiring. This is the first book in English which describes the process by which interested parties meet, plan, and build a community.

Ecocity Berkeley

Richard Register
North Atlantic
Berkeley, 1987

Richard Register's book tries to cover a lot of ground. In the first section, he defines, describes, and contextualizes the ecocity concept—with mixed results. In the second, more successful, section of the book he transforms present-day Berkeley into a working ecocity.

Residents and those familiar with Berkeley will especially appreciate Register's suggestions and the sketches which accompany them. His plan for creek restoration leading to a canaled city is particularly provocative, as is his proposal for an Ecology/Peace Center/Museum. Register's book is best when dealing with a specific problem and solution, such as a garbage mountain built from unrecyclable waste. He includes some fanciful plans, like one for a giant redwood forest in downtown San Francisco.

Shading Our Cities

Gary Moll & Sarah Edenbeck
Island Press
Washington DC, 1989

Do you want to start a tree-planting program in your town? This book is a complete resource guide for "urban and community forests," including many readings on the benefits of urban greening and information on how to turn barren, concrete cityscapes into lush, healthy treescapes.

The beginning of the book is a primer on urban forestry, with articles on various aspects of urban forestry, like *Branches and Wires:The Conflict Above* and *In Search of an Ecological Urban Landscape.* The latter part of the book is more action-oriented. Profiles of several programs, including Friends of the Urban Forest in San Francisco— help bring the vision alive for readers, as does the engaging and moving prose style. Extensive appendices, with resources from pamphlets to computer software to tree-planting organizations will help bring the vision to fruitation.

The Ohlone Way: Indian Life in the San Francisco-Monterey Bay

Malcolm Margolin
Heyday Books 1978
P.O. Box 9145 Berkeley CA 94709

This moving book evokes a past that will haunt you. Margolin has done a meticulous reconstruction from the few historical accounts available of the lives and ways of the Ohlone people, the original dwellers by the Bay.

Prior to the arrival of the Spanish, the Bay Area supported the densest population of Native Americans north of Mexico. Ten thousand people, members of approximately forty different groups, co-existed peacefully with each other and the natural wonders of the Bay. The Ohlone people lived a harmonious life in their villages, and amongst their neighbors. Their systems of cooperation and resource sharing, seasonally appropriate sustainable harvest, and the spiritual framework observed in daily life give modern bay inhabitants tenets for behavior and survival. *The Ohlone Way* draws a picture of people that are still an integral part of their environment. The book presents a temporal facet of our region that lasted much longer than the current one has–perhaps for good reason.

The San Francisco Bay Area: A Metropolis in Perspective

Mel Scott
University of California Press
Berkeley, 1985(out of print)

"The work was a pioneer effort to trace the influence of time and geography on the planning and re-planning of cities and to delineate the emergence of an interdependent, closely-linked metropolitan complex extending into the several valleys whose water courses drain into the bay," wrote Mel Scott in the preface to the 1985 edition of this book. Originally published in 1959, Scott's work wove social, economic, and political threads into a history of the region which was both entertaining and perceptive. A new chapter in this latest edition adds a most intelligent discussion of the factors involved in the struggle for regional rule.

JOURNALS

Journal of the New Alchemists

The New Alchemy Institute
Falmouth MA

This Journal is almost too good to be true. For the last twenty years the New Alchemy Institute has been doing R & D for a sustainable future. They have articles like "Greasing the Windmill" next to "Scale and Diversity in Energy Systems," "Surveying and Grafting Local and Antique Fruit Trees," and "Toxic Materials in the Bioshelter Food Chains and Surrounding Ecosystems." The work these people have done over the last twenty-five years has demonstrated that we are nowhere near the limits of the earth to support us if we learn to design, produce and live within natural processes. They have grown papayas in solar heated greenhouses in Massachusetts. They are most famous for their ark which is a self–contained ecology for growing food. It houses plants at several levels (depending on their heat requirements) fish, water plants, beneficial insects, and is heated in part by the compost bins which are strategically placed under the building. They have many other interesting, projects including a sustainable development project in Costa Rica.The institute is in Falmouth Massachusetts and well worth a visit if you should happen to be in the area. Call them at (508) 564-6301

NEWSPAPER

You Can't Get There From Here: Why Bay Area transportation is such a mess — and how we can get things moving again

October 10, 1990
The San Francisco Bay Guardian
520 Hampshire St.
San Francisco, CA 94110

This issue of the Bay Guardian is devoted to the transportation woes of the Bay Area. Just about every voice and organization in the area that is trying to address the issue gets a chance to speak. The Guardian shows how little we've progressed over the last twenty years, and how much we will lose in quality of life if things don't change in the near future. One of the most thorough and clear explanations of the unbelievably complex , splintered, and self-serving array that controls our transportation condition. The link between transportation and land use is made clear, and any solution we find must include this link, and be carried out regionally. Back Issues may be ordered for $3.00 or picked up for $1.00.

REPORTS

MTC Regional Transit Guide
MTC Publications
Oakland Metrocente,r Oakland, CA1989

Until the Bay Area has an integrated mass transit system, this is the best resource available for getting around on buses, trains, and boats. It is designed to help plan short or long trips with many maps, time tables, and phone numbers of all major transit agencies listed by counties. The list of popular destinations, such as Tilden Park and Chinatown, accompanied by all of the available transportation alternatives, is particularly useful for both tourists and residents.

The MTC is an agency which analyzes transportation problems in the Bay Area, and the 1989 guide is their third edition. A new guide is expected to be released in the late fall, and a form is included in every book for suggestions or comments for the next guide.

Projections 90
Association of Bay Area Governments
PO Box 2050, Oakland CA 94604, 1989

Do you ever wonder what the median income of Contra Costa County is, or will be in the year 2005? If so, this is the book for you. *Projections 90* is a forecast of what life could be like in the years 1990, 1995, 2000, and 2005. Including county-by-county data for nine counties , ABAG's report covers subjects like jobs, housing, migration patterns, land development, income, and population. It is interesting information, and great cocktail conversation. If you are a fan of LAFCOs and census data, however, run to your local library and check this out.

Food

BOOKS

Diet for a New America
John Robbins
Stillpoint
Box 640, Walpole, NH 03608 800-847-4014

People become vegetarians for all kinds of different reasons, ranging from politics to aesthetics to health. The arguments presented by Robbins basically fall into the "gross me out" category. Throughout this 450-page tome on the evils of eating flesh, Robbins discusses why we should love animals, why we shouldn't raise and slaughter them the way we do, and what we should eat instead. His thorough documentation of the chicken and beef industries, and his graphic descriptions of what their products do to our bodies, are enough to make even the most staunch of carnivores feel pretty bad. Whether or not they'll be inspired to become vegetarians is another issue. Robbins's propagandatistic tone overshadows the value of his information and analysis. One might feel more irritated than enlightened after reading this book.

Pesticide Alert: A Guide to Pesticides in Fruits and Vegetables
Laura Mott and K. Snyder
Sierra Club Books
San Francisco, CA, 1987

Scientists from Natural Resources Defense Council reviewed federal government documents and examined data from the California Department of Food and Agriculture's pesticide residue monitoring program to compile this guide.

Oriented to consumers, *Pesticide Alert* can be a valuable resource, as it offers both a product-by-product description of pesticide residues commonly found on fruits and vegetables, and the health hazards posed by these residues. Furthermore, it lists which pesticides can and cannot be removed by washing and offers alternative methods to reduce residues. Also of interest in the guide are a dozen short takes on pesticide-related issues, including Integrated Pest Management, environmental pollution, FDA and EPA regulation, and exportation of banned pesticides to developing countries.

The Catalogue of Healthy Food in America
John Marlin, Phd, Domenick Bertelli
Bantam Books 1990.

What is healthy food and where can it be found? To answer these questions, Domenick begins with a review of food families, and then continues to an easy-to-read description of food contents, healthy diets, fast foods, and organic farming. This comprehensive catalogue concludes with a guide to natural foods and products.

Domenick offers tips on selecting truly healthy foods by alerting the reader to labeling hypes and to food hazards such as pesticides and irradiation. Other helpful tidbits range from advice on "How to Organize a Buying Club or Food Co-op" to hints on "How to Identify Organic Produce." The resource guide is substantial, comprising more than half of the book, and includes state-by-state listings of natural food restaurants, products, producers, distributors, retailers, and mail order sources.

The Fast Food Guide

Michael F Jacobsen, PhD and Sarah Fritschner
Workman Publishing
New York, New York 1986

This guide presents facts that most of us would guess but couldn't document about fast food. The Center for Science in the Public Interest extensively researched each of the major fast food chains and all the products they offered. This book lists the nutritional values of every type of burger, french fries, soft drink, or dessert that is available. Charts with titles such as "Fast Foods with Highest Gloom Factors" and "Fast Foods Highest in Calcium" make the information accessible and amusing.

This book was compiled in 1987. Since then many changes have occurred in the fast-food industry. Partly because of consumer demand, chains such as McDonald's have announced that they no longer cook french fries in beef tallow. Healthier products such as salads have been added, but continued change depends on the input of informed consumers— such as those who have read this book.

The Flavors of Home: A Guide to Wild Edible Plants of the San Francisco Area

Margit Roos-Collins
Heyday Books
P.O. Box 9145, Berkeley, CA 94709 1990

This inspiring book is a general field guide to our region's edible wild plants. It describes both the wild native and non-native "weed" plants of the area whose food value is often neglected by busy urbanites. After an introduction to foraging, Roos-Collins gives descriptions and illustrations of poisonous plants, edible blossoms, berries, nuts, mushrooms, seaweed and herbs, as well as where and when these can be found. The appendix contains charts of harvest times and a map of plant locations.

The Flavors of Home evokes a remembrance of the connections that humans have to the land. The reader is brought back to the time when honeysuckle sampling was a required part of a child's daily life. The adult is allowed to renew connections as well. Foraging for wild plants conjures up a time before either agriculture or supermarkets existed. The gathering of the earth's gifts, especially if they are "just weeds," is wealth in its most tangible form.

The Tassajara Recipe Book

Edward Espe Brown
Shambhala Publications
Boston, MA 1985

Food is life. Life is food. This culinary maxim is found in an autobiographical passages of *The Tassajara Recipe Book.* Brown, a founder of Greens restaurant in San Francisco, compiled this collection of vegetarian recipes from the Zen Mountain Center at Tassajara.

The recipes, inspired by the favorites of the Center's summer guests, are as inviting as Brown's holistic approach to cooking: Lentil Tomato Mint Soup, Vegetable Bechamel, Jack Cheese and Onion Tart with Cumin Seeds, Chili Rellenos Souffle, Vanilla Creme Anglaise and Grapefruit Champagne Ice.

Brown won't have you scouring gourmet ghettos looking for saffron threads but instead will have you devote that extra time to preparing the dish itself. The novelty of recipes such as Brown's Tofu Gumbo or Spinach With Strawberries comes not from imported ingredients but from the creative combination of indigenous items. Unlike some vegetarian cookbooks, Brown follows the Zen tradition that relies on the "essential and vibrant flavors" of vegetables, not butter and cream, to enrich soups and entrees.

The Vegetarian Connection

Joel Rose
Facts on File Publications
New York, N.Y 1985

For those curious about vegetarianism, *The Vegetarian Connection* is a comprehensive reference source. The book contains descriptions of a variety of vegetarian diets, Lacto-ovo, vegan, macrobiotic, and even a breatharian regimen whose adherents attempt to live on air, sunlight and water alone. Rose also discusses nutritional facets of vegetarianism; how it may function as preventative medicine, how women can maintain a diet during pregnancy, how children at various stages of growth may be affected, and the value of supplemental "power foods" such as ginseng or spirulina.

As a veritable vegetarian yellow pages, this book also includes nationwide listings of health food products, retailers, restaurants, spas, health farms, newspapers and magazines, international and regional organizations, university courses, and even cookbooks. Some listings, such as restaurants and health food stores, are limited for the Bay Area; therefore, the extensive bibliography may prove to be the most useful offering.

Whole Foods Encyclopedia: A Shopper's Guide

Rebecca Wood
Prentice Hall Press
New York, N.Y

Are you spaced out? Uptight? Perhaps you've been eating too many expansive or contractive foods. The role food plays in our lives, author Wood explains, is as complex and varied as what we eat. For example, sweets can nourish our spleen, pancreas and stomach, whereas sour foods soothe the liver and gallbladder.

Wood discusses the nutritional value and medicinal properties of foods ranging from more mundane items like milk and flour to more exotic choices such as umerboshi, seitan, quinoa, and yautia. She also offers tips on the selection and storage of these foods.

Intermingled among these descriptions is an assortment of charts, recipes, and anecdotes. Among these are such gems as a test for rancid almonds, pesticide information, a ranking of sweeteners, a recipe for Lemon Drop Cookies with rice flour, tips on using arrowroot for diaper rash, and excerpts from those literary giants, Walt Disney and Laura Ingalls Wilder.

Gardening

BOOKS

How to Get Your Lawn and Garden Off Drugs
Carole Rubin
Friends of the Earth
251 Laurier Ave.#701 West, Ottowa ON K1P, Canada

This handy booklet will get you on the organic gardening path. Produced by the Canada chapter of Friends of the Earth, the booklet takes you through the steps of maintaining a pesticide–free lawn and garden, describes common pests and weeds that plague lawns, and discusses safe ways of destroying them. Organic fertilizer ingredients are listed so that you can make your own blends. A troubleshooting section helps you identify and deal with problems as soon as they arise, so you can avoid further complications. Great illustrations bring the information home.

How to Grow More Vegetables
John Jeavons
Ten Speed Press
Berkeley, CA, 1982.

John Jeavons, a member of Ecology Action of the Midpeninsula, uses his experience working in the organization's Common Ground Garden to compile this primer on the biodynamic/French Intensive method. A holistic approach that stresses harmony with, rather than domination of, soil and insects, this gardening style is unique as it requires using raised beds rather than rows, transplanting with a "breakfast, lunch and dinner" technique to minimize shock, and planting by the phase of moon.

Telling readers to "think like a seed," Jeavon explains that empathizing will enable them to better imagine what their seeds may require to flourish. In case this doesn't offer sufficient inspiration, he also provides nearly 30 pages of planning charts, which contain item-by-item listings of flat and bed spacing, weeks to maturity, harvesting period, and seed yield. Also included in the book is a simple recipe for manure-free biodynamic/French compost.

This guide is suited to both beginning and more advanced gardeners. It offers garden plans which begin with a one-person plot and increase in size and complexity to conclude with a garden plan for a family of four.

Shepherd's Purse
Pest Publications
The Book Publishing Company
Summertown, TN 1987

Sheperd's Purse is an invaluable resource and how-to booklet on organic pest control. The booklet outlines alternatives to the use of poisons and toxic chemicals, cultural and biological preventative measures, and natural pest controls already operating in your garden. Often, no control is the best control, since it is usually the imbalance of bugs and garden arthropods caused by the introduction of chemicals which destroy the balances of beneficial insects, pests, and their natural predators. The Sheperd's Purse warns against attempts at total eradication of pests. Biological control refers to the maintenance of pest populations within tolerable levels, not the total elimination of any one insect, since this is expensive, dangerous, and often impossible.The booklet includes large color pictures and descriptions of many pests, their life cycles, their habits and effects, and possible biological and cultural controls of their infestation. This book will get you started on a whole new relationship to pests and pest management.

Strawberries in November: A Guide to Year-Round Gardening
Judith Goldsmith
Heydey Books
Berkelely, CA, 1987

Strawberries in November? With her "calender guide," author Goldsmith maps out unique gardening possibilities in the East Bay for each month of the year . After a discussion of the particular climate, soil and water, and rhythms of the "bioregion," Goldsmith describes specific planting times, cultivars, and monthly "bloom lists" for the East Bay in particular. As the author points out, *Strawberries in November* is a "when to" not a "how to" gardening guide. When should deciduous fruit trees be fertilized or California poppies be watered? Should potatoes be planted in April or September? When can seeds be gathered for next season's harvest? When can gourmet mushrooms be found under East Bay live oaks?

Answers to these questions, tips on monthly gardening events such as flower shows, sales, and festivals, and seasonal recipes like "Pumpkin and White Bean Soup" can all be found in the "calendar guide." The author also provides a resource list for East Bay gardeners, which includes local nurseries, sources for seeds, soil amendments, organic gardening supplies, and gardening organizations.

The Earth Manual: How to Work on Wild Land Without Taming It

Malcolm Margolin
Heyday Books
P.O. Box 9145 Berkeley CA 94709, 1985

If you don't have your own garden, or if you just want to be adventurous, you can try to plant on wild land. This unconventional approach to gardening aims at making your land grow as naturally as possible. Malcolm Margolin draws from his experience at Oakland's Redwood Regional Park where he ran a park conservation program. The methods of digging erosion control ditches, pruning wild trees, collecting and scattering seeds, and laying down leaves as mulch can be done in anyone's backyard. This book provides interesting reading, and is required reading for anyone engaged in restoration work.

The Organic Gardener's Complete Guide to Vegetables and Fruits

Rodale Press
Emmaus, PA, 1982.

The editors of Rodale Press, a longtime leader in the publication of organic gardening literature, are the authors of this comprehensive guide. Divided into two sections,Vegetables and Fruit, the book meticulously maps out necessary steps to organic gardening. The reader is guided through crop selection, site planning, soil preparation, planting, cultivating, harvesting, and storage.

Advice on crop-specific planting times, diseases, and pests are included, as are charts for fruit pollination, and crop "companions" and "antagonists." This guide nicely complements *Strawberries in November*, as it offers very specific "how-to" tips for gardeners. "Tricks of the Trade" sidebars offer helpful advice on such tasks as drought-proofing a garden and building homemade drip irrigators, strawberry pyramids, or seed flatbeds.

Water-Conserving Plants & Landscapes for the Bay Area

Barrie Coate
East Bay Municipal Utility District
Alamo, CA, 1990

This beautiful guide, which catalogues hundreds of trees, shrubs, grasses, groundcovers, and perennials, is essential for homeowners seeking plants suited to the Bay Area's climate. The guide indicates the particular tolerances of each plant, such as coastal conditions, heat, wind, and drought. Full-color photographs help you easily identify each plant. An introductory section explains how to use drip-irrigation systems and how to adapt an existing one to use as little water as possible.

PERIODICALS

Conservation Trees

National Arbor Day Foundation
100 Arbor Ave., Nebraska City, NE 68410
(402)474-5655

The National Arbor Day Foundation's Conservation Trees program encourages Americans to plant, manage, and preserve trees to conserve soil, energy, water, wildlife, and the atmosphere. Their *Conservation Trees* brochure, printed, unfortunately, on unrecycled paper, contains everything you wanted to know about trees but were afraid to ask. A spread in the beginning describes the many wonderful functions of trees, from "sheltering farmsteads" to "shading homes and streets." Subjects such as planting, pruning, and planning tree placement for maximum energy gains are covered in detail; furthermore, every section lists other resources in a "Where to Get Help" box.

This booklet is free, and $10.00 memberships include a subscription to a bimonthly publication, *Arbor Day*, and ten free Colorado Blue Spruces.

General

BOOKS

30 Simple Energy Things You Can Do To Save the Earth

The Earth Works Group
San Diego Gas and Electric
Berkeley, CA 1990

If you're still operating under the impression that all utilities are bad, this book should help you develop a more sophisticated world-view. As the authors note in the introduction, most utilities have a number of reasons for promoting energy-efficiency, from lowering their own costs to helping the planet. While we don't want to err on the side of 'polly-annishness', this book shows that corporations can act responsibly. This short, concise guide, modeled on the best-selling "50 Simple Things You Can Do to Save the Planet," covers everything from furnace tune-ups to attic insulation, from efficient use of appliances to lowering energy costs for lawns. The format follows that of its predecessor, with background and tips for each action. The only difference is that the "What's Happening" summary of global environmental problems is at the back, not the front, for no apparent reason.

50 Simple Things You Can Do To Save the Earth

Earthworks Press
The Earth Works Group
Berkeley, CA 1989

This slim volume, one of the first to tell citizens how to address global environmental problems, hit the big-time fast, and no wonder. After a brief synopsis of global ills, the authors present myriad practical tips on environmentally-responsible living, including: stopping junk mail, recycling, conserving gas, picking up beach litter, eating low on the food chain, snipping six pack rings, boycotting styrofoam, and using energy-efficient lighting. Each "Simple Thing" includes background on the problem, and several actions to take to solve it. The style is light, informative, helpful, and convincing. Every ecologist wanna-be should have this book on the shelf; every page should be dog-eared from use.

Blueprint for a Green Planet

John Seymour and Herbert Girardet
Dorling Kindersley
9 Henrietta St. London WC2E 8PS, 1987

This ambitious book fulfills the promise of its title. It covers the full range of green issues, from energy to recycling to health and pollution. It is beautifully illustrated and clearly written. The presentation of the information is organized in an "ecological" manner with many of the chapters starting with something simple like housekeeping, water garbage, and then following all the connections which branch out from that beginning.

The merit of this book is that it talks about the solutions as well as the problems. It ends up convincing us that although the situation is bad there are practical and even elegant solutions waiting for us to implement them.

Call To Action: A Handbook for Ecology, Peace, and Justice.

Brad Erikson, Ed.
Sierra Club Books
San Francisco, CA, 1990

"The old order is dying; a new one struggles to be born," writes Jesse Jackson in the preface to this invaluable primer on citizen action. Chapters on everything from the ozone hole to feminism to multi-cultural alliances to species rights to toxics to rainforests to renewable energy make this a nice representation of what the new order needs to include. While the book's diversity by necessity excludes detail, you'll get a range of viewpoints that is both inspiring and enlightening. Articles by Claire Greensfelder, Larry Bensky, Mike Rosell, Randy Hayes, Bella Abzug, Brian Wilson, Margo Adair, Karl Linn, Nancy Skinner, Gar Smith, Carl Anthony, Peter Berg, and other leading Bay Area thinkers make this a community-based guide to action in more ways than one.

Guide to the California Environmental Quality Act (CEQA)

Michael Remy, Tina Thomas, Sharon Duggan & James Moose
Solano Press
PO Box 773, Point Arena, CA95468, Annual

This 450 page book is a complete guide to the California Environmental Quality Act. It is intended for use by lawyers, activists, planners, developers, and other people who need to have a good understanding of CEQA. It is updated every year and is the standard reference work on the subject. In spite of its technical subject matter it is well written and easy to understand.

The substance and procedures of CEQA are examined in detail which provides an overview of the Act's requirements for "adequate environmental review" and the preparation of an environmental impact report (EIR). It also includes a discussion of the guidelines for implementing CEQA and an appendix with summaries of the most frequently cited cases.

The Ages of Gaia: A Biography of Our Living Earth

James Lovelock
Bantam Books
New York

This is Lovelock's second volume describing the Gaia Hypothesis, and it's as good, if not better than *Gaia: A New Look at Life on Earth,* his first book. Lovelock describes the Earth as a living organism— rather than an inanimate piece of rock— in an elegant style which combines poetic language with scholarly insight, and passion with scientific detachment. Where the first volume delineated the chemical and biological bases for the Gaia Hypothesis, this volume etches out a theory of evolution which explains how the earth, its inhabitants, and its atmosphere came to be. It's Darwinism with a New Age twist, and, like the first book, it doesn't treat the subject lightly. Lovelock's clear, personal, and heart-felt style makes it an easy, engaging, and thought-provoking book to read.

The Animal Rights Handbook

Laura Fraser,
Living Planet Press
Venice, CA 1990

This book provides an overview of animal rights issues: the threats facing animals and what caring people can do to protect them. It covers how the everyday choices we make about the way we eat, how we dress, and how we educate our children, affect the animal kingdom. *The Animal Rights Handbook* provides consumer guidance on how and where to buy cruelty–free products; advice on educating your children to respect and value animals; and awareness exercises to heighten the reader's awareness of animal rights and protection.

The Complete Guide to Environmental Careers

The CEIP Fund
Island Press
Covelo, CA

Aspiring to become an environmental professional? This guide takes the student or concerned citizen through the steps that could lead to an environmental career. Planning, environmental education, solid waste management, hazardous waste management, air quality, water quality, land and water conservation, fishery and wildlife management, parks, and forestry are just a few of the fields that are offered. Information regarding internships, salary, trends, and entry requirements are provided. Case studies of professionals are included to give the reader an idea about what a typical day at work would include. Descriptions of jobs in the federal, state, and local government, and in private and non-profit sectors give the reader a good overview of their employment opportunities.

The Global Ecology Handbook: What You Can Do About the Environment

Ed. by Walter H. Corson
Beacon Press Boston MA 1990

The Global Ecology Handbook, produced by the non-profit Global Tomorrow Coalition, provides valuable tools, information, and access to resources which allow the individual to address environmental issues and contribute to a beneficial impact. It is a practical guide to active and effective modes of promoting conservation and restoration on individual, communal, and international levels.

The Global Ecology Handbook includes information on everything you need to know about the current state of the environment and linkages between economic development, energy policy, population growth, tropical rain forests, oceans and coasts, biological diversity, garbage and hazardous waste, global warming, and efficient agriculture. The Handbook also lists resources, books, films, and organizations and how they can be accessed. It describes simple changes you can make in your own home to save water and cut down on energy consumption and garbage production, how to change your yard to support and help protect valuable animal and plant life, how to set up a recycling program in your community. A valuable resource.

NEWSLETTERS

Green Calendar

GAIA/Earth Island Institute
300 Broadway, Suite 28
San Francisco, CA 94133

The Green Calendar is exactly that - a monthly list of events, lectures, seminars, workshops, benefits, and fundraisers that support environmental campaigns ranging from peace marches to theater pieces on environmental issues. The Calendar will keep you up to date on issues, where to get involved, and who to contact. It demonstrates the links between all of these issues which can otherwise become somewhat fragmented and overwhelming. Now not hearing about events is no longer an excuse for not participating. No day goes by in the Bay Area without something going on to save the planet.

PERIODICALS

Amicus Journal

Natural Resources Defense Council
40 West 20th Street , New York, NY 10011.

This quarterly publication, which focuses on national and international public policy issues, is free for members of the Council, but also can be obtained by the general public. Articles are well-written and informative, and often authored by noted writers and leading environmentalists. The book reviews and poetry give it a slightly literary air, but in general this is hard-hitting analysis and policy recommendations.

Business Ethics

1107 Hazeltine Blvd., Ste. 330, Chaska, MN 55318

Not too long ago, "Business Ethics" seemed like an oxymoron to most of us. This bimonthly publication, which claims its mission is to "promote ethical business practices, to serve a growing community of professionals striving to live and work in responsible ways, and to create a financially healthy company in the process" is positive proof that business and ethics can co-habitate in the office. Regular features include "Trend Watch," which contains a variety of useful management ideas on subjects like recycling and reducing toxics in the office. Business Ethics is a valuable resource for all business people with a social conscience and social activists just entering the business world.

Buzzworm: The Environmental Journal

PO Box 6853, Syracuse, NY 13217

This magazine is an accumulation of short articles and briefing concerning recent environmental policies and actions throughout the country. Other stories probe various threats to wildlife and natural habitats. The concerned consumer will find interesting articles about natural beauty products, recycled products, and computer networking. There is good coverage of citizen action groups, such as the community effort to save the Great Lakes. The magazine brings together profiles of interesting products and activities to convey a sense of how creatively people can respond to environmental crises.

E: The Environmental Magazine

P.O.Box 5098 Westport CT 06881
(203)854-5559
Subscriptions: P.O. Box 6667, Syracuse, NY 13217

E magazine was "formed for the purpose of acting as a clearinghouse of information, news, and commentary on environmental issues for the benefit of the general public and in sufficient depth to involve dedicated environmentalists." So far they are fulfilling their mission statement very well. Extra points for being printed on recycled paper.

The Earth Island Journal

Earth Island Institute
300 Broadway, #28, San Francisco CA 94133
(415) 788-3666

Hard hitting eco-news from around the world. The Earth Island Journal is an excellent resource for activists and concerned environmentalists. It covers global environmental, and events and the various projects of the institute. Its coverage is timely, with good information and sharp analysis. It is worth joining Earth Island Institute just for the Journal.

Garbage: The Journal for the Environment

Old House Journal Corp.
435 Ninth St., Brooklyn, NY 11215

This new journal, with its useful articles on transforming your house, garden, and living and shopping habits, is one of the best general environmental magazines on the market. Gardening tips range from how to practice pesticide-free pest-control to what kinds of drought-free plants are available and where to find them. Articles about over–packaging and degradable plastics are oriented towards consumer education in the marketplace. Occasional product ratings (such as low-flow toilets and gas-conserving cars) are an ecologist's dream. Articles about natural solutions to our waste-disposal problems are both practical and inspiring. This magazine is a definite must for folks who want to be on the cutting edge of a number of environmental trends— not just garbage.

Race, Poverty and the Environment
A newsletter for social and environmental justice

Earth Island Institute
300 Broadway, #28, San Francisco CA 94133
(415) 788-3666

"Toxics, pollution and pesticides especially affect poor people and people of color. We as environmentalists must build bridges to people affected by those hazards if our movement is to succeed." So say the editors in the first issue.

For the environmental movement to come of age it must break out of its narrow race and class constituency. We need to begin to see the inner cities and abandoned industrial areas as "damaged land" and that people in these areas are subjected to serious enviromental hazards. As Jessie Jackson says in the first issue: "...unless I have the right to good breathe, the right to good drinking water no other right can be realized. Environmental justice is a fundamental human right." We must develop a widely shared vision of a just, peaceful and sustainable society. This newsletter is setting out to do just that.

Utne Reader

Box 1974, Marion, OH 43305

The *Utne Reader* is a bimonthly which bills itself as "the best of the alternative press," and it is, for the most part. A compendium of articles from a variety of publications, most of them leaning to the left, the *Utne Reader* arranges articles around various themes, which allows the reader to absorb several points of view. The themes are usually provocative: television, education reform, the coming of the millennium, rootlessness, gender issues, the family, and third party politics. The "Gleanings" section includes non-thematic articles, and the media reviews are usually thoughtful, intelligent, and multi-cultural. The *Utne Reader* merits a review in this publication because its coverage of environmental issues has been consistently excellent.

World Watch

The Worldwatch Institute
1776 Massachusetts Ave., NW, Washington, DC. 20036

This magazine contains excellent coverage of international development and environmental issues that are often glossed over or trivialized by other publications. Stories on problems such as pesticide use, hunger, environmental refugees, drought, timber logging, and human rights are well-written, analytically sophisticated, and packed with facts. Doom and gloom forecasts are mitigated by reports on positive and creative developments including organic farming, slow-growth policies, and local progressive movements. While the magazine has been criticized for being overly–moderate by some environmentalists, it is probably one of the more important resources available for developing a thoughtful, well-rounded approach to world problems.

Various

Public Information Center
Environmental Protection Agency
401 M St. SW PM-211B, Washington DC 20460
(202) 475-7751 or (202)382-2080

The purpose of PIC is to distribute a wide variety of general nontechnical information about EPA to private citizens, federal, state and local agencies, industry academia, and civic or environmental organizations. Publications available include: 1990 Gas Mileage Guide: EPA Fuel Economy Estimates; Be an Environmentally Alert Consumer;

Citizen's Guide to Pesticides; Earth Trek... Explore your Environment; Environmental Enforcement: A Citizen's Guide; A Family Guide to Pollution Prevention; Glossary of Environmental Terms and Acronym List; The Inside Story: A guide to Indoor Air Quality; Is Your Drinking Water Safe?; Lead and Your Drinking Water; Protecting Our Ground Water; Superfund: Looking Back, Looking Ahead; and many others.

Green Consumerism

BOOKS

The Green Consumer
John Elkington, Julia Hailes and Joel Makower
Penguin Books
New York, NY, 1989

Did you realize that by taking your car in for repairs you may be contributing to global warming and increasing your chances of skin cancer? The authors of this guide illustrate how this and other purchasing decisions are a "series of never-ending votes for or against the environment" and explain to consumers that they can "buy products that don't cost the earth."

In this brand-specific handbook, the authors use two criteria in describing a products "greenness": the product must have either environmentally sound contents or packaging. Products discussed in the guide include: fuel efficient cars, pesticides, recycling containers, organic baby foods, garden and pet supplies, green travel, cruelty-free cosmetics, biodegradable detergents, and many others.

The final chapter of the book, "How to Get Involved," describes how consumers can conduct a boycott, confront local polluters, form a community recycling program, make their businesses "green" and also contains a detailed list of helpful publications and organizations.

MAGAZINES

In Business
The JG Press Inc.
Box 323, 18 South Seventh St. Emmaus, PA 18049.
(215) 967-4136.

This magazine has shifted its focus from small business concerns to ecologically-sound business concerns. It has tracked the surge of environmental work and enterprise throughout the country. It describes some of the more interesting solutions, carried out by members of the business sector, for environmental problems across the board. The magazine covers such diverse topics as the natural food industry, native plant consultants, recycling, waste management, green financial work, as well as practical tips on financing your business, organizing your time, and other more prosaic tasks of running a business. *In Business* publishes a directory of green entrepreneurs annually. It is a magazine of business ideas that are practical because they are based on and motivated by an understanding of the earth's limits.

Green Economics

BOOKS

Bridging the Global Gap: A Handbook to Linking of the First
Medea Benjamin and Andrea Freedman
Seven Locks Press, 1989
Washington, D.C.

Since the mid-1970s, the primary routes for citizen action and exchange were east to west, from the U.S. to the U.S.S.R. Motivated by a fear of nuclear war, and inspired by the thrill of making "lay" contact despite a political Cold War, citizens from both continents visited each other in groups of everything from grandmothers to nuclear physicists. With perestroika on the rise, and nuclear threats on the wane, it's time to turn our attention to new directions: that is, north and south.

Chapters in "Bridging the Global Gap" cover a variety of issues, ranging from sister cities and tourism to human rights to consumer and corporate responsibility and working with Third World governments. The general message of the book, however, is probably best summarized by the late President of Mozambique, Samora Machel, who once said: "International solidarity is not an act of charity. It is an act of unity between allies fighting on different terrains toward the same objectives. The foremost of these objectives is to aid the development of humanity to the highest level possible." This is a moving and educational account of how ordinary citizens can help the social, economic, and environmental state of the world in the late 20th century. The 100 page resource guide in the back of the book is essential to anyone doing citizen action.

Economics as if the Earth Really Mattered
New Society Publishers
Susan Meeker-Lowry
4527 Springfield Ave. Philadelphia PA 19143

This book is a no holds barred manual on bringing ethics back into economics. It covers the theoretical underpinnings for an new approach to "Gaean" economics. It explains the ins and outs of affecting change in corporate decision–making through boycotts and socially responsible investment. It discusses the issues which arise when setting up small businesses and ways to avoid recreating the same problems that occur in the economy at large. This section includes the well known example of *Ben and Jerry's Homemade*. The final section of the book is called "Creating the economy we want," a combination of sound arguments and inspiring examples.

Susan Meeker-Lowry has found lots of examples of people and organizations that are making economics work for the good of people, community and environment. She shows that it is possible to live happy, productive, and rich lives with an economy which is based on sharing, mutual support, community, and earth preservation. This book is required reading if you have money to invest or if you are thinking of starting a business.

Green Philosophy

BOOKS

Deep Ecology
Michael Tobias
Avant Books
San Marcos, CA 1988

Recycling, public transit, non-toxics, and energy efficiency may all help save the world as we know it today. But until we change the fundamental ways in which we look at and act in the world, nothing much will change. By integrating science, religion, philosophy, and economics, deep ecologists seek to transcend the limits of the human ego and forge greater bonds with the rest of nature. That is powerful stuff and this book is an excellent introduction to it. Chapter topics range from species extinction and development issues to Heidegger and environmental ethics. But it is the middle section, called "Heartlands" and devoted to essays on poetry, shamanism, and nature, that will give the readers the sense of passion, compassion, and creativity that lies beneath the provocative analyses that infuse the other chapters.

Dharma Gaia: A Harvest of Essays in Buddhism and Ecology
Allan Hunt-Badiner, Ed.
Parallax Press
Berkeley CA1990

In an era of elaborate, expensive, and unrealistic technical solutions to the world's environmental problems, this book suggests a far more profound strategy for healing: changing our ways of perceiving and relating to the planet. By integrating Buddhist philosophy with environmental science and politics, these essays guide the reader towards seeing the elegant interconnectedness of all beings, and show how vitally important it is for us to develop wisdom and compassion in our dealings with them. Essays by Joanna Macy, Jeremy Hayward, David Abram, and Peter Levitt provide intellectually engaging analyses of systemic thought, Buddhist ethics, and the nature of perception. Poetry and first person accounts by Gary Snyder, Thich Nhat Hanh and Allan Ginsberg provide a more personal expression of both the pain of environmental degradation and the power of envisioning better world. Readers looking for a primer on Buddhist thought should not turn to this book first. It assumes some accquaintance with Buddhist theory and practice. However, readers struggling to bring the profundity of spiritual practice to the political movement of environmentalism, which too often is riddled with the self-centeredness and confusion that caused problems in the first place, will find these essays inspiring and stimulating.

Green Politics

BOOKS

Building the Green Movement
Rudolf Bahro
New Society Publishers
4277 Baltimore Ave, Philadelphia PA 19143 1986

This collection of Bahro's writing and speeches from the early eighties gives you a direct encounter with one of the leading minds of the German green movement. Bahro is steeped in European political history and its theoretical debates about how and what we should learn from the past. It is a rare thing to find such a combination of intellectual breadth and acumen being brought to bear on the immediate practical problems of building the green movement and creating a sustainable future.

Bahro was exiled from East Germany in 1979 after being imprisoned for writing a far-reaching, ecologically based critique of the Eastern European societies. Immediately after entering West Germany he joined the nascent Green Party and was quickly elected to its Federal Executive.

The essays in this book cover the period from the spectacular rise of the Greens in the heyday of anti NATO protests; sparkling debates with the "realos" who want to compromise the fundamentally ecological stances of the greens to gain more access to state power; incisive arguments for the creation of alternative economic structures and communities which do not depend on the "life destroying international economic system"; to Bahro's resignation from the Greens when he felt that the "realos" were trying to save the party at the expense of its purpose.

Bahro's intelligence and commitment are inspiring and his integrity is contagious. This book will provoke the reader to question more deeply what we take for granted as "normal life", and it is highly recommended for anyone active in environmental politics in the US.

Resource Manual For a Living Revolution

Virginia Coover, Ellen Deacon,& Charles Esser
New Society Publishers
4277 Baltimore Ave, Philadelphia PA 19143 1981

The Resource Manual is an invaluable tool for nonviolent activists. It offers suggestions on everything from how to facilitate meetings and use consensus decision-making, to how to develop a training session for nonviolent monitors at rallies or actions, to how to help a group think through its long term strategy.

That's not all. You can also find ideas on how to cook for large groups, how to improve the vitality and group process of meetings so they energize people instead of burn out. It combines case histories of effective use of nonviolent action, community building exercises, and extensive bibliographies of further resources in all of those areas.

The German Greens: A Social and Political Profile

Werner Hulsberg
Verso Press 29 W 35th St, N.Y.C. 10001-2291 1988

This is a in-depth, blow-by-blow account of the West German Greens, from the conditions which gave rise to the Greens in the confusion and splinter–group politics of the mid- to late–70's, to the debates which emerged as the Greens gained electoral prominence. Hulsberg is obviously sympathetic to the Greens, and in a characteristically European way, is not afraid to polemicize. He chronicles the Greens' progress against a well-woven background of the social, political and economic history of West Germany. This makes it a useful book for American Green activists because one can see in some detail how the Greens in West Germany reacted to and created their particular historical situations. If you are looking for a thorough and precise left-oriented history of the German Greens, this is one of the best sources.

The Green Alternative: Creating an Ecological Future

Brian Tokar
R&E Miles
P.O. Box 1916, San Pedro, CA 90733
(213)833-8856

Brian Tokar has written an eloquent introduction to the emerging green alternative. He examines the roots of the basic assumptions of the Greens, develops these into a green historical critique of western "civilization," and outlines the prospects for a green future. The four pillars of the green program are Ecology, Democracy, Social Justice, and Nonviolence. Each of these topics is explored in depth along with their important interconnections. Tokar's green critique challenges conventional environmentalism to take a deeper view of the problems confronting our planet. He argues that the ecological problems cannot really be solved without addressing the inequities of power and wealth which are the manifestations of the same structures. What is exciting about the green program is that it offers the possibility of bringing together many of the movements and issues of the last twenty years under a cohesive program for a more just and greener world.

Home Hazards

BOOKS

Nontoxic, Natural & Earthwise

Debra Lynn Dadd
Jeremy P. Tarcher, Inc.
Los Angeles, CA,.1990

As consumer awareness grows and companies scramble to affix "green" labels to their products, it is becoming increasingly difficult to determine which products are truly environmental sound. It is easy to fall for misleading ads for "biodegradable" trashbags which actually only degrade into smaller, more dangerous bits of plastic, and which produce toxic waste when manufactured.

This excellent book can help consumers avoid such purchasing pitfalls. Dadd has done an environmental evaluated of hundreds of products. She evaluates them according to five criteria: ingredients, packaging, energy use, compassion to animals and social responsibility. Dadd suggests commercial products such as pesticide-free coffee, "earthwise" seafood, and natural fiber clothing as well as providing homemade alternatives to toxic cleaning and pest control chemicals. Brand name products are listed under each category (eg.. shoes, blankets) and a directory of mail-order houses can be found at the end of the book.

PAMPHLETS

How Green is My Home?
Household Tips For Saving the Planet
Michael Belliveau
Mother Jones Magazine
Home Audit, c/o Mother Jones Magazine,
1663 Mission,San Francisco 94103 1990

This pamphlet lists and describes various ways the general public and individuals can effect direct actions in pollution prevention, energy conservation, ozone preservation, indoor air quality, toxic household products, and other toxic hazards found in the common household. It covers water treatment devices, conscientious disposal of chemicals and toxins, and the application of political, consumer, and public pressure on local water utilities to improve water quality.

How Green is My Home? also describes some of the major environmental hazards caused by the cumulative impact of millions of families, and describes a few methods by which individuals can reduce their contribution to the problems. Some of these hazards include: global warming; destruction of the ozone layer; effects on air and water resources; solid waste production, management and disposal; local, regional, national and international trends in food production and distribution; impacts on wildlife and habitat; and the identification of toxics, hazards, and cancer-causing radon gas sources within the home and workplace. Information on accessing the necessary tools to effect hazard reduction is sometimes excluded or forgotten.

The pamphlet also presents information on the uses of research, policy advocacy, litigation, grassroots organizing, and education on the reduction and prevention of pollution in your home and the public environment. *How Green is My Home?* definitely offers useful information and ideas. $5.00 by mail.

Asbestos Ombudsman Clearinghouse
Office of Small & Disadvantaged
Business Utilization
Environmental Protection Agency
401 M St. SW (A149C), Washington DC 20460

The mission of the Asbestos Ombudsman is to provide to the public sector, including individual citizens and community services, information on handling and abatement of asbestos in schools, the workplace, and the home. In addition, interpretation of the asbestos-in-school requirements and publications are provided to explain recent legislation. The duties of the Ombudsman include: receipt of complaints and requests for information; rendering assistance with regard to complaints and requests for information; and making recommendations to the administrator of the EPA as deemed appropriate.

Pollution/Toxics

BOOKS

Fighting Toxics
National Toxics Campaign
Island Press
Covelo, CA, 1990

Fighting the production of toxic pollutants by industries can be frustrating and exhausting. This book teaches individuals and communities how to garner support for their cause. It details the steps of organizing, fundraising, gathering information, and using the media to lead a successful campaign. A small group of people can push an issue into high visibility and force the government to take action. This book is worth reading before you take on any sizeable pollution problems. It offers plenty of useful tips on dealing with the inevitable resistance you will encounter.

Silent Killers
Kathlyn Gay
Franklin Watts Inc.
New York, N.Y. 1988

Over the past few decades, hundreds of natural and manufactured substances found inside buildings and outdoors have been identified as health hazards. This book clearly explains which common substances are hazardous and how to avoid being contaminated by them. The author points out how a variety of cleanup and preventative steps are being taken by individual citizens, private organizations, government agencies, industries, and research groups. If you're worried about radon, asbestos, lead, carbon monoxide, or any other potential hazard in your home or workplace, this book is well worth reading.

The Toxic Cloud:The Poisoning of America's Air
Michael H. Brown
Harper & Row, Publishers
San Francisco, CA 1987

The chemical industry has been silently poisoning our air supply through long-term toxin leaks all across the country. These emissions collectively are greater than one or two big accidents. This book brings the leakages— and their threat— to light by chronicling the stories of both victims afflicted by leukemia and cancer and the companies' absolute disregard for them and their communities. The book also narrates the struggle between survivors who want to halt the leakages and an apathetic industry and regulatory structure. Good read, bad news.

NEWSLETTERS

Journal of Pesticide Reform
Northwest Coalition for Alternatives to Pesticides
P.O. Box 1393, Eugene, OR 97440

NCAP fights the unnecessary use of pesticides.They challenge pesticide dependence in forest, agriculture, road side, aquatic and urban settings. The group strives to educate the public, support community-based pesticide-alternative programs, and form policy for all of these areas of pesticide use. Other publications include: a school program on integrated pest management;a how-to pamphlet on contracting for pest control work; pesticide drift information; grassroots anti-spraying campaigns that have been successful; information packets on specific pesticides and chemicals (e.g. PCPs and 2,4-D); farmworker issues; school maintenance without pesticides; and other informative subjects.

Rainforest

BOOKS

Amazonia: Voices from the Rainforest
Angela Gennino, Ed.
Rainforest Action Network
301 Broadway, Ste. A, SF, CA 94133 1990

This publication is a directory to most of the international organizations that are fighting to save the rainforest. One of the main emphases of the book is the acknowledgement that indigenous people of Amazonia are very active in the struggle for the preservation of their land. The book is arranged to cover major issues such as First World industry in the forest, traditional peoples' and the new colonists' effects on the land, and global support groups. The directory gathers interviews of many groups about their efforts, gives a contact person, address and phone number, and is arranged by continent. The book is intended as a companion to the film *Amazonia: Voices from the Rainforest* by Monti Aquirre and Glenn Switkes.

The Rainforest Book
Scott Lewis/Natural Resources Defense Council
Living Planet Press
Venice CA 1990

This book offers the reader a non-technical perspective on the problems facing rainforest habitats around the globe. The book guides one through the biodiversity of the area, the destructive pressures on the ecosystem, and the possible solutions and actions that might make a difference. It gives the basics, and enables the reader to make intelligent decisions regarding rainforest products and practices, and where to get involved.

Recreation / Travel

BOOKS

A Bicyclist's Guide to Bay Area History
Fair Oaks Publishing Co.
C. O'Hare
Sunnyvale, CA1989

Carol O'Hare's purpose is "to guide you to the interesting historic sites that abound in the Bay Area, while also providing you with a pleasant day of cycling." Her method is straightforward and successful. She describes eighteen rides throughout eight counties of varying length and difficulty, and describes them in terms of distance, rating (easy, moderate, moderately strenuous, and strenuous), historic highlights, and a description of the ride itself.

There are an abundance of maps and photographs to accompany O'Hare's descriptions, which are both accurate and witty, and include historical background along with practical information, like the location of restrooms and traffic conditions. She covers both the familiar, such as Point Reyes and San Francisco rides, and the offbeat, like Pescadero and Benicia. This is also a good guide for walkers and anyone else who likes to explore local historic sites.

Cyclist's Route Atlas: The Delta, Farm, and Wine Country
Randall Grey Braun
Heyday Books
Berkeley CA

"The purpose of this book is to provide you with high-quality route information: how far or hilly it will be, what the terrain will be like, how good the road conditions will be, and where you can get food and water along the way," writes Braun in the introduction. The routes he has mapped range from 9 to 62 miles, and cover the Napa, Solano, Yolo, and Lake counties.

Tips on safety and detailed maps will aid the cyclist, and Braun's descriptions of the scenery will inspire her. Each route is eloquently narrated, with little details like where the "fertile fields" and egrets are, and restaurant recommendations.

Making the Most of the Peninsula
Lee Foster
Tioga Press Palo Alto CA

This great guide to San Mateo, Santa Clara, and Santa Cruz Counties, indexed by subregion and type of activity (e. g. camping parks), combines history, personal observations, and quirky activities in an amusing and highly readable book. Foster is clearly a fan of this area, and has researched its history quite thoroughly. He also provides a comprehensive bibliography of books about the region and various resources and visitor centers on the Peninsula .The appeal of Foster's book lies in its humor and offbeat suggestions, like a visit to the San Francisco airport as an evening on the town. His knowledge of the area is impressive: he explains everything from the love of horseracing in San Mateo County to the philosophy of James Lick, who endowed the observatory near San Jose. Coverage of the region's numerous parks and wildlife preserves in this region is impressive, and there are lots of maps to get you where you want to go. After browsing through this book, even Gilroy will sound like a romantic and exciting destination for a family trip.

Sacred Places
James A. Swan
Bear and Company Publishing
Santa Fe, NM 1990

Sacred sites are pilgrimage destinations such as Mecca, Ayer's Rock, Mt Fuji and tourist spots like the Grand Canyon or Mt. Tamalpais. This book explores various sacred sites and describes some of the traditional lore surrounding a site, whether it is Native American or Buddhist. It gives the reader an overview of the extremely influential nature of sacred places and attempts to explain some of the fascination that sacred places have held for people of all times.

The Bay Area at Your Feet: Walks with San Francisco's Margot
Margot Patterson Doss
Lexikos Press
San Francisco, CA 1988

Fans of Margot Patterson Doss' series in the San Francisco Chronicle are sure to love this book compiled and revised from her newspaper writings. Doss' introduction is a plea for readers to get out of the car, beyond the "profit-motive" mindset, and walk. Walking is the surest way to relearn your place with nature, and Doss feels that the Bay Area "is a more pleasurable place to walk than most metropolitan places." "However," she warns, "it gets less so every day."

The situation is not as dire as her introduction implies. There are many wonderful places still left to explore, from Oakland's Chinatown to the San Andreas Fault Trail to Loch Ness in Marin. Doss is bursting with history, ideas, and places to fish, lounge, people watch, eat, and use the bathroom. She also has suggestions about what to bring and directions of how to get to the walking sites by car or public transit.

The Bicycle Commuting Book
Robert Van der Plas
Bicycle Books
Mill Valley, 1989

"Using the bike for utility and transportation" is the subtitle of this practical manual for potential or current bicycle commuters. Part 1 covers selecting a bike, buying tools, fixing flats, brakes, lights, proper saddle height, gears, rain gear and other wardrobe tips. "Handling the Bike," discusses riding techniques and skills. There are some excellent safety tips and ideas on how to divert accidents, and a section called "Bicycling Health" which outlines common injuries, like backaches and saddle sores, and reccommends effective treatments. "Bike Commuting as a Way of Life," covers things like expressway cycling and traffic dodging. "The Politics of Bicycle Commuting," discusses bike lanes and parking. This is a thorough guide to bike commuting, with many photos and diagrams which enhance the straightforward text.

The East Bay Out: A Personal Guide to the East Bay Regional Park District
Malcolm Margolin
Heyday Books
P.O. Box 9145 Berkeley CA 94709 1988

The East Bay Regional Park District is very lucky to have a fan of the parks like Malcolm Margolin. Margolin describes the parks and their better features, and relates the essence of one man's relationship to his treasured wild areas. Margolin does not gloss over the wear and tear that is evident in the system's parks. His perspective includes urban pressures on the land, and ponders the future profiles of areas long harassed by a non-native presence. He covers the human history and the diversity of park users. Margolin excels in his descriptions of glimpsed wildlife and landscapes. His appreciation of the land has developed over a twenty year history of tramping, looking, sitting, and absorbing the messages he finds. *The East Bay Out* will remain a classic, a tribute, and a gift to our area.

Directory of Alternative Travel Resources
Diane Brause
One World Family Network
Dexter, OR

This is an invaluable resource for anyone trying to plan a socially-responsible vacation. Diane Brause, an alternative travel expert, has compiled this list of over 250 world organizations which are involved in alternative travel. Cross-referenced by geographical region and categories (including adventure, scientific expeditions, workcamps, citizen diplomacy, and ecology), there is a vacation in this book for everyone from hippies to medical professionals, ranging from Asian treks to Nebraska pioneer life to vision quests. The particularly useful "Other Travel Resources" section lists responsible travel agencies, books and guides, directories, networks, and publishers.

Recycling / Waste

BOOKS

The Economic Feasibility of Recycling: A Case Study of Plastic

T. Randall Curlee
Praeger Publishers
New York, N.Y. 1986

The vast majority of work done on plastics recycling has focused on technological questions, largely ignoring the economic aspects of plastics in the marketplace.

T. Randall Curlee presents a detailed critical analysis of the economic incentives and barriers of the plastics recycling industry and gives projections for resin production in the U.S. through 1995. He also includes, for the non-technical reader, a glossary of characteristics and uses of selected plastic resins.

The Recyclers Handbook

The Earth Works Group
Berkeley CA

If you're not already an avid recycler this book will make you into one. The recycler's handbook is full of facts about recycling. It is in the same format as *50 SIMPLE THINGS YOU CAN DO TO SAVE THE EARTH.* Each page takes on a different recycling issue and includes suggestions on how to make it easier to do "fascinating facts", and sources of more information.

The authors have done a really good job of presenting the issues in a simple and clear fashion.

War on Waste

Louis Blumberg and Robert Gottlieb
Island Press
Covelo, CA, 1989

Blumberg and Gottlieb explore the problem of garbage disposal and how it has changed over the years. They chronicle the controversy over waste incineration which took off in the early 80's. The public outcry surrounding many proposed waste-to-energy plants stalled further plans for the expensive and potentially hazardous technology. However, alternatives to incineration, such as recycling and composting, have met with heavy industrial and legislative resistance. The book concludes with a discussion of waste reduction at the source, which entails a total shift in the way we package and consume goods.

PERIODICALS

Biocycle: Journal of Waste Recycling

JG Press, Inc.
Box 351, 18 South 7th St., Emmaus, PA 1804
215/967-4135

Originally known as *Compost Science*, the nationally-distributed *Biocycle* caters to public works officials, consulting engineers, professionals in waste management, and environmental activists concerned with the latest technologies and business opportunities in recycling and composting. Concise, non-technical articles discuss issues ranging from plastics collection to the environmental impact of yard waste composting and feasible recycling systems for sludge. This monthly publication covers a nice variety of waste issues, yet, often too technical for its readers. Still, private sector subscription is growing as public awareness spurs the waste management industry.

Resource Recycling: North America's Recycling Journal

Resource Recycling, Inc.
PO Box 10540, 1206 N.W. 21st Ave, Portland, OR 97210
800/227-1424

Primarily designed for policy makers, waste management professionals, and private sector people interested in recycling and composting, this monthly national publication remains on the cutting edge of the developments of the waste management crisis on a state-by-state basis.

In search of the most practical and economical alternatives to landfilling, *Resource Recycling* investigates issues from regional beverage container reuse to the feasibility of creating products from mixed plastics. This clearly-written journal is a valuable source of recycling information for both the professional and the consumer.

World Watch Paper 76—Mining Urban Wastes

Cynthia Pollock
Worldwatch Institute
Washington, D.C., 1987

Half of the cities in the U.S. are scheduled to close their landfills within the next three years, so the need for alternatives waste disposal solutions is urgent.

Cynthia Pollock discusses the causes of the garbage glut and offers ways to solve it through recycling. She explains the different potentials for various recyclable materials and presents examples of successful recycling programs from around the world, emphasizing their enormous economic and environmental benefits.

Worldwatch papers are written for a worldwide audience of decision makers, scholars, and the general public, and they are usually well worth reading.

Water

BOOKS

Cadillac Desert

Marc Reisner
Penguin
40 west 23rd St New York, New York 10001 1986

This is unquestionably the best book on Western water issues written to date. Beginning with the great water rip–off of the Owens Valley and ending with the current debate over the fate of the Sacramento-San Joaquin Delta, Reisner uses anecdotes, government documents, letters, and personal interviews to develop an entertaining and disturbing account of the ways in which the government and water retailers colluded to destroy much of the West's land. The personalities are well-researched and well-rendered, and the hydrological, biological, and engineering issues are well-documented. This is a must-read for anyone interested in Western water, or for anyone interested in good reporting and splendid story-telling.

Overtapped Oasis:Reform or Revolution for Western Water

Marc Reisner and Sarah F. Bates
Island Press
Covelo, CA,1989

Water is a highly political issue in the Western states. As the current drought continues, household water conservation programs have become common in the Bay Area. However, agriculture consumes 85 percent of our water diversions. How can we use water more efficiently? Reisner and Bates suggest water marketing. They propose various steps to reform our current, wasteful system. Selling water rights and forcing people pay the true cost for their water may fundamentally alter our allocation of water. Many of the needed changes must take place in outdated bureaucratic institutions. This book explains how we, as citizens, can influence and pressure them to change.

RESOURCES

Hydrologic Information Unit

U.S. Geological Survey
419 National Center, Reston VA 22092
(703)648-6817

The USGS is one of the most prolific and useful research departments of the government. Renowned for their maps and earthquake studies, they are not as well known for their water-related reports. This unit will answer questions about water resources in general and about the water resources of specific areas of the United States. This office also provides information about the availability of water-resources investigations and reports.

Wildlife

BOOKS

California Wild Heritage: Endangered and Threatened Species in the Golden State.

Peter Steinhart
California Department of Fish and Game
Natural Heritage Division - 1416 Ninth St. 12th flr
Sacramento, CA 95814

A gorgeous book that shows just how beautiful and priceless our natural heritage and endangered wildlife really is. The book will inspire people to consider their beleaguered native neighbors. Each species is represented with a picture, range description, and basic natural history. It is a shame that there are enough threatened species to fill an entire book, but at least this book will insure that our wildlife will not disappear unnoticed.*California Wild Heritage* is now available in most bookstores but you can call the Natural Heritage Division of the Fish and Game Department to order a copy.

Biodiversity

E.O. Wilson, Ed.
National Academy Press
Washington DC 1988

An excellent overview of the status of biological diversity around the world. This book collects together leading studies of the threats, techniques of maintenance, and future prognosis for biodiversity in numerous habitats. Rainforests, oceans, temperate forest, and desert systems are thoroughly covered. This book is an excellent reference for current work in biodiversity in general, and current conservation biology directions. The manner in which people are addressing this critical turning point in global wildlife survival is as interesting as the areas covered. The book makes very clear why people should care about obscure species that may never be spotted in an individual's lifetime.

Preserving Comunities and Corridors

Defenders of Wildlife
1244 Nineteenth St. NW
Washington DC 20036 1989

Many of the conservation efforts for endangered and threatened species, wildlife management, and habitat preservation revolve around establishing preserves and parks. Though this process is essential in the face of enormous development pressures, it can be short-sighted if the need for wildlife corridors is not addressed. Wildlife corridors are critical for the survival of many species. They allow the animals to pass from one area of sizable habitat to another undisturbed. Predators have large range requirements. Some animals need access to water, forage and escape from predators. Others are sensitive to urban lights and noise. Each species has a need for refuge from our urban reality and access to other areas.

This book goes beyond just calling attention to the necisity of wildlife corridors. It makes a critical analysis of established game and species management practices. Efforts to save threatened wildlife usually employ a species approach. These focus on the requirements of one particular species rather than the availability of an adequate habitat base for survival of that particular ecosystem. This book takes the complex field of conservation biology and makes it accessible for the layperson. It outlines the mistakes we are making, the bureaucratic obstacles we face, and the directions that new programs are taking.

PERIODICALS

Backyard Wildlife Habitat

National Wildlife Federation
1400 16th St., NW Washington, DC 20036-2266
(800)432-6564

The National Wildlife Federation's Backyard Wildlife Habitat Program is a really exciting way to connect your household to the local eco-system. The information includes a booklet which expounds the virtues of having a backyard popular with flora and fauna, and contains a plan to design a space with special places for animals to eat, drink, mate, and find shelter. There is also a good bibliography of books about various plant possibilities, such as edibles, wildflowers, and natives, as well as on attracting birds, mammals, butterflies, and reptiles to your backyard. Application for certification as a Backyard Wildlife Habitat is included, something the young ones, especially, will appreciate.

Section Three

EDUCATION

Whether your interest is birdwatching, herbal studies, sustainable building techniques, marine mammals, ecology, or any other of myriad environment-related subjects, the Bay Area is rich with opportunities for expanding your understanding and horizons.

This section lists many of the environmental educators in our region. It describes courses ranging from weekend seminars to graduate degree programs for people of all ages, as well as resources educators can utilize in planning their curricula.

The offerings are organized by the following types: Adult, degree, K-12, resources, and short courses. Adult courses usually offer material that is more complex and detailed than most children can handle, but this is not a hard and fast rule. Degree listing educators provide either a degree or some other form of certification. The K-12 listings in some way support, develop, or provide curricula for shool-age children. Resources include museums, educational expeditions, visitor centers, and environmental libraries. The short courses have something for everyone, young or old — except a degree.

If you want to see what's available on a particular subject, find it in the index at the back of the book. Otherwise, just browse the listings and discover the remarkable array of opportunities for environmental education in the Bay Area.

Adult

American Lung Association of San Francisco

562 Mission Street #203
San Francisco, CA 94105
(415)543-4410
Fax: (415)543-3682

Contact: Jolie Pearl, Program Director for Occupational and Environmental Health

Description: The mission of the Association (ALA/SF) is the conquest of lung disease and promotion of lung health in the City and County of San Francisco. ALA/SF provides educational programs on lung diseases such as asthma, tuberculosis, emphysema and chronic bronchitis, as well as educational and advocacy programs to mitigate the factors that contribute to these ills: cigarette smoking, air pollution, toxic substances, and on-the-job lung hazards. Current program emphasizes in occupational and environmental health and includes public information and referral, and community and professional training in the following areas: asbestos, art hazards, on-the-job lung hazards, indoor air quality, and outdoor air pollution.

Resources: Small library on occupational health (especially lung hazards), Asbestos information packet and referrals, many pamphlets and brochures on occupational and environmental lung hazards, newsletter: *IT'S VITAL.*

Volunteer Opportunities: Needs volunteers for help on specific program projects, Speakers Bureau, Health Fairs and Community Fairs, office work.

Funding Sources: Donations, bequests, grants (government and private)

Annual Budget: $700,000

Employees: 13

Audubon Society - Marin

P.O. Box 599
Mill Valley, CA 94941
(415)456-5311

Description: General Membersip Meeting and Program: The first Friday of the month from October through June. Guest speakers provide information regarding their special fields; for instance, South American rainforests, raptors, bats, sea turtles, and so on. Generally programs cover topics of interests to birders and conservationists.

Resources: Classes: Usually two per year, ranging from 2 to 4 evenings; bird identification lectures and slide presentation; and field trips on Saturdays. The Society sponsors more than 24 field trips per year, primarily in the SF Bay Area to locations where there is "good birding." A few field trips are held each year outside the Bay Area to places like Monterey/ Carmel, Mono Lake, Gray Lodge, Sacramento Wildlife Refuge, etc.

Funding Sources: Marin Audubon publishes a monthly newsletter, *The Redwood Log*, which is mailed to each member and is available for a subscription fee of $8.00 per year by non-members.

Bay Area Mountainwatch

P.O. Box AO
Brisbane, CA 94005
(415)524-5609

Description: The environmental education program begins with year-round, weekly free hikes of all levels (gentle to rough, class learning to summertime research, and monitor survey work.) Also available are flyers and various other publications, verbal presentations, slide shows, photo, art and materials exhibits, not only to the general public, but also to official bodies that have jurisdiction on San Bruno Mountain. A compilation of written and oral histories of the mountain and its surrounding communities is in progress.

California School of Herbal Studies

P.O. Box 39
Forestville, CA 95436
(707)887-7457

Description: This organization offers a hands-on learning experience. Certificates are given for long programs. The California School of Herbal Studies offers programs ranging from one-day workshops to three day seminars, and programs up to five months for the beginner to the advanced, serious herbal student. For a catalog and more information, send $2.00.

Center for Science Information

4252 20th Street
San Francisco, CA 94114
(415)553-8772
Fax: (415)566-1048

Contact: Elizabeth Atcheson -V.P.

Description: CSI was founded in 1981. It educates decision-makers and journalists about the environmental and public policy issues created by biotechnology.

Resources: Briefbooks, Briefsheets, and Background Briefs.

Volunteer Opportunities: Need researchers and general office workers.

Funding Sources: Grants, contributions

Employees: 7

Clem Miller Environmental Education Center

Pt. Reyes National Seashore
Point Reyes, CA 94956
(415)663-1200
Contact: Fawn Bauer
Description: Established in the early 1970s, the Clem Miller Environmental Education Center is an overnight outdoor facility which can accomodate up to 80 people. The main building is a new, 4600-square foot cedar building with two large classrooms, various teaching aids and study materials, an extensive library, and some nature exhibits. The main building also houses a large kitchen . Aside from the main house there are also six dormitory cabins and a central restroom.

Environmental Traveling Companions (ETC)

Fort Mason Center, Building C
San Francisco, CA 94123
(415)474-7662
Description: For eighteen years ETC has been successfully providing innovative outdoor adventures and environmental education programs for people with special needs. ETC's sea kayaking, whitewater rafting, and nordic skiing programs provide therapeutic recreation, education and inspiration to over 1,000 participants annually. These include youth-at-risk, students, and people who are visually or hearing impaired, physically or developmentally disabled, and those with a terminal illness such as AIDS. Participants are challenged to succeed in ways they never thought possible. They network with over 65 established social service agencies and schools who look to ETC as a vital extension of their services. ETC offers a wide variety of "Benefit" trips, on a regular basis, to the general public and corporate groups. The funds from these trips (whitewater rafting and sea kayaking) go to a scholarship fund to subsidize the outings for the disabled and disadvantaged participants. These trips take place all year long, alongside those that support our mission.

Institute for the Study of Natural Systems

P.O. Box 637
Mill Valley, CA 94942
(415)383-5064
Description: A non-profit, tax-exempt education and resource organization concerned with increasing harmony between people and nature through programs and projects which bridge traditional cultural wisdom about place and nature with modern science, design, and the arts.

Primary programs are:

1) The Spirit of Place Symposium series — a 5 year series of annual public symposiums featuring 50 to 70 speakers, all of whom address how to develop and use place consciousness. Begun in 1988 at U.C. Davis, in early 1991 Quest Books will publish an anthology of selected papers from the 1988 and 1989 programs.

2) Development of Recycle Man, the Marin County recycling mascot.

3) Each year the institute offers a two-term environmental design internship program which focuses on the psycho-spiritual roots of design for people from design fields and the creative arts. The program is directed by James A. Swan, an environmental psychologist and author of *Sacred Places*, and Roberta A. Swan, former professor of architecture and design at the University of Oregon.

4) Celebration for Mother Earth Concerts - Begun in 1989, these public performances blend ancient mythic themes with modern musical artists to celebrate the earth and the wisdom of indigenous peoples.

International Wildlife Rehabilitation Council

4437 Central Place, Suite B-4
Suisun City, CA 94585
(707)864-1761
Contact: Rose Green
Description: The International Wildlife Rehabilitation Council offers classes to teach people who are interested in wildlife rehabilitation the basic and advanced skills necessary in a standardized approach to wildlife work. Classes are held on demand.

Naturalist Associates

107 Palm Ave
Corte Madera, CA 94925
(415)924-3572
Description: Naturalists Associates is a network of professional naturalists in various fields of natural history and ecology. They provide a number of environmental education programs, field seminars and trips,give slide presentations, and are involved in photography , publishing, and research.

Owner Builder Center

1250 Addison, #209
Berkeley, CA 94702
(415)848-6860
Contact: Charles Kim
Description: The Owner Builder Center offers over forty workshops and seminars on building and remodeling topics, with special emphasis in hands-on learning and consumer education. The non-profit Center has taught courses in the Bay Area since 1978. Energy and environmental issues are covered in special courses, including "Passive Solar Design/Title 24", "Photovoltaic Systems", and "Non-Poisonous Structural Pest Control". The Center also stocks and mail-orders 60 book titles and has a full-time consultant and drafting team. Free catalog for California residents; $4/per year out-of-state.

Peninsula Humane Society

12 Airport Blvd.
San Mateo, CA 94401
(415)340-8200
Description: Adult Education Classes for 1991

1/8 The Cruelty—Free Face
1/15 Consumer Beware
1/22 All Creatures Great and Small
1/29 Conservation of the Bay
2/5 Laws, Legislators, Lobbyists and You
2/12 Animals in Disasters
2/19 Integrating Animals into the Curriculum
2/26 Teaching Animal Issues to Children

Planet Drum Foundation

P.O. Box 31251
San Francisco, CA 94131
(415)285-6556
Contact: Crofton Diack
Description: Planet Drum was founded in 1974 to provide an effective grassroots approach to ecology that emphasizes sustainability, community self-determination, and regional self-reliance. In association with community activists and ecologists, Planet Drum developed the concept of a bioregion: a distinct area with coherent and interconnected plant and animal communities, often defined by a watershed. A bioregion is a whole "life-place" with unique requirements for human inhabitation so that it will not be disrupted and injured. Through its publications, speakers, and workshops, Planet Drum helps local organizations and individuals find ways to live within the natural confines of bioregions. Peter Berg offers a variety of courses, from lectures about bioregionalism in Australia to urban sustainability. Judy Goldhaft is an accomplished and well-known performer. Her performance piece 'water' is educational as well as inspirational.
Resources: The best bioregional library on the West Coast, open to the public with preference to members. Newsletter "Raise the Stakes", bi-annual.
Volunteer Opportunities: Cultivating the native plant garden, help in the production of "Raising the Stakes" (writing, proof-reading, copy editing), updating new and non-member files.
Funding Sources: Grants, membership, contributions from individuals.
Employees: 3

Tree Project—Circuit Rider Productions, Inc.

9619 Old Redwood Hwy
Windsor, CA 95492
(707)838-6641
Contact: Karen Kubrin
Description: Tree Project conducts a job training program for adults in landscaping, tree care and resource conservation. The project is sponsored by the Private Industry Council of Sonoma County. Emphasis is on hands-on work activity and integrated classroom training. This program is beginning its thirteenth year and has been selected by the National Association of Counties as an exemplary job training program. Tree Project won the 1983 National Arbor Foundation Award for an educational project that promotes the understanding and care of trees.
Volunteer Opportunities: In development stage.
Funding Sources: Grants.
Annual Budget: $92,000
Employees: 3

Shearwater Journeys

P.O. Box 1445
Soquel, CA 95073
(408)688-1990
Description: Dedicated to bringing marine birds, mammals and people together on one day boat excursions on Monterey Bay, Shearwater Journeys has operated educational and conservation oriented trips for 14 years. Charters for school groups are available.

Degree

Agroecology Program, University of California, Santa Cruz

Agroecology Program
Santa Cruz, CA 95064
(415)429-4140
Description: The Agroecology Program is a research and education group working toward the development of sustainable agricultural systems. These systems conserve natural resources, control pests with biological and ecological measures, provide employment and a safe and stable food supply, while ensuring the same for future generations. Undergraduates at UC Santa Cruz can pursue an Agroecology Concentration as part of the Environmental Studies major. Master's and Ph.D. work is done in conjunction with the Agroecology program by working through the Board of Study in Biology and the Committee on Education. Each April through September, the Agroecology Program/University of California Extension at Santa Cruz offers a full-time residential apprenticeship in Ecological Horticulture at the UCSC Farm and Garden. Emphasis is on hands-on learning of environmentally-sound horticultural methods, soil management, and marketing.

Cal State, Hayward Environmental Education Program

CSUH/ Department of Teacher Education
Hayward, CA 94542
(415)881-3016

Contact: Contact the university for qualifications and more information.

Description: Master of Science degree: Environmental Education Program. The program has three parts: environmental education courses, core courses, and electives. The environmental specialization consists of Environmental Education in the Curriculum, Community Resources for Environmental Education, and Field Study in Environmental Education. The program is designed for teachers who want to enrich their teaching or who want to expand their assignments and for other interested professionals. All courses are open as electives and through Extended Education.

Resources: Core courses include the Foundations of Curriculum Development, Advanced Educational Psychology, and Research in Education, with a choice of a Department Thesis, University Thesis, or Project. The Environmental Education courses are also available as electives through the open university.

College of Marin

Kentfield, CA 94904
(415)485-9506
Fax: (415)485-0135

Contact: Dona Boatright, Dean, Professional, Vocational, and Community Education

Description: This program prepares the student for employment in the field of commercial landscape management with a focus on design, development, and site maintenance. Courses include: Introduction to Commercial Horticulture, Home Gardener, Principles of Landscape Design, Turfgrass Management, Pest Prevention and Control, Landscape Management Practices, Water Conserving Plants, and many more.

Institute in Culture and Creation Spirituality

Holy Names College
3500 Mountain Boulevard
Oakland, CA 94619
(415)436-1046

Description: The Institute in Culture and Creation Spirituality (ICCS) offers a 9 month M.A., a 9 month Certificate, and 4 month Sabbatical experiences. ICCS re-awakens the Western spiritual tradition and makes connections between spirituality, native traditions, the arts, science, and concern for justice and the plights of the earth. Rooted in a reverence for creation and respect for the environment, ICCS's approach balances traditional classroom teaching with art-as-meditation courses.

John F. Kennedy University

12 Altarinda Rd.
Orinda, CA 94563
(415)253-2228

Description: JFK University offers a Bachelor of Arts Degree with emphasis in Environmental Studies. The approach to Environmental Studies stresses the interconnections between natural science and social science. It emphasizes ethical values and the role of consciousness in a practical approach to the global crisis.

Napa Valley College

Napa Valley, CA 94558
(707)253-3259

Contact: Stephen Krebs

Description: The school provides a two year program designed to prepare students for careers in viticulture and/or winery-related jobs. The courses helps individuals gain entry-level positions in the wine grape industry or improve career advancement opportunities for persons already employed in the industry. The program emphasizes the application of theory for decision-making in actual wine industry situations.

New College of California

50 Fell Street
San Francisco, CA 94102
(415)626-1694

Contact: Vicki Brezeale or Matt Kumin

Description: New College of California offers an Ecological Studies emphasis within the Humanities B.A. Program. This emphasis embraces a variety of perspectives pertaining to human interactions with the environment and life on earth— political, social, scientific, biological, economic, historical, spiritual and cultural. The intent of this multi-perspective approach is to cultivate a comprehensive, acute understanding of the complexity of environmental discourse, problems, and solutions. It is also congruent with the college's committment to relevant, interdisciplinary, multi-cultural education. The emphasis involves students in critical dialogue, creative problem solving, and real-world activity in both classroom and community. The program provides students with the ability to take part in internships and independent studies with environmentally-minded nonprofit organizations in the Bay Area that will aid them in their particular field of study. Through these learning opportunities, students are able to tailor their own particular focus within the emphasis and their relationship to the larger community. Upon graduation from New College, students will be prepared to enter environmentally-concerned jobs or graduate programs.

Romberg Tiburon Center for Environmental Studies

3150 Paradise Dr.
Tiburon, CA 94920
(415)435-0479
Contact: Dr. Tim Hollibaugh
Description: A bayside facility of San Francisco State University dedicated to the study of the natural and human-altered environments of the San Francisco Bay. Courses and lectures on estuarine ecosystems are offered to University students and the public. Research is conducted on the San Francisco Bay and other estuaries, and information is disseminated to scientific, governmental and public groups.

San Francisco State University

1600 Holloway Ave.
San Francisco, CA 94132
(415)469-2141
Fax: (415)338-6136
Contact: Dr. Peggy Fiedler
Description: SFSU offers a Master's degree program in biology with a concentration in conservation biology. Applicants must have an undergraduate degree in biology or a related field. The program covers a broad range of biological issues in the field of conservation, including, wetlands ecology, marine mammal biology and conservation, rare and endangered species, andsystematic biology.

San Jose State University, Environmental Studies Dept.

Dept. of Geography & Environmental Studies
San Jose, CA 95192
(408)924-5450
Contact: Gary Klee
Description: Environmental Studies provides well-structured opportunities for education and enrichment in the areas of environmental awareness and environmental problem-solving. Affiliated with the School of Social Sciences, the program complements technical offerings with analysis of the social, political, economic, legal, ethical, and cultural implications of resource management and ecological problems. A graduate of this program is able to make a tangible and useful contribution toward the protection of the environment. It provides the legal and political tools to function as environmentally responsible citizens, and training for public service in those environmental technologies now being defined by industry and public protection agencies.
Resources: Environmental Research Center, Center for Development of Recycling.
Volunteer Opportunities: Internship Program.
Funding Sources: San Jose State University
Employees: 10

K-12

Alameda County Office of Education

313 W Winton Ave.
Hayward, CA 94544
(415)670-4168
Contact: Dr. Loretta Chin
Description: The Alameda County Office of Education has a variety of resources available by mail for educators and parents, including lesson plans dealing with toxics and water usage, videos about hazardous waste, and the "Tripping with Elizabeth Terwillinger" series, a nature travelogue program for grades K-6.

Audubon Canyon Ranch

4900 Shoreline Highway, Rt. 1
Stinson Beach, CA 94970
(415)868-9244
Fax: (415)868-1699
Contact: Nancy Angelesco
Description: Audubon Canyon Ranch is a natural wildlife preserve which offers educational programs to both school children and adults in the area of natural history. There are on site classes and school programs available. Please contact for further information.

Audubon Society—Golden Gate

2530 San Pablo Ave., Suite G
Berkeley, CA 94702
(415)843-2222
Contact: Barbara Rivenes
Description: 1) The Golden Gate Audubon provides local school districts and classrooms with a program called "Audubon Adventures." These are materials sent to the classroom with a teacher's guide, and focus on environmental concepts. 2) Offers their membership monthly programs, featuring knowledgeable speakers, focusing on local, regional, state, national and international wildlife issues. 3) Offers several field trips per month , primarily bird-oriented hikes and walk. 4) Provides the Northern California Rare Bird Alert describing the location of rare and unusual bird sightings in California.
Resources: "The Gull"

Bay Area Science Associates

P.O. Box 431
San Francisco, CA 94101
(415)383-8099
Contact: Terry Sullivan
Description: BASA works to promote increased awareness of the need to preserve California's natural heritage. It creates and operates outdoor science education programs for elementary, high school and adult groups. It also offers environmental and historical tours of California for foreign students. Schools consult BASA in setting up science laboratories and curriculum.
Funding Sources: Fees and grants.

Berkeley—East Bay Humane Society

P.O. Box 2222
Berkeley, CA 94702
(415)845-7735
Contact: Nancy Frensley, Education Dept.
Description: They offer free programs to classrooms and youth groups which stress spaying and neutering of pets, and keeping pets from chasing wildlife or harming the environment in other ways. While not strictly environmental, humane treatment of pet, farm and wild animals is an important part of conservation.

Camp Unalayee

2448 Watson Court
Palo Alto, CA 94303
(415)493-3488
Contact: Phyllis Bolchalk
Description: Wilderness camp located in the Trinity Alps of Northern California for boys and girls ages, 10-17. Teaches respect and love for the environment. Non-profit, multicultural. Backpacking and hiking emphasized.
Resources: Two newsletters annually.
Volunteer Opportunities: Some summer staff are volunteers.
Funding Sources: Memberships, fees, corporate & foundation donations & grants.
Annual Budget: $100,000
Employees: 30

Coyote Hills Regional Park

8000 Patterson Ranch Road
Fremont, CA 94536
(415)795-9385
Contact: Norm Kidder — Supervising Naturalist
Description: K-12 Education: Coyote Hills offers field trips for school classes, youth groups, and individuals in the areas of cultural and natural history. Fourth Graders and older may schedule a trip to a 2400-year-old Ohlone shellmound to learn about the first inhabitants of the Bay Area.

Adult and Family: On weekends a wide variety of walks and workshops are offered for adults and familieson natural history topics and Native Americans —specifically relating to the local Ohlone Indians. A monthly publication, The East Bay Log, lists all these programs.
Resources: The East Bay Log; a small museum featuring displays and interactive materials on the Ohlone Indians, and the natural history (birds, deer, wildlife, marshlands, etc.) of the park.

East Bay Schools Energy Fitness Center

440 Grand Avenue, Ste. 425
Oakland, CA 94610
(415)261-8422
Contact: Ken Eagle
Description: Energy education, both in the classroom and in terms of co-curricular activities for students, is strongly encouraged so that the school truly becomes a laboratory to link energy education with energy management.
Resources: See "California Energy Extension Service" in the Organizations listings for publications available.

El Camino Real Energy Extension Center

c/o San Jose Unified School District, 250 Stockton Avenue
San Jose, CA 95126
(408)280-0990
Contact: Tom Tvetnes, or Andrea Denver at (415) 841-4552.
Description: The California Energy Extension Service has established regional energy extension centers to serve school districts. The centers strive to produce measurable energy savings in schools by involving everyone at the school site—from administrators and staff to teachers and students. Energy education, both in the classroom and in terms of co-curricular activities for students, is strongly encouraged so that the school truly becomes a laboratory to link energy education with energy management.
Resources: See "California Energy Extension Service" in the Organizations Listing for publications available.

4- H Program-Sonoma County

University of California Coop. Ext.
2604 Ventura Ave, Rm 100
Santa Rosa, CA 95403
(707)527-2681
Fax: (707)527-2022
Contact: Carol Kaney, 4- Program Coord.
Description: 4-H provides research and educational programs for youth and adults in marine science, natural resource management and agriculture. They also provide experiences aimed at building self-esteem and critical thinking skills.

Inner City Outings (ICO)

730 Polk Street
San Francisco, CA 94109
(415)776-2211 x5628
Fax: (415)776-0350
Contact: Debra Asher
Description: Inner City Outings (ICO) is a community outreach program of the Sierra Club. Volunteer leaders, trained in recreational and safety skills, provide wilderness adventures for people who wouldn't otherwise have them — including urban youth of diverse cultural and ethnic backgrounds, seniors, hearing or visually impaired individuals, and the physically disabled.
Volunteer Opportunities: run by volunteers.

Life Lab Science Program

1156 High Street
Santa Cruz, CA 95060
(408)459-8942
Contact: Rebecca Levy
Description: Life Lab is a Kindergarten to Sixth grade, garden-based science program, used by schools throughout the country. Two-day training workshops can be arranged and curriculum can be obtained through the program office.

Math/Science Nucleus

3710 Yale Way
Fremont, CA 94538
490-6284
Contact: Dr. J. Blueford
Description: Non-profit, science education group that developed the Integrating Science, Math, and Technology Curriculum for Kindergarten through Eighth Grade. Environmental education, both built and natural environments, are an integrative portion of the curriculum. Hands-on labs and lesson plans are available. Free scope and sequence is also available on request.

Natural Science Docent Program
Calif. Native Plant Society

P.O. BOX 2284
Yountville, CA 94599
(707)224-4747
Contact: Vivian Tucker
Description: The Natural Science Docents, volunteers from the Napa Chapter of the California Native Plant Society, provide environmental education to Napa upper elementary classes. The core presentation consists of a four part ecology suitcase program. Upon a teacher's request, a team of three docents visits the classroom for four one hour sessions on consecutive weeks. They bring games, hands-on activities, and exhibits which illustrate habitats, food and energy relationships, various plant functions, invertebrates, and vertebrates. They also make available a Native American suitcase program, several slide shows, and an interdisciplinary film strip series from the Cousteau Society.
Resources: Docents who visit classrooms and present information, bring games and projects, provide educational activities; films.

Pacific Energy and Resources Center

Building 1055, Ft. Cronkhite
Sausalito, CA 94965
(415)332-8200
Description: The Pacific Energy and Resources Center (PERC) is a nonprofit policy research and public education center committed to sound natural resource management, and to energy conservation and alternative development. It works to assess the status of local and global resources and environmental conditions, analyze emerging issues, develop innovative policy responses, and educate policy makers and the public about these issues. PERC has education programs that address global environmental issues.

PLAE, Inc. (Playing and Learning in Adaptable Environments)

1802 Fifth Street
Berkeley, CA 94710
(415)845-7523
Fax: (415)845-8750
Contact: Sally McIntyre, Susan Goltsman or Mi-Yung Rhee
Description: Traditionally, the training of recreation leaders has lacked a vital component — the use of the environment for play, learning, and integration. The physical environment is a critical element in the development of all children. It can also become a key element to promote the integration of disabled and non-disabled children. PLAE has developed leadership techniques to help recreation staff use simple materials to expand their curriculum and change their every day environment into a new adventure for children.

Project Learning Tree
California Department of Forestry

P.O. Box 944246
Sacramento, CO 94244
(916)445-9920
Description: Project Learning Tree (PLT) is an award-winning environmental education program designed for teachers and other educators working with students in K -12. PLT is a source of interdisciplinary instructional activities, and provides workshops and in-service programs for teachers, foresters, park and nature center staff, and youth group leaders. Children learn HOW to think, not WHAT

to think, about our complex environment; discover how subjects and skills taught in the classroom relate to the world around them; develop skills in creative problem solving, critical thinking, evaluation and research; and have fun while learning. Teachers receive a ready-to-use PLT guide, resource materials and a complimentary subscription to PLT's newsletter filled with teaching ideas and activities; discover over 175 activities that help teach science, mathematics, language arts, social studies, humanities, and other subjects; and participate in a creative workshop that helps improve classroom skills.

San Francisco Bay Girl Scout Council

P.O. Box 2389
San Leandro, CA 94709
(415)562-8470

Contact: Kim S. P. Campbell

Description: Girl Scouting is an informal educational program, serving girls in grades K-12. It is carried out in small groups with adult leadership, and provides a wide range of activities developed around the interests and needs of girls. Our environmental education activities include program packets on recycling, astronomy, water conservation and pollution, and shoreline exploration; several ecology related badges and "try it" patches for elementary school girls, an interest project for junior high school girls, and a nationally sponsored program called Earth Matters. For information on getting involved, call 1-800-447-4475.

Resources: Girl Scoutlook Newsletter as well as teaching materials on recycling, water polution, care for trees, and more are available to members. Membership is open to anyone by phoning 1-800-447-4475 or (415) 562-8470.

Volunteer Opportunities: contact Karen Kaho, Volunteer Coordinator, (415) 562-8470 or 1 (800) 447-4475.

Funding Sources: United Way, support and contributions.

Annual Budget: $1,000,000

Employees: 60

San Francisco Friends of the Urban Forest

512 2nd Street, 4th Floor
San Francisco, CA 94107
(415)543-5000

Contact: Clifford Janoff

Description: Friends of the Urban Forest (FUF) is a non-profit organization committed to the belief that trees are a critical element contributing to a liveable urban environment. Trees in the city, our urban forest, represent contact with nature in the midst of hard surfaces. They satisfy our basic human desire to live among growing things that both nurture and inspire us. FUF also believes that city trees provide an exciting laboratory in which students can explore their physical environment and community from many perspectives. *City Trees: A Curriculum Guide of Our Urban Forest and Community* is FUF's attempt to bring our belief into action. The activities in *City Trees* link in-depth classroom study with the excitement of hands-on learning. By using the urban forest as a textbook, teachers can provide a fresh approach to a variety of disciplines: math, social science/history, literature and fine arts, as well as science and ecology. When this guide is used in conjunction with a tree-planting project, the students will find a rewarding end to the early learning, and as the trees mature, they will provide an ongoing laboratory for study.

Santa Clara County Office of Education Walden West Center

15555 Sanborn Road
Saratoga, CA 95070
(408)867-5950

Contact: Dave Aikman

Description: Walden West Environmental Center is owned and operated by the Santa Clara County Office of Education. Its mission is to provide a week-long Environmental Education experience for fifth and sixth grade students. The children are taught an appreciation for the environment and their interdependence with it. The program operates from mid-September to mid-June.

Volunteer Opportunities: Naturalists, guest speakers, gardeners

Funding Sources: Santa Clara County Office of Education and user fees

Annual Budget: $800,000

Employees: 23

Slide Ranch

2025 Shoreline Hwy.
Muir Beach, CA 94965
(415)381-6155

Contact: Suzanne Connoly, Program Director; Ross Herbertson. Administrative Director

Description: Slide Ranch is a non-profit demonstration farm and environmental education center, operating as a park partner with the Golden Gate National Recreation Area. It has been offering educational programs since 1970 to school children, the learning and physically disadvantaged, inner-city kids, and multi-ethnic groups, girl and boy scouts, and other organizations. Slide Ranch programs reconnect participants to the plants and animals that feed and clothe them. The programs teach principles of ecology using the farm, wildlands and ocean environments, and provide the opportunity to explore the wonders of the natural world in a non-threatening manner. The Ranch offers programs Wednesday through Sunday to group programs and, on weekends offers programs for families. All programs need reservations. Scholarships are available, and all animal areas and garden are wheelchair accessible.

Funding Sources: Program fees; contributions from individuals; foundation, corporation, and government grants.

Annual Budget: $163,000

Sonoma Energy Extension Center
c/o Sonoma State University
Department of Environmental Studies
Rohnert Park, CA 94928
(707)664-3145
Contact: Irv Peterson
Description: The California Energy Extension Service has established regional energy extension centers to serve school districts. The centers strive to produce measurable energy savings in schools by involving everyone at the school site—from administrators and staff to teachers and students. Energy education, both in the classroom and as part of co-curricular activities for students, is strongly encouraged so that the school truly becomes a laboratory to link energy education with energy management.
Resources: See *California Energy Extension* Service in Organizations Listings for publications available.

Sulphur Creek Nature Center
1801 D Street
Hayward, CA 94541
(415)881-6747
Fax: (415)881-1716
Contact: Mike Koslosky, Hayward Shoreline
Description: Sulphur Creek Nature Center is a wildlife education center dedicated to introducing the visitor to the diversity of animal life in the greater Bay Area. The center houses over 100 different species of local wildlife in educational displays. There is also an animal rental program which allows children to have a pet for a week. A wildlife rehabilitation program treats injured and orphaned wildlife, with the intent of releasing them back into the wild once they have recovered from their injuries and are self-supporting. Programs include tours, field excursions, in-class presentations, visits to convalescent hospitals and rest homes, weekend interpretive programs, volunteer opportunities, seasonal special events, a wildlife discovery center with exhibits and displays, wildlife garden, nature trail, birthday party area (with program), picnic grounds, seasonal stream.
Volunteer Opportunities: general clerical, cleaning cages.
Employees: 750

Terwilliger Nature Education Center
50 El Camino Dr.
5 Gamma Bldg.
Corte Madera, CA 94925
(415)927-1670
Description: Terwilliger Nature Education Center is dedicated to furthering the work of Elizabeth Terwilliger who pioneered tactile/multi-sensory techniques for teaching children about the environment. Based on the belief that 'what people love, they take care of,' TNEC programs encourage children and families to learn about and love nature and nature's creatures. Programs include touching and careful examinaton of plants and taxidermied animals, circle time for questions, games and songs, sharing information in a fun, non-technical manner, and instilling in the participants a confidence in being in nature.
Volunteer Opportunities: Field trips, general clerical.
Employees: 9

Youth Science Institute
296 Garden Hill Drive
Los Altos, CA 95030
(408)356-4945
Contact: Dave Johnston, Executive Director
Description: The Youth Science Institute has been providing Santa Clara Valley children and families with the opportunity to discover the wonders of science since 1953. Its programs and exhibits are designed to interpret science and to instill an appreciation and an understanding of the natural environment. The Youth Science Institute, a non-profit organization, was founded by the Woman's Service Club (now called the Junior League of San Jose) and the National Science Foundation for Youth to augment science instruction in schools. YSI reaches more than 40,000 children through its programs taught in schools and at YSI's three discovery centers located in Alum Rock Park, San Jose, Vasona Lake Park, Los Gatos; and Sanborn-Skyline Park, Saratoga.

Resources

BAEER Fair (Bay Area Environmental Education Resources)
P.O. Box 391
Cupertino, CA 95015
(415)657-4847
Contact: Ken Hanley
Description: The BAEER Fair is and annual event which brings people in touch with the resources and issues surrounding our natural world. The fair will be held on Saturday, January 19, 1991 from 10 AM to 5 PM at the Marin Civic Center (San Rafael). Tickets will be $5 at the door.The exhibitors range from park services, museums, zoos, aquariums, maps, compasses, labatory equipment, live animals, environmental singers, and much more. Registration for workshops is now open. The first workshop will be at the Monterey Bay Aquarium and be on "Pollution Off Coast". Other workshops on animals, interpretative programs, and earthquakes will be held at the Yosemite Institution, and at the Richardson Bay Audubon Preserve.

California State University, Hayward Environmental Education Lab
CSUH/ Department of Teacher Education
Hayward, CA 94542
(415)881-3016
Contact: Dr. Esther Railton-Rice
Description: The Environmental Education Laboratory is designated by the Alliance for Environmental Education as one of four network centers in the U.S., providing teacher education, research, program development and community outreach. The resource center houses over 4,000 books, curriculum guides, periodicals, pamphlets, teaching units, activity cards, posters, games, and newsletters. The cataloged collection encompasses all topics of environmental concern. Further resources include files providing extensive information about research, philosophy, school site development, field trip sites, public and private agencies, and resident outdoor programs.
Resources: The calendar and bulletin board provide current information on workshops, conferences, classes, job openings, and other environmental education around the state. An assortment of free materials is available to visitors. The EEL produces topical bibliographies of Lab acquisitions. These computer printouts are available by mail and soon by EcoNet.
Funding Sources: University, State Department of Education Env. License Plate Grant, Industry, Public and Private Agencies.

California Academy of Sciences
Golden Gate Park
San Francisco, CA 94118
(415)750-7142
Fax: (415)750-7346
Contact: David Shaw
Description: The Academy consists of a natural history museum, aquarium and planetarium. The Steinhart Aquarium offers over 1,000 species include fresh and saltwater fishes, plus an impressive array of reptiles, amphibians, marine mammals, and penguins. The Morrison Planetarium offers educational and entertaining sky shows, call (415) 750-7141 for show times. Another feature is Laserium, an laser light and music show presented live Thursday through Sunday evenings. Call (415) 750-7138 for show times. Exhibits include: Life Through Time is an evolutionary display; Wild California features indigenous plants and animals; Wattis Hall of Human Cultures contains scenes of various groups adaption to climatic environments; African Safari consists of dioramas showing African animals in their natural surroundings; the Far Side of Science Gallery features 159 of Gary Larson's cartoons; Hohfeld Earth and Space Hall offers a ride on the Safe-Quake which simulates two of San Francisco's tremors; the Gem and Mineral Hall has over 1,000 specimens from gold to granite; the Discovery Room for Children is a hands-on-everything place designed for K-6 grades.

Center for Marine Conservation
312 Sutter Street, Suite 606
San Francisco, CA 94108
(415)956-7441
Contact: Kate Patterson
Description: Center for Marine Conservation is a national non-profit membership organization dedicated to protecting marine wildlife and their habitats, and to conserving ocean and coastal resources. Our Pacific Coast office is located in San Francisco and focuses on marine pollution prevention, and establishing marine protected areas along the Pacific Coast. The Center achieves its goals by conducting policy-oriented research, promoting public awareness through education, and involving citizens in public policy discussions. Marine educational activities, materials, and curriculum available.

Committee for Sustainable Agriculture
P.O. Box 1300
Colfax, CA 95713
(916)346-2777
Contact: Otis Wollan
Description: The CSA is an educational organization focused on educating farmers, consumers, retailers, manufacturers, and the public on organic farming and sustainable agriculture. Events sponsored by the CSA include: Ecological farming conference, National Organically Grown Week, Hoes Down Harvest Faire.
Resources: Quarterly publication, *Organic Food Matters*
Volunteer Opportunities: Many opportunities available.
Funding Sources: Projects, memberships, foundations.
Employees: 6

Conservation & Resource Studies Resource Center
112 Giannini Hall, UCB
Berkeley, CA 94720
(415)642-6746
Description: The CRS Resource Center offers books, periodicals, reference books, job newsletters, and an extensive pamphlet collection from miscellaneous sources. This center focuses on collecting material that other campus libraries don't stock, and has extensive materials that are action, policy, and community oriented. The Conservation and Resources Studies Resource Center is open to the public 9-5, M-F. Hours may vary.

Coyote Point Museum for Environmental Education

Coyote Point
San Mateo, CA 94401
(415)342-7755

Description: Coyote Point Museum has interactive displays, an aquarium, hands-on computers, a live bee hive and fun games in a beautiful warm woodsy setting. The purpose of the Museum is to increase knowledge of and appreciation for the natural world, enabling people to make wise decisions for the future. The Museum is currently constructing a wildlife habitat exhibit that will house animals native to California in naturalistic settings. It will emphasize the need for habitat preservation. The Wildlife Center is scheduled to open in Spring 1991. Educational programs include: Day Camp: docent tours of tidepools and foothills, The Wildlife Center, the Museum's environmental exhibit, changing exhibits, school programs, lecturers and field trips. The Museum is open Wed. - Fri. from 10:00am - 5:00pm; Sat. and Sun. from 1:00pm - 5:00pm. It is closed Mondays, Tuesdays, Thanksgiving, Christmas and New Year's Day.

Funding Sources: Individuals, corporations, foundations, membership, admission and program fees.

Crow Canyon Gardens, Inc.

10 Boardwalk
San Ramon, CA 94583
(415)837-9387

Description: Crow Canyon Gardens, Inc. is a producing organic herb, vegetable and fruit garden/mini farm, with an adjoining country restaurant. Virginia Mudd, Palmer Madden, Kerry Marshall and Jim Fellows secured land for an organic garden/restaurant in 1978, and started to sponsor a Summer Nature Camp in the creek canyon in 1980. Since then, they have continued to expand the Camp and present a series of classes throughout the year. They offer classes in food, gardening, nature studies, and the arts within the setting of the gardens and the creek canyon nature park. In 1987, they joined forces with the City of San Ramon Park and Community Services Department to further expand the Summer Nature Camp and out-reach education programming to the Schools and Seniors.

Dolphin Network

3220 Sacramento Street
San Francisco, CA 94115
(415)931-2593
Fax: (415)931-0948

Contact: Edward Ellsworth, Director

Description: The Dolphin Network is a project of the Washington Research Institute, a non-profit organization. It is an educational and research organization that publishes a quarterly newsletter called Dolphin Net, with networking information about research and legislation on marine mammal and environmental issues. It contains a directory of cetacean organization, announcements from members, volunteer human resources within the network, videotape and equipment resources, land and ocean area available for projects, bibliographies, and lively editorial discussion on the direction of The Network. They educate about dolphin welfare and captivity issues, and the importance of studying whales and dolphins as related to evolution and long-term survival.

Resources: *Dolphin Net*, a quarterly newsletter ($15.00 annual subscription) and an electronic listing of the information, including marine mammal stranding conferences on EcoNet. Annual education conferences are offered along with free listings of existing education materials to teachers.

Volunteer Opportunities: Anyone interested should contact the network in San Francisco.

Funding Sources: Community-supported by Supporting and Associate members, and subscribers to Dolphin Net.

EarthSave

706 Frederick Street
Santa Cruz, CA 95062
(408)423-4069
Fax: (408)458-0255

Contact: Holly Gordon or Patricia Carney

Description: EarthSave is a non-profit environmental and health educational organization dedicated to helping the Earth restore its delicate ecological balance by reducing the destructive impact human activity has upon the Earth. EarthSave promotes conservation of natural resources, respect for all life forms, compassion for humans and animals, sustainable agriculture, and improved human health through good dietary choices. EarthSave's mission is to alert the public to how our food choices effect not only our own health but the health of the planet as well. EarthSave teaches that our current meat and dairy based diet has a major detrimental impact on our environment, our health, animals, and all other life. We encourage people to learn the facts about a meat based diet and make wise choices. We also respect the inherent dignity of all people, regardless of their choice. Our actions are peaceful, positive, compassionate, non-judgmental, and life-affirming. EarthSave's iscommited to living in harmony and peace with nature, animals and other humans so that all future generations may know the full beauty and wonder of our home—planet Earth.

Resources: Books on making the correct food choices for a better Earth. Present work is being done on a school nutrition program called *Healthy People, Healthy Planet.*

Volunteer Opportunities: Many volunteer projects are available. EarthSave action groups provide local involvement. Please call for information. There are numerous local action groups across the country.

Funding Sources: Memberships and Donations.

Annual Budget: $250,000

Employees: 8

Hayward Shoreline Interpretive Center

4901 Breakwater Ave.
Hayward, CA 94545
(415)881-6751

Description: The Hayward Shoreline Interpretive Center is an education and visitor center on a San Francisco Bay salt marsh. Naturalists offer free weekend programs on the human and natural history of the Bay and adjacent wetlands. Schools and colleges may contract for fee programs of one or more hours. Teacher workshops on wetlands and bay ecology are offered during the school year or by special arrangement. The Center is open 7 days/wk from 10am - 5pm

Resources: Newsletter subscriptions are available.

Volunteer Opportunities: Center administration, marsh monitoring, teaching (a 6 week docent program is required) and exhibit maintenance.

Hidden Villa

26870 Moody Road
Los Altos Hills, CA 94022
(415)948-4690

Contact: Robin Winston (415) 948-4690

Description: Hidden Villa Farm and Wilderness Preserve engages children and adults in hands-on, innovative programs promoting humanitarian values and environmental awareness. A 1600 acre wilderness preserve and working farm, Hidden Villa demonstates stewardship of the land, fosters cooperation between groups, and offers the community an incomparable natural, educational, historic, and recreational preserve. Hidden Villa is open to the public Tuesday through Sunday from dawn until dusk. Hidden Valley Hostel is open September through May, seven days a week. Environmental Education Programs operate September through May. Summer Camp Programs are in June through August.

Volunteer Opportunities: Volunteers are needed in many capacities.

Funding Sources: In the 1960's the Duveneck Family invited the community to join them in founding The Trust for Hidden Villa, a non-profit organization dedicated to preserving Hidden Villa as a farm and wilderness area and continuing the program traditions that had been established. In 1987, two years after Frank Duveneck's death, The Trust for Hidden Villa assumed full financial responsiblity for Hidden Villa which had previously been subsidized by the Duveneck Estate. The trust is currently run by a Board of Trustees and is supported by program fees and charitable donations.

Jasper Ridge Biological Preserve

Herrin Lab - Stanford University
Stanford, CA 94305
(415)723-1580

Contact: Alan Grundmann

Description: The Jasper Ridge Biological Preserve is a 1200-acre natural area protected for research and teaching. Public tours are available via the Jasper Ridge Tour Service at (415) 327-2277.

Labor Occupational Health Program (LOHP)

2521 Channing Way
Berkeley, CA 94720
(415)642-5507
Fax: (415)643-5698

Contact: Donna Jarvis

Description: LOFP is a labor education project affiliated with the Center for Occupational and Environmental Health at the school of Public Health, University of California, Berkeley. The program produces a variety of printed and audiovisual materials on occupational health, and conducts workshops, conferences and training sessions for California workers and unions. A catalog of materials and a brochure which describes training services are available upon request. Call for information on volunteer opportunities. We are a small university-based program, directed toward community outreach with a variable staff of 10 to 15 people.

Lawrence Hall of Science

University of California
Berkeley, CA 94720
(415)642-5132

Contact: Lynn Stelmah, Public Programs

Description: The Lawrence Hall of Science is the public science center of the University of California at Berkeley. Visitors to the Hall can learn more about the Earth, its inhabitants, and the universe through interactive exhibits, planetarium shows, Discovery Laboratories, special events, classes, and lectures. Open daily 10:00 a.m. to 4:30 p.m. The Discovery Corner Gift and Bookstore offers a complete selection of books, posters, audio tapes, games, and T-shirts, for all ages. Call (415) 642-1929 for store information.

The Lindsay Museum

1901 First Avenue
Walnut Creek, CA 94596
(415)935-1983

Contact: Steve Barbata

Description: The Lindsay Museum is a Natural Science Museum for children and adults interested in the natural sciences. Regional in focus, the Museum applies universal principles of biology and ecologty to its programs, which feature direct experiences with live animals for children and adults. It offers tours of the museum for school groups, special interest groups, physically and mentally disabled groups, as well as others. It has a Pet Library where anyone 6 years old and older can check out a rabbit, hamster, guinea pig or rat for a week. The purpose of this program is to teach young children respect for animals and a sense of the responsibility involved in caring for a live creature. The Lindsay Museum has more than 400 volunteers. The Docents lead school tours, take animals out to schools and other facilities, and represent the Museum at fairs and other events; wildlife volunteers (age 16 and up) care for the more than 8,000 animals brought to the museum each year; Interpretive Guides are young people 12-16 years old who care for the Pet Library animals, interpret the animals for the public, and represent the museum at various events, the Alliance manages the Museum Store, the Pack Rat volunteers work at the annual fundraiser to raise money for the museum's operating expenses. The Museum is open Wednesday through Sunday 1pm to 5pm. Admission is free. The Wildlife Department is open daily from 9 am to 5 pm to receive injured and orphaned wild animals. For wildlife information, call (415) 935-1988. Preregistration is required for all classes and trips. For information about classes, trips, and school tours, call (415) 935-1983 between the hours of 11 am and 5 pm.

Merritt College -Self-Reliant House

Environmental Education Center
12500 Campus Drive
Oakland, CA 94619
(415)436-2619
Fax: (415)436-2405

Description: Come help build an urban ecological house and garden! The Self-Reliant House features energy and resource-conserving home improvements, recycling, restoration, and edible and drought-tolerant landscaping. We offer semester length, short lecture and laboratory courses, workshops and tours. Self-Reliant House is developing a library and databases.

Oakland Museum

1000 Oak Street
Oakland, CA 94607
(415)273-3884

Description: The museum offers wide range of educational offerings, including gallery tours, classes, outreach activities, suitcase exhibits, special programs and events, curriculum materials and guides are available to teachers and students of all ages. All programs are designed to supplement the school curriculum and to offer students significant opportunities for learning about the state's cultural heritage and natural resources. K-12 classes in natural history include: A Walk Across California - in the gallery; What's for Dinner? K-6 - explore the animals' culinary perspective; Plant and Animal Homes K-6; Adaptations for Survival 4-9; Species and Communities at Risk 6-12; Shaping California's Landscape 6-12 - on the causes of California's terrain; Museum Magic 4-12 -for students with limited English proficiency; Lake Merritt Channel Walk 4-8; African-American Professionals in the Sciences 9-12 -science career presentation. Other interdisciplinary classes include: Indian Life 4-12; Life's Recipe: Just Add Water 4-8 - role of water in California. Lending Collection on bird and mammal adaptations for teachers. A Suitcase exhibits on the ecology of the San Francisco Bay Area is available to schools. The Museum offers a variety of Family Programs strarting in April. These workshops for families with children are free and explore such subjects as animal habitat's, California indian life, animal food, and the lay of the land. Call for brochure. The Hall of California Ecology was designed to be a gallery of ideas rather than objects, and contains exhibits which present concepts of ecological science and the grandeur of California's diverse natural environment.

Peninsula Conservation Center Library

2448 Watson Court
Palo Alto, CA 94303
(415)494-9301

Contact: Connie Sutton, librarian

Description: An environmental lending library containing books, periodicals, maps, newspaper clippings, pamphlets (including Environmental Impact Reports) and videotapes on topics ranging from acid rain to Yosemite, with special emphasis on Bay Area environmental news. The library contains the following special collections:

- The Environmental Volunteers Collection
- The Wildlife and Endangered Species File
- The Santa Clara Valley Audubon Society Collection
- The Backpacking and Trails Collection

Richardson Bay Audubon Center and Sanctuary

376 Greenwood Cove Drive
Tiburon, CA 94920
(415)388-2524

Contact: Hours: 9 to 5 Wed through Sun.
(Limited handicap access)

Description: The center, 900 acres of submerged tidelands managed as a sanctuary, provides a refuge for thousands of wintering waterfowl. The 11 acres of land include an environmental education center, historic Lyford House, and a classroom-office-bookstore facility. The variety of mini-habitats — including a rocky and sandy shoreline on the Bay, grassland, coastal scrub, freshwater pond, thicket and riparian woodland — make the sanctuary a wonderful outdoor classroom for teaching science and conservation. Teachers can make appointments to bring classes for either self-guided or conducted programs. A self-guiding nature trail booklet is provided to groups. Education programs include the Bay Shore Studies Program, a marine biology program for grades 4 through 8. Classes in natural history include Super Bug, Super Bird, Hawk Watch Over Marin, and Plants Along The Trail.

Resources: Education Programs for grades 4 through 8, classes in natural history, outdoor classes in environmental science and conservation. There are programs for the public on Sundays.

San Francisco Bay Model Regional Visitor Center

2100 Bridgeway Blvd.
Sausalito, CA 94965
(415)332-3870
Fax: (415)332-0761

Contact: Daphne Derven, Park Manager

Description: A tidal hydraulic model, 1.5 acres in size, used to conduct research in the water movement patterns of San Francisco Bay and Delta, the Visitor Center offers guided tours for 10 or more (by reservation), or is open for self-guided tours. The Center has hands-on exhibits, audio tours, and orientation programs offered in English, French, German, Spanish and Japanese. Hours:

Winter —Tuesday through Saturday, 9 am to 4 pm
Summer —Tuesday through Friday, 9 am to 4 pm
Saturday, Sunday, Holidays, 10 am to 6 pm

The staff provides educational programs to schools and general public on the natural and cultural history of the Bay Area. The model is an excellent environmental education tool to learn about the spectrum of marine, geographic, urban planning, water use and conservation issues.

San Gregorio Environmental Resource Center

P.O. Box 49
San Gregorio, CA 94074
(415)726-0565

Contact: George Cattermole—Director

Description: A non-profit, public- benefit educational institution, the center is designed to: (1) provide youth with the guidance, training and support needed for the crafting of a cleaner and safer world, (2) serve as a conference and resource center for groups and individuals concerned with coastal ecology, development and agriculture, and (3) facilitate the use and sale of environmentally safe products at the manufacturing, resale and consumer levels. The Farm Ecology Camp, on 8 acres of prime agricultural land in the San Gregorio Valley in San Mateo County, is teaching the basics of sustainable organic farming, energy and natural resource conservation, coastal ecology, camping skills and crafts. By involving visitors in activities such as the growing and preparation of their own food, the running of a recycling center, erosion control and reforestation projects, and tidepool ecology, it is hoped they will become more concerned about their environment and, contribute directly to the preservation and enhancement of our coast's natural resources.

Resources: The Farm Ecology Camp

Volunteer Opportunities: fundraising s and maintaining the environmental garden.

Funding Sources: Contributions, benefits, grants

Annual Budget: $5,000

San Jose State University Environmental Resource Center

1 Washington Square
San Jose, CA 95192-0116
(408)924-5467

Contact: Steve Shunk

Description: The Environmental Resource Center is a student- run Environmental Studies library and information center providing resources on a huge range of environmental issues and organizations. Some of our areas of focus include alternative transportation, library information system and community outreach. We assist those doing specific research or just looking for answers to the many problems facing society and our planet.

Wildlife Rescue Incorporated

4000 Middlefield Rd., Bldg. V
Palo Alto, CA 94303
(415)494-7283

Description: Wildlife Rescue Inc. tries to instill awareness and appreciation for local wildlife. We provide a shelter for animal rehabilitation and classes.

Short Courses

Audubon Society—Mt. Diablo

P.O. Box 53
Walnut Creek, CA 94596
(415)376-8732

Description: The Mt. Diablo Audubon Society offers field trips to local areas of birding interest, geared to birders of all levels of expertise. They occur four to seven times a month. Monthly meetings highlight specific topics of natural history and bird information. Educational programs are available to local groups and classes interested in general bird information.

Bay Area Global Education Program

312 Sutter Street #200
San Francisco, CA 94108
(415)982-3263
Fax: (415)982-5028

Contact: Carol Marquis

Description: Since 1979, BAGEP, a consortium of organizations, has operated a comprehensive program to assist schools strengthen teaching about international affairs, global issues, and people from other national and cultures. It conducts in-depth teacher education programs and organizes local school and district programs that implement major curriculum change.

Resources: A resource center and teacher library is operated at the World Affairs Council of Northern California. Consulting services and conference presentations are available and a free newsletter *Colloquy*.

Volunteer Opportunities: Working in centers, reviewing, cataloging.

Funding Sources: Foundations and participating institutions.

Employees: 3

Body Tales

698 46th Street
Oakland, CA 94609
(415)547-4467

Contact: Olivia Corson, Artistic Director

Description: Eco-Theater informed and inspired by Earth conciousness. Experience storytelling at its best—imagination and information combined with original music and movement—drawing together facts and fantasy from deep ecology. Corson and company offer classes, workshops and performances tailored to the particular needs, the age range and the size of your group. Bring new life to your meeting, gathering, celebration, or personal path.

East Bay Conservation Corps

1021 3rd Street
Oakland, CA 94607
(415)891-3900
Fax: (415) 272-9001

Contact: Al Auletta, or Jim Sternberg, Education Coordinator

Description: East Bay Conservation Corps offers summer and year-round programs serving the needs of the community and public land managers. Youths, between ages 16 and 24 (age requirement dependent on program), who like to work outdoors, are out of school, need job experience and assistance in job-holding skills, basic skills, and life and survival skills and show a commitment to self-development are hired for participation in four days of conservation in crews of 6-12. Education program offers individualized teaching in math, reading and writing as well as lifeskill workshops, English as a second language, career development assistance, environmental education and conservation awareness, internships, and various other educational opportunities. Placement in college or vocational positions is also provided.

Volunteer Opportunities: Great demand for tutors in the learning center.

Funding Sources: Public and private sector grants, contracts with public land agencies.

Annual Budget: $4,300,000

Employees: 50

Ecological Heritage Foundation

3318 Doral Ct.
Walnut Creek, CA 94598
(415)945-7560

Contact: California-Mary Ann C. Simonds, Educational Director

Description: The Ecological Heritage Foundation is a conservation and educational organization. The Foundation's focus is to educate and enlighten people throughout the world concerning their intrinsic value and inter-connectedness with Nature. The emphasis is on human-nature interaction, perpetuating biodiversity, and developing ethical and wise policies and programs that will support sustainable ecosystems and cultures. The Foundation offers educational seminars, lectures, and "shareshops," designed to stimulate ideas and creativity concerning how humans can interact positively with nature. Research study programs on the Kiger Mustangs, developing ethical eco-tourism programs for public lands, and the Free-Spirit Horse Education Center are several of the current projects.

Funding Sources: Private donation, membership, grants and programs.

Green Gulch Farm

1601 Shoreline Hwy.
Sausalito, CA 94115
(415)383-3134

Contact: Emila Heller, Guest Manager

Description: Green Gulch Farm Zen Center is a Zen Buddhist community which includes an organic farm and garden as well as retreat facilities (day and overnight use) for groups of up to 30. Flowers, herbs, vegetables and fruit plants started at Green Gulch are available for sale, and produce from the farm is sold locally as well. Weekend organic gardening classes and events (some of which are specially planned to include children and families) are offered in subjects ranging from composting techniques and reforestation to flower arranging and wreath-making. Caring for the environment and conservation of natural resources are traditionally part of Buddhist practice, and are a vital part of daily activity at Green Gulch Farm. Schedules and information are available from the office.

Magic Incorporated

P.O. Box 5894
Stanford, CA 94309
(415)323-7333
Contact: David Schrom
Description: Magic offers a variety of educational services, from one-on-one instruction and counseling in various aspects of healthful living, to speakers for schools and community groups, to series of classes or seminars. For the remainder of 1990 and during 1991, we plan to offer a quarterly 10-session class, "The Nature of Value," in which participants learn how the methods and principles of ecology may be brought to bear upon fundamental questions like, "To what purposes do I live?" "By what paths do I aim to realize these purposes?" and "On what basis have I chosen these purposes and paths?" In this class we also carefully scrutinize, from the perspective of ecology, concepts commonly deemed within the realm of economics, like production, wealth, consumption, investment, capital, savings, GNP, and so on. We also coach individuals, usually middle-aged adults without particular athletic aptitude or training, in running and cycling more effectively (i.e. with less injury and more comfort through greater awareness in movement). Periodically we conduct workshops, usually video-feedback based, for swimmers and runners. Near each solstice and equinox we offer life-planning workshops.
Resources: Community/cooperation discussion groups, restoration projects, instruction and supervision , and outstanding opportunities for personal growth and contribution to others through internships.

Marin Discoveries

11 First Street
Corte Madera, CA 94925
(415) 927-0410
Contact: Ruth Catlin
Description: Marin Discoveries is committed to introducing people to educational, recreational, and inspirational experiences in the outdoors. Established in 1974, Marin Discoveries offers over 500 programs each year ranging from day walks, to rock climbing and family outings in the Bay Area to backpacking in the Sierras and canoeing the Colorado River. The programs are designed to provide instruction in low-impact, fun outdoor activities that lead people to a better understanding and appreciation of our natural areas, and to provide educational opportunities regarding environmental issues.
Resources: Newsletter
Volunteer Opportunities: Participation is open to everyone.

Marine Science Institute

P.O. Box 7142
Redwood City, CA 94063-7142
(415) 364-2760
Contact: S. James Taggart, Ph.D.
Description: The Marine Science Institute offers hands-on science education programs which encourage exploration and discovery. Under the guidance of the Institute's graduate-level faculty, participants operate oceanographic and marine research equipment to collect and analyze water, sediment, flora, and fauna samples from the living San Francisco Bay. Discovery Voyages are designed to promote critical thinking and awareness about the San Francisco Bay's unique estuarine ecosystem. The Marine Science Institute inspires learning by providing participants with hands-on positive learning experiences which ignite an interest in the field of science.

Merritt College Environmental Studies Program

12500 Campus Drive
Oakland, CA 94619
(415)436-2619
Contact: Charles Ford, Chairperson
Description: Merritt College's Environmental Studies Program offers many short courses on evenings and weekends that are primarily field-based and hands-on. Courses range from creek and landscape restoration to field trips in locations such as the Bay Area and the Everglades. The courses emphasize biology, geology, ecology, and architectural and environmental design. In the Fall of 1991 the college will offer a four-year transfer major in Environmenal Studies, and an Associate degree in Natural Resources Restoration and Management. In the Fall of 1992 the college plans an Environmental Hazardous Material Technology program, and an Environmental Technology and Planning program.

Natural Encounters

3318 Doral Court
Walnut Creek, CA 94598
(415)945-7560
Contact: Mary Ann C. Simonds, Principal
Description: Natural Encounters is an ecological, education and human resource consulting organization. Blending ecological and human values with science, economics and business, the firm promotes positive interaction with nature by developing ecological awareness and understanding in people through education and training programs. The programs focus on helping people perceive themselves as "holistic ecologists," intimately connected with instead of separate from, nature. The firm's emphasis is directed at education of organizations, particularly those in the natural resource, engineering, and develop-

ment fields.Programs on the subject of: conservation biology, ecology, energy and resource management, environmental studies, sustainable forestry, rangelands, and agriculture.

Resources: Private consulting, lectures, seminars, "shareshops," and training programs are offered and customized for various groups and organizations.

Funding Sources: Funding is through private fees.

Northern California Solar Energy Association

P.O. Box 3008
Berkeley, CA 94703

Description: The NCSEA promotes sustainable, affordable living through renewable energy. It is a non-profit educational society (a regional chapter of the American Solar Energy Society) incorporated to foster the development and application of solar energy through the exchange of information. NCSEA publishes the quarterly *Northern California Sun* and hosts periodic seminars; many short 1-day courses are offered in the uses of solar energy. Past topics have included: solar water heating, code compliance, biomass, solar in developing countries, wind energy, and greenhouse design.

Resources: Short courses, *Northern California Sun* a quarterly newsletter.

Point Reyes Field Seminars

Point Reyes National Seashore
Point Reyes Station, CA 94956

Contact: Fawn Bauer

Description: The program provides a variety of environmental education opportunities for people of all ages. They operate the Clem Miller Environmental Education Center—an overnight outdoor facility which can accommode up to 80 people.

Resources: The Education Center is available as a resource to groups. They offer field seminars in natural history, environmental education, and artistic expression of the surroundings. In the summer they host a natural history summer camp for kids to explore and learn about the Pt. Reyes National Seashore environment. They also custom design programs and trainings for special groups.

Saso Herb Gardens

14625 Fruitvale Ave
Saratoga, CA 95070
(408)867-0307

Description: The Saso Herb Gardens offer a variety of workshops on growing herbs, using herbs medicinally, and herb crafts. Upcoming workshops include wreath -making, garland-making, introduction to culinary, medicinal and ornamental herbs, herbal medicine, and aromatherapy

Strybing Arboretum Society

9th Ave at Lincoln
San Francisco, CA 94122
(415)661-0668

Contact: Bob Hyland, Director of Education

Description: Strybing Arboretum Society, a non-profit corporation, is the membership organization which supports the development of the Strybing Arboretum and Botanical Gardens. The Society also provides all the educational programs at the gardens, including lectures, workshops, publications, plant labelling and interpretive signs, and docent-led tours of the gardens for both children and adults. The Society also operates the Helen Crocker Russell Library of Horticulture, a library containing over 11,000 titles and open seven days a week, free to the public.

Resources: Education Programs: children's tours, The Flower Walk, Plant Travelers Walk, and The Indian Walk. The Redwood Trail takes children through an authentic recreation of a redwood forest to study plant adaptations in a natural plant community, entitled A First Look at Plants; also, an Adult Quarterly Program Calendar and a library.

University Research Expeditions Program (UREP)

University of California
Berkeley, CA 94720
(415)642-6586

Description: International expeditions in which participants assist University of California research teams in tasks such as collecting plant specimens in the Ecuadorian rain forest, monitoring bird mating rituals in Kenya, and testing non-toxic farming methods in Malawi. No previous experience is necessary. Scholarships for teachers and students are and a free brochure of expeditions are available.

Vision Quest—Wilderness Transitions

70 Rodeo Ave.
Sausalito, CA 94965
(415)331-5380
Fax: (415)432-9558

Contact: Marilyn Riley, Betty Warren

Description: Our week-long Vision Quest trips take small groups to a remote, safe place in the wilderness following four preparation meetings. Two days of community living at base camp deepen bonds of love and trust within the group. Each participant then finds a place apart where he spends three days and nights alone, usually fasting. Experienced guides and the companionship of a few others with the shared intention of finding new horizons combine to make this a unique opportunity to slow down, look within, and grow! To better understand the spirit of the Quest, you are welcome to visit one of our slide shows, presented at the first meeting for each trip. Trips take place in California about every two months. Groups include a maximum of ten women and men, age fifteen and up. We have over ten years of experience and are non-profit. Please call or write for brochure.

Wildlands Studies

3 Mosswood Circle
Cazadero, CA 95421
(707)632-5665

Description: Wildlands offers a year-round series of field study programs in North American and international wilderness locations. Participants join back country research teams in a search for answers to important environmental problems concerning wildlife populations and/or wildland habitats. Participants can earn 3-14 units of University credit.

Section Four

ORGANIZATIONS

The following list of organizations is an introduction to the huge number of groups in the Bay Area fighting to preserve the planet. There may well be more such groups here than anywhere in the world.

No group is exactly like any other. They range widely in their missions and scales of operation. Some groups operate out of members' living rooms and have a single local goal on their agenda, such as preserving or restoring a nearby stream. Others more closely resemble large corporations and campaign internationally on a broad spectrum of issues. Many fall somewhere in between.

This diversity of groups offers ample opportunity for every would-be activist, regardless of interests, to find a niche in which they can be comfortable. Many of these groups have an urgent need for skilled volunteers, and we encourage you to help out.

The descriptions presented here were for the most part provided by the organizations themselves, in answer to questionnaires we sent to them. Environmental organizations are often pressed for time and short on personnel, so ommissions may have occurred when other more pressing matters (like saving the planet) precluded questionnaire-answering. The fact that a group is not included in the *Green Pages* does not necessarily mean they've folded or disappeared, but more likely that the Green Pages somehow failed to connect with them. We apologize for any such ommissions, and ask unlisted groups to contact us for inclusion in next year's update.

We commend both the listed and unlisted groups for their efforts on behalf of the planet and everyone's future.

Please use the subject index at the back of the book to find groups working in your particular areas of interest. These descriptions are organized alphabetically by the first word of the organization's name.

Abalone Alliance
2940 16th Street #310
San Francisco, CA 94103
(415) 861-0592
Contact: Roger Herried
Description: The Abalone Alliance is a statewide organization of about 20 community anti-nuclear/pro-safe energy groups. Through public education and non-violent direct action, they are committed to:
- Stopping the construction and operation of all nuclear power plants in California.
- Promoting the realistic alternative of safe, clean, and renewable sources of energy.
- Encouraging responsible community control of energy production.
- Supporting efforts to eliminate nuclear weapons.

Resources: Radiation and Alternatives Bulletin newsletter
Volunteer Opportunities: Run by volunteers. From computer-database operations to staffing tables, newsletter production, research, and fundraising.
Funding Sources: Memberships $15-25, subscriptions to newsletter $10, fundraisers and donations.
Annual Budget: $15,000

Alameda-Contra Costa Regional Parks Foundation
P. O. Box 21074
Oakland, CA 94620
(415) 531-9300 x2201
Contact: Janet S. Cobb, Executive Director
Description: The Alameda-Contra Costa Regional Parks Foundation is a non-profit,the sole purpose of which is to supports the East Bay Regional Park District. This Foundation is dedicated to protecting and restoring regional park lands and trails for public use, as well as preserving critical wildlife habitat. It funds education programs which encourage enjoyable and responsible park access for special groups. It supports the Regional Park directors and staff in securing parklands and funding special facilities, programs and publications.
Resources: Monthly newsletter, *Regional Parks Log* published by EBRPD.
Volunteer Opportunities: Tree planting, Coast Cleanup, Statewide Trail Days, and other volunteer projects. Contact EBRPD Volunteer coordinator (415) 530-4875.
Funding Sources: Private contributions, in-kind gifts.
Annual Budget: $250,000

American Farmland Trust
California Field Office
512 Second Street
San Francisco, CA 94107
(415) 543-2098
Contact: Regina Macias
Description: The American Farmland Trust is a national nonprofit organizaton dedicated to the conservation of agricultural land. AFT seeks to accomplish its farmland conservation objectives by, 1) increasing public awareness of agricultural resource issues, 2) providing technical assistance to governmental agencies and private organizations on farmland preservation techniques and, 3) engaging in individual farmland conservation real estate transactions. AFT represents common ground upon which agriculturalists and conservationists, business people, scholars, public officials and private citizens can work together to safeguard the land that feeds America.
Resources: Quarterly newsletter, information packets, lobbying.
Volunteer Opportunities: Internships.
Funding Sources: Grants, private corporations, foundations.
Annual Budget: $84,000
Employees: 3

American Friends Service Committee
2160 Lake Street
San Francisco, CA 94121
(415) 752-7766
Contact: Tony Henry, Executive Secretary
Description: The American Friends Service Committee is committed to social change through nonviolent means. Programs focus on peace, disarmament, homelessness, the death penalty, and feeding the hungry. Environmental programs include energy, land use, water use and toxics (Land Use Alliance - Stockton). AFSC has programs all over the world, including a Middle East peace program, and a Pacific Rim disarmament project.
Volunteer Opportunities: Office work, program work abroad, maintenance of information files.
Funding Sources: Individual contributions, foundation grants.
Annual Budget: $1,100,000
Employees: 20

American Lung Association San Mateo County
2250 Palm Ave.
San Mateo, CA 94403
(415) 349-1111
Contact: Renee Bates
Description: The American Lung Association works in the general areas of air pollution, tobacco smoking, asthma and lung disease. Catalogs are available listing all materials available.
Resources: In-house work with hospitals such as Sequoia, Mills, and Kaiser.
Volunteer Opportunities: Volunteer center of San Mateo.
Funding Sources: Fundraising events.
Employees: 5

American Lung Association Santa Clara - San Benito Co.

1469 Park Avenue
San Jose, CA 95126
(408) 998-5864
Contact: Janet Ghanem
Description: The group acts as an advocate for programs to obtain and maintain air and indoor air quality. It provides expert testimony and consultation to governmental bodies and individualsand coordinates with BAAQMD activities.It also promotes various events for car care, and Clean Air Week. and Give a clean air award to local business. Consult with local business and individuals regarding smoking in the workplace and survey current trends about company smoking policies. Distributes clean air literature.
Resources: Quarterly newsletter, *Environmental Health* . *Considerate Smokers* booklet summarizing city smoking ordinances in Santa Clara County.
Volunteer Opportunities: Environmental Health Committee members and task force workers. Events for fundraising. Helpers for Clean Air and Car Care Weeks.
Funding Sources: Private donations from Christmas Seals, fundraising.
Annual Budget: $700,000
Employees: 7

American Youth Hostels Central California Council

P.O. Box 1241
Santa Cruz, CA 95061
(209)722-6101
Contact: Joe Madonna, Book Store Manager
(415) 863-9939
Description: American Youth Hostels operates hostels, promotes hosteling, and sponsors outdoor activites for people of all ages. AYH - Golden Gate Council operates or charters 12 hostels in Northern California. Outdoor activities include bicycling, hiking and skiing. Information is available on hostels world wide.
Resources: Library of reference materials on travel
Volunteer Opportunities: "Hostel Adventure", a program of environmental education school or community groups for disadvantaged, and information desk staffed by volunteers at Fort Mason.
Annual Budget: $1,000,000
Employees: 35

Americans for Nonsmokers' Rights

2530 San Pablo Ave, Suite J
Berkeley, CA 94702
(415) 841-3032
Fax: (415) 841-7702
Contact: Judith Edmonds
Description: ANR is a non-profit organization created to pursue a coordinated, action-oriented program of legislative and legal activities to assure that nonsmokers can avoid involuntary exposure to secondhand tobacco smoke in the workplace, restaurants, public places, and on airplanes and other public transportation vehicles. The American Nonsmokers' Rights Foundation is the educational arm of ANR and creates comprehensive educational programs for school children on issues of smoking prevention, and their right to breathe smoke-free air. It also Provides educational materials to aid adults in their quest for a smoke-free environment.
Resources: National Resource Center —provides information on passive smoking, tobacco and the tobacco industry to government agencies, local advocates and the media.
• UPDATE is a quarterly newsletter available to members.
• Also available is the *Smokefree Travel Guide*.
Funding Sources: Funded by membership.
Annual Budget: $287,000
Employees: 7

Animal Legal Defense Fund

1363 Lincoln Ave., Suite 7
San Rafael, CA 94901
(415) 459-0885
Contact: Joyce Tishler
Description: Provides assistance to students and teachers who object to dissecting animals in classrooms. The hotline—1-800-922-FROG—gives information on student's rights, educational alternatives, guidance on negotiating with school officials, and attorney references.
Volunteer Opportunities: Attorneys needed, clerical support.
Funding Sources: Memberships, foundation grants.
Annual Budget: $3,000,000
Employees: 3

Appropriate Technology Project, Volunteers in Asia

P.O. Box 4543
Stanford, CA 94305
(415) 326-8581
Contact: Ken Darrow
Description: The primary function of the project is the production and sale of the Appropriate Technology Microfiche Library, which contains the complete text (on Microfiche) of 1,000 of the best books and documents relating to appropriate technology. The project also publish the *Appropriate Technology Sourcebook*, which is a stand-alone reference, as well as serving as the index to the Microfiche Library. The *Sourcebook* contains reviews of, and ordering information for, paper copies of all the books in the microfiche collection, as well as additional books.
Resources: Microfiche library and the *Appropriate Technology Sourcebook*.
Employees: 3

Association of Environmental Professionals

52120 Overbrook Way
Sacramento, CA 95841
(916) 344-4136

Description: A California-wide association founded in 1974 for environmental improvement through professional involvement in all environmentally-related fields. The Association has programs and interests that span all aspects of environmental research, law, analysis, assessment, and education. Open for membership to active environmental professionals, with affiliate memberships available to others interested in the environmental management process, student memberships for currently enrolled college or university students and sponsoring memberships for interested public and private organizations are available.

Resources: AEP promotes education and training in environmentally related fields, conducts conferences, publishes newsletters and bulletins, and promotes public awareness and involvement in the environmental review process. AEP also collects and makes available environmental information on an open and nondiscriminatory basis of the public and to decision makers.

Funding Sources: Memberships and Donations, Conference Fees.

Annual Budget: $85,000

Employees: 3

Audubon Society –Madrone Chapter

P.O. Box 1911
Santa Rosa, CA 95402
(707) 938-5238

Contact: Karen Nagel - Membership Chairman

Description: The Audubon Society of Madrone sponsors monthly meetings with speakers lecturing not only on birds but also conservation issues and natural history. It sponsors field trips and boat trips to observe birds in their natural environment. The Audubon Society of Madrone participates in the nationwide bird count every Christmas season. Furthermore, it is one of the four chapters that sponsors Audubon Cannon Ranch. It also contributes to the nationwide project of the *Breeding Bird Atlas*.

Resources: Newsletters are available for members.

Funding Sources: Membership and fundraisers.

Audubon Society –Marin

P.O. Box 599
Mill Valley, CA 94941
(415) 383-1770

Contact: Harrison Karr, President

Description: Please see the *Education* section for information on classes offered by this environmental education and conservation organization.

Resources: Newsletter, *The Redwood Log*.

Funding Sources: Memberships, dues and bequests.

Audubon Society –Mt. Diablo

P.O. Box 53
Walnut Creek, CA 94596
(415) 945-1785

Contact: Barbara Vaughn, Board Member

Description: Mt. Diablo Audubon Society is dedicated to preserving the environment by conservation activities, field trips, and educational meetings. Conservation activities include active involvement in habitat protection in Contra Costa County of San Francisco Bay and Sacramento/San Joaquin Delta water quality preservation. TheP-eregrine Falcon Project is an attemptto re-introduce the endangered peregrine falcon to nesting sites around Mt. Diablo. Monthly meetings are highlighted by specific topics of natural history and bird information. Please see the *Education* section for a description of classes and field trips.

Funding Sources: Dues, donations (individual memberships).

Annual Budget: $2,000

Audubon Society –Santa Clara Valley Chapter

415 Cambridge Ave, #21
Palo Alto, CA 94306
(415) 329-1811

Contact: Cecily Harris, Managing Director

Description: The chapter's mission is to promote conservation of our natural resources, wildlife, plants, soil, water and to foster their intelligent treatment and wise use for the public welfare, and to cooperate with the Natonal Audubon Society in these pursuits.

Resources: Newsletter, *Avocet*. Library collection- 500 books at Peninsula Conservation Center, monthly general meetings in Palo Alto.

Volunteer Opportunities: Work as office liason on other nonprofit environmental committees, do internships, serve on the Field Trip, Program, Education, Library, Environmental Action, Fundraising, Photo Club, Garden, Habitat, or other Committees.

Funding Sources: Membership dues, grants, Environmental Federation gifts.

Annual Budget: $70,000

Employees: 1

Bay Area Action

P. O. Box AA
Stanford, CA 94309
(415) 321-1994

Contact: Michael Winkler or Jeff Hoover

Description: Founded by the organizers of Earth Day 1990, Bay Area Action is a grassroots network of citizens dedicated to preserving and improving the natural environment of the Bay Area. BAA is a direct action organization working to carry environmental issues beyond discussion into positive change. BAA's vision is that each individual in the community be an informed, participating citizen. To achieve this goal, BAA headquarters is an information center providing a bi-monthly newsletter, action alerts, television and radio broadcasts, a book and video library, and guest lecturers. BAA also takes an active role in educating the community through programs in workplaces and schools. BAA initiates and co-organizes a variety of actions and events, including Alternative Transportation Tuesday, the Workplace Program, the Schools and Youth Program, the Media Project, and the Phone Tree.

Resources: BAA has a newsletter, Phone Tree hotline, environmental video library, numerous fact sheets, environmental audits for workplaces and schools.

Volunteer Opportunities: Workplace Program, Schools and Youth Program, the Media Project, and the Phone Tree, Communication Committee.

Funding Sources: Grants, private donations, benefits.

Employees: 1

Bay Area Mountainwatch

P.O. Box AO
Brisbane, CA 94005
(415) 467-6631

Contact: David Schooley

Description: Bay Area Mountainwatch is a citizens' organization dedicated to the preservation of San Bruno Mountain, its landforms,and its unique ecosystems particularly its rare and endangered species. They are trying to protect the remaining open spaces of the habitat and small range running east-west on the border between San Francisco and San Mateo County threatened by development.

Resources: Free newsletter published and mailed to members and supporters (every two months) San Bruno Mountain calendar. Informational flyers are frequently issued.

Volunteer Opportunities: Writers, monitor survey & photo work on rare and endangered species on San Bruno Mountain, and non-native plant removal in special areas.

Funding Sources: Members, classes and some hikes. Sale of calendars, postcards, T-shirts, and other special publications, drawings. native plants, etc. Small grants have occasionally been contributed.

Employees: 6

Bay Institute of San Francisco

10 Liberty Ship Way #120
Sausalito, CA 94965
(415) 331-2303

Contact: William T. Davoren, Executive Director

Description: A non-profit organization with the purpose of halting the decline of the Bay-Delta ecosystem as an estuary. The primary goals are to obtain a water "right" to streamflows of the Sacramento-San Joaquin River for the Bay, to stop discharge of toxics to the bay-river systems by farms, industry and cities, and to establish local and state protection for 51,000 acres of diked lands containing 16,000 acres of ponds, lagoons and marshes. The goals are pursued by developing new scientific, engineering, biological and economic justifications to support, by advocacy, better enforcement of existing laws.

Volunteer Opportunities: Need volunteers with scientific and professional degrees.

Funding Sources: Foundation, corporate, and individual donations status, small government contract.

Annual Budget: $150,000

Employees: 2

BayKeeper - San Francisco Bay-Delta Preservation Association

Building A, Fort Mason
San Francisco, CA 94123
(415) 567-4401

Contact: Courtney Desio

Description: San Francisco Bay is under unprecedented pressure from oil spills, illegal filling of wetlands, point and non-point source pollution, dredging, and the diversion of fresh water. Fish, wildlife, and wetlands, each of which has commercial and recreational values to the citizens of the Bay Area, are at risk. Potential public health problems resulting from polluted fish, shellfish and water fowl, and possible economic losses from salinity intrusion into agricultural irrigation and drinking water sources, are of great concern to those living in and around the Bay.

Although the presence of many regulatory agencies created by landmark legislation of the 1960s and 70s gives the impression that the Bay is adequately monitored, in fact none of these agencies has any regular monitoring presence on the Bay or conducts routine on-the-water environmental enforcement programs.

The San Francisco Baykeeper project was designed in the summer of 1989 as a high-visibility, hands-on water quality enforcement and public-awareness-raising program by scientist/activist Dr. Michael Herz. Based on similar East Coast projects like the Hudson Riverkeeper, the project's mission is to protect, preserve, and restore the San Francisco Bay and Delta ecosystem by supplementing the activities of the regulatory agencies.

The Baykeeper and its corps of trained volunteers patrol the Bay and Delta with boats, planes, and on foot to detect and document violations of these laws and collect data to assist agencies and advocacy groups in bringing enforcement actions. We also use the results of our investigations to raise awareness concerning threats to the Bay and Delta through the media and public education.

Our Hotline (see below) serves both as an antenna for citizen complaints and reports, and appears increasingly to act as a deterrent to illegal activities. In one year, BayKeeperhas found over 170 pollution incidents or other illegal activities, which have resulted in fines, citations, abatement orders and one ongoing criminal investigation.

Resources: BayKeeper Hotline 1-800-Keep-Bay. Volunteer Training videotapes for potential volunteers. Media clips, with or without Baykeeper speakers, for public showing.

Newsletter - *BayKeeper Log*

We have limited educational and other resource materials available including a database summary of the incidents which have been reported to us since July, 1989.

Other environmental groups, funders, the media and trained volunteers are often invited to join the BayKeeper boat during one of its regular patrols.

Volunteer Opportunities: In addition to the patrol work, incident discovery and reporting requires prior research and significant follow-up investigation. We need volunteers with skills in: science and other technical areas; boat repair and maintenance; public speaking; organizational development and fundraising; graphics, writing and editing; office and computer support; media relations, law and politics, and construction. Volunteers meet regularly in strategy sessions focusing on specific projects as well as Bay -wide educational and social gatherings.

Funding Sources: Private and community foundations, individual donors, contracts with regulatory agencies, and business groups.

Annual Budget: $180,000

Employees: 5

Berkeley Creators Association Educational Foundation

2526 Shattuck Ave.
Berkeley, CA 94704
(415) 848-5713

Contact: Robin Freeman

Description: BCAEF provides educational research and support services in the environment, culture and public policy. They currently offer partially-paid internships and apprenticeships in ecological design, construction,and owner-designed communities. BCAEF also provides subsidized artist-in-residence live-work studios, career counseling in the arts and environmental work, and is developing a marine ecology education program, a wooden sailing ecology education vessel, and a photography internship. The organization hosts the Institute for Sustainable Policy Studies, a think tank which considers the social impacts of environmental issues from a broad collaborative point of view. The Institute gives lunch seminars. Call for schedule.

Resources: Reprints available of interviews with well-known environmentalists and pacifists, originally published in the Sierra Club *Yodeler*.

Volunteer Opportunities: Internships (see above), occasional work parties,marine ecology and education, and wooden boat maintenance.

Funding Sources: Contracts, contributions.

Annual Budget: $40,000

Berkeley TRiP

2033 Center Street
Berkeley, CA 94704
(415) 644-7665

Contact: Diane Sutch, Project Manager

Description: The primary goal of TRiP (transit , ridesharing, parking) is to reduce the adverse impacts of the single-occupant automobile in Berkeley for the work commute. TRiP operates a Commute Store in downtown Berkeley where assistance is provided on transportation alternatives to driving alone. The Commute Store offers tickets for all Bay Area transit systems; transit schedules and bicycle maps: on-the-spot carpool and vanpool matching through RIDES for Bay Area Commuters; and personalized planning for trips on transit. TRiP works with employers to provide incentives to their employees, such as offering monthly bus passes to employees at a discount and providing preferential carpool and vanpool parking at work sites.

Volunteer Opportunities: None.

Funding Sources: U.C. Berkeley Campus and Office of the President: Associated Students of the University of California, City of Berkeley, Lawrence of Berkeley Laboratory: Rides for Bay Area Commuters; Berkeley Chamber of Commerce.

Annual Budget: $300,000

Employees: 6

Beyond War

222 High St.
Palo Alto, CA 94301
(415) 328-7756
Fax: (415) 328-7785

Description: As an educational foundation, Beyond War's activities include presentation, seminars, discussion groups, speakers, conferences, and the production of audio and videotapes. The annual Beyond War Award honors outstanding contemporary contributions of individuals, groups, or organizations who are demonstrating cooperation in addressing survival issues. It is a non-profit nonpartisan educational foundation established in 1982, with volunteers in more than 40 states and six countries.The organization is run by a Board of Directors consisting of 15 active volunteers The mainstay of Beyond War's educational efforts is individual communication and initiative. In local communities, people work with neighbors, friends, business colleagues, religious groups, elected officials, and professional and volunteer groups to discover and implement solutions to contemporary problems. Various symposiums, conferences , publications, and projects include "National Security A Global Imperative," "A Planet in Every Place," "Earth Vision: The Search for a New Earth-Human Relationship," "Human Rights Initiative," "Helsinki 2000 Appeal," "Middle East

Project," "Afghanistan Initiative," "Philippines Project," "By Wonder Are We Saved," "Everything Has Changed."
Resources: Presentations, seminars, discussion groups, speakers, conferences, and audio and videotapes. *On Beyond War,* monthly newsletter.
Volunteer Opportunities: 30 to 40 at present, please contact for more information.
Funding Sources: Individual donations which average under $100 per contribution.
Annual Budget: $1,200,000

Bio-Integral Resource Center, The IPM Practitioner

P.O. Box 7414
Berkeley, CA 94707
(415) 524-2567
Contact: Sheila Daar
Description: BIRC is a non-profit organization undertaking research and education in ecosystem management. A monthly newsletter for professionals, *The IPMpractioner,* describes non-toxic pest management programs for urban and agricultural settings, reviews books, reports on research, and provides up-to-date news. IPM stands for integrated pest management— the selection, integration, and implementataion of pest control based on predicted economic, ecological and sociological consequences. BIRC also publishes *The Common Sense Pest Control Quarterly.* Memberships, which include the *Quarterly,* are $30/year for individuals or $50/year for institutions.
Resources: *The IPM Practitioner* monthly. Additional pest control pamphlets.*Common Sense Pest Control Quarterly*
Volunteer Opportunities: Library and database assistance. Must have library or computer skills.
Funding Sources: Publication subscriptions, grants, donations.
Employees: 8

Biological Urban Gardening Services

P.O. Box 76
Citrus Heights, CA 95611
(916) 726-5377
Contact: Steven Zien
Description: Biological Urban Gardening Services (B.U.G.S.) is an international membership organization devoted to reducing the use of potentially toxic pesticides in our highly populated urban landscape environments. B.U.G.S. informs consumers as well as landscape and horticultural professionals about the many natural horticultural techniques currently available to maintain a healthy, pest-free landscape and garden. Experimentation is under way at our research landscape and garden on a variety of sustainable horticultural techniques and products that could prove beneficial to landscape and garden plants and their stewards.
Resources: *The Voice of Ecological Urban Horticulture*-quarterly newsletter, catalog of informative brochures.
Funding Sources: Membership fees & catalog brochure sales.
Employees: 2

Buckhorn Canyon Legal Defense Fund (BCLDF)

P.O. Box 5856 College Ave. #110
Oakland, CA 94618
(415) 841-7679
Contact: Tom Hedges
Description: The mission of the Buckhorn Canyon Legal Defense Fund is to protect Buckhorn Canyon, in the Oakland hills, from a dam project proposed by the East Bay Municipal Utilities Districts. It sued on grounds of inadequate EIR coverage of seismic safety, wildlife impacts, growth-inducing effects, traffic impacts during construction, along with other issues. It was joined in the suit by California Department of Fish and Game, Sacramento County, and the Preserve Area Ridgelands Committee. BCLDF settled out of court in early 1990. EBMUD agreed to do additional research and studies on the contested issues for a new EIR/EIS, expected to be ready in 1991. At that time it will organize public response at a public hearing which EBMUD will be required to hold. If BCLDF's concerns are not adequately dealt with it may sue again.
Resources: Newsletters are occasionally mailed to donors and interested parties when the need arises.
Volunteer Opportunities: Once the new EIR/EIS is released in 1991, BCLDF could use help in fundraising, publicity and mailing parties. A background in land use, water policy, or seismic safety would be helpful.
Funding Sources: Donations.
Annual Budget: $10,000

Business for Environmental Action

3255 Broderick #2
San Francisco, CA 94123
(415) 921-0617
Contact: Elizabeth Fetter
Description: BEA provides a bridge between business and environmental groups in order to develop workable solutions to environmental problems. Its purpose is to encourage business to take an appropriate leadership role in solving local and global environmental problems by 1) implementing profitable and environmentally sound practices, and 2) demonstrating to the public and to other corporations that the complement of ecologically sound practices is necessary to insure the survival of our planet.
Resources: Newsletter; special issues on the Rainforest
Volunteer Opportunities: Need committees for upcoming events and freelance articles for the newsletter.
Funding Sources: All contributions.
Annual Budget: $25,000

California Association of Cooperatives

1563 Solano Avenue, #243
Albany, CA 94707
(415) 524-8826
Contact: Matt Kumin
Description: This is an organization set up to provide assistance to the State's coop community by way of resources, networking and education. At present they are working on legislation within the State legislature for the furtherance of Cooperatives. The association provides grants to members for the Peer Consultancy, whereby Cooperatives can interact with other Cooperative's management. A regular newsletter is provided to members outlining current activities of the organization, legislative news, and membership activity.
Resources: Referral service and newsletter.
Volunteer Opportunities: Many.
Funding Sources: Membership dues and grants.
Annual Budget: $4,000

California Certified Organic Farmers

P.O. Box 8136
Santa Cruz, CA 95061-8136
(408) 423-2263
Contact: Bob Scowcroft
Description: CCOF's purpose is to promote and support a healthful, ecologically-accountable, and permanent agriculture in California and elsewhere; to develop standards and certification programs for organic farming and processing of organic foods; to provide verification of adherence to those standards for distributors, retailers, and consumers; to provide educational forums and materials relating to sustainable agriculture and, generally, to share ideas and information.
Resources: Quarterly newsletter - free to members, or $2.00 per sample. Publications: *Certification Handbook, Retailers Guide to Organic Food and Farming, Growers List and Crop Index, Retailers and Wholesalers List, Farm Inspection Manual.*
Volunteer Opportunities: There are many projects at this statewide office and the fourteen regional chapters. We welcome people with talents in marketing, organic techniques, general office work, etc.
Funding Sources: Dues and assessments from members, grants, sale of literature and promotional items.
Annual Budget: $386,000
Employees: 6

California League of Conservation Voters

965 Mission Street, #750
San Francisco, CA 94103
(415) 896-5550
Fax: (415) 541-9253
Contact: Lucy Blake
Description: The California League of Conservation Voters (CLCV) is the non-partisan campaign arm of the environmental community in California. The League works to protect the environmental quality of our state by electing candidates to office and by passing ballot propositions. With 45 Congressional districts and 120 state legislative seats , California clearly represents a formidable challenge to any grassroots organization. To meet this challenge, CLCV conducts early research on candidates for office and concentrates on races where environmental resources might make the difference in the outcome of the race. CLCV backs their political endorsements with campaign expertise, assisting candidates with the media, fundraising, and grassroots strategies they need to win their races. Each year the league assigns experiencd campaign organizers to the very closest environmental contests. On Election Day, they comb the precincts, getting environmental voters to the polls for our candidates. League organizers also communicate directly with hundreds of thousands of Californians every year. In addition to providing information on legislators' environmental records, they register voters, recruit members and volunteers, generate letters to targeted representatives, and identify "conservation voters."
Resources: Quarterly newsletter - *The Conservation Voter*. Annual state and federal legislative voting records (scorecards of the environmental performance of our representatives).
Volunteer Opportunities: Campaign volunteers: phonebanks, precinct walks, mailings, general office support. Paid Staff Positions: CLCV has ongoing recruitment for out campaign staff. Full and part time positions available. Excellent pay and benefits.
Funding Sources: CLCV is supported solely by membership dues and individual contributions.
Annual Budget: $2,000,000
Employees: 120

California Marine Mammal Center

Marin Headlands
Fort Cronkhite, CA 94965
(415) 331-7325
Description: The center rescues, rehabilitates, and studies sick or injured marine mammals, and releases recovered animals back to the wild. Findings are shared with scientific associates and the public. Open every day of the year from 10 am to 4 pm. Admission free.
Resources: Call the rescue line at 331-SEAL. Their newsletter is sent with membership.
Volunteer Opportunities: Great volunteer and internship opportunites (95% of employees are volunteers).
Funding Sources: Membership, private foundations and corporations.
Annual Budget: $750,000
Employees: 18

California Native Grass Association

231 Escondido Drive
Martinez, CA 94553
(209)727-5319

Contact: Robert Delzell, Chairperson, or David Dyer.

Description: Membership in the association is open to all persons actively engaged or interested in the development, cultivation, application and marketing of California native grasses and associated species. The members represent the diverse interests of public and private land managers, technical agencies and organizations, the seed production and marketing industry, conservation organizations, universities and individuals. The goals of the association are: 1) to develop the technology to restore and/or rehabilitate ecosystems using native grasses and associated species for the purpose of soil stabilization and improvement, sustained productivity, conservation of biodiversity, and exotic weed control; 2) to cooordinate and support the production and marketing of commercial quantities of seed and other plant material and; 3) to educate people on the values of native grasses and associated species. Specific funding objectives and agenda have been developed to achieve these goals. A major objective of CNGA is to coordinate common garden experiments and field evaluation plantings with native grasses, to determine degrees of variability and range of adaptability.

California Native Plant Society

909 12th St., Suite 116
Sacramento, CA 95814
(916) 447-2677

Description: The California Native Plant Society is dedicated to creating a conservation-conscious world. It is an educational organization committed to an understanding of flora and preservation. Membership is open to all and includes a wide range of pursuits, including photography, hiking, drawing, natural history, etc. The California Native Plant Society conducts monthly lectures, field trips, and plant sales. It also conducts educational programs on native flora and supporting an environment for these plants.

Resources: *Fremontia* - jounal of Native Plants. *Bulletin* - on wild flowers. *The Inventory of Rare and Endangered Vascular Plants of California* - information on rare plants and threatened flora.

Volunteer Opportunities: Many volunteer opportunities available.

Funding Sources: Membership, grants, publications.

Employees: 4

California Public Interest Research Group (Cal PIRG)

1912 Bonita Avenue
Berkeley, CA 94704
(415) 644-3454

Contact: Jim Petruzzi

Description: California Public Interest Research Group (CalPIRG) does research and advocacy work in a variety of areas such as energy alternatives, toxics use reduction, pesticides control, food safety, environmental quality and consumer protection. They have 10 offices in California.

Resources: Quarterly newsletter.

Volunteer Opportunities: Summer volunteer jobs and fundraising are available.

Funding Sources: Citizens funding.

California Tomorrow

Fort Mason Center Bldg. B
San Francisco, CA 94123
(415) 441-7631

Description: California Tomorrow is a diverse group of citizens committed to California's future as a fair, working, multi-racial, multi-cultural society. At times, California Tomorrow focuses on environmental issues as they relate to population demographics.

Resources: *California Tomorrow: Our Changing State.*

Funding Sources: Tax deductable contributions, corporate and individual foundations.

Employees: 5

California Trout

870 Market St., Suite 859
San Francisco, CA 94102
(415) 392-8887

Contact: James Hamilton

Description: The main functions of California Trout are to restore and protect wild trout, native steelhead, the waters that nurture them and, to provide high quality angling adventures for the public to enjoy. Specific projects include:

- Tributaries to Mono Lake flow regulations and habitat recovery.
- North Coast androgenous fisheries protection-watershed protection, timber sale appeals, co-op work with the uses, CDF&G and USFWS.
- National Forest LMP Review & Mediation.
- McCloud River Camp, the Wild & Scenic program.
- Hat Creek Bank restoration.
- Apanscio Creek, Central Coast Steelhead advocacy.
- Calaveras/Alameda Creek Steelhead restoration.
- Malibu Creek Steelhead restoration.
- Mono County Trout Park.
- East Walker River Flow Regs. and habitat recovery.

Resources: *Streamkeeper's Log.*

Volunteer Opportunities: As needed.
Funding Sources: Membership, individual gifts, foundation and corporate Grants.
Annual Budget: $250,000
Employees: 21

CEIP Fund

512 Second Street - 4th Floor
San Francisco, CA 94107-1483
(415) 543-4400
Contact: Elizabeth Eckel, Regional Director
Description: The CEIP Fund is the nation's environmental careers organization. Founded to help people interested in environmental careers gain experience, and to supply new talent for organizations and companies in need of environmental professionals. CEIP, a non-profit organization, offers short-term, paid positions in various environmental disciplines to college students, recent graduates and career-changers. These internships are with private industry , government, and non-profit organizations. CEIP offers a wide spectrum of project types in the fields of environmental protection, public policy, waste management, natural resource management, and enviromental health and safety.
Resources: Quarterly newsletter, reference desk in office.
Volunteer Opportunities: Opportunities exist in the California Regional office to assist with project development, recruitment, and office administration.
Funding Sources: Management fees and grants.
Employees: 30

Center for Economic Conversion

222 View Street, Suite C
Mountain View, CA 94041
(415) 968-8798
Fax: (415) 968-1126
Contact: Michael Closson, Executive Director
Description: The Center for Economic Conversion was founded in 1975 and is dedicated to creating a sustainable future and building a society of peace, jobs, and justice. It educates the public about the need for positive alternatives to excessive miltary spending. Economic Conversion facilitates the process of converting to an economy responsive to both human and environmental needs. With a vision and a strategy for obtaining real security, nationally and globally, Economic Conversion includes: reorientating national priorities; rebuilding productive capacities to meet critical needs; revitalizing the economies of military-dependent communities; and transforming defense plans and military facilities to productive civilian uses.
Resources: Quarterly newsletter *Positive Alternative* for Fall of 1990, New and improved from *Plow Share Press* 1976-Summer 1990. *Base Conversion News* - focusing on the conserving and closure of bases.

Volunteer Opportunities: Great volunteer opportunities and internships for anyone interested in making the world a better place.
Funding Sources: Membership, donations, foundation grants.
Annual Budget: $260,000
Employees: 4

Center for Investigative Reporting

530 Howard Street, 2nd Floor
San Francisco, CA 94105-3007
(415) 543-1200
Fax: (415) 543-8311
Contact: Sharon Tiller, Program Director
Description: In existence for thirteen years, CIR is the only organization in the country entirely dedicated to investigative reporting. A core staff of twelve writers oversees several projects, the largest of which has an environmental emphasis. CIR publishes approximately one book per year, covering topics such as the export of banned pesticides to the Third World (*Circle of Poison*), the poisoning of the domestic water supply (*Troubled Water*), and the hazardous conditions of American factories here and in the Third World (*The Bhopal Syndrome, The Electronic Sweatshop*). CIR's most recent book is called the *Global Dumping Ground* and deals with the dumping of US toxic waste onto lands held by poor nations. In addition to books, CIR publishes reports for magazines and newspapers, produces video segments for local, national and international news broadcast, and is in the process of developing a documentary unit.
Resources: Books - see above, newsletter - *In House*, which recapitulates, in brief, published stories, and allows the reader to order reprints. (Bi-annual), *Nuclear California* - produced in conjunction with Greenpeace, this newsletter describes the extent of California's nuclear industry.
Volunteer Opportunities: Intern program, six months long, to promote educational and editorial development. CIR encourages people from all backgrounds to apply.
Funding Sources: Foundation and individual grants, fundraising. Commercial income from publications and television segments.
Annual Budget: $500,000
Employees: 12

Center for Third World Organizing

3861 Martin Luther King, Jr. Way
Oakland, CA 94606
(415) 654-9601
Fax: (415) 654-5862
Contact: Rinku Sen.
Description: The Center for Third World Organizing is a non-profit support and training institute for organizations fighting for the rights of communities of color. Major projects include: the Campaign for Accessible Health Care; the Saturday school for leaders of the community and labor

organizations; the Residential Minority Activist Apprenticeship Program for new organizers; one-, two- and three-day trainings in organizing and organizational development; and technical assistance to organizations. This is a national organization and includes a speakers' bureau.
Resources: Quarterly newsetter, *Minority Trends* $20/yr; *Activists' Guide to Religious Funders*, $25; *Fundraising for Social Change* $20.
Volunteer Opportunities: Internships in organizing, fundraising, research.
Funding Sources: Private and church foundations, publications, training and consultant fees.
Annual Budget: $400,000
Employees: 8

Central Coast Conservation Center

790 Main Street, Suite E
Half Moon Bay, CA 94019
(415) 726-3613
Fax: (415) 726-9708
Electronic Address on EcoNet: central coast
Contact: Kathleen Van Velsor, Executive Director
Description: The Central Coast Conservation Center, a project of the Montara Institute of Citizen Research, Planning and Action, works to protect, restore, and enhance the coastal environment of the central coast region from San Francisco to Monterey. The Center is part of a growing network of regional conservation centers that provide focal points and support for citizens in education and action. The Center offers educational forums, training seminars, and conservation research. It also informs the public through the news media, policy analysis, advocacy, and provides information and referrals. The Central Coast Conservation Center is currently working on projects that include offshore water quality, wetlands protection, habitat and land conservation, and facilitation of electronic networking among environmentalists concerned with coastal issues. Current projects include: Designation of the Monterey Bay National Sanctuary, Wetlands Protection Project - including Pacifica's San Pedro Creek, and Apanolio Creek, a riparian rights handbook, and the Associate Program, to gather environmental services donated by environmental, planning and legal professionals. The center also operates CoastNet, a computer network information exchange for the preservation of the coastline's environmental integrity.
Resources: An informal library of conservation organization profiles. Telephone assistance regarding land-use planning and resource issues.
Volunteer Opportunities: Office support, campaign assistance, research, interns, CoastNet. Associates: We need the skills and talents of a legal advocate, resource or urban planner, librarian, biologist, writer, graphic designer, media expert or photographer.
Funding Sources: Grants, fees.
Employees: 2

Circuit Rider Productions

9619 Old Redwood Highway
Windsor, CA 95492
(707) 838-6641
Contact: Karen Gaffney, Environmental Project Manager
Description: CRP is a non-profit service corporation dedicated to the enhancement of environmental and human resources. Active since 1976, CRP operates innovative programs in environmental restoration and vocational training. CRP's Environmental Division provides a full range of revegetation, erosion control, and stream enhancement services.
Funding Sources: Fees for service, grants.

Citizens Committee to Complete the Refuge

453 Tennessee Lane
Palo Alto, CA 94306
(415) 493-5540
Contact: Florence La Riviere
Description: The Citizen's Committee was formed in 1985 to save the Bay's remaining wetlands by incorporating them into the San Francisco Bay National Wildlife Refuge. Through education, legislation and legal action, the committee strives to protect the remaining wetlands of the Bay by placing them in public ownership. Land acquisition is the committee's major task, carried out while fending off developers who would destroy these lands before they can be purchased.
Resources: Periodical newsletter, monthly mailings to activists.
Volunteer Opportunities: The Committee depends on its small localized groups of volunteers. Activists watch over their local stretch of the Bay, and keep others informed of threats or opportunities in their locality.
Funding Sources: Foundation grants and direct mailings.
Annual Budget: $15,000
Employees: 1

Citizens for a Better Environment

501 Second Street Suite 305
San Francisco, CA 94107
(415) 243-8393
Contact: Hannah Creighton
Description: The mission of Citizens for a Better Environment is to translate an ecological and democratic vision of California's future into practical, effective advocacy and policy analysis. CBE's objective is to prevent and reduce toxic hazards to human health and the environment, specifically from pollution of air, water and land in major urban areas of California. CBE uses four tools to achieve its objectives - in-depth technical research; watchdogging regulatory agencies and industrial polluters; carefully planned litigation based on sound technical data; and public education and participation.A statewide organization, Citizens for a Better Environment is headquartered in San

Francisco, with local offices in Los Angeles and Berkeley. In 1990-91, Citizens for a Better Environment will focus its attention on the following program in the S.F. Bay Area:

- A new campaign to win the phase-out of chlorinated solvents, industrial chemicals that are destroying the Earth's atmosphere and, at the ground level, poisoning the environment and causing public health injury.
- The campaign to protect San Francisco Bay, focuses on three sources of Bay pollution — industrial wastes released through municipal sewer systems, dredge disposal, and reduction of fresh water flows into the Bay.
- In 1990-91, CBE will continue to focus on air quality, building on the 1990 victory in federal court, which required public agencies in the Bay Area to finally enforce the Clean Air Act of 1982.
- Citizens for a Better Environment's ongoing work in the community of Richmond, California - a low-income, predominantly minority community which is home to more than 300 polluting facilities - was designed to provide a model response to the fact that three out of five Black and Hispanic Americans live in communities with uncontrolled toxic sites.

Resources: Referrals for local groups fighting toxic battles. Publishes a quarterly newsletter for members.
Volunteer Opportunities: Office work, work on newsletter, research, and legal intern opportunities.
Funding Sources: Door-to-door canvass, memberships, foundation grants.
Annual Budget: $1,150,000
Employees: 61

Citizens for Urban Wilderness Areas

4325 Mountain View Ave.
Oakland, CA 94607
(415) 530-7547
Description: Citizens for Urban Wilderness Areas works for promotion and protection of parklands and natural preserves in East Bay areas, including the Bay shoreline as well as hills and woodlands. Citizens of Urban Wilderness Areas offers letters and occasional meetings supporting promotion and protection of open space, natural preserves, park land, and improved qualities for inner city living.
Funding Sources: Contributions.

Citizens Opposing Polluted Environments

Berkeley, CA
(415) 548-0861
Contact: Jamie Caseber, Acting Director
Description: A local environmental group based in Berkeley dedicated to preventing the construction and operation of hazardous waste incinerators. The group promotes toxics-use reduction and is proposing an ordinance to regulate importation of hazardous waste into Berkeley. It pinpoints local sources and producers of hazardous waste, and has formed a tough issues questionnaire for local politicians.
Volunteer Opportunities: Volunteers needed for contacting media, speaking engagements, writing letters, re-writing, and outreach to other communities.

COAAST Californians Organized to Require Access To State Tidelands,

P.O. Box 3284
Santa Rosa, CA 95409
(707) 539-0153
Contact: Chuck Rhinehart or Carol Vellutini
Description: COAAST is primarily concerned with matters related to land and resource use on the California Coast. A major function is to assure that the general public has access to the publicly-owned tidelands. COAAST advises and supports agencies with their planning for access, acquisition, development, and controls on coastal lands. It has several on-going projects such as: leasing and offshore drilling, oil spills, waste water disposal, and the fouling of the offshore environment, transportation, and waste problems as they relate to coastal development, with particular emphasis on the Sonoma County Coast. This organization provides support for environmental groups in the form of minor financing, experts in coastal matter, and consulting. COAAST has been established as a non-profit educational corporation since 1968. Their 1991 projects include the Santa Rosa Wastewater Disposal project, and the Sonoma County Transportation Plan.
Resources: *COAAST NEWS* - Newsletter, bi-annual; referral through Sonoma County Conservation Council's Environmental Center
Volunteer Opportunities: Board of Directors, Legal consultancy, biology, hydrology, geology, and experts of all resource planning types.
Funding Sources: Membership dues, contributions, recycling, annual banquet.
Annual Budget: $2,200

CoastNet

725 Main Street
Half Moon Bay, CA 94019
(415) 726-3613
Description: CoastNet is an association of education and action organizations dedicated to the protection and restoration of the natural environment of coastal regions. CoastNet's purpose is to facilitate communication and cooperation among environmentalists through electronic networking for computer-to-computer information exchange. CoastNet was initiated in 1988 by a group of environmental organizations with the aid of a computing equipment grant from Apple Computer. Participation in CoastNet is open to organizations and individuals who share our commitment to the environmental integrity of the coast.

Resources: E-mail, conferences (electronic bulletin boards), databases of information that can be searched with keywords for easy retrieval of everything on a particular subject.
Volunteer Opportunities: Technical help, outreach.
Funding Sources: Monthly fee plus connect time charges, grants.
Annual Budget: $200,000
Employees: 10

Committee for Green Foothills

2448 Watson Court
Palo Alto, CA 94303
(415) 494-7158
Fax: (415) 494-7158
Contact: Barbara Brown, Coordinator
Description: The Committee is concerned with political land issues in the Bay Area. Founded in 1962, the Committee now represents nearly 1300 members in San Mateo and Santa Clara Counties. The Committee has provided a consistently effective voice in promoting integrated land use planning. CGF has two legislative advocates who research issues, testify before public agencies, and work with City and County decision-makers. CGF supports park and open space acquisition, preservation of agricultural lands, protection of natural resources, and long-range growth planning to maintain the region's quality of life. Members receive a quarterly newsletter, *Green Footnotes*, which gives special insights into local issues and includes announcements of field trips and slide presentations.
Resources: Peninsula Conservation Center Library and *Green Footnotes*, a quarterly publication.
Volunteer Opportunities: Office and fieldwork
Funding Sources: Memberships ($35) and fundraising
Annual Budget: $60,000
Employees: 3

Commonweal

P.O. Box 316, 451 Mesa Rd.
Bolinas, CA 94924
(415) 868-0970
Contact: Nadine Parker
Description: Commonweal is a non-profit institute, founded in 1976, engaged in service, research, and demonstration programs in health, education, and human ecology. It serves as an instrument through which projects of value to humanity and to the earth can be implemented. Commonweal has three major areas of interest: children and families; patient-centered medicine and health care; and environmental security for a sustainable future. The patient-centered health promotion programs for cancer patients and their families include retreats, yoga, and discussion groups focusing on diet and nutrition, exercise, and other factors that contribute to optimum health. Retreat center facilities are available for use by other groups when not in use for Commonweal programs (consists of three furnished houses). Commonweal also has a biodynamic/"French intensive" garden open to visitors.
Resources: Paper available: *World Order and Environmental Security*
Volunteer Opportunities: None.
Funding Sources: Foundation grants, private and corporate donations, fees for services and products.
Annual Budget: $500,000
Employees: 6

Community Action Marin

408 Fourth Street
San Rafael, CA 94901
(415) 457-2522
Contact: Richard McKee, Program Director
Description: CAM works onommunity, energy, and poverty issues. This program addresses disproportionate energy expenses for low-income residents of Marin County through direct weatherization, energy education and direct financial energy asistance.
Volunteer Opportunities: Volunteers for outreach needed.
Funding Sources: State and federal programs, foundation grants.
Employees: 6

Consumer Pesticide Project

425 Missippi Street
San Francisco, CA 94107
(415) 826-6314
Contact: Craig Merilees
Description: The Consumer Pesticide Project is a network of environmentalists, consumers, and labor groups working to promote practical pesticide reduction, and in the process, promote excellence in American agriculture.

Design Associates Working With Nature (DAWN)

Spinnaker Way
Berkeley, CA 94710
(415) 644-1315
Contact: Jane Andrews
Description: DAWN contracts with the City of Berkeley to provide native plants for city gardeners and parks. Members of DAWN may purchase natives from the nursery. Projects range from a large restoration plant supply, to appropriate plantings for school environmental projects.
Resources: Periodic *Land Steward* newsletter, brochures on membership and publications on thistle removal, and native grass vegetation. North Waterfront Park is a display garden of native plantings placed by DAWN.
Volunteer Opportunities: We need volunteers to weed, learn to propagate properly, water and other chores.
Funding Sources: Grants, fees for seeds, consultation.
Employees: 1

Earth Action Network

1711 Martin Luther King Jr. Way, Suite. D
Berkeley, CA 94709
(415) 843-4306
Fax: (415) 649-1895
Contact: Stephen
Description: EAN is a continuation of the Earth Day Action Coalition which organized the Pacific Stock Exchange demonstration and Blockade in April of 1990 in conjunction with the Wall Street action on the same day. Their goal is to continue supporting and organizing just actions involving civil disobedience, direct action, and creative resistance. They are seeking to build a diverse coalition of environmental, labor, cultural, peace and social justice, anti-intervention, commintiy and minority organizatons and groups. They believe that any entrenched system of hierarchical authority and control should be placed on the extinction list, whether it be western capitalism or state socialism. Earth Action Network desires to create an ecologically sustainable future predicated on non-violence and the building of an alternative community.
Resources: Community Action Network Bulletin Board at (415) 843-8788, and the monthly *Earth Action Network* paper.
Volunteer Opportunities: Many. Please call.
Funding Sources: Monthly benefits and individual contributions.
Annual Budget: $8,500

Earth Day 1990

P.O. Box 192026
San Francisco, CA 94119-2026
(415) 986-5140
Contact: Blaine Townsend
Description: Earth Day 1990 marked the beginning of a long-term commitment to building a safe, just and sustainable society. Earth Day will launch a "decade of the environment,"promoting biological diversity, human health and regenerative agriculture. It will involve a broad cross-section of society in creating a groundswell of support for environmentally sound products, investments and policies. Among the goals are: a worldwide ban on chlorofluorocarbons, sustained reductions in carbon dioxide emissions through higher standards for automobile fuel efficiency, adoption of a transportation system not powered by fossil fuels, preservation of old-growth forests, a ban of non-recyclable or non-biodegradable packaging, a transition to renewable energy resources, increases in energy efficiency, hazardous waste minimization emphasizing source reduction, heightened protection for endangered species and habitats, a powerful international agency with authority to safe-guard the atmosphere, the oceans and other commons from international threats, and a new sense of responsibility for the protection and preservation of the planet by indiviuals, communities, and nations.

Earth First! San Francisco

P.O. Box 411233
San Francisco, CA 94141
(415) 824-3841
Contact: Daniel Barron
Description: Earth First! is a loose-knit national network with many local 'chapters'. Earth First! espouses a philosophy of action which includes a full spectrum from letter-writing to nonviolent direct action to 'monkeywrenching' (ecotage).
Resources: Hotline for information: (415) 824-3841.
Volunteer Opportunities: Volunteers welcome: Office work, fundraising, organizing action.
Funding Sources: Individual donations, grants.

Earth First!- East Bay

P.O. Box 83
Canyon, Ca 94516
(415) 376-7329
Contact: Karen Pickett
Description: See San Francisco listing.
Resources: Sporadically published newsletter.
Volunteer Opportunities: Production of press releases, banner makers, tree and rock climbers, etc.
Employees: 23

Earth Island Institute

300 Broadway, Suite 28
San Francisco, CA 94133
(415) 788-3666
Fax: (415) 788-7324
Contact: Kendra Ellis
Description: Earth Island Institute is a non-profit organization, founded by David Brower, which works on environmental issues in their global context by initiating internationally–oriented conservation and restoration action projects, publishing a quarterly international environmental news magazine, and developing a world-wide network of Earth Island Centers to provide support for grassroots action. They emphasize actions that can be taken by individuals, and build networks to involve other organizations and concerned individuals. Projects include: International Marine Mammal Project (IMMP); Environmental Project On Central America (EPOCA); Rainforest Health Alliance (RHA); Japan Environmental Exchange; Climate Protection Network; Information for the Public Trust; Brower Fund; conferences on the fate of the Earth; Environmental Litigation Fund; Friends of the Ancient Forest; Green Alternatve Information for Action (GAIA); Ben Linder Memorial Fund; San Bruno Mountain Preservation Project; Stewards of the Earth; Sacred Land Film Project; Radio Earth Island; Sea Turtle Restoration Project; US/USSR Environmental Exchange; Urban Habitat Program; and Watchfire Productions.

Resources: *Earth Island Journal* quarterly, Econet.
Volunteer Opportunities: Clerical wprk, and projects.
Funding Sources: Memberships ($25), grants, donations.
Annual Budget: $1,100,000
Employees: 15

Earth Regeneration Society, Inc.

470 Vassar Ave.
Berkeley, CA 94708
(415) 525-4877
Contact: Alden Bryant
Description: The ERS has recognized that a massive emergency program is necessary to halt the increase in carbon dioxide and commence its reduction. The means to accomplish this are three–fold: 1) stop the production of carbon dioxide by developing and using non-fossil fuel (non–nuclear), 2) stop deforestation and begin a massive reforestation program, and 3) remineralize the soils to provide healthy and rich soils for proper plant and tree growth. They argue that funds for CO-2 control and climate stabilization must be diverted from military uses. This would create employment in soil and forest restoration, and the development of alternative energy sources.
Resources: California CO2 budget, newsletter
Volunteer Opportunities: Please call; at present , we need volunteers to prepare documents for education and the media.
Funding Sources: Personal contributions, memberships.
Annual Budget: $5,000

EarthSea Institute

P. O. Box 2164
Sausalito, CA 94966
(415) 331-7060
Fax: (415) 331-8278
Contact: Terry Carlisle
Description: The EarthSea Institute's purpose is to help single individuals have an impact on global concerns. Tree Life Connection is one of EarthSea's main projects, with future plans to include planting sites throughout the world. EarthSea's other focus is Antarctica and the current efforts to preserve the continent in its pristine state. EarthSea has strong ties to the music industry, and in January 1991 will release Polar Shift, a compilation album featuring major instrumental artists as a benefit to The Cousteau Society in its efforts to save Antarctica. The Tree Life Connection is a service where a tree will be planted in honor of the person you choose. We will send that person a hand-lettered certificate, a card, and a color photo of the site. Their name will be placed in a book at the headquarters of the Cosumnes River Preserve. The cost is $20. Your gift is tax deductible. To order call (800) 326-6575
Funding Sources: Tree Life, and private donations.
Employees: 3

East Bay Area Trails Council

11500 Skyline Blvd.
Oakland, CA 94619
(415) 531-9300 x2316
Contact: Steve Fiala
Description: The EBATC is a grass roots organization representing individuals, trail user groups, and land management agencies involved in the preservation and expansion of trail opportunities in the East Bay Area. Members assist in volunteer projects, participate in planning of trails in local jurisdictions, and watchdog potential problems related to trails and access to open space areas. The organization represents hiking, running, equestrian, and bicycle groups in the East Bay Area.
Resources: Monthly newsletter including minutes of the meetings.
Volunteer Opportunities: Trail–related projects including: planning, advocacy, maintenance and construction.
Funding Sources: Membership dues (175 members at present).

East Bay Bicycle Coalition

P.O. Box 1736
Oakland, CA 94604
(415) 452-1221
Contact: Alex Zuckermann
Description: The Coalition promotes bicycle use and safety as an every day means of transportation and recreation. They actively support programs which coordinate bicycle travel with public transportation. Major accomplishments include Bike-on-Bart, Bay Bridge Bicycle Shuttle, East Bay Bicycle Commute Route Map, regional bike routes, and racks on AC Transit buses. Meetings are the third Tuesday of every month at 7:30 . Please call for meeting place.
Resources: Publications: *Ride On* quarterly, *Bicycle Commuter Guide* with information on bicycles, ferries, bridges, BART, buses and rail.
Volunteer Opportunities: Clerical, special projects, and tabling (setting up tables at conferences) positions.
Funding Sources: Memberships starting at $12.00.

East Bay Citizens for Creek Restoration

2721 Stuart Street
Berkeley, CA 94705
(415) 486-1742
Contact: Barry Waldman, Outreach Coordinator or John Steere, President (415) 849-1969.
Description: East Bay Citizens for Creek Restoration (EBCCR) is a grassroots environmental action and education group dedicated to renewing creeks in the urban setting. They develop and implement creek revegetation and restoration projects as well as educate and encourage people to treat creeks in a conscious and environmentally sensitive manner. It is their long-term intent to reclaim our creeks as corridors of nature, running like green spines

through our neighborhoods. They seek to work and form coalitions with other community and environmental organizations whenever the occasion arises. Since its inception in 1988, EBCCR has coordinated four "Creek Weeks," semi-annual series of stream restoration and awareness events that synthesize art, action, and community participation. They have also initiated a number of planting projects, and co-sponsored a curbside creek identification project in Berkeley. We are beginning a restoration-needs inventory for local urban streams; assisting in the development of an elementary school-age creek curriculum; and developing an interpretive map of East Bay Creeks—illustrating their present status, as well as recommending walking tours and opportunities for restoration.

Resources: Quarterly newsletter, *Creeks Speak;* small library of books and technical information on riparian restoration.

Volunteer Opportunities: Hands-on restoration activities; creative arts as part of awareness building; research and writing for newsletter and restoration inventory; information booth and other outreach; administrative responsibilities; public education; map–making and landscape design for creeks.

Funding Sources: Membership dues; donations; proceeds from EBCCR-sponsored events.

Annual Budget: $2,000

EcoAct!

438 Paris Street
San Francisco, CA 94112
(415) 587-5372

Contact: John Isom, SFSU Geography Dept. 1600 Holloway Avenue, SF,CA 94132

Description: After a year of organizing for Earth Day in April, our group has started this school year by affirming that we do not wish to be affiliated with Earth Day either as an anniversary, or with the Earth Day group in Palo Alto. We are a small-scale organization that is planning several projects for the San Francisco State University campus:

- Conducting an environmental audit, expanding recycling, and introducing composting on campus.
- Transportation issues: tying SFSU fees to MUNI fees and getting a reduced fast pass rate.
- Promoting bicycle riding to the campus and around the city.
- Establishing a farmer's market on campus.
- Outreach to schools in San Francisco and the Bay Area for a second Kid's Day, itself to be a part of an Alternatives Week (with films, Energy Day, Alternative Transportation Day, etc.).

Resources: We can help you work out ideas that are similar to our own.

Volunteer Opportunities: This is an all volunteer organization.

Funding Sources: Student government.

Annual Budget: $100

Ecology Action
Common Ground Garden Supply

2225 El Camino Real
Palo Alto, CA 94306
(415) 328-6752

Description: Ecology Action has a mini-farm for research, demonstration and education. There is also a community garden and an educational center. Common Ground Garden Supply is a non-profit project of Ecology Action of the Mid-Peninsula. They have been in business since 1972 and sell organic gardening supplies.

Resources: Extensive gardening, agricultural library, mail order, publications and books.

Ecology Center

2530 San Pablo Avenue
Berkeley, CA 94702
(415) 548-2220

Description: The Ecology Center is a non-profit educational and recycling organization which has served the Bay Area since 1969. We operate the oldest on-going recycling program in the state, as well as a comprehensive environmental information service. The Ecology Center is working to develop a more responsible society by identifying environmentally destructive practices and demonstrating sound alternatives. Our programs include a citywide weekly curbside recycling program and two weekly farmers' markets.

Resources: Monthly environmental newsletter, telephone information switchboard, recycling hotline, referral service, environmental bookstore, ecological living and organic gardening supply center, organic gardening and simple living classes, and an environmental library and clipping files.

Volunteer Opportunities: Assistants needed in the following areas: library, information resources, newsletter, farmers' market, fact sheet development, and outreach.

Funding Sources: Members, bookstore, farmers' market sponsorships, and a contract with the city for curbside recycling.

Annual Budget: $500,000

Employees: 23

EcoNet

3228 Sacramento Street
San Francisco, CA 94115
(415) 923-0900
Fax: (415) 923-1665

Contact: Bill Leland, Director, or Jill Small, Asst. Director

Description: EcoNet is a computer-based communication system helping the environmental movement throughout the world to communicate and cooperate more effectively and efficiently. A large minicomputer, based in Northern California, is connected to SpringNet (a common carrier), and enables EcoNet users to communicate globally, usually through a local phone call. EcoNet is part of the Institute for Global Communications (EcoNet, PeaceNet, ConflictNet). EcoNet is compatible with nearly any personal computer or computer terminal outfitted with 300 or 1200 baud modem. It has a full range of telecommunications capabilities including electronic mail, electronic conferencing, and databases - all covering environmental issues.

Funding Sources: Memberships. $15 sign-up fee and $10 monthly subsription fee.

Elmwood Institute

P.O. Box 5805
Berkeley, CA 94705
(415) 845-4595

Contact: Philippa Winkler or Lee Summerell

Description: The Elmwood Institute is an ecological ("green") think tank founded by Fritjof Capra (*The Tao of Physics, The Turning Point*) with Ernest Callerbach (*Ecotopia*), Hazel Henderson (*Politics of the Solar Age*) and others. The Institute promotes the ecological world view and systemic solutions that reflect the multiple manifestations of life. Its programs include the *Elmwood Newslet*ter, "think-and-do" Elmwood Circles; symposia, conferences and *Global File* reports on ecological practices in business and government worldwide, edited by Ernest Callerbach. Membership starts at $25 ($15.00 for those living lightly)

Resources: *The Elmwood Newsletter*

Volunteer Opportunities: Writing, research, publicity, fundraising, administrative duties.

Funding Sources: Membership, major gifts, foundations, fees for services.

Annual Budget: $250,000

Employees: 3

Environmental Action Center

P.O. Box 410563
San Francisco, CA 94144
(415) 647-9160
Fax: (415) 647-9175

Contact: Daniel Barron or Alan Van Tress

Description: The Environmental Action Center is an alliance of grassroots activists serving as an information clearinghouse and support network for Bay Area activism. We focus on local, regional, and global ecology issues and on the related struggles of oppressed peoples. The Action Center is a resource for organizing non-violent direct action and creative educational projects, and provides a place for people to learn about current issues and get involved.

Resources: Resources for organizing and networking; current information on ecology issues and upcoming events.

Volunteer Opportunities: Volunteers welcome: office work, research, organizing, sign and banner making.

Funding Sources: Donations, grants, member organizations

Environmental Council of Santa Cruz County

P.O. Box 1769
Santa Cruz, CA 95061
(408) 426-2286

Contact: Becky Luening (chairperson thru 12/90).

Description: The Environmental Council was founded in 1978 to protect, maintain, and restore the Santa Cruz County environment, including the quality of its air and water, its wildlife habitat, and all those values inherent in the preservation of open space and a sense of community. The Council addresses these concerns through public education, environmental issues forums, dissemination of news through publishing a regular newsletter, and the support of environmental events and initiatives. The Council actively nurtures an environmental network and invites environmentally concerned residents to join us in this urgent and rewarding work.The Environmental Council was the umbrella organization for the 1990 Earth Day Santa Cruz, a week-long series of educational and celebratory events culminating in a festival.

Resources: Monthly newsletter; a telephone referral and networking service.

Volunteer Opportunities: We have an ongoing need for volunteers to do research, writing, attend other groups and ad hoc meetings, and to do organizational work (organize volunteers, events, etc.).

Funding Sources: Donations, business sponsors.

Annual Budget: $10,000

Environmental Federation of California

116 New Montgomery Street, Suite 231
San Francisco, CA 94105
(415) 882-9330

Contact: Lara Casby, Office Manager

Description: EFC is a coalition of 33 environmental groups based in California. They do work-place payroll deduction to support the work of those groups who are fighting toxics at work and at home. They are involved in preserving open space, planting trees, cleaning up our air and water, working for a healthy ocean and desert, and sending out alerts for individual environmental action. EFC is running more than 70 work-place campaigns this year and making over 600 presentations to employees all over California.

Resources: Newsletter under production, referrals.

Volunteer Opportunities: Making presentations to employees representing EFC, office administrative assistance, referral resource, information research.

Funding Sources: Work-place campaigns.

Employees: 5

Environmental Forum of Marin

P.O. Box 74
Larkspur, CA 94939
(415) 924-0320

Contact: Virginia Souders-Mason, President

Description: The purpose of The Environmental Forum of Marin is to improve the quality of our environment through citizen education. We are a non-profit organization, operated since 1972 on a voluntary basis by dues-paying members. Membership is gained by graduation from the Forum Training Program. The Forum has two goals: 1) To provide an intensive training program and public educational services that increase understanding of ecology, environmental issues, and the planning process. 2) To support and advocate informed citizen action for the environment.

Resources: Five-month training program, and many other public educational services.

Environmental Project On Central America (EPOCA)

300 Broadway, Suite 28
San Francisco, CA 94133-3312
(415) 788-3666

Contact: Dave Henson or Jane McAlevy

Description: The Environmental Project On Central America (EPOCA) promotes environmental protection and social justice in Central America. In general EPOCA outlines the connections between the U.S. government and military and the environmental degradation of Central America. Areas of interest include military impacts in Honduras, Nicaraguan political strife and its impact on the environment, the difficulties and repression encountered by Central American environmentalists, the trade-offs between popular movements and the environment, agriculture and land reform.

Resources: *EPOCA Update* - quarterly. Greenpapers - periodical. Environment Under Fire -video.

Volunteer Opportunities: Call for information.

Environmental Traveling Companions (ETC)

Fort Mason Center, Building C
San Francisco, CA 94123
(415) 474-7662

Contact: Shawn Haber

Description: For eighteen years ETC has been successfully providing innovative outdoor adventures and environmental education programs for people with special needs. ETC's sea kayaking, whitewater rafting, and nordic skiing programs provide therapeutic recreation, education and inspiration to over 1,000 participants annually. These include youth-at-risk, students and people who are visually or hearing impaired, physically or developmentally disabled, and those with a terminal illness such as AIDS. ETC's programs have a dramatic impact on the quality of their participants' lives. Participants are challenged to succeed in ways they never thought possible. They network with over 65 established social service agencies and schools who look to ETC as a vital extension of their services. ETC offers a wide variety of "Benefit" trips, on a regular basis, to the general public and corporate groups. The funds from these trips (whitewater rafting and sea kayaking) go to a scholarship fund to subsidize the outings for the disabled and disadvantaged participants. These trips take place all year long, alongside those that support their mission.

Resources: Newsletter

Volunteer Opportunities: Over 200 volunteer guides, volunteer raft guide training in March, kayak guide training in August, training in wilderness skills, first aid, and disability awareness. Office volunteer opportunities also available; summer internships available for raft program to live at site on the American River.

Funding Sources: Raffle ticket drives for outdoor gear prizes, grants, benefit trips, auctions, memberships, fundraisers (special events such as surf contest, Angel Island Picnic, films and speakers.

Annual Budget: $250,000

Employees: 5

Farallones Institute Rural Center

15290 Coleman Valley Rd.
Occidental, CA 95465
(707) 874-3060

Contact: Salli Rasberry, Director

Description: The Farallones Institute Rural Center is a non-profit organization which seeks to encourage community self-reliance. Located on 80 acres of coastal farmland 70 miles north of San Francisco, the Center focuses on educational programs which includes residential training and workshops on appropriate development, and on integration of technical expertise and skills for activists and international development workers. The Center is working in the following areas: renewable energy technologies, small scale food systems, edible landscaping, and development strategies.

Resources: Training materials and an annual report.

Funding Sources: Contracts, grants, educational programs, publications, memberships, consulting fees.

Friends of Creeks in Urban Settings

1778 Sunnyvale Ave.
Walnut Creek, CA 94596
(415) 938-6323

Description: Their mission is to work for the preservation and enhancement of urban streams in Contra Costa County. Projects include: Referrals and information to individuals and groups seeking to preserve urban streams; meetings with the Flood Control District and the Army Corps of Engineers concerning preservation of Murderer's, Matson, and Grayson Creeks in Central Contra Costa; application for nonpoint source water quality implementation grant; distribution of native plants raised from local

seedstock for planting along streams in Central Contra Costa County.
Resources: Free brouchure available with SASE, entitled *The Creeks in Your Backyard.* Slides available on loan with script entitled, *Urban Creeks: A Problem or an Asset.*
Funding Sources: Membership, grants.

Friends of Fitzgerald Marine Reserve

P.O. Box 451
Moss Beach, CA 94038
(415) 728-3584
Contact: Virginia Welch, Bob Brien
Description: An organization dedicated to the protection of the James V. Fitzgerald Marine Reserve. Supports educational, scientific and conservation projects at the Reserve.
Resources: Newsletter - *Between the Tides.*
Volunteer Opportunities: Roving Interpreters, Docent program through the Coyote Point Museum in San Mateo.
Funding Sources: Memberships, donations and workshops, government.
Annual Budget: $80,000
Employees: 4

Friends of Islais Creek

6 Hillview Court
San Francisco, CA 94124
(415) 826-5669
Contact: Julia Viera
Description: The group consists of citizen volunteers working to rehabilitate a section of urban waterway in the industrial section of San Francisco.
Volunteer Opportunities: Wide open— no regular meetings, no dues, no officers, only core groups of 25-50 professionals from widely diverse fields. All volunteers welcome.
Funding Sources: State Department grant for urban streams restoration for $48,000.
Annual Budget: $48,000

Friends of the River - San Francisco

Bldg. C, Fort Mason Center
San Francisco, CA 94123
(415) 771-0400
Fax: (415) 771-0301
Contact: David Rouslin
Description: Friends of the River is a non-profit membership organization dedicated to protection of western rivers and to conservation of water and energy resources. Through research, public education, lobbying, political action, and the legal process, FOR works to secure state and federal Wild and Scenic River recommendations and protections as well as conservation-oriented reforms of state and federal water and energy policies. The 100 Rivers Campaign is a multi-year effort to secure protection of rivers and streams within national forests in California under the Wild and Scenic Rivers Act. The New Water Agenda is a multi-faceted program to prepare concise, accessible documentation on river ecology, to develop a consensus for water policy reform in California, to prepare public school curricula about water, its sources, destinations, uses and abuses.
Resources: *Headwaters*, a bi-monthly newspaper; *Cross-Currents*, a quarterly newsletter; a booking service for whitewater rafting trips; and guide training for whitewater rafting.
Volunteer Opportunities: In San Francisco and Sacramento offices including support for special events, periodicals indexing, data entry, and general office support.
Funding Sources: Membership, donations, special events, corporate support, grants.
Annual Budget: $650,000
Employees: 15

Fund For Animals

Fort Mason Center
San Franciso, CA 94123
(415) 474-4020
Contact: Virginia Handley
Description: The Fund For Animals aims to protect endangered species along with the fight against cruelty, and the protection of all animals through education, legislation and litigation.
Resources: Animal Rights Actionline: (415) 474-4202, *Animal Rights Resources Catalog, California Animal Rights Legislative Action Alerts.*
Volunteer Opportunities: Office work, Demos, meetings, projects and Legislation.
Funding Sources: Donations and memberships.
Employees: 21

Golden Gate Council of American Youth Hostels

425 Divisadero St, #307
San Francisco, CA 94117
(415) 863-1444
Contact: Bob Leone
Description: The Golden Gate Council of American Youth Hostels promotes world peace through educational and recreational travel. Programs include:

- Hostel Adventure Program/Inside Out Academy-provides overnight trips for disadvantaged youth to a Northern California hostel. Activities include hiking, biking, tidepooling, and learning about the environment while building self-esteem.
- A new youth hostel in Sacramento - this project involves moving and renovating a historic Sacramento Victorian that will open in the Spring of 1991.
- Council-sponsored biking and hiking clubs which offer 1 and 2 day Bay Area trips.
- The Great San Francisco Bike Adventure - a 15 mile non-competitive ride on the streets and highways of San Francisco. The June 1990 event attracted nearly 10,000 riders.

- World Adventure Trips Program - Hiking, biking and van trips throughout the world.

Resources: Quarterly newsletter for Northern California, *Golden Gate Hosteler*. Travel Services (store): Sells AYH memberships, Eurail pass, budget international flights, travel gear, books and maps. Also has a reference library of travel books.
Volunteer Opportunities: Hostel Adventure Program, Great San Francisco Bike Adventure (see above). Marketing intern program.
Funding Sources: Overnight fees from council hostels, donations, grants and annual Bike Adventure.
Annual Budget: $1,000,000
Employees: 38

Green Alternative Information for Action (GAIA)

Earth Island Institute
300 Broadway, #28
San Francisco, CA 94133
(415) 530-4935
Fax: (415) 788-7324
Contact: Terrie Schultz or Roger Picklum
Description: GAIA publishes the Green Calendar, a monthly listing of Bay Area activist events: fundraisers, meetings, demonstrations, fairs and other activities. Available for $10/year. Event information may be submitted to the above address or numbers.
Resources: Green Calendar.
Volunteer Opportunities: The Calendar is distributed by volunteers in many neighborhoods throughout the Bay Area.
Annual Budget: $7,000

Green Party –East Bay Chapter

East Bay Green Alliance, P.O. Box 20999
Oakland, CA 94609
(415) 549-1011
Description: The East Bay Green Party organizing group is one of 24 statewide, seeking 80,000 California voters in 2 years to gain ballot status. Although the Green Party in California has been active only since February, 1990, it is a direct descendant of the various European Green parties which began in the early 1980's, and which now hold parliamentary seats in a number of countries around the world. They presently meet as a general body in Berkeley twice a month and in a number of committees including: platform, orientation, tabling, publicity, finance, and speaker series. The party is culturally very diverse, young and old, female and male; sharing common concerns for the environment, peace, and justice, and frustration with the unresponsiveness of the present political system. They invite everyone to join and work toward a future for all. Your vote, ideas, opinions, and contributions of time and money can make a difference. Are you contributing to the problem or its solution? They're already having fun - join. (Note: voter registration forms can be obtained at your local post office or by calling the State toll-free number: 1-800-345-VOTE). The East Bay Green Alliance, from which The East Bay Green Party evolved, is the 5 year old non-profit educational branch, which brings together those interested in a variety of subjects such as Green theory, community, tree planting, spirituality, energy, etc.; and which sponsors occasional lectures, beach clean-ups, and other events.
Resources: Newsletter mailings, small library, sale items (books, T-shirts, buttons, stickers, etc.), information sheets, Green group contacts for many other California counties and U.S. states
Volunteer Opportunities: Tabling, registration, publicity, telephoning, networking, fundraising, organizing events, writing articles, etc.
Funding Sources: donations, merchandise sales, contributions from speakers series.
Annual Budget: $8,000

Green Party –Marin: See Marin Greens

Greenbelt Alliance

116 New Montgomery, Suite 640
San Francisco, CA 94105
(415) 543-4291
Fax: (415) 543-1093
Contact: Larry Orman, Executive Director
Description: The Greenbelt Alliance is the Bay Area's region-wide land conservation organization. Founded in 1958, it is dedicated to the protection of the region's greenbelt of farms, parks and other open lands. Greenbelt Alliance is also an acknowledged leader in Bay Area metropolitan planning aimed at enhancing the livability of the region's cities. Current projects include:

- Protecting the Greenbelt in fast-developing counties (Alameda, Contra Costa, Santa Clara, Solano, and Sonoma) through land use regulation and, in some cases, acquisition.
- Outreach program: Reaching the Bay Area public through presentations, festival appearances, and showings of the award-winning film, *Treasures of the Greenbelt*.
- Media Access Program: Ensure in-depth coverage of key open space issues.
- Greenbelt Mapping and Assessment Program: An advanced computer program that charts land use trends in the Bay Area.
- Sustainable Metropolis Program: Works with local officials, developers, housing advocates to encourage more compact, environmentally sustainable development. Also cultivates regional leadership to better manage regional land use programs.
- Bay Area Ridge Trail: 400 mile loop around the Bay Area's ridges now one quarter completed.

Resources: Quarterly newsletter: *Greenbelt Action*; quarterly news coverage on land use issues, *Newsclips*; Numerous reports and technical guides on land use in the Bay Area an California; send SASE to Greenbelt Alliance for publications list.

Volunteer Opportunities: Public education; presentations and appearances at festivals; Land use monitoring and activism; Office support (typing, reception, mailings); research; writing and graphic design assistance. As well as arranging and guiding hikes and bike trips into the Greenbelt.
Funding Sources: Memberships, foundations, business sponsors, federal government funding for Ridge Trail, and contract work on San Francisco Estuary Project.
Annual Budget: $650,000
Employees: 13

Greenpeace

139 Townsend St. 4th Floor
San Francisco, CA 94107
(415) 512-9025
Description: Greenpeace resolves environmental problems through the use of nonviolent direct action, environmental education, and media exposure. Greenpeace's many areas of attention include the preservation of marine mammals; the nuclear industry; toxic pollution of our air, land and water; and problems of waste. The group publishes a bi-monthly magazine, *Greenpeace* which covers articles on these topics, as well as green advertising hype, and international environmental degradation.
Resources: *Greenpeace* a bimonthly magazine.
Volunteer Opportunities: Call for information.

Half-Cost Carpool Transit Systems

2720 Martin Luther King Jr. Way
Berkeley, CA 94703
(415) 845-1769
Contact: J. Ralph Richards
Description: Half-Cost Carpool Transit Systems is an efficient way to get current information on ride sharing. In an effort to save energy and establish low cost transportation, this organization provides daily updates to members on listings of available carpool opportunities. Non-members may submit their name, phone number, and ride situation for no fee, but they will not receive information on other carpoolers. Information on four means of travel are given out: You can ride with someone else and pay only half the gas costs; you can drive your own car and collect half of the gas costs from riders; you can use a driveaway car and earn cash to take the car to your destination; and you can buy a "Cheap Heap" or sell your car for not over $500. Membership cards are transferable, meaning that more than one person can share privileges, and institutions such as schools may purchase membership and allow others to use the service at no cost. Operates 9am - 3pm, Monday through Saturday.
Resources: Lists of riders, drivers, and Driveaways.

List of buyers and sellers of autos that do not cost over $500, published daily.
Funding Sources: Memberships, self-funded

Harbinger Communications

250 Homestead Trail
Santa Cruz, CA 95060
(408) 429-8727
Contact: Bill Leland
Description: *The Harbinger File* is a directory of citizen groups, government agencies, and environmental education programs concerned with California environmental issues. Published every other year, the 1990-91 edition is 288 pages and describes 959 organizations including: name, address, phone, contact person, publications, one-to-four paragraph description, table of contents, and indexes. *The Harbinger File* has become the standard reference of its kind. [The *Green Pages* staff found it invaluable throughout our production] Price: $15.25, includes postage and handling.

Hayward Area Planning Association

2787 Hillcrest Avenue
Hayward, CA 94542
(415) 538-3692
Contact: Sherman Lewis - Chair
Description: The Hayward Area Planning Association was formed in October 1978 by citizens concerned with planning in the Hayward area. A small steering committee sets policy and does the work. HAPA supports managed compact growth, open space protection, mass transit, downtown revitalization, jobs-housing balance and citizen participation. In 1984 HAPA sponsored Measure M to limit hill development, but was outspent more than 25 to 1 by foreign and corporate developers. HAPA nevertheless won 46% of the votes. HAPA sued successfully to stop a new law which guts the Williamson Act. HAPA now faces massive freeway and ridge development proposals.
Resources: HAPA News
Funding Sources: Member dues ($10), contributions, newsletter subscriptions

In Vivo: Radiation Response

P.O. Box 31958
Oakland, CA 94604
(415) 763-8837
Contact: Geoffrey Sea, Director and Editor
Description: In Vivo publishes an international journal and provides educational and consulting services on issues of radiation, health, and the environment. They're a project of the Tides Foundation. In Vivo sponsors regional conferences on radiation around the world. They are affiliated with organizations that work with radiation survivors, and on radiation issues, in the Soviet Union, Europe, Japan, and other countries. They provide expert testimony on the health physics and medical effects of radiation, in support of radiation compensation litigation in nuclear-free-zone campaigns. Consulting services include research and

provision of educational materials. Electronic Mail - Econet igc: invivo.

Resources: *In Vivo: Radiation Response* : a quarterly journal in English and Russian which covers radiation science issues, and is an independent monitor of the radiation industry. It carries news, research and commentary concerning the impact of ionizing and non-ionizing radiation on life, health, and the environment. The library conatins radiation issues and maintains files of speakers and experts on these issues.

Volunteer Opportunities: Clerical work, writing , research, publications, computer work, translation.

Funding Sources: Grants, and subscriptions.

Annual Budget: $200,000

Employees: 4

INFACT/Northern California

2414 B Telegraph Ave.
Oakland, CA 94612
(415) 272-9522

Contact: Kirsten Cross

Description: INFACT is making the connections between nuclear weapons work and the overwhelming environmental contamination it causes. They are the organizers of the G.E. boycott.

Inquiring Systems, Inc

3111 Deakin Street
Berkeley, CA 94705
(415) 848-5212

Contact: S. Loren Cole, Ph.D.

Description: Inquiring System has been helping non-profit organizations, small businesses and public agencies operate more effectively. The experienced staff provides expert assistance. ISI believes that the survival of all non-profits will depend on their abilty to gain independence from grants and other soft money. Hence, ISI services are designed to help organizations become more self-sustaining.

Resources: Workshop materials.

Volunteer Opportunities: Apprentices and interns.

Funding Sources: Fees for service.

Annual Budget: $180,000

Employees: 5

Institute for Food and Developement Policy, Food First

145 Ninth Street
San Francisco, CA 94103
(415) 864-8555
Fax: (415) 864-3909

Contact: Marilyn Borchardt

Description: Food First's goal is to promote participatory, equitable, and sustainable development. Current projects include building citizen democracy through democracy dialogues with students and adult commmunity groups; exploring environmental and human rights, and democratic developments in the Third World. The goal with all of their educational work, which includes public speaking and media work, is to provide an awareness of the interconnectedness of social problems, such as Third World debt, that lead to both environmental destruction and starvation, and to encourage people to act together to find solutions-locally, nationally, and globally. Food First is a non-profit organization supported by donations, book sales, and public speaking. Donors and members receive a quarterly newsletter. Recent action alerts include "Brazil's Debt and Deforestation - A Global Warning" and a paper on South Korea's labor and pollution problems. Write for free information. Books recently released include: *Rediscovering America's Values*, by Frances Moore Lappe, author of *Diet for a Small Planet*; *Dragons in Distress: Asia's Miracle Economies in Crisis*; *The Philippines: Fire on the Rim*. Upcoming books include: *Liberation Ecology*; *Child Alive*; and *Development that Works*.

Resources: Directories include: *Alternatives to the Peace Corps* ($6.95), *Education for Action - a guide to progressive graduate programs* ($6.95), Quarterly newsletter - *Food First News and Action Alerts* ($25)

Volunteer Opportunities: Volunteers and interns are involved in all aspects of work from research to book-packing, from telephone referral to member relations, from publicity to word processing.

Funding Sources: Donations (56%) book sales (18%) public speaking (5%), grants (11%), royalties (4%), other (6%).

Annual Budget: $750,000

Employees: 8

Institute for the Study of Natural Systems

P.O. Box 637
Mill Valley, CA 94942
(415) 383-5064

Contact: James A. Swan, Ph.D.

Description: This is a non-profit education and research organization seeking to produce innovative programs to increase harmony between people and nature. Projects include: 1) The Spirit of Place symposiums a five year program of annual symposiums bringing together representatives of traditional cultures to see how their earth wisdom may guide modern design, law, architecture, and planning to create sustainable communities; 2) the Annual Celebration for Mother Earth Concert; 3) creation of "Mr. Recycleman", the Marin County recycling mascot.

Resources: A selection of presentations from the 1988 and 1989 Spirit of Place programs will be published as *The Spirit of Place: Sacred Places and Spaces* (Quest Books) in 1991. A personal report on 15 years of research on the scientific significance of sacred places, *Sacred Places: How The Living Earth Seeks Our Friendship*, by James A. Swan, was published in 1990.

Funding Sources: Gifts, grants, registration fees.

Employees: 2

International Rivers Network

301 Broadway
San Francisco, CA 94133
(415) 986-4694
Fax: (415) 398-2732
Contact: Owen Lammers, Director
Description: The Rivers Network researches and publishes information concerning threats to rivers, watersheds, and local populations from destructive development projects. It focuses on World Bank, IMF, and government economic development policies. Current programs include reforming World Bank policies, and stopping several high dam projects in the Third World and Eastern Europe. Supports grassroots water projects in Africa.
Resources: Bi-monthly newsletter *World Rivers Review*; directory of water resources, professionals and organizations around the world; library.
Volunteer Opportunities: Office work, writing of the newsletter, research, database maintenance.
Funding Sources: Foundation grants, private donors and members.
Annual Budget: $120,000
Employees: 2

League for Coastal Protection

P.O. Box 190812
San Francisco, CA 94119-0812
(415) 777-0220
Fax: (415) 495-5996
Contact: Ann Notthoff
Description: The League is a coalition of public interest and environmental organizations and individuals created in 1981 to re-ignite a strong defense of coastal resources. The League is the only statewide organization concentrating all its efforts on protecting California's coast. The League works to establish coastal protection advocates as an effective political constituency, ensure adequate funding of and balanced appointments to the California Coastal Commission, monitor Coastal Commission hearings, renew efforts to improve enforcement of Coastal Act policies at all levels of government, and work for passage of Congressional legislation to uphold the state's role in federal consistency decisions under the Coastal Zone Management act.
Resources: *Coastlines*, quarterly newsletter
Volunteer Opportunities: Office administrative work, mailings, issue research, etc.
Funding Sources: Memberships, subscriptions, grants.
Annual Budget: $25,000
Employees: 1

League of Women Voters - Sonoma County

2421 Coddington Center
Santa Rosa, CA 95404
(707) 546-5943
Contact: Lynn Camhi
Description: The League of Women Voters is a nonpartisan political organization that encourages the informed and active participation of citizens in government and influences public policy through education and advocacy. The League promotes an environment beneficial to life through the protection and wise management of natural resources in the public interest by recognizing the interrelationship of air quality, energy, land use, waste management, and water resources.
Resources: Voter newsletter published monthly.
Volunteer Opportunities: All volunteers.
Annual Budget: $10,350

Lifeweb

P.O. Box 20803
San Jose, CA 95160
(408) 972-0348
Fax: (408) 927-0348
Contact: Rick Bernardi
Description: Lifeweb's main interest is in animal rights, deep ecology, and social justice. They are a group of people whose purpose is to recognize and act upon a boundless set of ethics which will extend the circle of human compassion to include the Earth and all beings. Lifeweb understands that all life exists as part of a biocentric system, and therefore, other beings do not exist to serve the interests of an anthropocentric heirarchy. All beings have a right to live free, under their own self-determination, and according to their own self-interest. Since we recognize that human animals are capable of making moral choices, it is our responsibiity not to infringe upon the inherent rights of others. We are opposed to the oppression of all beings, human and non-human, and seek to create a world in which humans live in harmony with other beings and the environment. Acknowledging that such a world does not now exist, Lifeweb shall encourage the transition to such a world, for the mutual benefit of all life. Current projects include: Animal- based agriculture, an anti-vivisection campaign, animals in entertainment, biodiversity, environmental center, anti-fur campaign.
Resources: Quarterly newsletter- "Connectons" - published on solstices and equinoxes; speakers bureau; slideshows; events calendar hotline.
Volunteer Opportunities: Grassroots direct action- all campaigns are run by volunteers- volunteers are welcome for any existing campaign, or they may devise new campaigns for Lifeweb.
Funding Sources: Donations, grassroots fundraising.
Annual Budget: $5,000

Local Solutions to Global Pollution

2121 Bonar, Studio A
Berkeley, CA 94702
(415) 540-8843
Fax: (415) 540-4898

Contact: Nancy Skinner or Beth Weinberger

Description: Local Solutions to Global Pollution is a clearinghouse and technical assistance organization to assist grassroots groups, citizens, and local elected officials wishing to initiate solutions to environmental issues on the local level. LSGP is an outgrowth of an informal clearinghouse established in Berkeley by City Council member Nancy Skinner's office in response to requests for information on polystrene foam/plastics legislation and other innovative enviromental programs.

Resources: Write for informational packet request form.

Volunteer Opportunities: Volunteers needed to help answer information requests, research issue areas, and help develop information packets.

Funding Sources: Sales of publication, grants.

Annual Budget: $20,000

Employees: 2

Magic, Incorporated

P. O. Box 5894
Stanford, CA 94309
(415) 323-7333

Contact: David Schrom

Description: Magic, Incorporated is an educational, public- benefit corporation. The people who are creating Magic are joined in an intention to become more loving. By this we mean we are expanding our self-interest to encompass the well-being of our fellow humans, of other life, and of the abiotic elements of the the environment on which life depends. Magic also shares a perception that the methods and accumulated information of the science of ecology are among the most valuable of tools for realizing an intention to love. The activities of Magic are highly integrated. Through research and study, by teaching and publishing, and in the example of our everyday corporate operations and personal lives, we aim to discover and to disseminate a more accurate picture of the relationships of people to one another and to the other living and non-living elements of the environment, and to apply this understanding to identify and contribute to the common good. Magic has four overlapping programs in ecological philosophy, personal awareness, cooperation, and environmental Protection. we are developing an ecological approach to the concept of value, redefining health, organizing ourselves and others to make peacefulness and mutual aid more universal, and encouraging stewardship of the earth. We provide a variety of ecosystem protection and restoration services, including ecological analysis of land use alternatives, reforestation of grazed lands, comprehensive urban forest planning, forest modeling software, and redesign of streets to improve the safety, quiet, and stability of residential neighborhoods.

Resources: A modest library on land use, aspects of urban forestry, and permaculture. We have a section on personal awareness, health, cooperation, and community. We have a school speaker program.

Volunteer Opportunities: An all-volunteer organization. Immediate needs: clerical, correspondence, document production, database maintenance, library research, field work in permaculture/restoration, drawing and photography.

Funding Sources: Fees for services, contributions.

Marin Agricultural Land Trust

P.O. Box 809
Point Reyes Station, CA 94956
(415) 663-1158
Fax: (415) 663-1099

Contact: Robert Berner, Executive Director

Description: Marin Agricultural Land Trust (MALT) is a member-supported, non-profit organization the purpose of which is to help protect and preserve Marin County's extraordinarily beautiful and economically productive agricultural land. They do this by acquiring, in voluntary transactions with landowners, agricultural conservation easements. A conservation easement is a written agreement between the landowner and MALT which prohibits subdivision and development, and preserves the land's agricultural character and use. The land remains in private ownership, while MALT assures that the terms of the easement are observed in perpetuity. Since 1980 MALT has acquired conservation easements on 12,500 (13 ranches) of the approximately 135,000 acres of privately owned agricultural land in the county. MALT easements protect not only the agricultural future of the land, but also preserve the creeks and streams, wildlife habitat, scenic values, and social and cultural diversity represented by these lands. In 1987 and 1988 MALT participated in developing, qualifying and passing the California Wildlife, Coastal and Park Land Conservation Bond Act (Proposition 70) which provides $15, 000,000 for agricultural land preservation through easement acquisition in Marin County. Using these funds, MALT expects to acquire easements on an additional 25,000 to 35,000 acres (20-30 ranches) over the next 5-7 years. Proposition 70 funds cannot be used for the costs of operating MALT's conservation easement program or for monitoring the easements already held. Contributions from individuals provide the majority of MALT's operating support.

Resources: Tri-annual newsletter, information on agricultural land conservation.

Funding Sources: membership, grants, Bond Act (Prop 70).

Annual Budget: $250,000

Employees: 4

Marin Conservation Corps

P.O. Box 89
San Rafael, CA 94915
(415) 454-4554
Fax: (415) 454-4595
Contact: Lisa Bereny or Jeff Lumb

Description: MCC's mission is to help the youth of Marin County develop by providing meaningful employment, education, and training opportunites through projects that conserve natural resources, deliver human services, and respond to public emergencies. Some current projects are stream clearance and flood control, recycling, erosion control, trail construction and maintenance, and cutting of fire breaks.

Resources: Job training and employment sources for youth.

Volunteer Opportunities: Volunteer program pending.

Funding Sources: Contracts, Marin Community Foundation, Regional Occupation Program.

Annual Budget: $1,200,000

Employees: 60

Marin Conservation League

35 Mitchell Blvd., Suite 11
San Rafael, CA 94903
(415) 472-6170
Contact: Karin Urquhart, Executive director

Description: Marin Conservation League is dedicated to preserving and protecting the natural assets of Marin County. The league has studied environmental issues, testified at public hearings, met with elected officials, and educated Marin's citizens to critical concerns. In recent months, the group has spearheaded efforts to ban styrofoam and reduce solid waste in Marin County, and initiated efforts to establish a countywide planning agency to coordinate development relative to regional infrastructure and transportation needs. The league has active committees that work on issues related to land use, water, bayfront and wetlands, parks and open space, transportation, hazardous materials and solid waste , and rural land use. Marin Conservation League has a public education program which provides a variety of resources including issue-oriented public forums and teachers' workshops, resource files, a small library, and newsletters focused on key environmental topics.

Resources: Members newsletter, issues-oriented newsletter for students and teachers, extensive clip file from local newspapers on key environmental issues, modest video and reference library, referrals.

Volunteer Opportunities: Clerical

Funding Sources: Memberships & grants

Employees: 4

Marin Greens

P.O. Box 3908
San Rafael, CA 94912
(415) 459-3765
Contact: Ken Smith

Description: Marin Greens is a local, grassroots organization affiliated with the regional, national and international Green Movement. Their basis of unity is the "Ten Key Values" - non-violence, community-based economics, grassroots democracy, decentralization, feminism, respect for diversity, social justice, personal and global responsibility, ecological wisdom, and future focus. The Greens recognize the common root causes of diverse problems, and seek change at that deep, structural level. They use non-violent direct action, public education, community organizing, coalition-building, lobbying, and electoral politics to effect change. The Marin Greens stress the interdependence of personal and global change, of means and ends, and of understanding and action. Projects include: Protest actions (Redwood Summer, and in El Salvador and The Middle East), initiative campaigns (NFZ, Forests Forever , and Big Green), Green Party organizing and voter registration, tree plantings, public lectures and dialogues, public events and potlucks, tabling, newsletter production, advocacy and lobbying on diverse local issues, and more. Green Party Marin is the county chapter of the statewide Green Party organization, and acts to promote green values in electoral politics.

Resources: Event Line: 45-EARTH; quarterly newsletter, *Green Calendar.*

Volunteer Opportunities: Internships and volunteer training for: community organizing, electoral campaigning, direct action.

Funding Sources: Donations, memberships events, literature, potlucks.

Annual Budget: $1,000

Mt. Diablo Peace Center

65 Eckley Lane
Walnut Creek, CA 94596
(415) 447-7300
Contact: Andy Baltzo, or Jean Boverd

Description: For 1990-91 their major focii are: 1) conversion of our economy from military to environmentally sound civilian production and; 2) continued efforts for peace with justice in Central America. They offer a 700 book and videotape lending library and the peace forums speaker series. Recurring events include: May and November "Celebration" pot-lucks with a program. The annual end-of the-summer International Peace Festival features music, games, and sales, and information booths for peace, justice and environmental groups, as well as craftspeople.

Resources: *Business Directory, Peace Gazette* newsletter - 10 times per year, annual calendar and lending library.

Volunteer Opportunities: Everything from stamping envelopes and help with mailings to fundraising, newsletter, and the Board of Directors.
Funding Sources: Appeals, events, personal donors, grants, recycling bins.
Annual Budget: $70,000
Employees: 2

National Toxics Campaign

425 Missippi Street
San Francisco, CA 94107
(415) 826-6314
Contact: Craig Merilees
Description: The National Toxics Campaign is a grassroots nationwide network of environmental activists with headquarters in Boston and offices in twelve states. Their emphasis is in grassroots organizing, aggressive toxics-use reduction, pesticide policy, and oversight or testing of products or services that claim to be environmentally sound.

Natural Building Network

P.O. Box 553
Occidental, CA 95465
Contact: Lauren Halling
Description: The Natural Building Network promotes healthful home and workplaces, land and communities through the use of nontoxic building materials and techniques. The Network is a group of professionals who have adopted a building ethic. Architects, contractors and other interested builders are encouraged to join.
Resources: Resource guide, information and referral, newsletter (on recent legislation, techniques and new ideas).

Natural Resources Defense Council, Inc.

Western Office: 90 New Montgomery, Suite 620
San Francisco, CA 94105
(415) 777-0220
Fax: (415) 495-5996
Description: NRDC is a non-profit membership organization with nearly 168,000 members nationwide. Since 1970, NRDC scientist and lawyers have been working to preserve and enhance the integrity of the environment, protect public health, and provide for the conservation and wise management of land and natural resources. NRDC pursues these goals through shaping government action at the federal, state, and local levels and through extensive public education and outreach. NRDC's major public accomplishments have been in the fields of public health, resource conservation protection, and stemming the proliferation of nuclear weapons. The San Francisco office is engaged in a variety of on-going efforts including: energy conservation and the battle against global warming; working to protect America's own rainforests in Hawaii, Puerto Rico and the U.S. Virgin Islands; preserving the country's last remaining old growth forests; examination and reform of water-use policies in California; coastal protection, both from on-shore development and off-shore oil exploration; conservation of public lands; challenging the use of cancer-causing chemicals by agriculture, and reducing the poisons in our food stream.
Resources: Bi-monthly newsletter, quarterly magazine, extensive publications list.
Volunteer Opportunities: NRDC's San Francisco office relies on volunteers from time to time to assist with the planning and carrying out of special events. We welcome offers of assistance.
Funding Sources: NRDC receives nearly half of its funding from memberships. The other half is made up primarily by foundation grants and augmented by fees, contracts, and other income.
Annual Budget: $16,000,000
Employees: 30

News from Native California

P.O. Box 9145
Berkeley, CA 94709
(415) 549-3564
Contact: Malcolm Margolin, Publisher
Description: Publishes a quarterly magazine concerned with California Indians.
Resources: newsletter
Volunteer Opportunities: Interns for research and marketing.
Funding Sources: Subscriptions, consignment sales, grants.
Employees: 4

North Bay Wetlands Coalition

c/o Save San Francisco Bay Assoc., P.O. Box 925
Berkeley, CA 94701
(415) 452-9261
Contact: Mark Holmes
Description: A citizens group interested in protecting wetlands in the North Bay. Of seventy-five members, 25-30 are organization representatives. Each group or member keeps track of their particular wetland of concern. The Coalition meets once a month.

Northern California Land Trust

3120 Shattuck Ave
Berkeley, CA 94705
(415) 548-7878
Contact: Dale Becknell or Mary Carleton
Description: The Northern California Land Trust is an organization that provides for the leasing of lands that are held in trusts for low income persons and communities at low rates. NCLT provides affordable housing to low-income people, office and retail space for non-profit and

community businesses, and land for small farmers. NCLT serves as a consultant for affordable ownership structures, and productive, ecological land use and conservation. The leaseholder is able to own improvements — buildings, crops, etc. Donations of land or money are tax-deductible. Loans or investments are also administered for socially responsible leaders/investors to finance projects. NCLT also provides educaton and assistance for groups who want to start independent, locally-administered community land trusts. They also give workshops on solar energy use for affordable living, land use policies to prevent urban sprawl, and sustainable agriculture. The NCLT owns 9 residential units in Berkeley and Oakland, a five-storefront building in Berkeley, and a 40 acre farm near Lodi. Membership in the NCLT is $20 per year, or 4 hours of work per year. Members help with workshop planning, property acquisition, community planning and organizing,and mailings at the office in Berkeley, and with sheepshearing, egg collection, demolition and building at the farm. Please join, and join in.

Resources: Referral to other farming land trusts, distributor of the Community Land Trust Handbook, The Directory of Collectives, collected CLT information.

Volunteer Opportunities: Many –four hours of service per year constitutes membership. Please call.

Funding Sources: Private donations, grants, memberships, $20 regular, $8 low income.

Funding Sources: , Memberships.

Annual Budget: $100,000

Employees: 1

Northern California Recycling Association

P.O. Box 5581
Berkeley, CA 94705
(415) 548-6659

Contact: Portie Sinnott

Description: The Northern California Recycling Association is a trade association of recyclers, including community groups, corporations, and government entities. The association is run through individual memberships, at $35 per year. The association is in its tenth year of operation and is organized through monthly board meetings and quarterly membership meetings.

Resources: Facility tours, special programs, debates, speakers, Newsletter.

Volunteer Opportunities: This is almost entirely a volunteer-run organization.

Funding Sources: Membership dues and grants.

Employees: 1

Oakland Recycling Association

P.O. Box 11492
Oakland, CA 94611-0492
(415) 655-5373

Contact: Susan Bluestone, General Manager

Description: The Oakland Recycling Association is a non-profit corporation providing educational and consulting services concerning waste reduction and recycling in the Oakland area.

Volunteer Opportunities: Opportunities to learn about recycling.

Funding Sources: Donations, grants, and contracts.

Ocean Alliance

Fort Mason Center, Bldg. E
San Francisco, CA 94123
(415) 441-5970

Contact: Craig Strang, Education director

Description: Ocean Alliance is a non-profit marine education organization dedicated to fostering an understanding and appreciation of the marine environment, especially among young people. Ocean Alliance promotes the protection, preservation and restoration of the marine environment. Programs include: Project Ocean, a national model for curriculum reform in science education which brings the lessons of the sea to schools, both in ocean-side and inland communities nationwide. The Whale Bus brings an exciting marine mammal program to classrooms, libraries and community groups. Sea Camp is a summer marine science and conservation camp for kids, Adopt-a-Whale allows people to protect whales through symbolic adoptions, and the Conservation Project helps establish marine sanctuaries and assists in other conservation efforts.

Resources: Resource library for teachers and an extensive marine mammal library.

Volunteer Opportunities: Assistants are needed for beach cleanup activities and in the office. Research internships are available.

Funding Sources: Government grants, foundation grants, corporate grants, membership dues, and individual donations.

Annual Budget: $1,000,000

Employees: 15

Pacific Coast Federation of Fishermen's Associations, Inc.

P.O. Box 989
Sausalito, CA 94966
(415) 332-5080

Contact: Joyce Larson, Elizabeth Stewart Thomson

Description: The Pacific Coast Federation of Fishermen's Associations, Inc. (PCFFA), is a California non-profit corporation organized in March 1976. Its membership is composed of 24 commercial fishermen's associations in California and Alaska. Fishermen belonging to the PCFFA member associations fish for Chinook and Coho salmon, Dungeness and rock crab, swordfish, sea bass, Pink shrimp, abalone, sablefish, rockfish, herring , sea urchin, and California lobster. PCFFA provides the local associations with a full time staff to address fisheries education, communication, promotion, and legislation. PCFFA represents its member associations at the local, state, and regional lev-

els on fisheries issues before commission, councils, and the legislature.
Resources: Bi-weekly newsletter
Volunteer Opportunities: Please inquire.
Funding Sources: Assessments, individual and business contributions.
Employees: 4

Peninsula Conservation Center Foundation

2448 Watson Court
Palo Alto, CA 94303
(415) 494-9301
Contact: Joan Van Gelder, Coordinator of Volunteers at the Peninsula Conservation Center.
Description: P.C.C. is a member- supported organization devoted to preserving and protecting the environment of the San Francisco Peninsula. Through conferences, workshops, lectures, and information services, the P.C.C. actively promotes the protection of our natural resources (coastline, wetlands, and open space lands) and advocates long range, environmentally safe planning on all fronts. Current projects include:
- The Downtown Environmental Action Project is organizing a comprehensive recycling and resource conservation program for local businesses.
- South Bay Wetlands Coalition is encouraging and training activists in how to preserve our remaining wetland resources.
- A 24 hour glossy paper (magazine) recycling program has served the community for over a year, reducing landfill and providing a "new" product.
- The Brown Bag Lunch is a weekly lecture series on environmental issue.
- The Speakers Bureau provides speakers on environmental issues upon request.
- The PCC is home to a number of other environmental organizations—Environmental Volunteers, Sierra Club, Camp Unalayee, Committee for Green foothills, Peninsula Greens, and Midpeninsula Environmental Education Alliance.

Resources: Largest environmental resources lending library on the Peninsula; over 1,000 cards in PCC environmental information rolodex, plus resource directory; store with environmentally-responsible products and books
Volunteer Opportunities: Multiple volunteer opportunities—receptionist, store assistant, research projects, library assistant, special projects, etc. Create your own project: PCC welcomes people with ideas and initiative to use the Center's resources to create new solutions to environmental problems. Call a meeting, offer a class, etc. PCC will help organize and address the environmental issues you want to work on.
Funding Sources: Memberships ($25— tax deductable), grants and donations.
Annual Budget: $125,000
Employees: 5

Peninsula Humane Society

12 Airport Blvd.
San Mateo, CA 94401
(415) 340-8200
Fax: (415) 348-7891
Contact: Sherry Richert, Public Information Manager
Description: The Peninsula Humane Society is an animal welfare organization dedicated to improving the quality of life for all animals. With an "open door" policy, PHS provides shelter for all types of creatures, and never turns away an animal in need. Domestic or wild, feathered or furry, stray or surrendered, all are cared for at the Peninsula Humane Society. PHS provides adoption services and counseling to assure that animals are placed in suitable home environments to become companions for life. This organization also offers a lost and found service to help connect lost animals with their guardians. PHS has the only wildlife rehabilitation facility in San Mateo County, so they are called upon to rescue a vast array of animals injured by everything from habitat destruction to oil spills or the plastic waste generated by people. The Wildlife Care Center has been able to sucessfully treat and release a large portion of these animals back into their natural environments. The community and its animals also benefit from other services, such as low-cost spray and neuter operations, vaccination clinics, adult education classes, school programs, animal behavioral training classes, the Pet Supply Shop (which carries a large selection of of cruelty-free personal care items), and the Pick of the Litter Thrift Store.
Resources: *Pawprint Magazine*, animal welfare library, many animal-related brochures, an animal-behaviorists hotline (340-7022 ext. 306), and a speaker's bureau with experts on wildlife animal control, domestic animals, legislation.
Volunteer Opportunities: Wildlife care center, shelter, special events, vetenary clinic, front office, adoption, and docents. There are currently 300 volunteeers.
Funding Sources: Memberships and donations, Animal Control funds from County government.
Annual Budget: $4,000,000
Employees: 90

Peninsula Open Space Trust

3000 Sand Hill Rd., Bldg. 4, #135
Menlo Park, CA 94025
(415) 854-7696
Fax: (415) 854-2803
Contact: Audrey C. Rust
Description: Peninsula Open Space Trust (POST) is a regional land conservancy, the objective of which is to preserve the beauty, character and diversity of the San Francisco Peninsula landscape. POST creates dedicated open space through acceptance of gifts, private purchase, and cooperative efforts with public agencies. It encourages the use of land for recreation, wildlife habitat, agriculture, and natural resource protection. Currently, POST is involved in the acquisition of marsh and agricultural land along the

coast, restoration of wetlands near Alviso, and acquisition of scenic Skyline properties.
Resources: Quarterly newsletter, *Landscapes.*
Volunteer Opportunities: Special events, pro-bono technical assistance.
Funding Sources: Private donations, foundation grants, land transactions.
Annual Budget: $6,500,000
Employees: 6

People for a Golden Gate National Recreation Area (PFGGNRA)

3627 Clement St.
San Francisco, CA 94121
(415) 752-2777
Contact: Amy Meyer
Description: PFGGNRA is a coalition of conservation and civic-minded groups and individuals who are concerned with the protection and use of land within the boundaries of GGNRA, Point Reyes National Seashore, Muir Woods National Monument, and Fort Point National Historic Site. They are also concerned with the timely acquisition of land included in the boundary, but not yet acquired. PFGGNRA's current projects include conversion of the Presidio of San Francisco from an Army post to a National Park, protection and enhancement of native habitat, the status of the FAA domes and antennae on Mt. Tamalpais, access to parklands in northwest Marin County, preservation of the integrity of the San Francisco watershed in San Mateo County, or anything of a policy-nature relating to the above-named national parks.
Resources: They network with groups and individuals.
Volunteer Opportunities: Volunteers needed to work on issues concerning The Presidio of San Francisco, native fauna and flora, removal of exotic species, and policy issues in the GGNRA and Point Reyes.
Funding Sources: Donations.
Annual Budget: $600

Permaculture Committee

P.O. Box 101
Davis, CA 95617
(916) 679-2729
Contact: Guy Baldwin
Description: Permaculture Communications publishes pamphlets on permaculture subjects; conducts retail/mail order sales of permaculture books and publications; provides mailing-list services (see below); collects native plant-seed (specializing in nitrogen-fixing (actinochizal) plants and perennial grasses) and provides permaculture design, education and referral services on an hourly fee basis. Referral is to qualified designers and teachers working in other areas of North America. Emphasis is on the Keyline Design method —water harvesting, soil fertility development, pond siting and design— developed by P. A. Yeomans. Consulting and implementation services are available.
Resources: *The Permaculture Activist* - quarterly, Database-Mailing list of permaculture, organic farmers, sustainable agriculture interests, tree crops, and botanic gardens . Over 26,000 entries. Lists are for sale to business and organizations.
Funding Sources: Self-funded, self-employed entrepreneur.

Physicians for Social Responsibility San Francisco Bay Area

2288 Fulton St., #307
Berkeley, CA 94704
(415) 845-8395
Contact: Joan Ali
Description: Physicians for Social Responsibilty is concerned with the health effects of the nuclear arms race and nuclear war. The group has educational materials, lecture series, referral library, and speaker's bureau that addresses these issues. Individual members present data at public hearings, and can suppport local groups and scientists with medical information on these issues.
Volunteer Opportunities: Newsletter, computer work, public speaking, head committee work, office work, community outreach.
Funding Sources: Donations, grants , memberships.
Annual Budget: $100,000
Employees: 2

Planned Parenthood - Contra Costa

1291 Oakland Blvd
Walnut Creek, CA 94596
(415) 935-4066
Contact: Kathy Simpson
Description: Operating with the credo that all children have a right to be wanted and that all people have a right to reproductive freedom and services, Planned Parenthood provides high-quality, affordable family planning services, including reproductive health care, education, and counseling. Planned Parenthood also monitors domestic and international family planning issues, maintains a resource library, and does advocacy work on political battles around the rights of choice and privacy.
Resources: Library; Journals: *International Family Planning Perspectives* and *Family Planning Perspectives;* quarterly newsletter; speaker's bureau on family planning, sexually transmitted diseases, choice, menopause, and sexuality in the 90's; books for sale or loan; films; and other educational materials.
Volunteer Opportunities: Clinic Assistant; Pregnancy Counselor; Picket Escort; Parent/Child Education; Program Assistant; clerical support; fundraising; Auxiliary/Guild members.
Funding Sources: Private pay, Medi-Cal, Office of Family Planning, and private donations.
Annual Budget: $5,700,000
Employees: 180

Quail RidgeWilderness Conservancy

25344 County Road 95
Davis, CA 95616-9735
(916) 758-1387

Contact: Frank Maurer, Jr., Executive Director

Description: Established as a nonprofit organization in February 1989, QRWC is dedicated to promoting biological diversity, native habitat research, and education about environmental stewardship. They are focusing efforts at present on the protection of a near-pristine peninsula—Quail Ridge Wilderness Preserve (1200 acres)—located on Lake Berryessa in Napa County, though the scope of the Conservancy will eventually broaden to include other wild areas in need of protection. The Conservancy provides guided, interpretative tours to the Preserve during the dry months of the year. There is a three-mile nature trail with 200 staked sites identifying native flora or other natural items. Research projects anticipated include censuses of the bird, mammal, and reptile populations, and an exploration of the dependence of oak tree regeneration on native bunchgrasses (Quail Ridge Wilderness preserve still harbors vigorous populations of various species of oak trees and bunchgrasses). The results of this research will be especially important for future management of about 8.5 million acres of the state's Engleman and blue oak populations (which are dwindling away).

Resources: Newsletter in the planning stage; *QRWP Trail Guide* (nearly complete).

Volunteer Opportunities: Volunteers needed for various preserve-related projects—e.g. scat-gathering and analysis, census work on native fauna, work on the newsletter, and staffing tables at conferences, fairs, etc.

Funding Sources: Memberships, donations, grants to be sought.

Annual Budget: $12,500

Rainforest Action Network

301 Broadway, Suite A
San Francisco, CA 94133
(415) 398-4404

Description: The Rainforest Action Network is a grassroots organization working to stop multi-national corporations and international development agencies from destroying tropical rainforests. Through action, education, consumer campaigns, and letter-writing, the Rainforest Action Network informs the public and private sectors of the critical importance of rainforests and what action can be taken to preserve them. The network works closely with indigenous peoples and grassroots groups in tropical forest areas to help protect their homelands.

Resources: *World Rainforest Report* (quarterly), *Action Alert* (monthly), *Rainforest Action Guide.*

Volunteer Opportunities: General office support, campaign work, membership department.

Annual Budget: $900,000

Employees: 15

Regional Bicycle Advisory Committee (REBAC)

3313 Grand Ave.
Oakland, CA 94610
(415) 452-1221

Contact: Alex Zuckerman

Description: The Regional Bicycle Advisory Committee was formed in1986 for the purpose of improving conditions for bicycling in the nine-county Bay Area. One of the major tasks is to create a regional planning map of bicycle routes covering the entire region. Some of the key issues and problems that REBAC is dealing with are 1) Eliminating barriers to bicycle travel, such as bridges, tunnels, freeways, and limited access roads; 2) improving road conditions; 3) working with city, county, and state agencies on their bicycle plans and policies; 4) having REBAC become a center for regional bicycle concerns in the Bay Area, receiving input, giving advice, suggestions and recommendations regarding bicycle projects and policies. They have meetings every third Wednesday of every odd-numbered month, at 7:30 PM at the MTC Metrocenter, 101 Eighth Street in Oakland, directly across from the Lake Merritt Bart Station.

Resources: Newsletter - *The REBAC Reporter*, bi-monthly. Gives a summary of accomplishments and problems to be tackled and provides agendas for upcoming meetings.

Restoring the Earth

1713C Martin Luther King Jr. Way
Berkeley, CA 94709
(415) 843-2645

Description: Restoring the Earth (RTE), a project of the Tides Foundation, was founded in 1985 to focus national attention on creative solutions to environmental problems by means of ecological restoration. RTE's overall mission is to promote the repair of environmental restoration, providing information on how restoration can be accomplished, and providing large numbers of people with opportunities for participating in restoration work. RTE recently published an inventory of ecological restoration projects in the nine-county Bay Area. Current projects include Project Restore, a volunteer action program RTE is collaborating with the East Bay Regional Park District and other local land managers to identify and facilitate the restoration of damaged sites on public lands in the Bay Area. They are also seeking funds to produce a public television series on restoration, and conduct a national restoration needs assessment.

Resources: *Ecological Restoration Directory for the San Francisco Bay Area: A descriptive directory and sourcebook*; information about restoration volunteer opportunities in the Bay Area; quarterly news bulletin, beginning Fall, 1990.

Volunteer Opportunities: Hands-on restoration fieldwork every other Saturday from 9 a.m. to 3 p.m. A variety of tasks and projects in office.

Funding Sources: Foundations and individual donations.

Annual Budget: $80,000

Employees: 6

RIDES for Bay Area Commuters, Inc.

60 Spear Street, Suite 650
San Francisco, CA 94105
(415) 861-7665

Description: RIDES For Bay Area Commuters began in 1977 as energy resources and the environment became growing concerns for the State of California and Bay Area residents. Today, RIDES is the best ridesharing resource in the Bay Area for both commuters and employers. We are a private, nonprofit corporation dedicated to providing assistance to commuters, and to helping relieve traffic congestion and air pollution throughout the ten-county region. RIDES has helped more than 120,000 commuters join carpools and vanpools. We've also provided commute-program assistance to more than 1,200 employers, developers, and local governments. Funding for services and operations is principally provided by the California Department of Transportation and the Metropolitan Transportation Commission.

S.A.F.E.R
Student Affiliation for Environmental Respect

335 South Eleventh #2
San Jose, CA 95112
(408) 280-0418

Contact: David Lombard

Description: S.A.F.E.R. is a group of San Jose student activists seeking environmental reform, direct action, and education. They have an intensive recycling program in place, currently recycling approximately eight tons monthly during the school year, and continuously expanding. They deliberate with campus Food Services in an effort to encourage the use of reusable products, and work with Campus Administration to reform the Bike Ban policy. They are conducting a campus environmental audit in an effort to evaluate the impact of the campus and make recommendations. Their goal is to raise student and campus awareness of environmental issues.

Resources: Workshops on environmental issues.

Volunteer Opportunities: On-campus recycling; eco-restoration field trip.

Funding Sources: Recycling program.

San Francisco 2000

130 Kearny, 33rd Floor
San Francisco, CA 94108
(415) 394-3489
Fax: (415) 362-8609

Contact: Kathy Lu

Description: San Francisco 2000 is a committee formed by Mayor Agnos as a "special forum to bring business and community together to find common ground and unite disparate interests as we move towards the year 2000." S.F. 2000 is envisioned as a process lasting 18-24 months that will yield agreements on the City's future and new ways of working together on civic tasks. Current projects include developing contacts and creating the Work Program for S.F. 2000, which will involve large numbers of San Franciscans in creating a shared vision of the City's future.

Volunteer Opportunities: Future opportunities for involvement will include volunteering in the SF 2000 offices, assisting in with community outreach, and working toward a consensus-based decision-making process. Specific duties may include data entry, survey-taking, discussions, forum organization and participation.

Funding Sources: Corporate, public, and private contributions.

Annual Budget:

Employees: 2

San Francisco Bay Bird Observatory

P.O. Box 247
Alviso, CA 95002
(408) 946-6548

Contact: Donald S. Starks

Description: S.F.B.B.O. is a non-profit volunteer agency that is dedicated to the management and preservation of the wildlife of the south San Francisco Bay. They have an ongoing schedule of classes dealing with wildlife biology and identification as well as nature tours and a class in museum methods. Current projects include: (1) monitoring for the prevention of avian botulism; (2) censusing shorebirds in salt ponds to determine patterns of usage throughout the year; (3) banding and observing rufous-crowned sparrows to study their breeding biology.

Resources: Bi-monthly newsletter; library for member use.

Volunteer Opportunities: Participation in research

Funding Sources: Primarily membership dues, supplemented by contracts and grants.

Annual Budget: $35,000

Employees: 1

San Francisco Bay-Delta Aquatic Habitat Institute

Richmond Field Station
1301 South 46th St.
Richmond, CA 94806
(415) 231-9539
Fax: (415) 231-9520

Contact: Kathy Kramer

Description: The San Francisco Bay-Delta Aquatic Habitat Institute is an independent non-profit corporation with a legislative mandate to coordinate all pollutant-related research and monitoring programs in the Estuary, and to develop a data management system that will allow easier access to information on these research programs and their results. The Institute has developed a computerized Bay Information Network (described below) that serves as a centralized clearinghouse for scientific informatnion about the Estuary. Access to the BIN is available free of charge by calling (415) 643-7485 (modem line). A VT-100 emulation package is necessary. In addition to the computerized databases, the Institute is reviewing current monitoring efforts with the aim of developing a long-term, comprehensive, and coordinated pollutant monitoring program for the Estuary. The Institute is also developing an outreach program.

Resources: Library with 1000 volumes containing information on the San Francsico Bay Delta, pollution and contaminant monitoring. Other resources include a quarterly newsletter and three BIN data bases, the Estuarine Data Index, the Bay-Delta Hearing Testimony and Exhibits Database, and the Bay-Delta Bibliography. Other databases include a San Francisco Estuary Effluent Monitoring Database and a public education display.

Volunteer Opportunities: Internships are available. Interns should have a background, or interest in, the natural sciences, library science, communications, marketing, or education.

Funding Sources: Federal, state, and local government, foundations and industry.

Annual Budget: $500,000

Employees: 7

San Francisco Community Recyclers

701 Amador
San Francisco, CA 94124
(415) 285-0669

Contact: Beryl Magilavy, Executive Director

Description: San Francisco Community Recyclers is a non-profit organization that has been recycling materials in San Francisco since 1980. They operate four centers in the city and provide the only convenient way for many neighborhoods to recycle. Planned projects include a demonstration recycling facility at Amador Street and Cargo Way, which will include composting, an organic garden, and displays.

Resources: A recycling program for students in San Francisco schools as part of a community recyclers/private business partnership, newsletter *ReNews.*

Volunteer Opportunities: Membership in the Friends of Recycling volunteer group. On-site help, graphic design, grass-roots organizing, data-entry, research and writing.

Employees: 23

Funding Sources: Membership donations, contributions, grants

Employees: 15

San Francisco Greens

New College, 777 Valencia Street
San Francisco, CA 94110
(415) 255-2940

Description: The San Francisco Greens are a broad-based group interested in environmental and political issues, pacifism, and social justice. The group is based on the U.S. Green Movement that deals with ecological wisdom, non-violence, equality, and diversity efforts. They sponsor lectures, "Green Talks", in Fall and Spring. There are working groups on recycling, transportation, toxics, and political action.

Resources: Bimonthly newspaper, *Green Consensus*

Volunteer Opportunities: Call for information.

Funding Sources: Dues and donations.

San Francisco League of Urban Gardeners

2540 Newhall Street
San Francisco, CA 94124
(415) 468-0110

Contact: Cynthia Hall, Director

Description: The San Francisco League of Urban Gardeners (SLUG) promotes ecologically-sound urban gardening in the San Francisco Bay Area. They believe anyone should be able to garden in San Francisco. They promote gardens in containers, backyards, schools, and on private and public vacant land and roof tops. They work with individuals, families, senior citizens, and youth, and strengthen communities through gardening and greening activities. Their Garden for The Environment. features self-guided tours, youth activities, workshops on home composting,and water conservation.

Resources: Horticulture library for members, quarterly newsletter, telephone consultation, fact sheets on Bay Area gardening techniques (organic gardening), workshops.

Volunteer Opportunities: Board of Directors - 3 year committment, people of color encouraged; Learning Garden (offers workshops and demonstrations); Newsletter, office support.

Funding Sources: S.F. Community Development Block Grants, private foundations, grants, donations.

Annual Budget: $250,000

Employees: 4

San Francisco Planning and Urban Research Association (SPUR)

312 Sutter Street, Room 500
San Francisco, CA 94108
(415) 781-8726

Contact: James Fussell, Jr., Executive Director

Description: SPUR conducts public policy analysis regarding all aspects of San Francisco City planning and policy. As a SPUR member, you'll be invited to join one of our standing committees: Housing, Transportation, Regional Issues, Education, City Planning & Development, City Management & Finance, Northern Waterfront, Southern Waterfront, and Social Issues. There you'll work closely with other SPUR members and staff, reviewing plans, policies, and ideas, and help create the position statements that government officials consider before they make decisions.

Resources: Standing Committees give Monthly Lunchtime presentations.

Funding Sources: Private and business.

Annual Budget: $230,000

San Francisco Tomorrow

942 Market Street, Room 505
San Francisco, CA 94102
(415) 566-7050
Contact: Andy Nash, President
Description: In 1970, neighborhood activists and concerned residents from throughout the City joined together to form San Francisco Tomorrow. SFT is dedicated to promoting environmental quality, protecting neighborhood livability, and encouraging good government in San Francisco. SFT is an all-volunteer effort that operates soley on the talent, hard work and support of its individual members. Most of the work is done in issue-oriented committees. Board of Directors meet monthly to set policies and approve actions. The main issues include: reducing impacts of downtown development; preserving the waterfront; improving public transportation; expanding open space; preserving and enhancing the city's life; reducing toxics; and encouraging affordable housing.
Resources: A monthly newletter is sent to members and friends.
Volunteer Opportunities: Political involvement on environmental issues.
Funding Sources: Contributions and memberships.
Annual Budget: $10,000

San Francisco Vegetarian Society

1450 Broadway #4
San Francisco, CA 94109
(415) 775-6874
Contact: Dixie Mahy, President
Description: The purpose of the society is to promote vegetarianism for ethical, health, and ecological reasons. The Society has monthly meetings at Fort Mason Center, every second Sunday at 3:30 PM (none in July and August). They have a packet of information for $2.00 available through the mail. They also offer memberships for $12.00 per year.
Resources: Newsletter –Fall/Winter
Volunteer Opportunities: Staffing tables at fairs.
Funding Sources: Membership.

San Jose Conservation Corps

2650 A Senter Road.
San Jose, CA 95111
(408) 998-5884
Fax: (408) 288-6821
Description: The San Jose Conservation Corps is a youth employment and development organization. We hire unemployed men and women ages 18-23, provide them with 32 hours of work and 8 hours of education each week. The work consists of 2 major categories: park maintenance type projects with local municipalities and recycling projects consisting of both residential outreach and education, and commercial recycling.

SJCC's mission is to develop our greatest natural resource, our young people, while at the same time protecting and preserving the environment in the South Bay. Corps members earn minimum wage for 32 hours of work each week. The education is unpaid and an investment in their own futures. After one year, permanent-job placement assistance is also provided.
Volunteer Opportunities: Assistance with ESL and GED classes (M-Th nights,after work from 5-7 PM, and on Fridays from 8 AM-noon).
Funding Sources: Fee for service contracts, foundation grants.
Annual Budget: $1,500,000
Employees: 100

Santa Clara Valley Bicycle Association

P.O. Box 831
Cupertino, CA 95015
(415) 651-6472
Contact: Bill Michel, President
Description: The objectives of the Association are to promote the use of bicycles, to encourage legislation favorable to bicycles at all levels of government, to preserve the identity of the bicycle as a vehicle, to improve bicycle safety, to educate both cyclists and non-cyclists regarding proper use and enjoyment of bicycles, and to encourage the provision of improved facilities for bicyclists. The Santa Clara Valley is one of the world's premier areas for bicycle commuting, because of its good weather, relatively flat topography, and good roads. Studies show that close to 50% of the residents of Santa Clara County live within five miles of work. This translate into about 30 minutes ride, for the average rider. We offer the bicycle as one of the most effective components in the solution to the problems of pollution and traffic congestion. Meetings are held every other month in the Cupertino Library, 10400 Torre Ave., Cupertino.
Resources: Newsletter, *The Spinning Crank* (available in bike shops).

Save Mount Diablo (SMD)

P.O. Box 25
Concord, CA 94522
(415) 549-2821
Contact: Seth Adams, Project Director
Description: Save Mount Diablo, a local citizens organization, raises funds and works with landowners who may be interested in selling their property. Many landowners who have kept their family's land in a natural condition now face pressure to subivide. SMD works with these landowners to acquire property before it's too late. SMD was formed in 1971 by a group of Contra Costa County residents who were alarmed at the rapid rate of urbanizaton taking place around the mountain and the lack of recreational open space. They track the development status of every parcel of land on and around Mount Diablo,

and work with willing sellers when opportunities arise to act quickly to preserve more land. They alert government agencies to important natural and recreational values. Thanks to the generous support of thousands of people, Save Mount Diablo has succeeded in almost tripling the size of the protected areas in less than 20 years. The Mount Diablo State Park consists of more than 19,000 acres, and that's not counting other parts of Diablo now protected by the East Bay Regional Park District and the Walnut Creek and Concord Open Space areas.

Save Our Shores (S.O.S.)

P.O. Box 1560
Santa Cruz, CA 95061
(408) 425-1769
Contact: Dan Haifley, Chairperson
Description: Save Our Shores is a nonprofit citizen's organization acting to preserve and protect the quality of the marine and coastal environments of the Central California coast. Their concern is not limited to an ecological viewpoint, but extends to the local economies and civic pride which also depend upon a healthy coastal environment. They are particularly concerned about outer continental shelf lease sales and their potential environmental impacts.
Resources: Quarterly newlsetters and fact sheets.
Volunteer Opportunities: General clerical and outdoor work.
Funding Sources: Fundraisers, membership donations.
Annual Budget: $5,000
Employees: 1

Save our South Bay Wetlands

784 Danforth Terrace
Sunnyvale, CA 94087-1225
(408) 720-1955
Contact: Tom Espersen
Description: A sub-group of the Citizen's Committee to Complete the Refuge, the SOSBW is very active in the preservation of wetlands in Santa Clara, San Jose, Sunnyvale, Alviso, and Milpitas. In addition to watching over those lands already earmarked for inclusion in the refuge, SOSBW also monitors activities on private property. Volunteers equipped with cameras and binoculars look for signs of disturbance on their assigned wetland. Whether the destructive activities are the result of illegal or misguided motives, the SOSBW puts "heat and light" on the situation. Working closely with the press, city government, the District Attorney's Office and local politicians, SOSBW allows no wetland to disappear unnoticed.
Volunteer Opportunities: This all-volunteer group encourages people to join and become watchdogs for the Bay.
Funding Sources: Citizen's Committtee to Complete the Refuge, grants.
Annual Budget: $1,500

Save San Francisco Bay Association

P.O. Box 925
Berkeley, CA 94701
(415) 452-9261
Description: This non-profit, tax-exempt organization is dedicated to the protection and enhancement of San Francisco Bay's many natural and public values. Current priorities include: protecting wetlands, improving water quality, providing shoreline parks and trails, preventing additional fresh water diversion from the estuary. Membership over 23,000. SSFBA was established in 1961, and has more than 23,000 members.
Resources: *The Baywatcher*, quarterly newsletter; free informational fact sheets and materials; slide shows with scripts that can be borrowed,and several films.
Volunteer Opportunities: Occasional "baywatching" projects and other program-related matters, office work including typing, filing, data entry, envelope stuffing, etc.
Funding Sources: Membership dues and contributions, some foundation grants for special projects.
Employees: 6

Save the Dolphins Project

300 Broadway, Suite 28
San Francisco, CA 94133
(415) 788-3666
Contact: Todd Steiner, Director
Description: Save the Dolphins Project is working to stop the slaughter of more than one hundred thousand dolphins a year by the international tuna industry. They believe tuna can be caught without killing dolphins. With several of the dolphin populations already seriously depleted, they invite others to join with them to end the unnecessary killing (80% of all tuna is caught without killing dolphins).
Resources: *Dolphin Alert*, quarterly; *Earth Island Journal* article, "The Tragedy Continues: The Killing of Dolphins by the Tuna Industry" (position paper).
Volunteer Opportunities: General clerical and special talents.
Funding Sources: Memberships, donations, private foundations.
Employees: 4

Seeds of Peace

2440 Sixteenth Street, Box 241
San Francisco, CA 94103
(415) 420-1799
Description: Seeds of Peace is a collective dedicated to living together peacefully through the process of consensus, and in assisting the peace, social justice, and environmental communities in the endeavor to make the world a more peaceful place to live. Some of Seed's involvments include: the 300-mile Florida Peace Pilgrimage to protest the Trident 2 D-5 Missile test, the Brotherhood/Sisterhood March for civil rights, the Via Crucis March in Mas-

sachusetts, the 700-mile Michigan Peace March, the New England Walk for Nuclear Disarmament, assisting the American Peace Test in organizing the Reclaim of the Test Site Action at the Nevada Test Site. They have participated in Habitat for Humanity campaign, and the International Peace Walk from Washington, D.C. to San Francisco along with 100 Americans and 220 Soviets. They have provided assistance to the Housing Now! New Exodus Walk from N.Y. to D.C., which included 600 homeless activists, and helped the Red Cross in feeding the victims of the Loma Prieta earthquake.
Resources: Extensive movement network, newsletter : *Across The Lines,* ongoing referrals, and much more.
Volunteer Opportunities: Ongoing opportunities to plug into various peace, social justice, and environmental actions and campaigns.
Funding Sources: Donations , grants, collective funding.

Sempervirens Fund

Drawer BE
Los Altos, CA 94023
(415) 968-4509
Contact: Kristie Knutson
Description: The Sempervirens Fund is a non-profit conservation organization established to acquire park land in the Santa Cruz Mountains and secure them for state ownership. The Fund is interested in land conservancy and completing Big Basin Redwoods, Castle Rock Park in the Santa Cruz Mountains. The Fund sponsors much parkland restoration. It is currently in the process of purchasing a thirty-nine acre parcel of the upper San Lorenzo watershed, and is carrying out numerous volunteer restoration projects here and in the Waddell watershed.
Resources: Membership newsletter, *Mountain Echo.* Brochures, maps, and Redwood videotape available at the office.
Volunteer Opportunities: Many volunteer oppurtunities, especially in restoration programs that meet every other Saturday.
Funding Sources: Public, foundations, donations, reimbursed by California for parks.
Annual Budget: $1,500,000
Employees: 11

Shared Living Resource Center

2375 Shattuck Ave.
Berkeley, CA 94704
(415) 548-6608
Contact: Ken Norwood, Executive Director
Description: The Shared Living Resource Center is committed to creating supportive, shared living communities that integrate quality-designed housing with cooperative self-management, ecological design, and affordability. The creation of successful cooperative shared living communities requires SLRC to focus on both the innovative design of housing and the social organization of community. Through a participatory process, the shared living designs unfold with the essential qualities of vitality and timelessness. User participation in the design and development of the community, along with resident control and management of the completed project, are prime objectives throughout the entire process.
Resources: Bimonthly newsletter, *Shared Living Community;* library with books on housing, shared living design and site planning, personal and global transformation, the environment, alternative economics, and more. Books and publications on sale. Architectural design services for the creation of shared living communities and households.
Volunteer Opportunities: Extensive opportunities, particularly for those with strong writing, drawing, architectural, and/or computer skills. SLRC also offers an Apprentice/Internship program for one to two individuals per year.
Funding Sources: Grants, consultations, architectural design services, workshops, seminars.
Annual Budget: $200,000
Employees: 5

Sierra Club - Headquarters

730 Polk Street
San Francisco, CA 94109
(415) 776-2211
Contact: Public Affairs
Description: This is the main administrative office of the Sierra Club. See other Sierra Club entries in this listing for regional chapters. The Sierra Club is active throughout the U.S. in nearly all areas of environmental concern. Their major activity categories include studying and influencing public policy, grassroots organizing, publications, and outdoor activities.
Resources: Wide range of pubications. Write for listings.
Volunteer Opportunities: General clerical.
Funding Sources: Memberships ($33 annual), donations, sales.
Annual Budget: $35,000,000
Employees: 250

Sierra Club - Loma Prieta Chapter

2448 Watson Court
Palo Alto, CA 94303
(415) 494-9901
Description: The Chapter for the Sierra Club members of San Benito, Santa Clara and San Mateo Counties.

Sierra Club - Redwood Chapter

Box 466
Santa Rosa, CA 95402
(707) 544-7651
Contact: Shirley McGovern
Description: The Redwood Chapter is concerned with a wide range of environmental issues. It works with the national Sierra Club to protect and preserve the environment. The chapter supports clean air and clean soil, and

the preservation of the wilderness. The chapter leads hikes, backpacking trips etc. Region includes those counties north of Marin to the Oregon border and east to Solano County. Office is located at 825 College Ave., Santa Rosa.
Resources: Newsletter, *Redwood Needles* .
Volunteer Opportunities: General clerical, outdoor leadership, call for details.
Funding Sources: Memberships and donations.

Sierra Club - San Francisco Bay Chapter

6014 College Ave
Oakland, CA 94618
(415) 653-6127
Contact: Jackie McCort
Description: San Francisco Bay Chapter is an organization aiming to explore and protect our natural heritage. They use continuing efforts to protect wildlands and to improve the quality of the urban environment. Talks and films are periodically scheduled. A program of local outings is maintained. A variety of conservation campaigns are constantly underway.
Resources: Scheduled talks and films, *The Yodeler*, newsletter-monthly, *Schedule* activities quarterly, *Sierra* magazine-bimonthly.
Volunteer Opportunities: Clerical, writing and research, lobbyists.
Funding Sources: Memberships: $33 regular, $15 student.
Employees: 8

Sierra Club - Santa Cruz Regional Group of the Ventana Chapt

P.O. Box 604
Santa Cruz, CA 95061
(408) 426-4453
Contact: John Stonum
Description: The group's goal is to restore the quality of the natural environment and to maintain the integrity of ecosystems. Educating the public to understand and support these objectives is a basic part of the Club's program. All are invited to particiate in its activities, which include programs to study, explore, and enjoy wildlands.
Resources: Format books and handbooks, *The Ventana*, chapter newsletter.
Volunteer Opportunities: Organization is completely volunteer.
Funding Sources: Memberships, donations, publications, and sale of merchandise.

Sierra Club - Ventana Chapter

P.O. Box 5667
Carmel-by-the-Sea, CA 93921
(408) 624-8032
Contact: Mary Dainton
Description: The purpose of the Sierra Club is to explore, enjoy, preserve and protect the nation's forests, waters, wildlife and wilderness. To accomplish these objectives they have field trips, lobby for state, local and national bills that fit their goals, and have speakers at their meetings who can keep them informed on problems in the environment. An important facet of Club activities are the outings (hikes, backpack trips, bicycle trips, ski trips, etc.) that are led by the Club, one to three times every week.
Resources: Monthly newsletter: *The Ventana* and *Sierra* Magazine.
Funding Sources: Membership, donations, fundraising.

Sierra Club Legal Defense Fund, Inc.

2044 Fillmore Street
San Francisco, CA 94115
(415) 567-6100
Contact: Joanna Chestnut
Description: The Sierra Club Legal Defense Fund is a public interest law firm which provides legal services to environmental and conservation organizations or to concerned citizens in environmental matters of a broad public nature.
Resources: Annual report and quarterly newsletter, *In Brief*,
Funding Sources: Contributions.

Silicon Valley Toxics Coalition

760 North First Street
San Jose, CA 95112
(408) 286-0315
Contact: Ted Smith
Description: The Silicon Valley Toxics Coalition (SVTC) has become a recognized leader both locally and nationally because of its work is on issues ranging from groundwater clean up, to pollution and accident prevention. SVTC has documented hazards and promoted solutions that reduce problems at the source. By working with neighborhood groups, elected officials, and others, SVTC has played a key role in developing solutions. Current projects include: 1) monitoring of IBM and Moffitt field superfund sites; 2) continuing efforts to shift industry and government to priorities of pollution prevention by encouraging source reduction; 3) promoting rapid phase out of chloroflurocarbons (CFC's).
Resources: Newsletter, *Silicon Valley Toxic News* ($25 for one year).
Volunteer Opportunities: Filing/office work, phoning, computer input, fundraising events, artwork, research, newsletter writing and editing, computer skills, public speaking.
Funding Sources: Membership dues, foundation grants, events, and individual donors.
Annual Budget: $125,000
Employees: 3

Sonoma County Tomorrow

P.O. Box 5712
Santa Rosa, CA 95402
(707) 538-2589
Contact: Warren Watkins
Description: Sonoma County Tomorrow is a non-profit, public interest, local coalition of residents dedicated to responsible air, energy, land, and water use. They are active in public testimony and letter writing, providing seed money for conferences, initiating legal action when necessary, watchdogging environmental laws and regulations and supporting like-minded groups.
Funding Sources: Membership, fundraisers, donations.
Annual Budget: $3,000

Sonoma Land Trust

534 B. Street
Santa Rosa, CA 95401
(707) 526-6930
Contact: Dan Schurman, Executive Director
Description: The Land Trust is a non-profit organization founded by local residents to preserve significant agricultural, ecological, and scenic sites in Sonoma County. To secure these resource lands, the Trust works directly with landowners, offering income and estate tax benefits in exchange for gifts of land and conservation easements. It also strives to become more active in the market place by matching conservation buyers and sellers and by developing an emergency fund for purchasing threatened lands.
Resources: Brochure, newsletter.
Volunteer Opportunities: Monitoring easements.
Funding Sources: Donations, memberships, grants.
Annual Budget: $100,000
Employees: 3

SONOMoreAtomics

540 Pacific Ave
Santa Rosa, CA 95404
(707) 526-7220
Contact: Larry Harper
Description: SONOMore Atomics is committed to a nuclear-free future. They do outreach and education in their community, and organize and participate in non-violent direct action. Emphasis is on direct action and educational work and making the corporate connections to nuclear power and weapons.
Resources: Newsletter, *Peace Press*, research library and videotape of Rancho Seco action.
Volunteer Opportunities: Lecturing, meetings.
Funding Sources: Donations.

Tri-City Ecology Center

P.O. Box 674
Fremont, CA 94537
(415) 793-6222
Fax: (415) 793-6222
Contact: Donna Olsen
Description: The Tri-City Ecology Center is an all-purpose ecology center servicing Fremont, Newark, and Union City. Actually, it does not have a center or an office. Rather, Tri-City is an all-volunteer affiliation of 800 local residents concerned about recycling, wetlands preservation and hillside protection, and environmental education. Tri-City activists have donated thousands of environmental books to local libraries, initiated water conservation efforts, advised curbside recycling organizations, pressed for better sewage treatment facilites, and initiated an annual program for the disposal of household toxics.
Resources: Monthly newsletter, *Eco-Logic*.
Funding Sources: Memberships.

Trust For Public Land

116 New Montgomery Street, 4th floor
San Francisco, CA 94105
(415) 495-4014
Fax: (415) 495-4103
Contact: Susan Ives, Director of Public Affairs
Description: The Trust for Public Land is a national , non-profit land conservation organization working with public agencies and citizens' groups to acquire land with scenic, recreational, historical or cultural value for public use in urban, rural and wilderness areas in 37 states and Canada. It comprises a network of real estate, finance and law experts. Commmitted to sharing knowledge of non-profit land acquisition techniques and pioneering new methods of land conservation and enviromentally sound land use, the Trust provides training for and collaborates with, community land trusts— nonprofit community conservation groups that preserve cherished local landscapes
Resources: Tri-annual magazine, *Land and People*.
Volunteer Opportunities: No formal program, as arranged with National or Western Region office.
Funding Sources: 62% completed projects, 15% grants and contributions, 23% other (investment lands, fees, interest).
Annual Budget: $18,854,000
Employees: 178

Tuolumne River Preservation Trust

6529 Telegraph Ave.
Oakland, CA 94609
(415) 652-1600
Contact: Joe Daly
Description: The trust is involved in protecting the Tuolumne River and its tributaries. Its major current project is monitoring Clavey River Dam Project.

Urban Creeks Council

2530 San Pablo Avenue
Berkeley, CA 94702
(415) 540-6669

Contact: Bruce Van Allen, Executive Director

Description: The Urban Creeks Council is a tax-exempt organization formed to encourage the preservation, protection, restoration, and maintenance of natural streams in urban and developing areas. The Council's goal is to educate the general public, along with environmental and engineering professionals, on the aesthetic, recreational, and ecological values of urban streams. The UCC encourages alternative flood control project design, biological streambank stabilization methods, and riparian land use planning, to decrease damage from flooding without sacrificing environmental features. Central to the UCC's approach is community involvement and participatory planning. Members help organize creek clubs, citizen's groups, and neighborhoods who wish to carry out clean-ups and restoration projects. The three main functions of UCC are to educate the public, organize support for projects, and manage actual restoration.

Resources: Video (27 min) - *Appreciating Our Urban Streams* (covering clean-up and biotechnical aspects of stream restoration; quarterly, *Creek Currents;* manual on stream restoration, Urban Creek Database ; speakers bureau - made up of the board members willing to give talks on urban creek subjects.

Volunteer Opportunities: Participation in creek clean-up and restoration projects. Office work and newsletter work.

Funding Sources: Grants, donations, newsletter subscriptions, and service contracts.

Annual Budget: $150,000

Employees: 1

Urban Ecology

P. O. Box 10144
Berkeley, CA 94709
(415) 549-1724

Contact: Richard Register

Description: Urban Ecology aims to help build and rebuild cities in a way that makes them environmentally healthy and culturally vital. Ecological city building involves principals of access by proximity, mixed-use zoning, energy conservation, appropriate technology, and more. What this group is most interested in is the interconnection of all of these areas of interest. Projects include: bicycle and pedestrian oriented slow streets, public transit promotions, city energy planning, solar greenhouse construction, creek restoration, urban orchards, integral neighborhoods, joint action with other groups, lectures, classes, conferences.

Resources: Lectures, classes, conferences, quarterly newsletter,*Urban Ecologist* (free with membership), slide shows, posters, papers, books.

Volunteer Opportunities: Office and organization work, field projects: trees, creeks, planning issues.

Funding Sources: Membership dues ($25 , donations, conferences, slide shows, posters, papers.

Annual Budget: $50,000

Urban Habitat Program

300 Broadway, Suite 28
San Francisco, CA 94133
(415) 649-7877
Fax: (415) 788-7324

Contact: Ellie Goodwin, Victor Lewis

Description: The Urban Habitat Program was established in October of 1989 by Carl Anthony, Karl Linn, and David Brower as a project of Earth Island Institute. The purpose of the Urban Habitat Program is to encourage, nurture, and develop multi-cultural, urban environmental leadership in the Bay Area. Events for 1991: first Bay Area Eco-Rap Festival, May 1991; Bay Area people of color environmental summit; bi-monthly forums on social-environmental justice.

Resources: Newsletter, *Race, Poverty and the Environment;* speakers bureau; consulting service on cross-cultural sensitivity and alliance-building for environmental organizations; monthly career exchange letter, and career hotline for people of color seeking employment in environmental field.

Volunteer Opportunities: Writers and editorial assistance for every phase of production and distribution of newsletter; people to help produce public events, environmental outreach into disenfrancised communities; and fundraising.

Funding Sources: San Francisco Foundation. Other funding sources pending.

Annual Budget: $79,000

Employees: 2

West County Toxics Coalition

1019 MacDonald Ave.
Richmond, CA 94801
(415) 232-3427

Contact: Henry Clark

Description: The West County Toxic Coalition is a community organization to help people become aware of toxics and environmental hazards. It monitors environmental issues in the Richmond area, and the Bay. The Coalition promotes better assessment of health risks from local industry, monitors pollution and provides solutions, and is developing an early warning system for accidents. The coalition produces reports on their efforts to control the health and environmental risks of this industrialized area.

Volunteer Opportunities: Many.

Funding Sources: Newsletter.

Annual Budget: $85,000

Employees: 1

Whole Access

517 Lincoln Avenue
Redwood City, CA 94061
(415) 363-2647

Contact: Phyllis Cangemi, Executive Director

Description: Whole Access is a Bay-Area based non-profit organization which works to increase opportunities for participation in outdoor recreation and in the nature experience for people of all abilities; including persons with mobility related, sensory and communicaton, cognitive and learning and emotional disabilities. Whole Access provides consulting and educational services to the planners and providers of park and nature programs and facilities to help them to create and maintain facilities and programs - including environmental education and nature programs - which are accessible to all. Whole Access also offers (some) public education and speakers' programs.

Resources: Two newsletters, *Park Access Digest* for professionals, and the *Whole Access Quarterly* for members.

Volunteer Opportunities: Office work, resource library, data entry, mailings, computer work, newsletter, trainings, photo/video work, architectural skills for access.

Funding Sources: Memberships, grants, fees (training and consulting).

Annual Budget: $100,000

Employees: 2

Wild in the City

522 Albemarle
El Cerrito, CA 94530
(415) 525-1082

Contact: Mancy Morita, Director

Description: City people often get the mistaken impression that the world was created by and for humankind. Wild in the City's purpose is to help people understand the connection between the city environment and the natural one. The focus is on San Francisco. The goal is to help foster more conscientious city lifestyles and to teach an appreciation for the wild that is tapped to fuel the city. The Wild in the City poster compares San Francisco's present-day grid of streets and buildings with it's original (pre-1750) natural pattern of creeks, marshes, flora, fauna, and Ohlone village sites. Wild in the City is sponsored by Planet Drum Foundation.

Resources: Wild in the City poster (new edition pending).

Volunteer Opportunities: Many.

Funding Sources: Grant, poster sales, donations.

Annual Budget: $5,000

Employees: 1

Wilderness Society

116 New Montgomery Street, #526
San Francisco, CA 94105
(415) 541-9144

Description: The Wilderness Society is a non-profit membership organization devoted to preserving wilderness and wildlife, protecting America's prime forests, parks, rivers, deserts, and shorelands, and fostering an American conservation ethic. Current projects include: protecting Yosemite National Park, the California desert, an ancient forests.

Resources: The Wilderness Society publishes a quarterly magazine and a variety of other publications.

Volunteer Opportunities: The Society has an intern program and welcomes volunteers to help in the office.

Funding Sources: Memberships.

Employees: 4

Zero Population Growth - San Jose Chapter

P.O. Box 9393
San Jose, CA 95157
(408) 379-1213

Contact: Zara Van Wichen

Description: Zero Population Growth is a nonprofit citizens' organization with headquarters in Washington, D.C. The San Jose chapter covers both Santa Clara and Santa Cruz counties and is run by volunteers. ZPG members are deeply concerned over the growth of world population, a major cause of exacerbating hunger, poverty, political unrest, international strife, deforestation, and desertification. They urge that foreign aid be accompanied by family planning assistance where needed. The group advocates adoption of a national policy stating a goal of population stabilization through: voluntary limitation of family size encouraged by wide availability of reliable and safe methods of birth control, ensuring equal rights for women, discouraging teenage pregnancy, stronger control of illegal immigration, expansion of factual population/ resources/ environment education via the schools and the media, and curbing waste and extravagance. The San Jose chapter's major activities include public education through booths at environmental and community events, supplying informational materials to libraries, colleges and individuals; promotion of legislation favorable to ZPG's goals; media publicity and membership development. Meetings are held as required to plan for special projects. Call for information regarding dates and locations.

Resources: Monthly newsletter, *Reporter*, quarterly chapter bulletin.

Volunteer Opportunities: Call for information.

Funding Sources: Portion of dues from memberships in national organization (which start at $10) and donations.

Annual Budget: $2,000

Section Five

PUBLIC AGENCIES

The following list of public agencies is arranged by level of government oversight — local, county, regional, state, and national. This heirarchy tends to reflect the levels of responsiveness to public input or requests. An individual is usually best able to get answers from their local government, fastest emergency response from their own fire department, and wield the most political clout over local politicians.

County and regional government agencies act most effectively as the record keepers, report makers and the trend watchers of an area. State government is primarily regulatory, and therefore concerned with minute details and procedural requirements; it often takes much perserverance to get what you want from state agencies, and usually you must know ahead of time exactly what you are looking for. National government departments are primarily useful for their publications and their information hotlines.

There is too often a wide difference between what a public agency's stated mission is and the way it actually operates or what it is actually able to accomplish. As with the organizations, the *Bay Area Green Pages* has given most of these agencies free rein to describe their own purposes and resources. But some of the agencies charged with protecting the environment have at times been key players in many of the damaging processes that contribute to its degradation.

The reader should therefore question whether the agency actually carries out its mission, and know that many aspects of our bureaucracy are ineffective, outdated, and staffed by people with varying opinions as to the extent to which sacrifices should be made for the environment. (This is not to say there aren't some excellent services provided by some government agencies as well.)

Articles in The Green Digest section will give you a different perspective on many of these public agencies and their accomplishments.

Local

Not all of the environmentally-related offices for the many cities in the Bay Area have been included in this list. Those listed here were chosen because they have become models for similar programs across the country.

City governments are becoming increasingly sensitive to environmental issues, and responding with real programs and departments to deal with them. No city election goes by without candidates stating where they stand on local environmental problems and what they are going to do about them. The *Bay Area Green Pages* looks forward to hearing from more offices, departments or divisions that are addressing the issues, for inclusion in our next edition.

In general the cities listed below have established effective environmental units that deal with recycling issues, city planning or environmental health. If your city is not listed, it does not mean there is no corresponding department or agency. Call to find out. Also check the county office listings — jurisdictions and responsibilities are often shared between city and county governments.

Residents of any city can call their local fire departments for assistance with suspected toxic dumps or hazardous waste emergencies. Call the regional air pollution control watchdog, the Bay Area Air Quality Management District (800) 334-ODOR for air pollution complaints.

Berkeley

City of Berkeley—Energy Office

2180 Milvia Street
Berkeley, CA 94704
(415) 644-6309

Contact: Elaine Eisenstadt

Description: The energy office provides a wide variety of services to the Berkeley community to advance energy conservation and management. The services include program planning, legislation, and education. The office works with the local energy commission to prepare cost-saving projects for the city primarily through retrofitting of city buildings. Other programs for residents include free weatherization, and installation of conservation measures for low and moderate income dwellings (Call 644-8544). An energy crisis-intervention program assists low-income residents in paying P. G. & E. bills. Available for the school district use is an educational curriculum, supporting materials, and a videotape on energy .

Resources: *Berkeley Energy Resource Directory* a free guide to local resources for homeowners and businesses who wish to maximize their residential or commercial energy efficiency. *Energy Efficient Buildings in Berkeley,* - geared toward developers and renovators. An on-site library of useful tips on energy, many of which are available in pamphlet form.

Employees: 8

City of Berkeley—Environmental Health

2180 Milvia Street
Berkeley, CA 94704
(415) 644-6510

Description: Most cities in the Bay Area rely solely on their county government for environmental health services. Berkeley is rather unique in that it has its own division. The Environmental Health Office oversees a number of areas, dealing with issues of food facility health inspection, hazardous waste, noise pollution and vector control. This office handles issues on hazardous materials, including regulations concerning inspection and overseeing of underground tank installation and removal; waste generator storage, handling;, and transportation of hazardous waste materials. They have a waste reduction program that provides information on resources and advice for product substitution or elimination. The office also maintains spill response teams for emergencies. This office oversees disclosure compliance for Community-Right-To-Know requirements on hazardous materials. If you have a noise complaint or nuisance complaint (garbage pile-ups, etc.), a pest problem, or a Community-Right-To-Know request, these are the people to call. Yet, if you suspect a toxic spill, call the fire department first. They have the fastest response time, and they will call the Environmental Health Unit if the spill response team services are needed.

Resources: Pamphlets on hazardous waste, tank removal, and waste reduction are available for businesses.

Employees: 14

City of Berkeley—Recycling Office

2180 Milvia Street
Berkeley, CA 94704
(415) 664-6858

Description: The City of Berkeley currently contracts with three organizations to provide recycling services to Berkeley residents. Citizens may recycle glass, aluminum, tin and bundled newspaper in the citywide residential curbside recyclable collection program. A Buyback Center where citizens are paid for materials that they recycle, and a flea market for salvaged goods are located at the City of Berkeley Solid Waste Management Center.

- The City Refuse Division provides a commercial recycling service citywide, a pilot yard debris collection program, and a holiday tree recycling program. Recycling Education presentations are also available.

• The City of Berkeley Motor Oil Recycling Depot is scheduled to open at the Solid Waste Management Center by February, 1991.

Resources: *Where to Recycle in Berkeley* (Brochure), Pre-Cycling Poster, Environmental shopping product list, and a Commercial recycling brochure.

Employees: 9

City of Berkeley—Refuse Division

Recycling Operations
1201 Second Street
Berkeley, CA 94710
(415) 644-6465

Contact: Karen Hemphill

Description: The City of Berkeley Refuse department oversees regular garbage pick-up and disposal, and commercial recycling. Its Recycling Operations provides recycling carts that collect cans, bottles, newspaper, cardboard, and white office paper on-site at businesses. A Yard Debris pilot project has been serving approximately ten percent of the city, and may be expanded in the year 1991. This debris, taken to a private facility, ends up as mulch (70%) and boiler fuel (30%). Included in the commercial recycling operations are services to city and school facilities. By mid-1991, the refuse division will be opening a used motor oil depot at the Second Street site. Scrap metals, cardboard, and glass are routinely salvaged off of the transfer station's tipping floor, and approximately two percent of the 85,000 tons disposed of annually is recovered.

Resources: Newsletter; *Waste Watch* and a Commerical Recycling brochure.

El Cerrito

City of El Cerrito Recycling Program

10890 San Pablo Avenue
El Cerrito, CA 94530
(415) 527-6077

Contact: Susan Kattchee, Director of Community Services, City of El Cerrito

Description: The City of El Cerrito Recycling Program is one of the oldest city-run recycling programs in the nation, and serves as a model program for other cities. The program maintains a weekly curbside pick-up service of recyclables for El Cerrito residents. It operates a drop-off center open seven days a week: 8:00 am to 7:00 pm weekdays and 9:00 am to 5:00 pm weekends. Recycling slide show and tour of facilities for children K-6 grades available.

Funding Sources: Sale of material, monthly surcharge per household, and grants.

Palo Alto

City of Palo Alto Materials Recovery Program

P.O. Box 10250
Palo Alto, CA 94303
(415) 329-2495

Contact: Janet Foreman, Public Works Special Project Coordinator.

Description: The Program's goal is to reduce solid waste by 25% through source separation of recyclable materials. The program operates a curbside collection of these materials servicing single family dwellings. A municipal composting program of yard waste is in operation at the City landfill. Other projects include a wood recovery area, white paper and computer paper recycling program, and commericial/industrial recycling facilitation. The Recycling Center at 2380 East Embarcadero takes compostables, appliances, refrigerators, large scrap, and car batteries.

Resources: Pamphlets on curbside service, and multi-unit white paper,commercial recycling.

Volunteer Opportunities: Clerical, phone answering, data collection, report writing.

San Francisco

San Francisco Hazardous Waste Program

City Hall, Room 271
San Francisco, CA 94102
(415) 554-6194

Contact: Bill Quan

Description: The Office of the Chief Administrative Officer (CAO) is responsible for developing a city-wide plan to manage and reduce the usage of hazardous materials and generation of hazardous waste. The plan focuses on the implementation of active waste–minimization programs, involving education and, especially, technical assistance to reduce the potential problems associated with hazardous materials and wastes. In addition, the CAO office oversees the operation of the City's Household Hazardous Waste Collection Facility. The purpose of this program is to encourage reduction in the use of household toxics and their proper management so as to minimize the amounts of toxics in the City's refuse that goes to the landfills. As of 1990, some of the projects that the CAO's office includes are: a waste reduction recycling checklist for gasoline service stations; a grants program for the promotion of waste reduction techniques applicable to small business; a citywide oil collection and recycling program; and one hundred waste audits and waste–reduction workshops for San Francisco's small businesses.

Resources: Newsletter, *On the Safe Side* (for small businesses), Household Hazardous Waste Hotline: (415) 554-4333, and *Directory of Bay Area Hazardous Waste Management and Recycling Firms* (for small businesses).
Volunteer Opportunities: Many.
Funding Sources: Garbage bills.
Annual Budget: $625,000
Employees: 3

San Francisco Recycling Program
City Hall, Room #271
San Francisco, CA 94102
(415) 554-6193
Contact: Amy Perlmuttter, Recycling Coordinator
Description: The San Francisco Recycling Program promotes recycling through coordination and support of existing programs; through planning, research, design, and facilitation of new programs in all sectors; and through an intensive public education campaign. In April 1989 San Francisco began a commingled curbside recyling pickup service—the first of its kind in the State of California. Weekly, residents voluntarily recycle commingled glass, aluminum, tin, and bimetal food and beverage containers in a plastic bin provided to each home. Commingled newspaper, junk mail, cardboard, and magazines are bundled or placed in paper bags. Serving as an example for local corporations, all city departments participate in office paper recycling. Where feasible, city employees also set aside bottles, cans, and newspaper for collection and recycling by the Conservation Corps and Association for Retarded Citizens. Also under the Recycling Program's leadership, representatives of major purchasing departments have come together to push for increased government purchase of reusable, recycled, and recyclable products. In the private sector, the Recycling Program provides businesses with an office paper recycling guide, desk top sorters and waste reduction tips. San Francisco's Office Paper Recycling Coordinator holds "how-to" seminars and provides free consulting services. Bars and restaurants are also encouraged to recycle glass and cardboard and reduce waste. The Recycling Program distributes "Starter Kits," guides, fact sheets and newsletters to residents and businesses. A recycling hotline, 554-6193, operates in City Hall and receives over 300 calls per week. The Program's Education Coordinator has developed recycling curricula, conducts classroom demonstrations, and takes children on field trips to the garbage transfer station and recycling centers.
Resources: Environmental Shopping Guide, Home Recycling Starter Kit (guide, list of recyclers, booklet on waste reduction, etc.), Office Recycling Starter Kit-free to SF residents, $5.00 elsewhere, and K-5 Curriculum - free to SF residents, $10.00 elsewhere.
Volunteer Opportunities: Limited.
Funding Sources: Garbage bills.
Annual Budget: $900,000
Employees: 7

San Jose

Center for the Development of Recycling
Geography & Environmental Studies, San Jose State University (SJSU)
One Washington Square, Rm BT550
San Jose, CA 95192-0204
(408) 924-5453
Contact: John Thomson
Description: The Center researches and develops information on waste management, plans activities, and makes such information available to government, industry, non-profit groups, and individual citizens. The Center is currently documenting Bay Area recycling strategies, developing fact sheets on Bay Area recycling organizations and facilities, compiling market development research on recycled materials, and developing internships in solid waste management.
Resources: SJSU Environmental Resource Center library, computer network links to State of California Department of Conservation and City of San Jose Office of Environmental Management databases, and to both the California State University and the University of California library systems. It provides listings of other solid waste management organizations - governmental, non-profit, and trade - and various collections of reports on recycling, reuse, and source reduction.
Volunteer Opportunities: Paid and unpaid internships with government, non-profit, and for-profit organizations.
Funding Sources: State of California, Department of Conservation; City of San Jose, Office of Environmental Management; the San Jose State University Foundation; and industry sources.
Annual Budget: $100,000
Employees: 5

San Jose Office of Environmental Management
777 N. First Street, Suite 450
San Jose, CA 95112
(408) 277-5533
Fax: (408) 277-3606
Contact: Michelle Yesney, Director
Description: The mission of the office is to promote and encourage the efficient, non-wasteful use of natural resources. The office is divided into four major programs: Energy Management, Water Resources and Conservation, Integrated Waste Management, and Environmental Protection. The Energy Management "sustainable city" program aims to reduce citywide energy use 10% by the year 2000, including offering technical assistance to the building-design and construction community for energy-efficient design. The In-House Energy Management program institutes energy-saving measures within city infras-

tructure. The Water Resources and Conservation Program offers a technical assistance program, financial incentives program, drought respons, long-term drought planning, landscape audits, residential audits, waste minimization, and non-point source projects. The Integrated Waste Management residential curbside recycling provides; pilot programs for yard waste, mixed paper, multi-unit residential recycling; commercial recycling program; household hazardous waste collection programs including curbside pick-up of waste oil; source reduction; and management of garbage/rubbish collection. The Environmental Protection program oversees the City ordinance limiting CFC's and coordinates City response to toxics, air, and water quality issues.

Volunteer Opportunities: Volunteer program for community events, information assembly, and dissemination (contact Arleen Arimura).

Funding Sources: City funds, state and federal grants

Annual Budget: $9,100,000

Employees: 60

San Jose Office of Environmental Management—Energy Program

777 N. 1st St., Suite 540
San Jose, CA 95110
(408) 277-5533
Fax: (408) 277-3606

Contact: Mary Tucker

Description: The Office of Environmental Management aims to develop more efficient use of energy in municipal government and the community to achieve cost savings and reduce future resource uncertainty. Energy Management Programs include in-house energy management, innovative design and an energy analysis service for new commercial buildings, solar access and solar building design assistance, housing energy rebate service, residential information and referral, public school programs, lighting and appliance information, research, development and monitoring.

Resources: Technical library, energy clearinghouse and referral service, technical assistance to builders, architects on energy efficiency, and solar design.

Volunteer Opportunities: Internships available.

Funding Sources: City funding and federal grants.

Employees: 7

Santa Clara

City of Santa Clara—Street Department

1500 Warburton Ave
Santa Clara, CA 95050
(408) 984-5188
Fax: (408) 241-8291

Contact: Annie Horton, Recycling Coordinator

Description: The City of Santa Clara Street Department coordinates Household Hazardous Waste (HHW) Collection Days (Drop-off) in the Spring and Fall. Call 984-3080 for information. City of Santa Clara Street Department also provides a residential curbside recycling program (weekly, includes motor oil recycling), and commercial recycling program assistance.

Resources: Free copy to residents, *Consumer Guide to Safer Alternatives for Household Hazardous Products;* and a Recycling "Hotline" - (408) 984-5188, 8am-5pm.

Volunteer Opportunities: Recycling block leaders.

Funding Sources: Government, City general fund, and customer rate charges.

County

The following is a general description of county departments, divisions and districts with environmental responsibilities. These departments are similar from county to county, although differences do exist. Also included are some resources people might expect their county to handle, but which are actually handled by regional agencies.

Agricultural Commissioners

The county agricultural commissioner's office enforces agricultural codes set forth by the state legislature. These offices are the implementation arm for agricultural pesticide use, structural and residential pesticide use, and worker safety regulations. The county inspects produce brought in as nursery stock, checking truck shipments, bus stations and airports, and duplicates the quarantine work of border stations. The inspection of retail, wholesale, and farmer's market fruit, vegetable, and egg quality primarily emphasizes size and minimal disease standards only (the California Department of Food and Agriculture monitors pesticide residues, (415) 540-2910). These commissioners' offices monitor quarantine pests like the medfly, compile statistics on the type, value and acreage of crops in each county, and investigate health injuries due to pesticide misuse. The public is assisted with questions about pests, and pesticide misuse. The offices publish annual reports on county agricultural production, and usually have back issues that can provide perspective on changing land-use over time. Unfortunately, counties at this time do not distinguish organic from non-organic farm production in their reports.

Agricultural Commissioners (continued)

Alameda
(415) 670-5232

Contra Costa
(415) 646-5250
(415) 634-5682— East County.
This office is called the Agriculture Department.

Marin
(415) 499-6349
This office is called Weights and Measures.

Napa
(707) 253-4357
(707) 963-7146 — St. Helena Office

San Francisco
(415) 285-5010
San Francisco's short annual report covers the eleven and a half acres of agricultural land in the city.

San Mateo
(415) 363-4700

Santa Clara
(408) 299-2171 or (408) 779-0861
(408) 299-2338 — Morgan Hill Office, Insect Pest Detection Program. This number is for any inquiries regarding fruit fly traps that the county places in yards.

Santa Cruz
(408) 425-2162 — City of Santa Cruz.
(408) 761-4080 — Elsewhere in the county.

Solano
(707) 421-7465.

Sonoma
(707) 527-2371.
(707) 762-3059 — Petaluma Office.
(707) 938-5215 — Sonoma Office.

Air Pollution

Call the **Bay Area Air Quality Management District** for all suspected air pollution problems that stem from stationary sources, whether they are belching factories or overpowering smells from a bakery. Their number is **(800) 334-ODOR**. If your complaint is with a movable source like a bus or a truck, call the **States's Air Resources Board** (800) 952-5588. This number is a message machine, on which you leave your address and number. Responses can take awhile and are usually in the form of a letter. If your problem is in Santa Cruz county call the **Monterey Air Quality Management District** at **(408) 443-1135**. San Francisco has an Office Smoking Complaints line, (415) 554-2780.

Cooperative Extension

Cooperative Extension is part of the Division of Agriculture and Natural Resources of The University of California. It provides a direct, personal link between Californians and the teaching, research, and public services activities of the University. CE is a three-way partnership of the U.S. Department of Agriculture at the federal level, the University at the state level, and county government at the local level. University advisors extract useful and practical information dealing with the farm, the garden, or home from their ongoing research. Specific advice, given by staff with specialized knowledge, can be obtained on a wide range of subjects including nursery supply, organic gardening, nutrition, farm advising, marine biology or home economics. Educational programs are conducted through individual consultations, publications, seminars, field days and mass media. By calling your local Cooperative Extension number you can get a brochure on what subjects are commonly covered. Some branches have a Teletips number where you choose a prerecorded tape on a particular topic. You will find the touchtone dialing code for this electronic help system in their brochure.

Alameda
(415) 670-5200
(415) 763-9061 — Teletips.

Contra Costa
(415) 646-6540 — Pleasant Hill Office.
(415) 634-3012 — Brentwood Office, limited hours.
(415) 763-9061 — Teletips.

Marin
(415) 499-6352
(415) 499-8142 — Teletips.

Napa
(707) 253-4221

San Francisco
(415) 871-7559

San Mateo
(415) 726-9059

Santa Clara
(408) 299-2635
(408) 292-9054 — Teletips.

Cooperative Extension (continued)

Santa Cruz
(408) 425-2591 — Santa Cruz Office.
(408) 761-4056 — County-wide Office.

Solano
(707) 421-6790
(707) 427-3555 — Teletips

Sonoma
(707) 527-2621
(707) 575-8341 — Teletips

District Attorney's Offices—Consumer Fraud & Environmental Protection Units

These departments enforce state and local environmental regulations. They typically prosecute cases that have originated with the Environmental Health Office, Agriculture Commissioner's Office, or other agencies that are in a position to report on violations of environmental codes. The units can respond to citizen complaints or tips regarding consumer fraud or environmental violations. Typical cases include toxic dumping, pesticide misuse, or streambed alteration. If your county does not have a specific Environmental Protection Unit, your complaint will probably have to go through the police department first. They can determine whether the case is criminal, and refer it to the District Attorney's Office for assignment to an attorney.

Alameda
(415) 569-9281.

Contra Costa
(415) 646-4556 — Consumer fraud.
(415) 646-4620 — Environmental protection.

Marin
(415) 499-6450 — Consumer protection only.

Napa
(707) 253-4059

San Francisco
(415) 553-1814

San Mateo
(415) 363-4656

Santa Clara
(408) 299-8477 — Environmental Crimes Hotline -24 hours.

Santa Cruz
(408) 425-2054 — Consumer affairs only.

Solano
(707) 421-6859 — Consumer fraud.
(707) 421-6860 — Environmental protection.

Sonoma
(707) 527-3161

Environmental Health Office

These offices are usually departments under the larger county Health Care (sometimes called Public Health) service agencies. They may contain offices dealing with hazardous waste management, solid waste management, vector control, water supply, restaurant health inspections, and occupational health (not all of these topics are covered in each county). If you are looking for general information on environmental health issues and procedures it is useful to contact these offices, or the **State Department of Health Services** (see State Government section).

Alameda
(415) 271-4330 — North County.
(415) 670-5275 — South County.

Contra Costa
(415) 646-2521 — East-Central County.
(415) 374-3141 — West County.
(415) 646-2286 — Occupational Health.

Marin
(415) 499-6907

Napa
(707) 253-4471

San Francisco
(415) 554-2795

San Mateo
(415) 363-4305

Santa Clara
(408) 299-6060

Santa Cruz
(408) 425-2341

Solano
(707) 421-6770

Sonoma
(707) 525-6500

Flood Control

Floods are not often a feature of urban life due to the efforts of these departments. Flood Control departments construct systems for flood control that use reservoirs, channels, pipelines, levees, pump stations, and flood storage ponding facilities. They usually oversee some natural watercourses, and coordinate community or agency restoration projects of creeks and waterways in support of flood control. The departments conduct water resource studies, monitor runoff water quality, and issue permits for work or grading in watercourse areas. In some counties this department is responsible for the operation of water supply facilities including well fields, water treatment plants, groundwater recharge facilities, and pipelines (e.g., the Livermore area's water supply).

Alameda
(415) 670-5480

Contra Costa
(415) 646-4470

Marin
(415) 499-6528

Napa
(707) 253-4351

San Francisco
No office

San Mateo
(415) 363-4100

Santa Clara
(408) 265-2600 — Santa Clara Valley Water District.

Santa Cruz
(408) 425-2860

Solano
(707) 448-6942 — Transportation Office.

Sonoma
(707) 526-5370

4-H Program

A part of University of California's Cooperative Extension Program, each county's 4-H Program provides research and educational programs for youth and adults in marine science, natural resource management, and agriculture. The programs provide experiences aimed at building self-esteem and critical thinking skills for youth from the ages of nine to nineteen.

4–H Program

Alameda
(415) 670-5210

Contra Costa
(415) 646-6543

Marin
(415) 499-6352

Napa
(707) 253-4221

San Francisco
(415) 871-7559 — Only an informal program currently, call Cooperative Extension, or the San Mateo 4-H.

San Mateo
(415) 726-9059

Santa Clara
(408) 299-2630

Santa Cruz
(408) 425-2591 — Santa Cruz.
(408) 761-4056 — From elsewhere in the county.

Solano
(707) 429-6383

Sonoma
(707) 527-2681

Hazardous Materials

These departments handle hazardous materials issues, and oversee compliance with regulations for hazardous waste disposal and storage for businesses. They monitor underground tank permitting, installation, modification and removal. They also oversee clean-up of contaminated sites by approving consultant-designed site mitigation plans, and investigate waste and asbestos haulers and removers. These are the people to call if you wish to report any strange new storage techniques of a business neighbor, a potentially unpermitted underground tank removal, or other nefarious business doings. If you suspect that you have found a toxic minidump in your area, call the nearest Fire Department. They usually have the quickest response time, and can call the County emergency response teams if they are needed.

Alameda
(415) 271-4320

Hazardous Materials (continued)

Contra Costa
(415) 646-2286
(415) 646-1112 — Hazardous Materials Emergency Response Hotline.

Marin
(415) 499-7233

Napa
(707) 253-4269

San Francisco
(415) 554-2775.
(415) 554-2780 — Toxics complaints.

San Mateo
(415) 363-4305

Santa Clara
(408) 441-1195

Santa Cruz
(408) 425-2341

Solano
(707) 421-6770

Sonoma
(707) 525-6565

Household Hazardous Materials

So far, only San Francisco has institutionalized a permanent household hazardous waste drop-off site. Other counties occasionally hold household hazardous waste drop-off days for paint, pesticides etc., and several are in the process of establishing permanent drop-off sites. Many cities also hold occasional drop-off days. You can get the dates by calling your local city government or the county's Solid Waste Management Office. For advice in a situation that may be an emergency, call you local fire department. (For instance, if you have the ballast in an old flourescent light fixture blow up, and suspect it may contain PCBs). See the article The Hassles of Household Hazardous Waste on page 13 for more information.

Alameda
(415) 271-4320

Contra Costa
(415) 646-2286 — Department of Health.

Marin County
(415) 499-626 — County Planning Department.

Napa
(707) 253-4269 — Hazardous Materials Section.

San Francisco
(415) 468-2442 — Drop-off site. Open Thursday, Friday and Saturday.

San Mateo
(415) 363-4305 — Environmental Health.

Santa Clara
(408) 441-1195 — Toxics Control.

Santa Cruz
(408) 425-2341 — Environmental Health.

Solano
(707) 421-6770 — Department of Environmental Manage ment.

Sonoma
(707) 525-6565 — Environmental Health.

Local Agency Formation Commissions (LAFCOs)

Not well-known to the general public, these commissions play a large role in county planning. Charged with the task of establishing "spheres of influence" for cities and service districts (e.g. water districts and sewer districts), the commissions can have great impact on the control of urban or suburban sprawl. Each commission contains two members of the county board of supervisors, two city council members from each city in the county, and one public member. The different types of commissioners are appointed by their peers - in the first case by the entire board of supervisors, in the second case by the county's mayors association, and in the third (the public member) by the commission itself. Some commissions have service district representatives.

The decisions made by LAFCO can increase the size, and therefore the tax base of cities, taking land out of the county General Plan jurisdiction, and into the city's. Open space-maintenance goals of the county can be usurped by city rezoning. Service district increases allowed by the commission pave the way for the installation of infrastructure, that in turn becomes the basis for new development. Some county LAFCOs have the reputation of being "sprawl-busters," as in Marin County, while others (Contra Costa, for instance) tend to the other end of the spectrum. The LAFCOs can answer questions about individual requests for annexation into, for example, a sewer district, answer questions that city planners or environmental organizations might have, or tell a potential land developer the likelihood of district services in their area in the near future. Among resources sometimes available to the public are directories of county-wide agencies, complete with meeting schedules and rosters, and "Sphere of Influence" maps.

LAFCOs

Alameda
(415) 272-6984

Marin
(415) 499-7395

Napa
(707) 253-4421

San Francisco
Does not have a LAFCO since the city's borders are also the county's borders. There are no questions regarding expan sion left to be decided.

San Mateo
(415) 344-8592

Santa Clara
(408) 299-2424

Santa Cruz
(408) 425-2694

Solano
(707) 429-6599

Sonoma
(707) 527-2577

Mosquito Abatement Districts

These agencies take calls regarding any problems with mosquitoes at your business or home. If you have a backyard pond, they will put in mosquito fish to eat the mosquitoes' larvae. If your pond is already established, the district will probably spray with non-toxic BTI - a bacteria deadly only to mosquitoes, or an insect growth regulator that keeps larvae in the juvenile state. The San Francisco Bay Area is fortunate to have enough money to implement these "bio-rational," non-toxic techniques; some abatement districts in the state still use toxic pesticides.

Alameda
(415) 783-7744

Contra Costa
(415) 685-9301

Marin
(800) 231-3236

Napa
(415) 258-6044

San Francisco
No abatement district.

San Mateo
(415) 344-8592

Santa Clara
(408) 299-2050

Santa Cruz
(408) 425-2341 — Does not have an abatement district, but will give advice.

Solano
(707) 425-5768

Sonoma
(800) 231-3236

Vector Control Districts

Mandated to protect humans from any animals or insects that cause disease, discomfort, or injury, these districts monitor, report on, and eradicate "pests." They bait rats with poison, destroy wasp nests, make referrals to beekeepers to remove bee hives, and arrange for the trapping of mammals, if a bite has occurred. These districts compile reports on Lyme disease-bearing ticks, monitor rabies incidence and flea populations, and handle problems with flies and outdoor waste storage. They emphasize prevention through education, specifically encouraging the securing of homes against unwanted visitors, and respond to the public's questions.

Alameda
(415) 667-7557

Contra Costa
(415) 646-2521 — East Central County.
(415) 374-3141 — West County.

Marin
(415) 499-7805 — Communicable disease statistics (e.g. Lyme disease).
(707) 499-6873 — Rabies questions
(707) 499-7237 — Emergency weekend number.
(707) 883-4621 — Animal Services, Marin Humane Society.

Vector Control Districts (continued)

Napa
Check with the **State Department of Heatlth** for statistics on disease. For other questions regarding wildlife call the **Agricultural Commissioner.**

San Francisco
(415) 255-3610 — Rodent control.

San Mateo
(415) 363-4305

Santa Clara
(415) 299-2050

Santa Cruz
(408) 425-2341

Solano
(707) 421-6770

Sonoma
Check with the State's **Department of Health Services** for statistics on vector-borne disease. For other questions call **Animal Regulation** (707) 527-2471.

Worksite Hazards

If you suspect asbestos, PCBs, noxious carpet-glue fumes, or any dangerous condition at your worksite contact **Cal-OSHA**. Cal-OSHA (California Occupational Safety and Health Administration) responds to complaints made by employees regarding unsafe and unhealthy conditions at work. If you are an employer and wish to partake of some free consulation services on improving the safety conditions of your company, call the San Mateo consulting office (415) 573-3864.

Alameda
(415) 568-8602

Contra Costa
(415) 676-5333

Marin
(707) 576-2388

Napa
(707) 576-2388

San Francisco
(415) 557-1677

San Mateo
(415) 573-3812

Santa Clara
(408) 277-1260

Santa Cruz
(408) 277-1260

Solano
(415) 676-5333

Sonoma
(707) 576-2388

The following agencies are a short list of other types of county departments that work for the environment. There are many Resource Conservation Districts in ou area. Two of them chose to be included in the *Bay Area Green Pages*.

Evergreen Resource Conservation District

888 North First St., Rm. 203
San Jose, CA 95112
(408) 288-5888
Contact: Robin Daniels
Description: The Evergreen R.C.D. is a public agency organized under the State of California Public Resources code. It is authorized and directed to conduct research, and advise and assist private individuals and other public agencies in the field of conservation of soils, water, woodlands, wildlife and other natural resources. It is not a rule-making agency, but it can advise individual members of the public. Certain forms of federal assistance are available only through the R.C.D. The basic tenet of the district is the treatment of each acre according to its capabilities. Among those who are assisted by the district are farmers, ranchers, landowners, schools, educators, engineering consultants, planning commissions, and others who request conservation information. Current projects include: cost-shared flood control projects on Silver and Penitencia Creeks, irrigation workshops, range management workshops, and youth board activities.
Resources: Free Conservation education materials; film library; Soil survey reports; and technical informationon ranching, farming, biology, agronomy, engineering, soil conservation, range management, forestry, and wildlife.
Volunteer Opportunities: District needs help in the following areas: office work, technical assistance, education-information, etc. training provided.
Funding Sources: County Property Tax
Annual Budget: $82,520

Santa Cruz County Planning Department—Resource Section

Planning Office, 701 Ocean St. Room 406B
Santa Cruz, CA 95060
(408) 425-2858

Contact: Ken Hart

Description: The Resource Section of the Planning Department is involved with stream restoration, timber harvest review and policy development, water quality monitoring, long-range water planning, off-shore oil strategy, solid waste management planning, agricultural preservation, and flood prediction. Additional programs for 1991 include stream restoration and fish habitat improvement programs, and the ongoing design of the county's building-materials recovery center.

Resources: Stream Care Guide and Recycling brochures for Santa Cruz residents.

Employees: 10

Santa Cruz County Regional Transportation Commission

Bicycle Program, 701 Ocean St., Room 406-B
Santa Cruz, CA 95060
(408) 425-2951
Fax: (408) 458-7139

Contact: Jack Witthaus

Description: The SCCRTC Bicycle Program encompasses planning and policy making for bicycle facilities in the Santa Cruz County region, as well as provision of technical and other information to local governments and the public on bicycle issues. The program includes a bicycle safety element comprised of information dissemination and in-class instruction components. Staff for the program support a citizen's advisory committee on bicycling which advises the Regional Transportation Commission on bicycling issues in general, and makes recommendations to local jurisdictions on bicycle-related plans and improvements. The committee also recommends approval, disapproval, or amendment of TDA Article 8 (Transportation Development Act - bicycle and pedestrian) funding claims for bicycle projects by local jurisdictions. The Bicycle Program is coordinated with the SCCRTC Traffic Management Program, primarily through the provision of bicycle commuting information for local businesses, support of the annual Bike - To - Work Day event, and administration of a bicycle parking subsidy program.

Resources: *Santa Cruz County Bikeways Map*; bicycle safety presentations; bicycle parking subsidy program. (Santa Cruz County Region only.)

Volunteer Opportunities: Citizen Advisory Committee to Transportation Commission on bicycling issues.

Funding Sources: TDA funds.

Annual Budget: $23,000

Santa Cruz County Resource Conservation District

3233 Valencia Ave., Suite B-6
Aptos, CA 95003
(408) 688-2692

Contact: Sharon Corkrean, Administrative Assistant

Description: The Santa Cruz County Resource Conservation District is responsible for a comprehensive natural resource conservation program in the county of Santa Cruz, excluding the cities of Santa Cruz and Watsonville. The District is governed by five locally- elected or appointed directors that are familiar with county resource problems and issues. Land users within District boundaries can contact the USDA Soil Conservation Service (SCS) for free soil, water, and related resource conservation assistance. Current programs include: information & education campaigns for erosion control, water conservation, soil stewardships, technical assistance to farmers, wildlife enhancement, and many others.

Resources: *Santa Cruz County Soil Survey Report*, 1980; leaflets on land and water related problems; directory of certified professionals in soil erosion and sediment control in California.

Volunteer Opportunities: College student internships; "Earth Team" volunteer program.

Funding Sources: Small percentage of county land taxes and fundraising.

Annual Budget: $45,000

Employees: 4

Regional

Association of Bay Area Governments (ABAG)

P.O. Box 2050
Oakland, CA 94604
(415) 464-7900

Contact: Jose Rodriguez, Cathryn Hilliard - Director of Public Affairs

Description: ABAG is one of nearly 534 regional planning agencies across the nation working to help solve problems in areas such as environmental quality, housing, transportation and economic development. ABAG is owned and operated by the cities and counties of the San Franacisco Bay Area. It was established by them in 1961 to protect local control, plan for the future, and promote cooperation on area-wide issues. Through its role as an association of cities and counties, ABAG has been designated by the state and federal governments as the official comprehensive planning agency for the Bay Area. One of ABAG's vital functions is to provide demographics and a forum to resolve local differences through workable compromises. The association sponsors workshops and conferences where local officials, businesses, industries, special interest groups and private citizens can discuss programs, regulations, and legislation affecting their communities. ABAG also helps local governments prepare grant applications and reviews over $2 billion annually of locally-prepared federal grant requests. ABAG also has a Training Institute with a wide variety of classes ranging from first response to hazardous materials spills to computer training. It provides insurance and financing for member governments.

Resources: Bibliography and Catalogue.

Volunteer Opportunities: In specific projects such as the Bay Trail project, the SF Estuary project, and wetlands cleanup.

Funding Sources: U.S. Environmental Protection Agency, local governments; member dues, other state and federal agencies, private foundations, and other organizations.

Annual Budget: $6,000,000

Employees: 65

Bay Area Air Quality Management District

939 Ellis Street
San Francisco, CA 94109
(415) 771-6000 x210

Contact: Ted McHugh, Information Officer

Description: This agency regulates emissions from stationary sources in the San Francisco Bay Area. Its region of coverage includes parts of Sonoma and Solano counties, and all other counties that ring the bay. Santa Cruz is under the **Monterey Air Quality Management District** jurisdiction ((408) 443-1135). The Bay Area District has a complaint line ((800) 334-ODOR) for the public to call if they encounter a problem with some factory smoke, an unbearable odor, or any other air-related problem that emanates from a stationary source. After a complaint is registered the district sends out enforcement inspectors to look into the problem. The Bay Area Air Quality Management District's Public Information and Education division also includes among its services a limited program of speakers with an emphasis on two-way discussion rather than a lecture approach.

Resources: Monthly, free newsletter called *Air Currents* , *Air Quality Handbook*, publications and pamphlets on asbestos, radon, global warming, permits, violation notices, and the difference between statospheric and ground level ozone.

Funding Sources: Property tax, state and federal grants, permit fees.

Bay Conservation and Development Commission (BCDC)

30 Van Ness Ave. # 2011
San Francisco, CA 94102
(415) 557-3686

Contact: Steve McAdam, Assistant Executive Director

Description: A permanent state planning and regulatory agency charged with preparing and implementing the S.F. Bay Plan, under the McAteer-Petris Act, and the Suisun Marsh Protection Plan, under the Suisun Marsh Preservation Act. Its main responsibilities are: (1) regulating all filling and dredging in S.F. Bay, its slough and marshes, and certain creeks and tributaries that are a part of the estuary, salt ponds, and some other managed wetlands that have been diked from the Bay; (2) regulating shoreline projects which are subject to design review and public access; (3) minimizing pressures to fill the Bay by reserving shoreline areas that have been designated in the Bay Plan for water-orientated priority use; (4)carrying out the provisions of the Suisun Marsh Preservation Act.

Volunteer Opportunities: Interns: Planning projects, reviewing applications, etc.

Funding Sources: State General Fund, Federal Grant from Office of Ocean and Coastal Resources Management.

Annual Budget: $1,800,000

Employees: 27

Please see also the articles on pages 97 and 108.

Center for the Development of Recycling

Geography & Environmental Studies, San Jose State University (SJSU)
One Washington Square, Rm BT550
San Jose, CA 95192-0204
(408) 924-5453

Contact: John Thomson

Description: The Center researches and develops information on waste management, plans activities, and makes such information available to government, industry, non-profit groups, and individual citizens. The Center is currently documenting Bay Area recycling strategies, developing fact sheets on Bay Area recycling organizations and facilities, compiling market development research on recycled materials, and developing internships in solid waste management.

Resources: SJSU Environmental Resource Center library, computer network links to the State of California Department of Conservation, the City of San Jose Office of Environmental Management databases, the California State University and the University of California library systems. It provides listings of other solid waste management organizations - governmental, non-profit, and trade - and various collections of reports on recycling, reuse, and source reduction.

Volunteer Opportunities: Paid and unpaid internships with government, non-profit, and for-profit organizations.

Metropolitan Transportation Commission
MetroCenter - 101 8th St.
Oakland, CA 94607-4700
(415) 464-7700
Contact: Catalina Alvarado
Description: The Metropolitan Transportation Commission was created by the state legislature in 1970 to provide transportation planning for the nine-county San Francisco Bay Area. The Commission has been assigned responsibility for administering several important public transit funding sources in the Bay Area, including monies from the State Transportaition Development Act and the federal Urban Mass Transportation Administration. The MTC is responsibe for overseeing the efficiency and effectiveness of the region's transit operators. MTC monitors the operators' budgets, conducts performance audits and adopts a yearly productivity/transit coordination improvement program. MTC sets capital investment priorities for both transit and highways in the Bay Area. MTC is given policy direction by an eighteen-member panel. Fourteen members are appointed directly by local elected officials. Two members represent regional agencies—The Association of Bay Area Governments and the Bay Conservation and Development Commission. In addition, two non-voting members have been appointed to represent federal and state transportation agencies. The combined annual operating budget of the transit agencies is in excess of $700 million, placing this region among the top transit operating budgets in the nation. In addition, there are numerous specialized services for the elderly and handicapped.

Regional Water Quality Control Board
1111 Jackson St., Rm. 6000
Oakland, CA 94607
(415) 464-1255
Description: The Water Quality Board is a state regulatory committee responsible for surface and ground waters. It regulates all California laws under the Water Pollution Act as well as the Federal Clean Water Act defined by the E.P.A. They have a staff of about one hundred including engineers, geologists, and biologists. Members of the regulatory committee are appointed by the Governor and serve a fixed four year term. Two thirds of the organization are working on programs cleaning up solvent spill problems in the Silicon Valley. The rest are involved in surface water problems. The Board meets monthly on every third Wednesday at BART headquarters, 800 Madison Ave. in Oakland. This meeting is open to the public. Most major decisions or topics are discussed at the meetings.

State

The following state agencies have areas of environmental responsibility, or provide information that may be useful to environmentalists. For more information, the California State Telephone Directory will help you find your way around the State Government. You can find one in the reference section of your local library, or order one for $9.00, by mail only, from the General Services Publication Office, listed below.

Air Resources Board
P.O. Box 2815
Sacramento, CA 95812
(800) 952-5588
Description: The Air Resources Board's (800) 952-5588 statewide complaint number is available to anyone in California for reporting air pollution complaints on disagreeable odors, smoke and other emissions from industrial and agricultural burning sources, problems with vapor-recovery equipment at service stations, smoking vehicles, etc. Complaints received during the workweek usually are referred for investigation to the appropriate local air pollution control district (APCD) within a few hours of their receipt. However, since California's 41 ACPDs have primary responsibility for control of air pollution from stationary sources, the ARB recommends that these districts be called before calling the ARB in order to speed up investigation of the complaint. The Bay Area's ACPD can be reached at (800) 334-ODOR. The office number for the Air Resources Board–not the complaint line–is (916) 322-2990.

California Coastal Commission
631 Howard St., 4th Floor
San Francisco, CA 94105
(415) 543-8555
Contact: Jack Liebster, Director of Public Affairs
Description: The Commission is the state's permanent coastal zone management agency, and was established after the approval of the 1972 Coastal Protection Initiative (Proposition 20) and the Coastal Act of 1976. The Commission promotes coastal conservation and regulates coastal development. It aids local planning efforts concerned with land and water use, development, access, natural resources, off-shore oil development, agriculture, and issues affecting the coastal zone. In conjunction with the Center for Marine Conservation, the Commission uses

data acquired in its Adopt-A-Beach cleanup program to assess coastal pollution. It sponsors "Coast Weeks" in the fall, a statewide celebration of coastal resources that includes seminars, tours, canoeing, and tidepooling. It also sponsors the statewide Coastal Cleanup Day.
Resources: *Coastal Access Guide*, *Coastal Resource Guide*, *Adopt-A-Beach Manual*, *Adopt-A-Beach Curriculum*, and School curriculum material on coastal aspects.
Volunteer Opportunities: Adopt-A-Beach– a statewide program in which volunteers make a committment to a specific beach, agreeing to clean it three times a year, recycle materials found, dispose of unrecyclable matter properly, and collect data for the Center for Marine Conservation. This data will be examined to determine sources of pollution, whether it's beachgoers, boaters, sewage or offshore dumping. Various groups have adopted beaches - companies, city government employees, Girl Scout troops and environmental organizations.
Funding Sources: State and Federal Budget.
Annual Budget: $9,000,000
Employees: 110

California Coastal Commission—North Coast District

640 Capitola Road
Santa Cruz, CA 95062
(408) 479-3511
Contact: Dave Loomis, Assistant District Director
Description: Like the Coastal Commission's main office, this office covers regulatory issues, permitting, and planning for the coastal zones of San Mateo, Santa Cruz, and Monterey . The public can arrange for talks that describe the Commission's aims, general coastal issues, resource issues, etc.
Resources: Same publications as main office. The group offers a library of regional planning documents (California Environmental Quality Act, EIR's , project evaluatations) forming a planning history of the area, and a Monterey Bay Natural History Display
Volunteer Opportunities: Adopt-A-Beach, with EcologyAction. Paid internships, volunteer office work.
Funding Sources: State and Federal budget.
Employees: 10

California Conservation Corps (CCC)

1530 Capitol Avenue
Sacramento, CA 95814
(916) 445-6819
Fax: (916) 323-4989
Contact: Susanne Levitsky, Public Information Officer
Description: The California Conservation Corps (CCC) has a dual mission: the improvement and enhancement of California's natural resources and the employment and development of the state's youth. An innovative state agency, the CCC works throughout the state tackling natural resource projects and emergency work (fire fighting, flood control, earthquake recovery, oil spill clean-up etc.). Projects include trail-building, park development, landscaping, stream clearance, historical renovation work, and energy auditing. Young people between the ages of 18 and 23 are hired year-round for corpsmember positions.
Resources: Toll- free number for corpsmember applicants and others: 1-800-952-JOBS
Volunteer Opportunities: Adult volunteers are welcomed at any of our CCC facilities.
Funding Sources: State of California. Also, non-profit CCC foundation.
Annual Budget: $58,000,000
Employees: 400

California Endangered Species Preservation Program

1115 Merrill Street
Menlo Park, CA 94025
(415) 327-2251
Contact: Robert Caughlan
Description: This program, operated by the California Department of Fish and Game, allows individual taxpayers to help financially with the preservation of "rare", "threatened" or "endangered" species by making a tax-deductible contribution on Line 45 of their State Tax Form. The Program provides a supplemental source of funds for management and recovery of California wildlife species that are classified as rare, threatened or endangered.
Resources: *Wildlife Lines* newsletter
Funding Sources: Line 45 of state tax form, individual taxpayers
Annual Budget: $1,000,000
Employees: 4

California Energy Commission

1516 Ninth St.
Sacramento, CA 95814
(916) 324-3009
Description: The Commission has five major areas of responsibility carried out by five divisions. They are the Energy Forecasting and Planning Division, Energy Technology Development Division, Energy Efficiency and Local Assistance Division, Energy Facility Siting and Environmental Protection Division, and the Administrative Services Division. These divisions are responsible for: forecasting future statewide electricity needs; licensing power plants to meet those needs; promoting energy conservation; developing renewable energy resources, and alternative energy generating technologies; and planning for, and directing state response to energy emergencies.
Resources: To receive current publication lists, and to get on CEC mailing list call (916) 324-3014. Or write to California Energy Commission Publications Office, 1516 9th St., MS-13, P.O. Box 944295, Sacramento CA 94244-2950. The Commission publishes conservation, electricity, energy development, and petroleum reports. Appliance efficiency trends and incentives, certified appliance directo-

ries, biomass, building standards, cogeneration, geothermal, hydroelectric, insulation, natural gas, solar and wind topics are covered in great (legal and regulatory) detail.

Funding Sources: Electricity consumption surcharge collected by utilities, and Petroleum Violation Escrow Account (restitution by oil companies for overcharging customers from 1973-1981).

California Energy Extension Service

1400 Tenth Street
Sacramento, CA 95814
(916) 323-4388

Description: The California Energy Extension Service (CEES) is part of a nationwide effort to lower energy use for small-scale consumers. Through demonstration projects CEES staff plan and implement energy management projects serving many segments of California, (primarily small businesses, school districts, fishing fleets, Native Americans, educators, and nonprofits). The primary objective of the CEES program is to provide personalized technical assistance and information. The CEES assists organizations to deliver effective programs by passing on lessons learned from program evaluations. CEES oversees Regional Energy Extension Centers including those described in the Education Listings under Sonoma Energy Extension Center (page 179), the East Bay Schools Energy Fitness Center and the El Camino Real Energy Extension Center (page 176). These centers provide energy educational materials to school districts.

Resources: *Save Energy* - quarterly newsletter; *Animated Bibliography of Energy Curriculum; Energy Activities for the Primary Classroom; How to Organize and Coummunicate your Energy Data: A Guide to Energy Accounting*; and more. List of publications available upon request.

Funding Sources: United States Department of Energy

California Public Utilities Commission

505 Van Ness Ave.
San Francisco, CA 94102
(415) 557-2621

Description: The CPUC regulates the rates, service, and safety of the privately-owned gas and electric utilities in California. The CPUC does not regulate city or district-owned energy utilities. The PUC also regulates private shipping rates, bus and train rates, and water rates. The PUC oversees the safety and efficency of public utility operation.

Resources: *A Consumer's Guide to Gas and Electric Service*—gas and electic customer's rights , reponsibilities, andquestions answered. *Consumer Information Bulletin*—outlines services and programs offered by the California PUC.

Coastal Conservancy

1330 Broadway, Suite 1100
Oakland, CA 94612
(415) 464-1015

Description: The State Coastal Conservancy is neither the Coastal Commission nor the Nature Conservancy. Instead it buys or secures land or easements for parks, open space, wildlife protection, and agriculture conservation. It funds public access to the shore–by way of stairways, promenades, picnic areas, piers, docks, commercial fishing facilities, and hostels. It restores wetlands, streams and watersheds. They also secure piers, docks, commercial fishing facilities. The Conservancy does all of this by working with local governments, other public agencies, land trusts and other nonprofit organizations, community groups, and landowners. They feel that it is their job to realize the purpose of the Coastal Act of 1976, which declared "that the California coastal zone is a distinct and valuable natural resource of vital and enduring interest to all the people." The Conservancy is the main funding agency for the Bay Trail. Since 1981 it has provided $6 million for 46 Bay Trail projects in nine counties. It is also working with many cities to restore waterfronts, to build a park along the East Bay shoreline from Richmond to Oakland, and to buy and protect wetlands. Not long ago it approved $3.5 million to purchase 1,400 acres of wetland near Vallejo (Culllinan Ranch). Its members are also working with the Port of San Francisco to improve commercial fishing facilities and public access at Fisherman's Wharf.

Funding Sources: The Coastal Conservancy is funded mostly by state bond acts. Funds are either spent directly or through grants to eligible public sector and non-profit recipients. Since 1976, the Conservancy has received nearly $185 million from the various state park and wildlife bond acts and from other sources.

Seismic Safety Commission

1900 K Street, Suite 100
Sacramento, CA 95814
(916) 322-4917

Contact: L. Thomas Tobin, Executive Director

Description: The purpose of the Seismic Safety Commission is to improve earthquake safety in California by improving public policy, especially that related to reducing hazards and mitigating the effects of potentially damaging earthquakes. The Commission is responsible for setting goals and priorities, requesting state agencies to devise criteria to promote seismic safety, recommending program changes to state agencies, local agencies, and the private sector where such changes would reduce earthquake hazards, reviewing reconstruction efforts after damaging earthquakes, gathering, analyzing, and disseminating information, encouraging research, sponsoring training, and coordinating the seismic safety activity of government. The Commission performs policy studies, reviews programs, and conducts hearings on subjects important to earthquake safety. It issues special reports and findings, and it reports annually to the Governor and the Legislature on its findings, progress, and recommendations.

Resources: Speakers on seismic safety, training modules through the Bay Area Regional Earthquake Preparedness Project for use in schools and communities, the California Earthquake Project at the Lawrence Hall of science, University of California, Berkeley, with earthquake educational materials.

Volunteer Opportunities: General clerical

Funding Sources: State General Funds, Federal Emergency Management Agency (FEMA)

Employees: 12

State Lands Commission

1807 13th Street
Sacramento, CA 95814
(916) 322-4105
Fax: (916) 322-3568

Contact: Charles Warren, Executive Director

Description: The State Lands Commission is composed of the State Controller, the Lieutenant Governor, and the Director of Finance. The Commission has jurisdiction over, and management responsibility for State-owned sovereign tidal and submerged lands which extend from the mean high tide line to three miles offshore, the beds of navigable waterways and congressional school grant lands.

Resources: CCORS Newsletter

Volunteer Opportunities: Volunteer opportunities exist through the State Lands Commission for individuals with appropriate interests and backgrounds.

Funding Sources: State General Fund (primary source); Environmental License Plate Fund.

Annual Budget: $18,000,000

Employees: 250

State Water Resources Control Board

P.O. Box 100
Sacramento, CA 95801
(916) 322-3132

Description: The State Water Resources Control Board sets the standards that are designed to protect all uses of the San Francisco Estuary's water: agricultural, residential, recreation, drinking water, as well as its use as fish and wildlife habitat. They regulate to prevent pollutants, and insure fresh water quality. A 1988 plan would have increased freshwater flows into the Estuary during the spring to insure that fish, wildlife and the Bay's health in general would be maintained. Unfortunately, the plan was rejected and a new plan falls far short of effecting protection for our Bay's resources. The new plan does not set minimal quantity standards for flows into the estuary and thus leaves the Bay's ecosystem vulnerable to water demands from areas throughout the state.

Waste Management Board

1020 9th Street, Suite 300
Sacramento, CA 95814
(916) 322-3330

Description: The Integrated Waste Management Board was created by statute on January 1, 1990. The mission of the new Board is to help local government meet aggressive waste reduction goals of twenty-five percent by 1995 and fifty percent by 2000. The Board will provide technical assistance in the preparation of city and county waste reduction and recycling plans, review and approve countywide integrated waste management plans, and conduct a broad range of enforcement, research and development, and public awareness activities.

Departments

These usually enormous departments are the regulatory sections of the state government. Here policy guidelines for state laws that affect the environment are designed, implemented, and enforced. Occasionally, a department will have regional offices that deal with the local details involved in enforcing and implementing programs.

Department of Conservation

1416 9th Street, Room 1320
Sacramento, CA 95814
(916) 322-1080

Description: Charged with mission of serving as the state's earth resources steward, the Department of Conservation has various responsibilities in the following areas: mining and geology; recycling; land resources protection; and oil and gas. The department can refer the caller to specific people to deal with issues of earthquake information, mining reclamation plans, surface mining, beverage container recycling (only), the Williamson Act (agricultural land preservation), and oil or gas facility safety. The department now administers the Beverage Container Recycling Act. Other recycling issues should be addressed to the California Integrated **Waste Management Board** (916-322-3330).

Resources: Referral capability.

Funding Sources: State budget.

Employees: 410

Department of Fish and Game—Central Office

1416 9th Street, Room 1236-8
Sacramento, CA 95814
(916) 445-7613

Description: The California Department of Fish and Game (DFG) insures that fish and wildlife are preserved to be used and enjoyed by the people in the state, now and in the future. DFG provides for the maintenance of all species

PUBLIC AGENCIES

of fish and wildlife for their natural, ecological, economic, educational and recreational value. Information officers assist conservation and other environmentally-oriented organizations. Although welfare of fish and wildlife is the Department's primary concern, it is also interested in a broad range of environmental issues including pollution, land and water uses, development of energy resources, forestry, mineral rights and fish and game law enforcement. The Sacramento office has a staff of approximately sixty that reviews all documentation generated by the California Environmental Quality Act (Environmental Impact Reports, Negative Declarations, Categorical Exemptions etc.) Timber and Water leases, and Fishery work. Tens of thousands of documents pass through this upper review level (initial reviews are often done on the regional level-see Region 3 below). Decisions are made as to where to best spend the limited resources of the department. Projects that are reviewed include any that affect riparian corridors, native plant and animal species and fisheries (sport and commercial). The DFG serves on the committees that develop Habitat Conservation Plans (those plans that provide for exceptions, through mitigation, to the prohibition of the taking of endangered species).

Department of Fish and Game—Region 3 Headquarters

7329 Silverado Trail
Napa, CA 94599
(707) 944-5500
Contact: Brian Hunter
Description: This regional office oversees the thousands of California Environmental Quality Act related projects that affect our ten county area (the region actually covers fourteen and a half counties). The Environmental Services section reviews Environmental Impact Reports, Army Corps of Engineers plans, Coastal Commission documents, and Bay Conservation and Development Commission plans for accuracy and scientific validity regarding impacts on California wildlife. Only four wildlife biologists inspect development sites (as time and resources allow) for an area stretching from Mendocino to San Luis Obispo. The counties served by the regional headquarters rely on this office for wildlife protection, management, and environmental law enforcement. The department also oversees interpretive sites in the region, such as Grizzly Island and Elkhorn Slough. The promotion of the non-consumptive (meaning nohunting or fishing is allowed) enjoyment of the state's natural heritage is a recent attempt to raise revenue that is not tied to the more destructive outdoor pursuits. The Department maintains hunting areas, fish stocking programs, and other game-hunting programs.

Department of Forestry and Fire Protection

1416 9th St.
Sacramento, CA 95814
(916) 445-9920
Description: The purpose of the department is to provide for the prevention and suppression of fires occurring on state and privately-owned forest, brush, and grass covered lands; to provide land management programs; to administer and enforce forest practice rules; to assist in range improvement programs; and to conduct or cooperate in forest and fire research programs.
Resources: *Protecting Trees When Building on Forested Land.*

Department of Forestry and Fire Protection—Urban Forestry

1416 9th St. Room 1540-36
Sacramento, CA 95814
(916) 322-0109
Description: The Urban Forestry program directs its attention to the improvement of the management of trees in urban areas, assisting cities, counties, districts and non-profits with information, networking and distribution of materials. In addition to urban planting, management and care advice, Urban Forestry oversees the Project Living Tree educational program. See Education Listings for info on this K-12 curriculum
Resources: *Catalogue of Forestry and Tree Education Resources; State of the Urban Forest for California ; An Introductory Guide to Urban and Community Forestry; The Grass Roots City Tree Planner; Technical Guide to Urban Forestry; State of California's Tree Ordinances (1991).*

Department of Forestry and Fire Protection—Regional Headquarters

135 Ridgeway, P.O. Box 670
Santa Rosa, CA 95401
(415) 576-2275
Description: The implementation arm of the state Department of Forestry and Fire Protection, this office is responsible for fire fighting and detection, fire training and public education. The Fire Safe Program advises rural residents on "defensible space" needs for protecting their homes in high risk areas. The regional headquarters oversees San Francisco Bay's Dutch Elm Disease problem and can identify pests, and advise the public on pest control strategies. The Urban Forestry program helps cities make informed decisions about tree management practises through publications and specialists.
Volunteer Opportunities: Citizen volunteers work in five areas; child fire safety education, public information/education, home fire safety, arson patrols, and radio communications.

Department of Health Services

2151 Berkeley Way

Department of Health Services

2151 Berkeley Way
Berkeley, CA 94701

Description: The Health Services Department is the state research and policy formulator for health issues affecting Californians. A number of their departments specialize in the study or implementation of environmental issues, including lead poisoning among children, workplace hazards, environmental testing certification, and pesticide illness monitoring.

Resources: Childhood Lead Program (415) 540-3657. Epidemiology Studies Section–Environmental Toxicology (415) 540-2669. Pesticide Illness Reporting (415) 540-3063.

Department of Health Services—Office of Drinking Water

2151 Berkeley Way
Berkeley, CA 94704
(415) 540-2173
Fax: (415) 540-2181

Contact: Alexis M. Milea

Description: The main objective of the state Department of Health Services—Office of Drinking water is to ensure that water delivered by public water systems in California is pure, wholesome and potable at all times. The Technical Programs Branch develops monitoring and water quality regulations and is responsible for maintaining expertise on state-of-the-art technology, responding to public inquiries related to drinking water quality, and conducting special studies of contaminants in drinking water. This Branch also mantains a data base of drinking water monitoring results throughout the state, processes financial assistance applications from public water systems, and oversees the certification of water treatment plant operators. The Field Operations Branch monitors public water systems compliance with the drinking water quality regulations, issues operating permits, monitors water quality data, and takes action when systems are out of compliance. In addition to the above , ODW is developing a comprehensive Safe Drinking Water Plan mandated by Assembly Bill 21 which includes an analysis of the overall quality of California's drinking water and the identification of water quality problems.

Funding Sources: State of California General Fund and Federal Grant.

Employees: 160

Department of Transportation (CALTRANS)-District 4

P.O. Box 7310, 3333 California St.
San Francisco, CA 94120-7310
(415) 923-4444

Description: CALTRANS has three areas of concern that are of great interest to the Bay Area. 1) Transportation Planning–CALTRANS is responsible for analyzing transportation policy issues, developing a systems plan for the effective integration of the various modes of transportation, and coordinating development evaluation of regional plans and transportation improvement programs. 2) Highways–CALTRANS is responsible for developing,operating, rehabilitating and maintaining the state highway system, making improvements to improve traffic flow, and expanding the capacity of the system. 3) Mass Transportation–CALTRANS is responsible for administering various state-funded programs for transit operators, Amtrak, and the San Francisco Peninsula Commuter train; providing technical assistance to transit operators, and supporting measures to integrate transit facilities with other modes of transportation wherever feasible.

Department of Water Resources

P.O. Box 942836
Sacramento, CA 94236
(916) 445-9248

Contact: Alan Jones

Description: The Department of Water Resources major responsibilities are: operating, maintaining, and adding facilities to the State Water Project, an immense system of dams, reservoirs, aqueducts, and pumping and power plants, which delivers water to urban and agricultural areas throughout much of California; evaluating current and projected needs for water, and developing programs that assure the best use of the resource; protecting the public through water quality improvement, flood control, and dam safety programs; and assisting local water agencies with funds, expertise, and technical support to improve their water delivery systems and meet the increasing demands of their communities. Planning for the future, the DWR is investigating several new projects, including improvements in existing Delta channels in coordination with State and federal agencies to assure quality water supplies for California's two largest water delivery systems, the State Water Project and the federal Central Valley Project.

Government Committees

Some of the most useful resources for information on pending legislation are the Senate and Assembly Committees. These committees prepare analyses of pending legislation, draft bills on behalf of elected backers, and publish reports of major legislation affecting the environment in California. Typically, a small staff examines bills for their legality, cost, constitutionality, policy perspective, and technical integrity. They prepare analyses of bills for legislators, that distill the purpose and problems of a bill, recommending changes where necessary. The committees take input from environmental groups and other lobbyists, and attempt to create balanced arguments.

The public may write or call the Committees to request major legislative publications (also call Joint Publications (916-445-4874) or ask for the reports on a specific bill. These minimally-staffed departments are not able to describe which bills they are covering in detail, and it is imperative that the caller know what he or she wants sent. To find out the specific numbers or titles of bills contact any local environmental organization that is concerned with your issue (e.g. Sierra Club, Friends of the River), or your Legislator's local district office.

Assembly Committee on Agriculture

P.O. Box 942849, State Capitol
Sacramento, CA 94249-0001
(916) 445-1918
Contact: Mike Falasco, Principal Consultant
Description: Makes policy analyses of legislation on most issues surrounding the food and agricultural code of California. Bills include dairy, livestock, crop, marketing, pest eradication and control, agricultural veterinary bills, and some bio-engineering subjects.
Resources: Publishes reports on committee work for each session. Contact Joint Publications Department to order (916) 445-4874. Policy analyses on legislative bills if requested specifically.
Volunteer Opportunities: Internships
Funding Sources: State budget
Employees: 3

Assembly Committee on Environmental Safety & Toxic Material

P.O. Box 942849, State Capitol
Sacramento, CA 94249-0001
(916) 445-0991
Contact: Arnold Peters, Principal Consultant
Description: Makes policy analyses of legislation on all issues that deal with the use, management and disposal of chemicals. Topics include pesticide residue in processed food, additives in food, and toxic waste disposal.
Resources: Publishes reports on committee work for each session. Contact Joint Publications Department to order (916) 445-4874. Policy analyses on legislative bills if requested specifically.
Employees: 3

Assembly Committee on Health

P.O. Box 942849, State Capitol
Sacramento, CA 94249-0001
(916) 445-1770
Contact: Paul Press, Principal Consultant
Description: Makes policy analyses of legislation on issues of the health effects of social environments and the workplace, regulatory legislation for doctors, drugs, food, and public and personal health care. Environmental issues range from the reproductive effects of pesticides to safe levels of contamination of drinking water.
Resources: Publishes reports on committee work for each session. Contact Joint Publications Department to order (916) 445-4874. Policy analyses on legislative bills if requested specifically.
Funding Sources: State budget
Annual Budget: $300,000
Employees: 4

Assembly Committee on Natural Resources

P.O. Box 942849, State Capitol
Sacramento, CA 94249-0001
(916) 445-9367
Contact: Kip Lipper, Principal Consultant
Description: Prepares policy analyses of legislation on issues of air quality, the California Environmental Quality Act, California pollution control financing, California alternative energy financing, coastal and bay protection, energy research, forestry, fire suppression, small hydroelectric plants, solid waste, surface mining, state lands, LakeTahoe, Wild and Scenic rivers, and the Williamson Act (agricultural land use fees).
Resources: Publishes reports on committee work for each session. Contact Joint Publications Department to order (916) 445-4874. Policy analyses on legislative bills if requested specifically.
Volunteer Opportunities: Internships.
Funding Sources: State budget.
Employees: 50

Assembly Committee on Water, Parks and Wildlife

P.O. Box 942849, State Capitol
Sacramento, CA 94249-0001
(916) 445-6164
Contact: Linda Adams, Principal Consultant
Peggy Lusk, Committeee Secretary
Description: Prepares policy analyses of legislation on issues of water quality, water development, water rights, water reclamation, water conservation, parks and recreation, river recreation, urban creeks, commercial and sport fishing and hunting, hunting seasons, rare and endangered species, wetlands, wildlife habitat and marine mammal protection.
Resources: Publishes reports on committee work for each session. Contact Joint Publications Department to order (916) 445-4874. Policy analyses of legislative bills if requested specifically.
Volunteer Opportunities: Internships–students from around the state may spend a half year working with a committee and a half year working with an Assembly Member. Volunteer positions with school credit.
Funding Sources: State budget.
Employees: 4

Senate Agriculture and Water Resources Committee

State Capitol, Box 942849
Sacramento, CA 94249-0001
(916) 445-2206

Contact: Stephen Macola, Principal Consultant

Description: The committee makes policy analyses of legislation on issues of water and agriculture. This includes water reclamation, water allocation, and agricultural genetic engineering bills.

Resources: Publishes reports on committee work for each session. Contact Joint Publications Department to order (916) 445-4874. Policy analyses on legislative bills if requested specifically.

Funding Sources: State budget.

Employees: 2

Senate Health and Human Services Committee

State Capitol, Box 942849
Sacramento, CA 94249-0001
(916) 445-5965

Contact: John Miller, Senior Consultant

Description: Makes policy analyses of legislation on issues of public health, mental health, welfare, blind and disabled policy, alcohol and drug treatment. Environmental issues include workplace health hazards, pesticide effects, asbestos and drinking water standards. Contact Senate Toxics and Public Safety Management Committee, regarding toxic substance bills.

Resources: Publishes reports on committee work for each session. Contact Joint Publications Department to order (916) 445-4874. Policy analyses on legislative bills if requested specifically.

Funding Sources: State Budget.

Employees: 8

Senate Natural Resources and Wildlife Committee

State Capitol, Box 942849
Sacramento, CA 94249-0001
(916) 445-5441

Contact: Peter Szego, Principal Consultant

Description: Makes policy analyses of legislation on fish and game issues, ocean resources, marine wildlife, recycling, mines and geology, CFC's, parks and recreation, forestry, solid waste, state land conservation, and rare and endangered species.

Resources: Publishes reports on committee work for each session. Contact Joint Publications Department to order (916) 445-4874. Policy analyses on legislative bills if requested specifically.

Funding Sources: State budget.

Employees: 4

Senate Toxics and Public Safety Management Committee

State Capitol, Box 942849
Sacramento, CA 94249-0001
(916) 324-0894

Contact: Bob Fredenburg, Principal Consultant

Description: Makes policy analyses of legislation on regulatory issues of hazardous waste facilities, storage tanks, waste oil, oil spills, and earthquake safety.

Resources: Publishes reports on committee work for each session. Contact Joint Publications Department to order (916) 445-4874. Policy analyses on legislative bills if requested specifically.

Funding Sources: State budget.

Employees: 3

Publications

General Services Publication Office

4675 Watts Avenue
North Highlands, CA 95660
(916) 973-3700

Description: Publishes a selection of state department publications. You will be referred to this office by various state departments when ordering copies of their material.

Joint Publications

State Capitol Box 942849
Sacramento, CA 94249-0001
(916) 445-4874

Description: Joint Publications is the central clearinghouse for legislative documents including hearings, transcripts, summaries of legislative sessions, digests, reports, appointee backgrounds, proposition analyses, and district maps. JP does not handle state department publications (Call the specific State Department or General Services Publications Office (916) 973-3700), nor actual text of bills (Call the Legislative Bill Room (916) 445-2323). JP is a good place to call if you are lost in the bureaucratic maze, and need some friendly informal referral help.

Employees: 5

Legislative Bill Room

State Capitol, Room B32
Sacramento, CA 95814-4997
(916) 445-2323

Description: The Bill Room takes orders for copies of the text of various bills. This is the actual text of the bill, and includes little or no additional analysis (See Legislative Committees for policy analysis reports). Individuals may order the texts of up to five legislative bills per day over the phone. Written requests for bills do not have a limit.

The Bill Room provides an indexed legislative bill list, by subject. The Bill Room only handles bills by their numbers. Check with local environmental groups, or the State Law Library for guidance on Bill numbers.

State Law Library

(916) 445-8833

Description: Has similar resources to most law libraries. Can provide legislative bill numbers when asked. Caller needs to know at least the subject of the bill, or bills. Bill numbers are required if one is trying to order a bill's text from the Legislative Bill Room, or bill analyses from Joint Publications.

National

Environmental Protection Agency

Public Information Center, 401 M Street SW
Washington, DC 20460
(202)475-7751

Description: The EPA sets and maintains air and water pollution standards, drinking water standards, ambient radiation standards, toxic substance standards, regulates sale and use of pesticides, oversees techniques of solid waste management, researches and demonstrates pollution control methods. The public may order general information pamphlets from this main office or regional offices, on such topics as air or water pollution, waste water treatment, pesticides and other EPA programs.

Resources: EPA Emergency Planning and Community Right-To-Know Hotline 800-535-0202 provides information on hazardous chemicals used by local businesses. Many publications available–see Reviews section, General on page 156.

Environmental Protection Agency Region 9

Public Information Office
75 Hawthorne Street
San Francisco, CA 94105
(415) 774-1500

Description: Has the same publications, and regulatory duties as the Washington Headquarters. You can often expedite requests by calling your regional office. The EPA public information line answers your questions, and can refer you to various local authorities or publications that may help you with questions regarding environmental pollution. They can also arrange for the free loan of environmental educational films. The San Francisco office has a library (774-1510) with a comprehensive collection of EPA Reports and other environmental information that is available to the public for on-site use, or through an inter-library loan. Companies can establish borrowing privileges through a written application process.

National Park Service Western Region Rivers and Trails Conservation Assistance

450 Golden Gate Avenue, Box 36063
San Francisco, CA 94102
(415) 556-5751

Contact: Nancy Stone

Description: Rivers and Trails Conservation Assistance is a program of the National Park Service which cooperates with states, local governments, and citizens to protect and restore river corridors and establish trail systems. The goal of this outreach program is to share the expertise of the National Park service with community groups working to protect their river and trail resources.

Resources: *Riverwork Book*

Volunteer Opportunities: Graduate and undergraduate interns needed.

Employees: 7

U.S. Army Corps of Engineers— San Francisco District

630 Sansome, Room 710
San Francisco, CA 94111
(415) 744-3276

Contact: Michael Keuss

Description: As the appointed steward of this area's coastline, ports, harbors, rivers, lakes, and wetlands, the Corps maintains the navigation channels, develops water and energy resources, designs flood plain water management plans, and increases recreational opportunities for the public. The harbors of San Francisco, Oakland, Richmond, Redwood City, and the terminal and harbor facilities in San Pablo Bay, Mare Island Strait, Carquinez Strait and Suisun Bay combined account for about one quarter of the state's total annual waterborne commerce. To accomplish the maintenance of coastal harbors, and inland straits of the Bay, the San Francisco District uses the capabilities of both the private dredging industry and government-owned vessels. The Corps removes an average of 5.5.million cubic yards of dredge material every year which is transported to either open-water approved disposal areas or selected landfill sites. As an extension of its traditional responsibility for regulating navigable waterways the Corps today seeks to manage all dredging, filling, and construction activity in 131,000 acres of Bay Area wetlands.

Annual Budget: $28,000,000

Employees: 195

U.S. Bureau of Land Management

2800 Cottage Way
Sacramento, CA 95825
(916) 978-4754
Contact: Tony Stead, Public Affairs Chief
Description: The Bureau is a federal multiple-use land management agency in the Department of the Interior which manages approximately 17 million acres of public land in California. There are small undeveloped parcels of BLM land in Alameda, Napa, and Santa Clara, which are primarily landlocked amongst private holdings and have no public access.

U.S. Fish and Wildlife Service

Regional Director
1002 NE Holladay St.
Portland, OR 97232-4181
(503)231-6828
Description: The Fish and Wildlife Service enforces federal wildlife laws that protect endangered species, migratory birds, marine mammals, and fisheries. The Service carries out U.S. enforcement obligations under international agreements. The Service reviews permit applications and, when appropriate, issues permits and documents. The USFWS employs about 200 special agents and inspectors who help enforce wildlife laws and treaty obligations. Special agents investigate cases ranging from individual migratory bird hunting violations to large-scale poaching and commercial trade in protected wildlife.The USFWS is also the key agency for the establishment of Habitat Conservation Plans which can allow for the destruction of an endangered species' habitat in exchange for mitigating efforts to enhance the species chances of overall survival.

U.S. Geological Survey

345 Middlefield Rd.
Menlo Park, CA 94025
(415) 329-4006
Description: The USGS was established by an act of Congress on March 3, 1879, to provide a permanent Federal agency to conduct the systematic and scientific "classification of the public lands, and examination of the geological structure, mineral resources, and products of the national domain." An integral part of that mission includes publishing and disseminating the earth-science information needed to understand, plan the use of, and manage the nation's energy, land, mineral, and water resources. Currently, USGS is the Federal Government's largest earth-science research agency, the primary source of data on the nation's surface and groundwater resources, and the employer of the largest number of professional earth scientists. The USGS is a research organization, and has no regulatory function.

Programs of interest include:

- basic and applied research in hydrology, mapping and related sciences.
- produces geographic, cartographic, and remotely sensed information in graphic and digital forms.
- assesses energy and mineral resources and develops techniques for their discovery.
- collects and analyzes data on water use, quality and projected quantity.
- evaluates hazards associated with earthquakes, volcanoes, floods, droughts, toxic material etc.

Resources: USGS Map Sales - Box 25286, Denver C) 80225 for topographic maps and free indexes. *New Publications of the U..S. Geological Survey*, 582 National Center Reston VA 22092- updated list that is free. *Earthquakes and Volcanoes* –bulletin, Superintendent of Documents, Government Printing Office, Washington DC 20402. *Guide to Obtaining USGS Information*–describes the services, types of publications and information products of USGS. From USGS Book and Report Sales, Box 25425, Denver CO 80225.
Volunteer Opportunities: Volunteers needed to assist scientists in collecting field data.

U.S. Nuclear Regulatory Commission—Region V

1450 Maria Lane, Suite 202
Walnut Creek, CA 94596
(415) 943-3809
Description: The Nuclear Regulatory Commission was created by Congress in 1975 to license and inspect all civilian uses of radiaoactive materials and nuclear facilities. Their mandate is to insure that these facilities are conducted in a manner consistent with the public health, safety, environmental quality, national security and anti-trust laws. The NRC is supposed to assure that nuclear power plants adhere to these regulations, and are not taking short-cuts in operation or maintenance. The Region V Office is responsible for NRC-licensed activities in a region stretching from Alaska to Mexico, and Arizona to Hawaii.

BROWN PELICAN

Section Six

PARKS

Here is a list of some of our region's best parks. We are very fortunate to have well-organized regional park and open space districts — and a large number of vocal private proponents — that together have preserved a greenbelt unmatched in any other major metropolitan area. Much of the credit for the beauty of the Bay Area must be given to our parks and preserves — local, county, regional, state, and national.

The diversity and quality of outdoor recreation opportunities in the area undoubtedly accounts for the great number of inhabitants who are willing to fight to preserve these types of resources, both here and around the world.

The major park districts are covered here, and there are enough possibilities represented to keep you hiking, swimming, boating, camping, birdwatching, and simply enjoying the beauty of our natural setting until the next edition comes out with more detailed descriptions of each of the parks within the districts.

City

Berkeley Camps

2180 Milvia Street, 4th Floor
Berkeley, CA 94704
(415) 644-6520

Description: The City of Berkeley offers two summer camping programs: Tuolumne Family Camp located in the Stanislaus National Forest near Yosemite Park, and a Summer Day Camp in Tilden Park Regional Park and at the Berkeley Marina (for 6 to 13 year-olds).

Resources: Brochures are availabe for each camp.

Funding Sources: Fees, Charges.

Deer Hollow Farm (Mountain View Recreation Department)

7550 St. Joseph Avenue
Los Altos, CA 94022
(415) 966-6331

Contact: Mary Gilman

Description: Deer Hollow Farm is a working homestead and educational center on 976 acres of Midpeninsula Regional Open Space District land. Farm animals, gardening, wool spinning, birding, Ohlone Indians, trees, native plants and wilderness exploration are among some of the topics offered to school classes and youth groups. Summer camp is also available.

Resources: Newsletter, brochures (general & class descriptions)

Volunteer Opportunities: Volunteers are involved in many capacities: teaching, gardening, farm maintenance, summer-aides and research. An extensive training program is available to those who wish to teach at Deer Hollow.

Funding Sources: Donations, City of Mountain View.

Foothills Park Nature Interpretive Center

3300 Page Mill Rd.
Los Altos Hills, CA 94022
(415) 329-2423

Contact: Annette Coleman, Senior Ranger

Description: The Interpretive Center interprets the flora, fauna, and history of the park to Palo Alto residents. The center contains a lecture room, exhibits, and displays, and is the starting point for various walks and talks given by the Palo Alto naturalists. Foothills Park is open to Palo Alto residents and their accompanied guests only.

Resources: More than thirty different one-page *Nature Notes*, on various topics. Free upon request. A self-guided nature trail brochure.

Funding Sources: City of Palo Alto.

Larsen Nature Preserve

P.O. Box 308
San Geronimo, CA 94963
(415) 488-9399

Contact: M. Dale Lambert or Anna Mae Kondratieff

Description: The Larsen Nature Preserve is a special outdoor education area with seven self-guided study walks, large enough to contain three distinct ecosystems, yet small enough to walk from one end to the other in five minutes.

Resources: *Guide and Handbook to the Marietta Larsen Memorial Nature Preserve.*

Volunteer Opportunities: Yes.

Livermore Area Recreation and Park District (LARPD)

71 Trevarno Road
Livermore, CA 94550
(415) 449-3832

Contact: Mike Nicholson - Ranger/Naruralist Supervisor

Description: LARPD is a special district which provides for the recreational and park needs of the City of Livermore and the Murray Township. The District manages 400 acres of natural parkland, Sycamore Grove and Veterans Park with bike and horse trails, picnic tables, and hiking trails open during daylight hours to the general public. There is no entrance fee.

Resources: Brochures on the parks and the school programs are available.

Volunteer Opportunities: Yes

Funding Sources: Fees for school programs and District's budget.

San Francisco Recreation and Park Department

McLaren Lodge, Golden Gate Park
San Francisco, CA 94117
(415) 666-7106

Description: The world-renowned Golden Gate Park offers lush green areas and vast recreational opportunities to the visitor. Eleven lakes, ten formal gardens, numerous meadows and picnic sites, woodland areas, two waterfalls, and abundant wildlfe environmentally balance the 21 tennis courts, horseshoe pitches, casting pools, rental boats, baseball diamonds, soccer pitches, stables, handball courts, four children's play areas, miles of bike trails, lawn bowling greens, statues and monuments, archery range, golf course, bicycle track, football field, an exquisitely restored carousel, Sharon Arts Building, restored windmill, and the Music Concourse area (where the Academy of Sciences, the DeYoung Museum, the Asian Art Museum and the Tea Garden are located). Visitors marvel that there is no admission charge to enter this facility. The San Francisco Zoo, Coit Tower (which houses depression-era murals depicting California life), Lake Merced Boating and Fishing facility, the Marina Green and Harbor, municipal swimming pools, golf courses, bocce ball courts, The Randall Museum, Candlestick Park, and more than 225 district parks and city squares ensure specialized activities and neighborhood greenbelts for San Francisco and its visitors.

Santa Cruz Park, Open Space and Cultural Services

Santa Cruz, CA
(408) 425-2395
Description: Santa Cruz parks offer an array of facilities and recreational opportunities including picnic sites, wedding sites, Pinto Lake Park, trails, and soccer and ball fields. Some open space areas, like Quail Hollow, have been acquired, but access is currently restricted to selected weekends when docent-led tours are offered. The Park services department offers after-school activities (no classes). No camping is available.

County & Regional

Black Diamond Mines Regional Preserve

EBRPD, 5175 Somersville Road
Antioch, CA 94509
(415) 757-2620
Contact: Joan Dougherty
Description: Located in the northern foothills of Mount Diablo, this nearly 4,000-acre preserve was the site of 19th-century coal mining and 20th-century sand mining. At one time home for the Bay Miwok Indians, the area's ranching history is preserved in buildings, corrals, and the cattle that still graze the hills. Rare species of flora and fauna are found in its grassland, foothill woodland, and chaparral habitats. The park is a geological, botanical, and zoological preserve, with nearly 40 miles of hiking trails. The park offers naturalist- and historian- led talks and walks, a slide show on park history, and a slide show on the local regional parks.

Volunteer Opportunities: A Variety of positions.
Funding Sources: Taxes, grants, donations.
Employees: 15

East Bay Regional Park District

11500 Skyline Blvd.
Oakland, CA 94619
(415) 531-9300 x2201
Fax: (415) 531-3239
Contact: Janet S. Cobb, Assistant General Manager of Public Affairs
Description: EBRPD is a splendid system of public parks and trails in Alameda and Contra Costa Counties. The district was founded in 1934 by public-spirited citizens who had the foresight to preserve some of the region's unique natural resources for future generations to enjoy. Today the district operates 47 parks and 11 regional trails covering 65,000 acres. These parklands assure preservation of the natural beauty that has made the Bay Area such a desirable place to live. They also protect habitats of wildlife, including many rare and endangered species. The parks offer many choices of recreational and educational activities. Whether you enjoy fishing, swimming, hiking, picnicking, nature programs, or just relaxing in beautiful natural surroundings, there is a park for you.
Resources: Newsletter - Regional Parks "Log"; Regional Parks Brochure
Volunteer Opportunities: Docents, tree planting, trail rehabilitation. Parks Partners are volunteer groups active at Ardenwood (Fremont), Black Diamond Mines (Antioch),Sunol Wilderness (Sunol), Point Pinole, (Pinole), and Lake Chabot. Call 531-9300 ext. 2200 for a brochure.
Funding Sources: Property tax from Alameda and Contra Costa Counties.
Annual Budget: $30,000,000
Employees: 500

Marin Parks, Open Space and Cultural Services

417 Marin County Civic Center
San Rafael, CA 94903
(415) 499-6387
Description: Parks run by the county range from redwood forest areas to coastal beach. Recreation opportunities include picnicking, tennis, softball, equestrian and hiking trails, a fishing pier, a model-car race track, and playgrounds. No camping.
Resources: Brochures available at District Office.

Midpeninsula Regional Open SpaceDistrict

Bldg. C, Suite 135201, San Antonio Circle
Mountain View, CA 94040
(415) 949-5500
Contact: Mary Hale, Public Communications Coordinator
Description: The District's primary goal is to preserve a regional greenbelt of open space lands, linking District preserves with state, county, and local parklands. In just 18 years the District has acquired 33,000 acres of foothill and bayland open space, preserved the scenic backdrop

of the midpeninsula, and provided the district's population of 600,000 people with access to these lands for low-intensity recreation.
Resources: Visitor's guide with trail information available at district office.
Volunteer Opportunities: Yes.
Funding Sources: Property taxes, grants, gifts.

Sunol Regional Wilderness

Box 82
Sunol, CA 94586
(415) 862-2244
Contact: Paul Ferreira
Description: Nature walks 7 days a week, workshops for rock climbing, backpacking, camping, excellent hiking, bird watching, and a creek.
Resources: *East Bay Log* (free), *Bird Key*, *Tree Key* (free), *Magic World of Mushrooms* ($.35), *Splashes of Spring Color* ($.10), *Survival Plants* ($.25), *Parkland Discoveries* ($4.00), and many other books in stock.
Volunteer Opportunities: Yes.
Funding Sources: Taxes.

Tilden Nature Area

Tilden Nature Area
Berkeley, CA 94708
(415) 525-2233
Contact: Alan Kaplan
Description: The park sponsors environmental education programs and learning experiences for all age groups. It also offers a wide variety of counseling and advice, outdoor education programs and conservation projects.
Resources: Monthly newsletter, educator's guide, pamphlets.
Funding Sources: Government, tax revenues, grants.

Oakland Zoo

9777 Golf Links Road.
Oakland, CA 94605
(415) 632-9525
Contact: Martha Smith, Public Relations Director
Description: The Oakland Zoo is the home for over 330 animals from around the world, located on 525 acres of rolling green parkland, tucked away from the city's hustle and bustle. There is Children's Petting Zoo and a Skyfari Ride which takes vistors high into the hills above the African Veldt. Grassy lawns and barbeque facilities located throughout the Park invite families and friends to picnic.
Resources: Free quarterly newsletter.
Volunteer Opportunities: Yes.
Funding Sources: State, county, and city subsidies, and private donations.

San Mateo Parks and Recreation Department

590 Hamilton Ave.
Redwood City, CA 94070
(415) 363-4020
Description: The County Parks Department manages 14,000 acres of park land that varies from the tidepools of the Fitzgerald Marine Reserve, to redwoods at Pescadero Creek, to an historic Spanish California site, to the Bayshore at Coyote Point. There are natural preserves at San Bruno Mountain and Edgewood Park. The parks offer hiking and riding trails, a marina, camping, and a museum. Send for a brochure or pick one up at the main office. Call (415) 363-4021 for park reservations.
Resources: Brochures.
Volunteer Opportunities: Yes.
Funding Sources: State.

Santa Clara Parks and Recreation Department

298 Garden Hill Drive
Los Gatos, CA 95030
(408) 358-3741
Description: Santa Clara's Parks Department oversees parks, campgrounds, lakes, and trails that vary in habitat from the western to the eastern portions of the county. There are redwoods at Mt. Madonna; picknicking at Sanborn-Skyline; Grant Ranch's camping and equestrian trails; Levin Park hang-gliding; reservoirs; boating at Calero, Anderson and Coyote Lake Parks; a velodrome at Hellyer, and urban parks at Vasona and Rancho San Antonio. For reservations for facilities, sites, and camping call (408) 358-3751. Trails connect in some areas with those of the Midpeninsula Open Space District.
Resources: Brochures on each park, color brochure on entire system.

Solano County Regional Parks

603 Texas St.
Fairfield, CA 94533
(707) 421-7925
Description: The county park system consists primarily of three large parks; Sandy Beach on the Sacramento River in Rio Vista with launching facilities, two miles of beach, and 44 campsites; Lagoon Valley Park, 350 acres along I-80 in Vacaville with a 110 acre lake, fishing, and wind surfing (day use only); and Lake Solano between Winters and Vacaville with 110 acres along Putah Creek, 50 campsites amongst a mixed evergreen forest, a day use area with two swimming lagoons, boating, canoeing, and fishing. No reservations, first come first serve.

Sonoma County Regional Parks

2403 Profession Dr. Ste. 100
Santa Rosa, CA 95403
(707) 527-2041

Description: The Sonoma County Regional Park has fifty-five parks spanning from Cloverdale to Petaluma, and from Hudeman's Slough near Skagg's Island to Doran and Westside Park on Bodega Bay. There is swimming at Hillsburg Veterans Memorial Dam and Beach on the Russian River, and at the Spring Lake lagoons in the summer. There are many campgrounds at the parks, including Doran, Westside, Gualala, Stillwater Cove, and Spring Lake. There are visitor centers, picnicking areas, soccer and baseball fields, trails, bike trails, and parcourses at many of the parks. Habitats range from sloughs, to mixed-evergreen forests, to mountain terrain. The park system has also historically overseen the operation of eight veterans memorial buildings and now operates a 250-slip marina in Bodega Bay. More parks are in the master plan stage, including the recently acquired Shiloh, Foothill, Crane Creek and Kenwood Parks. Call for more information on the many parks in our county.

Volunteer Opportunities: Yes.

State Parks and Recreation

There are many state park districts in the ten-county area. Each district office oversees a number of parks, and can answer questions regarding camping, tour hours, reservations, mountain bike rules and trails, horseback riding, and concessionnaires. Camping reservations are made through the MISTIX reservation system at 800-444-PARK (7275). District offices are occasionally unable to answer your questions because the rangers are out ranging instead of answering the phone.

State Department of Parks and Recreation

P.O. Box 2390
Sacramento, CA 95811
(916) 445-4614

Contact: Call 445-6477

Description: California State Department of Parks and Recreation acquires, designs, develops, operates, and maintains units of state parks. It is responsible for administering federal, state, and local assistance programs.

Resources: *Guide to the California State Park System* - $1 at parks and $2 by mail, individual brochures.

Volunteer Opportunities: Eighty percent of park units have many volunteer opportunities.

Funding Sources: Sale of bonds, state funds.

Employees: 2330

Candlestick Point State Recreation Area

P.O. Box 34159
San Francisco, CA 94134
(415) 557-4127

Description: Candlestick Point is a State Recreation Area by the shore of the San Francisco Bay. Located on land reclaimed after years of abuse by garbarge dumps, and auto wreckers, the land is now dedicated to recreational, educational, and aesthetic use. This is California's first urban State Park. Activities include fishing programs for kids and seniors, picnicking, bird-watching, bayshore walks,and more. There is a visitor center and trails with environmental education programs for schools and other groups.

Resources: "Wild Walk" - publication.

Volunteer Opportunities: Yes.

Funding Sources: State budget.

State Parks - Diablo District

(415) 687-1800

Description: Mt. Diablo State Park offers camping which is first come, first serve in the summer's fire season, due to possible closures. Parks include Bethany Reservoir State Recreation Area, Benicia Capitol State Historic Park, and Benicia State Recreation Area.

State Parks - Gavilan District

(408) 623-4526

Contact: San Juan Bautista : (408) 623-4881
Freemont Peak: (408) 623- 4255
Henry W. Coe: (408) 779-2728

Description: The Gavilan District of State Parks consists of three parks: 1) San Juan Bautista State Historic Park; 2)Fremont Peak State Park, and 3)Henry W. Coe State Park. San Juan Bautista has structures from the old town and mission of San Juan that date back to the late 1800's. Fremont Peak is a 13-acre park on the top of a peak, with camping, hiking, and trails. On October 13 and 27 star shows are conducted within the park. Henry W. Coe is 6,700 acres of unlimited hiking with primitive camping and no fresh, drinkable water.

Resources: Brochures, maps, vistor centers at San Juan Bautista and Henry W. Coe.

Volunteer Opportunities: Many volunteer opportunites at all three parks, ranging from wearing costumes at San Juan Bautista to being a horseman at Henry W. Coe.

Funding Sources: Fees and general funds from the public.

Annual Budget: $107,000

Employees: 15

State Parks - Marin District

(415) 456-1286

Description: Mt. Tamalpais, Angel Island, China Camp, Samuel P. Taylor, and Tomales Bay State Parks, and Olompali State Historic Park.

State Parks - Napa District

(707) 942-5370

Description: Bothe Napa Valley State Park (camping); Bale Grist Mill State Historic Park; Robert Louis Stevenson Park.

State Parks - Pajaro District

(408) 688-3241

Description: New Brighton State Beach (camping); Natural Bridges State Beach, Natural Bridges Monarch Butterfly Nature Preserve; Seacliff State Beach (camping); Sunset State Beach (camping); Manresa Campgrounds and Uplands; Twin Lakes State Beach; Wilder Ranch State Park; and Rio Del Mar Beach.

State Parks - Russian River District

(707) 865-2391

Description: Fort Ross State Historic Parks (first come, first serve camping - no dogs); Salt Point State Park (camping); Sonoma Coast State Beaches - including Bodega Dunes, Wright's Beach (with camping); Armstrong Redwood State Reserve; and Austin Creek State Recreation Area (first come, first serve camping).

State Parks - San Francisco District

(415) 666-7200

Description: Candlestick Point State Recreation Area - day use only.

State Parks - San Mateo District

(415) 726-6203

Description: San Mateo Coast State Beaches; Francis State Beach (camping); three ranches - Mc Nee (Montara State Beach), Burleyigh-Murray (Half Moon Bay), Cascade Ranch (Ano Nuevo); Ano Nuevo State Reserve; and Pescadero Marsh State Preserve.

State Parks - Santa Cruz Mountain District

(408) 335-9145

Description: The Santa Cruz Mountain District is made up of six parks: Portola, Big Basin-Rancho Del Oso, Henry Cowell, The Forest of Nisene Marks, (Loma Prieta Quake Epicenter Trail), Butano , Castle Rock State Parks. Portola is a 2,500-acre redwood park in San Mateo County. Big Basin is the largest of the parks with 18,000 acres and a virgin redwood forest. Henry Cowell is 4,500 acres with 34 acres of virgin redwoods and is near the San Lorenzo River. The Forest of Nisene Marks is a 10,000 acre park with redwoods, oaks, and creeks. Butano is 3,000 acres with many Douglas Firs. Castle Rock is a 4,000-acre oak woodland park with outstanding views. All parks have picnicking, hiking, nature walks, and nature trails. Some have camping as well.

Resources: Brochures, folders on each park with maps and general information, and publications by the National History Association.

Volunteer Opportunities: All parks have many volunteer opportunities, ranging from leading nature walks and campfire circles to working in the visitor or information centers.

Funding Sources: State funds, bond acts, donations from Friends of the Parks.

Annual Budget: $1,200,000

Employees: 52

State Parks - Sonoma District

(707) 938-1519

Description: Sonoma Mission; Vallejo Home; Sonoma Barracks (Historic Park); Annadel (Camping); Sugarloaf Ridge; Jack London State Historic Park; Petaluma Adobe.

National

Elkhorn Slough National Estuarine Research Reserve

1700 Elkhorn Road
Watsonville, CA 95076
(408) 728-2822

Contact: Julie Baretti Heffington

Description: The ESNERR is one of the few relatively undisturbed coastal wetlands remaining in California. It is managed by the California Department of Fish and Game (CDFG) in cooperation with the National Oceanic & Atmospheric Administration (NOAA). The area is also part of the California Wildlands Program and is designated a State Ecological Reserve. All of these programs are in concert with the following: habitat preservation for native wildlife, research for the perpetuity of that wildlife, education of California's school children, and interpretation of the resource and conservation education. Ongoing projects include volunteer work and training; training teachers for field trips; special public events featuring natural or cultural history; periodic environmental or conservation workshops or symposia; and continued habitat restoration. Proposed projects include permanent exhibits for the nature center, a native plant interpretive garden, a boardwalk into oak canopy, and an environmental education lab in a restored barn.Reserve is open from 9 to 5, Wed.-Sun. Fee: $2.25 per person, or $11 per year. No fee with a valid California Hunting/Fishing license.

Resources: 1400 acre reserve, more than 5 miles of hiking (only) trails, bird blinds, heron and egret rookery, nature center, guided walks (weekends, 10am to 1pm), and a natural history bookstore operated by the Elkhorn Slough Foundation.

Volunteer Opportunities: Large volunteer group (eighty plus) trained yearly, usually in the spring, for interpretation and other resource enhancement work.

Funding Sources: Government, both state and federal, with support from the non-profit Elkhorn Slough Foundation.

Annual Budget: $65,000

Employees: 3

Fort Point National Historic Site

Box 29333
Presidio of San Francisco, CA 94129
(415) 556-1693

Contact: Charles S. Hawkins , Site Manager

Description: Fort Point offers various interpretive programs to the public, including 30-minute tours led by Park Service Rangers wearing replica Civil War uniforms and demonstrations in the loading and firing of a Civil War cannon. Museum exhibits feature displays on artillery, soldiers' uniforms and belongings, Civil War medicine, historic photos of the fort, and an exhibit on the history of the black soldier in the U.S. Army.

Resources: Brochures.

Funding Sources: Federal budget.

Golden Gate National Recreation Area

Ft Mason, Bldg 201
San Francisco, CA 94123
(415) 556-0560

Contact: David Harbert

Description: GGNRA represents one of the nation's largest coastal preserves. Its North Pacific Coast landscape includes sandy beaches, rugged headlands, grasslands, forests, lakes, streams, estuaries, and marshes. Important scientific resources range from small rare plants in the San Francisco and Marine Headlands, through towering redwood trees, to the diverse organisms of the intertidal zone. The recreation area is located in three coastal counties: San Francisco, Marin, and San Mateo. The total park area of 114 square miles of land and water includes approximately 28 miles of Pacific Ocean, Tomales Bay, and the San Francisco Bay coastline.

Funding Sources: State.

Gulf of the Farallones National Marine Sanctuary

Fort Mason, Building 201
San Francisco, CA 94123
(415) 556-3509

Contact: Angie Coleman, Sanctuary Coordinator

Description: The Gulf of the Farallones National Marine Sanctuary preserves a portion of the rich marine ecology of the Gulf of the Farallones just north of San Francisco. Unusual concentrations of marine wildlife and plants and a diversity of ocean habitats are found within this particular region of ocean, making it a nationally significant marine ecosystem. Sanctuary status provides increased protection and allows enhanced resource management to maintain the integrity of this unique marine environment. Public education and research programs aid theunderstanding of the value of our marine ecosystems.

Resources: General brochure, management plan, FEIS, Education/Interpretive Plan, research reports.

Volunteer Opportunities: Yes

Funding Sources: National Oceanic and Atmospheric Administration, U. S. Department of Commerce.

John Muir National Historic Site - National Park Service

4202 Alhambra Ave
Martinez, CA 94553
(415) 228-8860

Contact: Linda Moon Stumpff

Description: Congress has set aside the home of John Muir to commemorate Muir's contribution to conservation and literature. The eight-and-three-quarters acres of land have been restored as a miniature of the orchards and vineyards that once comprised the 2600-acre ranch.

Funding Sources;Federal Government Southwest Parks and Monuments Association

Kule Loklo - A Coast Miwok Cultural Exhibit

Point Reyes National Seashore
Point Reyes, CA 94956
(415) 663-1092

Contact: Lanny Pinola

Description: Kule Loklo, a Coast Miwok Indian Cultural Exhibit replica, offers a setting for teachers, students, and others to explore their environment and imagine life the way it was. It is a functional tool for environmental education and living history.

Resources: Program brochures.

Volunteer Opportunities: Many volunteer opportunities.

Funding Sources: National Park Service.

Lake Sonoma Park

Warm Springs Dam Office, 3333 Skaggs Springs Road
Geyserville, CA 95441
(707) 433-9483

Description: The Army Corps of Engineers oversees this recreational lake, camping site (first come, first serve), and educational Fish Hatchery. When staffing permits, the Corps leads tours for school children during school session.

Muir Woods National Monument

Mill Valley, CA 94941
(415) 388-2595

Contact: Mia Monroe

Description: Muir Wood is a beautiful and inspiring redwood canyon on the side of Mt. Tamalpais in Marin County. Six miles of trail bring you close to the giant redwoods, representatives of the tallest trees in the world, and give a chance to observe the unique features of a redwood forest. Redwood ecology, watershed and stream studies, early conservation history, and plant identification are popular programs for school groups. Rangers are available for guided tours, and pre-visit materials are sent ahead to scheduled groups.

Resources: Slideshows, pamphlets for self-guided trail, and a vistor center.

Volunteer Opportunities: Yes.

Point Reyes National Seashore

Point Reyes, CA 94956
(415) 663-1092

Description: The Point Reyes National Seashore encompasses most of the Point Reyes Peninsula, lying west of the San Andreas Rift Zone. Point Reyes exhibits tremendous diversity of plant and avian wildlife, many miles of remote beaches, an Earthquake trail, camping facilities, historic ranches, a lighthouse, and a large parcel of authentic wilderness very close to urban areas. Headquarters and visitor center are located on Bear Valley Road, .6 miles northwest of Olema.

San Francisco Bay National Wildlife Refuge

U. S. Fish and Wildlife Service
P.O. Box 524
Newark, CA 94560
(415) 792-0222

Contact: John Steiner

Description: San Francisco Bay National Wldlife Refuge protects over 20,000 acres of open space for wildlife, and provides opportunities for outdoor education and wildlife-oriented recreation. The Visitor Center, near the Dumbarton Bridge toll plaza, features exhibits about Bay Area wildlife, interpretive programs, and seminars. Exhibits along the Tidelands Trail interpret the natural history of the South Bay marshes. The Environmental Education Center in Alviso allows classes and groups to learn about the Bay through hands- on activities, available by reservation only. All activities at the Refuge are free. A system of hiking trails provides access to salt marshes, bay shore and good bird-watching areas. Trails are open every day during daylight hours.

Resources: Visitor center featuring exhibits, interpretive programs and seminars. The Environmental Education Center offers classes and groups hands- on activities, Tuesday-Friday, 9am-3pm.

Volunteer Opportunities: Visitor center, environmental center, maintenance.

Funding Sources: Federal Government.

Annual Budget: $1,000,000

Employees: 26

WESTERN WALLFLOWER

Section Seven

CONSUMER GUIDE

If we had only two words of advice we could give about consuming for a better environment, they would be these: consume less. For no matter how careful a manufacturer is to avoid the use of toxic substances, use recycled materials, and minimize environmental impacts, the fact remains that manufacturing always consumes energy and resources.

Of course, very few of us are capable of a fully ascetic lifestyle. So on those occasions when you find yourself lapsing into consumerism, you should try to buy goods made with the fate of the Earth in mind. That's the sort of product you'll find in the Green Pages' Consumer Guide.

There is much more here than a list of relatively environmentally-sound products, however. It is really a guide to living a cleaner, "greener" life, and contains many service-providers with the ability to help you do it. There are architects who specialize in passive solar building design, gardeners with a fund of knowledge about landscaping with drought-tolerant plants and without pesticides, and consultants who can advise you about everything from how to set up an office recycling system to avoiding toxic materials in your home.

Our research for this section revealed that the distribution of the types of businesses we wanted to include was quite uneven; in certain classifications we were unable to find any businesses operating in some counties of the Bay Area. We can only hope that some of our readers, noticing such gaps, will help us fill them for the next edition. Please tell us about any green businesses you know of that are missing.

For certain categories that have more application for businesses than individual consumers (printers, for instance), check the Business-to-Business Guide, which starts on page 337.

The Consumer Guide is organized by major categories with subclassifications. Energy consultants, for instance, are under the Building category. See the table of contents on pages 265-266 to orient yourself to the classification system. To quickly find a particular item or service, check the Consumer Guide subject index, pages 267-270. (This index also covers the classifications in the Business-to-Business Guide, section eight.)

C.G. CONTENTS

C.G. CONTENTS

BUILDING

The buildings in which we live and work have a profound effect on the natural environment. Many of them waste energy, use precious resources, and are built with materials that make us sick. But it doesn't have to be that way. We can use techniques, materials, and designs which are healthy both for us and for the environment . The individuals and companies listed here try to work with non-toxic, non-endangered, renewable resources. Whether you are expanding, converting, or building from the ground up, they can help you develop a home or building that will not harm the environment. (See related articles in the *Home and Garden and Workplace* chapters.)

We currently get most of our lumber from companies that are cutting down trees at twice the rate they are planting them. The companies which are listed in our lumber section select out or salvage wood only. The way most houses are sited and designed is dictated more by the need for short-term profit than the needs for community, open space, gardens, shared appliances, recycling, easy access to work and services, and energy effiency for heating, lighting, and transportation.

When you build a house or modify a existing one you have a chance to change the physical structure of your life, and you can change it so it is more environmentally sound.

Beyond the individual house is the block or community. Individuals and families in our society usually express their concerns about their neighborhoods in the form of worries about security, property values, noise, and quality of schools etc. But we are also beginning to see organized concern about community gardens, tree planting, recycling, toxics, traffic, and open space. As we come out of our individualistic cocoons at the prodding of larger social and environmental problems and rub the suburban dream out of our eyes, we may well want to engage the services of a planner specializing in community design. Cohousing, and other shared living alternatives like cooperative blocks and apartment buildings, as well as new developments designed in keeping with ecological and community values, help solve many of the tangled problems which are inextricably bound up with our consumer lifestyles and disposeable society.

ARCHITECTS & DESIGNERS

When choosing an architect or designer, ask to see examples of their solar designs, and ask them what they know about nontoxic building materials.

Ask about incorporating built-in recycling systems and areas, and be sure to specify that you want energy saving windows and appliances. The architects and designers listed here all take such factors into account in their work.

ADPFSR
Architects, Designers & Planners for Social Responsibility
P.O. Box 9126, BERK (415)937-9010

MICHAEL BLACK
146B N. Main St., Sebastapol (707)829-9319

CLINT GOOD ARCHITECTS
P.O. Box 143, Lincoln VA (703)478-1352

MARK GORRELL ARCHITECT
850 Mendocino Ave., BERK (415)528-1208

JACOBSON, SILVERSTEIN, & WINSLOW
3106 Shattuck Ave., BERK (415)848-8861
(408)423-5263

WESTIN MILES DESIGN
17400 Monterey Rd., Morgan Hill (408)779-6686

MORRIS & CLEANER
47 Sixth St., Petaluma (707)763-0152

JEFF OBERDORFER & ASSOCIATES, INC.
Santa Cruz (408)423-5263
See display ad on this page

STEVE SHELDON
303 N. Main St., Sebastapol (707)823-6331

THE COHOUSING COMPANY
48 Shattuck Square, BERK (415)848-0331

NANCY G. SIMPSON
P.O. Box 94, Willis (707)459-2595

SIM VAN DER RYN & ASSOCIATES
55-C Gate Five Rd., Sausalito, (415)332-5806
See display ad on this page

CAROL VENOLIA, ARCHITECT
P.O.Box 69, Gualala (707)884-4513

BUILDING CONSULTANTS

These building consultants are specialists in non-toxic construction, solar design, energy efficient construction or a combination of specialties. Even if you don't hire one of these people to design or build your project, a few hours of their time looking

BUILDING

over your plans would be money well-spent. Some intelligent and informed thought at the early stages of a project can make it possible to build your home (or make additions and renovations) with the least toxic building materials and the most energy efficient.

Fact: Trees naturally cleanse the atmosphere of carbondioxide (CO_2).They do so by each absorbing an average of 47 pounds of CO_2 a year.

NANCY G. SIMPSON
P.O. Box 94, Willits (707)459-2595

OWNER BUILDER CENTER CONSULTING
1250 Addison St., Ste. 209, BERK (415)848-6860
See display ad on this page

SIM VAN DER RYN & ASSOCIATES
55-C Gate Five Rd., Sausalito (415)332-5806
See display ad, page 271

BUILDING SUPPLY

The businesses listed in this section are here for three reasons: They offer products and techniques which are non-toxic alternatives to conventional ones, they offer solar and/or super energy efficient materials; or they sell recycled building materials. Some of these businesses may fit into more than one of these categories.

Although the nontoxic building movement is relatively new in this country, it is well-developed in Germany and Scandinavia. There are full lines of non-toxic finishes, sealers, paints, adhesives, caulking, insulation, and vapor barriers. (See the articles on solar energy on page 79 and The Integral Urban House on page 10 for a discussion of the merits of integrated design.)

Recycled building materials can be a cost saver as well as a resource saver. Consider taking your salvagable materials to a place which will recycle them. Reuse is the most efficient form of recycling, because it doesn't involve remanufacturing or redistribution.

AFM ENTERPRISES
1140 Stacy Ct., Riverside (714)781-6860
See display ad on this page

AURO ORGANIC PAINTS
P.O. Box 857, Davis (916)753-3104

BAUBIOLOGIE HARDWARE
207 16th St., Pacific Grove (408)372-8626

C AND M DIVERSIFIED
330 N. Montgomery St., San Jose (408)294-5185

DJ'S LOCKER, INC.
2076 Agnew Rd., Santa Clara (408)727-3919

ECO DESIGN CO.
Catalog of nontoxic producs for home (505)438-3448

FINE ADDITIONS
2405 Macoovia Ln., Santa Fe NM (505)471-4549

HENDRICKSEN FLOORCOVERING
8031 Mill Station Rd., Sebastopol (707)829-3959

LIVOS PLANT CHEMISTRY
Catalog of nontoxic building supply (800)621-2591

NON-TOXIC ENVIRONMENT
6135 N.W. Mntn. View Dr., Corvalis, OR (503)745-7838

PACE CHEM INDUSTRIES, INC.
779 S. La Grange Ave, Newbury Park (805)499-2911

REAL GOODS TRADING CO.
966 Mazzoni St.. , Ukiah (800)762-7325

RESOURCE CONSERVATION TECHNOLOGY
2633 North Calvert St., Baltimore MD. (301)336-1146

SHAKER WORKSHOPS WEST
5 Inverness Way, Inverness (415)669-7256

SINAN COMPANY
P.O. Box 857, Davis (916)753-3104

SOLUTIONS TO INDOOR POLLUTION
10565 Caminito Bayon, San Diego (619)271-6082

THE ALLERGY STORE
PO Box 2555, Sebastopol (800)824-7163

TURTLE HOUSE - NON-TOXIC MATERIALS
(707)823-7233

DISCARD MANAGEMENT CENTER
Urban Ore, 1231 2nd St., BERK (415)526-9467

URBAN ORE BUILDING MATERIALS
1325 Sixth, BERK (415)526-7080
Fashion Island Shop. Ctr., S. Mateo (415)578-9200
401 Bayshore Blvd., SF (415)285-5244
Willow Shopping Center, Concord (415)686-2270

WHOLE EARTH ACCESS
2990 Seventh St., BERK (415)845-3000

CONSTRUCTION

There are a wide range of considerations in choosing a builder, even for a small job. Some of the "green" questions you might ask about are:

- Do they use imported tropical hardwoods?
- Do they use toxic stains and finishes and do they know where to find nontoxic ones?
- What efforts do they make to recycle?
- Do they plant trees to replace the wood they use?

The following companies have addressed these issues and come up with some interesting approaches to the somewhat contradictory "environmentally-sound construction practice" notion.

JTD CONSTRUCTION
1165 Montgomery Rd., Sebastapol (707)829-5856

MALLEK CONTRACTORS, INC.
473 S. San Antonio Rd., Los Altos (415)948-0797

SONOMA CONSTRUCTION
(707)792-2055

ENERGY CONSULTANTS

These consultants will come to your home or business and advise you on tactics for energy-use reduction and energy cost savings. Many of the consultants will also contract for the work to be done, or carry it out themselves. Check your local library for many books on home energy projects to do yourself.

Tip: Insulating attic floors and weatherstripping windows and doors saves heat and money.

ANDREA DENVER ASSOCIATES
Energy Management, Consulting, and Contracting
2243 Browning St., BERK
(415)841-4552

BARAKAT & CHAMBERLIN, INC
180 Grand Ave., Suite1090, OAK (415)893-7800

BERKELEY ENERGY MANAGEMENT
3120 Shattuck Ave., BERK (415)841-4036

BETA ASSOCIATES
6000 Hollis, Emeryville (415)832-6000

BULES & ASSOCIATES
461 Bush St. SF (415)399-9812

CALIFORNIA ENERGY COMPANY
601 California, SF (415)391-7700

CALIFORNIA MICRO UTILITY
1195 Park Ave., Suite 202A, Emeryville (415)658-1273

COMMUNITY ENERGY SERVICES CORPORATION
1013 Pardee St., BERK (415)644-8546

CUNNINGHAM ASSOCIATES
512 2nd St., SF (415)495-2220

ENERGY CALC CO.
3255 Kerner Blvd., San Rafael (415)457-0990

ENERGY WEST
123 Townsend, SF (415)495-7505

ENTEK CORP.
1043 Stuart, Lafayette (415)283-4040

ENVIROTHERM INC.
18 Woodland Ave., San Rafael (415)268-9733

FMC ASSOCIATES
1068 Mission St., SF (415)864-6776

GIBBS & HILL INC.
44 Montgomery, SF (415)296-8300

GLUMAC & ASSOCIATES
275 Battery, SF (415)398-7667

GROUP Z
2055 Bush, SF (415)922-9777

INQUIRING SYSTEMS, INC.
3111 Deakin St., BERK (415)843-3135

LEWIS JOHN & ASSOCIATES
110 Sutter, SF (415)398-5003

NEWCOMB ANDERSON ASSOCIATES
151 Union, SF (415)434-2600

SMITH, OTTO J.M.
612 Euclid Ave., BERK (415)525-9126

SAVE ENERGY COMPANY
2410 Harrison St., SF (415)824-6010

TEM ASSOCIATES
1900 Powell St., Emeryville (415)655-6576

WATER & ENERGY MANAGEMENT
79 Hillmont Place, Danville (415)820-6603

FLOORING

Because flooring involves the use of glues and finishes, they can be a source of many toxic chemicals in a house. Wood floors are also notorious users of endangered tropical hardwoods. If you are planning to install a new wood floor, make sure to ask for farmed or ecologically-harvested domestic woods. As poorer nations rip down their forests in a mad scramble to pay their debts, sustainable techniques are ignored. American forestry is not much better; this country logs trees at a faster rate than any nation in the world. You should ask if the contractors and flooring suppliers plant trees to make up for the wood they use. The rare few do.

Fact: Most houses waste 50% of the energy they consume.

BUILDING

CAROUSEL CARPETS
1044 Fourth St., Santa Rosa (707)485-0333

COLLINS AND AIKMAN
5341 Port Sailwood Dr, Newark (415)791-1623

DESIGNER HARDWOOD FLOORS
206 Pomona Ave., BERK (415)528-4420
See display ad, page 273

HENDRICKSEN FLOOR COVERING
8031 Mill Station Rd., Sebastopol (707)829-3959
See display ad, page 273

INTO THE WOODS
300 N. Water St., Petaluma (707)763-0159

JORDAN AND BOWMAN
6520, OAKmont Dr., Santa Rosa (707)539-1606

INSULATION

There are a variety of different insulating materials: Foams, fiberglass, chopped and shredded paper, blow-in foams and rigid foams.

Problem insulation:

Blow-in foams have been known to "outgas" formaldehyde for one to two years after installation. They are generally made from polyurethane.

Though rigid foams of polystyrene (put in place before the walls go up) are very good insulators, they produce very toxic fumes in a fire, and use CFCs in their manufacture. Some rigid foams grouped under the term "Expanded polystyrene foam" do not use CFCs, but they do use pentane which is a smog-producing chemical.

Better Insulation:

For the best, environmentally-sound, least toxic insulation in the home, use fiberglass, either installed during construction or blown-in. Or use recycled cellulose fiber insulation which is fairly non-toxic. This paper insulation is frequently treated with either aluminum, boron, or ammonium sulfate to retard fires. These chemicals do not out-gas but if they get wet they can leach down. There are a few other nontoxic insulation materials such as Air-Krete, or Icynene, with slightly less insulating capability than the foams. Fortunately the Bay Area's climate is not so extreme as to warrant an environmental compromise.

The following companies can help guide you through the maze of choices.

Tip: Insulating hot water tanks and piping in the basement can greatly reduce energy use and heating costs.

DWYER INSULATION
8401 Baldwin St., OAK (415)430-8383

PALMER INDUSTRIES INC.
10611 Old Annapolis Rd., Frederick MD (301)898-7848

INSULATION CONTRACTORS

LANDSCAPE ARCHITECTS

These landscape architects specialize in appropriate siting, temperature control plantings, and drought -tolerant and native plant landscaping.

4 DIMENSIONS LANDSCAPE DEVELOPMENT
4121 Culver, OAK (415)261-5820

DON JENSEN & ASSOCIATES
20 E. 5th St., Morgan Hill (408)778-2495

NATIVE LANDSCAPES
1021 University, San Jose (408)379-1323

LUMBER

Large lumber companies have produced major green-washing campaigns explaining how benign their managed forestry processes are. What the companies claim is sustainable harvesting technique is, in reality, massive clearcutting, and the monoculture of fast-growing softwood trees.

It is, however, possible to grow trees and harvest them in ecologically sound ways. It is very important to be informed about the how wood is grown, harvested, processed, and used, especially if you are a frequent user. The following companies produce or sell lumber only from selective cuts or the harvest of dead wood. There is a growing alliance of selective lumberers, and their products should soon be more available, and visible in the Bay Area.

If you do consume wood, tithe some money back into a tree planting program to replace the trees you use. (See Index under Tree Planting

If you buy tropical hardwoods, chances are great that you are contributing to the destruction of the rainforest. Avoid buying these tropical hardwoods: African walnut, ebony, greenwood, rosewood, tulipwood

Cedar and Redwood – Even though they are a beautiful woods to build with, cedar and redwood should be avoided. The last ancient redwoods in Mendocino county are now being cut down. Even if the trees grow back in the distant future, the unique Redwood ecosystem they support may never return.

Wood you can buy: apple, ash, aspen, beech, birch, black walnut, cherry, chestnut, elm hickory, larch, maple, oak, pear, pine, poplar, spruce, sycamore, hemlock, and Douglas fir.

Fact: In the current rate of destruction, scientists have estimated that the worlds rainforests could be completely destroyed in 30-50 years.

WILD IRIS FORESTRY
Catalog, P.O. Box 1423, Redway (707)923-2344

Fact: In rainforests, nearly 117 acres are chopped down every minute. This comes to 60 million acres destroyed a year.

Fact: Hundreds of thousands of people are already affected by rainforest depletion, through flooding, erosion, drought, shortages of fuelwood, and society and culture displacement.

Fact: 70 percent of more than 3,000 plant species known to help in treatment of cancer are found in tropical rainforests.

Fact: As a result of rainforest destruction, at least one species of plant or animal becomes extinct every single day.

PAINT, NON-TOXIC

AFM ENTERPRISES
1140 Stacy Ct., Riverside (714)781-6860

ARMSTRONG PAINTING & WATERPROOFING
774 Harrison St., SF (415)777-1234

AURO ORGANIC PAINTS
PO Box 857, Davis (916)753-3104

LIVOS PLANT CHEMISTRY
Catalog of nontoxic home products (800)621-2591

MILLER PAINT
317 S.E. Grana, Portland (503)233-4491

PACE CHEM INDUSTRIES, INC.
779 S. La Grange Ave, Newbury Park (805)499-2911

SAGE: THE ENVIRONMENTAL STORE
(415)485-5441

SINAN COMPANY
P.O. Box 857, Davis (916)753-3104

THE COHERENCY COMPANY
P.O. Box 553, Occidental (707)869-0956

THE LIVING SOURCE
3500 Mac Arthur Drive, Waco, TN (817)756-6341

Fact: One gal. of gasoline can contaminate up to 250,000 gal. of water.

THE OLD FASHIONED MILK PAINT CO.
completely non-toxic, water based paint for interior use
P.O. Box 222, Groton MA
(508)448-6336

PAINTERS

ARMSTRONG NATURAL PAINTING
774 Harrison St., SF (415)826-9134

CAL-COAT PAINTING
complete non-toxic interior paints & painting
licensed & insured
415 Trowbridge Road, Santa Rosa
(707)545-8838

NON-TOXIC HOUSEHOLD WORKERS CO-OP
P.O. Box 22324, SF (415)681-6113

Fact: Twenty-five percent of hazardous waste comes from individual households.

WANTED:

a healthier San Francisco Bay

If you see:
- ☐ **Oil Slicks**
- ☐ **Fish or Bird Kills**
- ☐ **Filling of Marshes or Wetlands**
- ☐ **Floating Sewage**
- ☐ **Water Discoloration**
- ☐ **Leaking Land Fills**
- ☐ **Vessels Discharging Oil or Waste**

If you catch or find:
- ☐ **Deformed Fish**
- ☐ **Fish with Fin Rot, Ulcerations, Parasites**
- ☐ **Clams or Mussels with Tumors**
- ☐ **Crabs or Shrimp with "Burn" Holes**
- ☐ **Deformed Ducks, Shorebirds or Seabirds**

Record the date, time, location, a description; take a photo and **CALL 1-800-KEEPBAY**

If you care about the Bay and Delta:
- ☐ **Become a BayKeeper Hotline Volunteer**
- ☐ **Take our course which trains certified BayKeepers who will patrol the Bay and Delta**

The Baykeeper program is a project of the San Francisco Bay-Delta Preservation Association, a non-profit organization dedicated to the protection and enhancement of San Francisco Bay and Delta. Our work supplements that of the enforcement branches of regulatory agencies and will detect & document violations of environmental regulations.

The BayKeeper is designed to raise public awareness concerning threats to the the Bay and Delta by providing accurate information to agencies, advocacy groups & the media. The program will also serve as an antenna for citizen complaints and act as a deterrent to illegal activities on the Bay and Delta.

BayKeeper

CAREERS

You don't have to be a wildlife biologist or a trained ecologist to work for the betterment of the planet. Environmental organizations have many other needs, especially in these days of diminished public funding for non-profits. They need help doing everything from direct mail campaigns to computer programming. For a full list of socially-responsible job placement services, and a discussion of the how to get work with non-profits, see the article starting on page 139.

CA LEAGUE OF CONSERVATION VOTERS
965 Mission St., SF (415)896-5550
See Display Ad This Page

Employment

JOBS & CAREERS IN ENVIRONMENTAL PROTECTION

- ***Full and Part Time***
- ***Management Training***
- ***Political & Electoral Training***
- ***Progressive Workplace***

896-0665

California League of Conservation Voters

CHILDREN

The children in your life present a wonderful opportunity and reason for living by ecologically-sound principles, and teaching children about the means to a positive and sustaining future.. Children's wares have traditionally been the biggest reuse item. Buying used clothing saves resources for the planet, and money for you. Cloth diapers don't pollute, persist, or fill up our landfills. Their natural fibers are better for your baby and pocket book than a ton of disposables. Feeding your child organic baby food, rather than processed, supports organic farmers and your baby's health. An incredible array of toys are available that teach kids about the environment. (See the Book Review section for kid's publications about the environment too).

CHILDREN'S GOODS

BABY BOOM
1601 Irving, SF (415)564-2666

BABY BOOM II
1518 Grant Ave., Novato (415)892-4264

CRACKERJACKS
14 Glen Ave., OAK (415)654-8844

IT'S A BIG DEAL
918 San Pablo Ave., Albany (415)527-5533

KIDS AGAIN
11815, Dublin Blvd., Dublin (415)828-7334

KIDSYSTEMS TOUGH TRAVELER LTD.
1012 State St., Schenectady NY (518)377-8526
See display ad on this page

LOU'S BABY FURNITURE
221 Willow Ave, Hayward (415)581-6082

QUINBY'S
3411 California, SF (415)751-7727

SCHNEIDER EDUCATIONAL PRODUCTS
2880 Green St., SF (415)567-4455

THE CHILDPROOFER
P.O.Box 14718, Santa Rosa (707)545-1116

THE PUZZLE PEOPLE
22719 Tree Farm Rd., Colfax (800)443-4823

CLOTHING

BIOBOTTOMS
P.O. Box 6009, Petaluma (707)778-7945
See display ad on this page

KIDS AGAIN
11815, Dublin Blvd., Dublin (415)828-7334

KIDZOO
1201 C Solano Ave, Albany (415)525-3488

MOONFLOWER BIRTHING SUPPLY
8593 Hwy.172, Ignacio (303)884-2383

FOOD

Since the Alar scare in 1989, when a NRDC report revealed that children are being exposed to a greater risk of cancer because of pesticides on produce, many parents are seeking baby foods that are free of chemical residues. As these residues do seem to affect children more profoundly than adults - they are still growing, and eat proportionally more food with greater toxicological effects - close scrutiny of their food does seem wise.

EARTH'S BEST INC.
6786 Sebastopol Ave., Sebastopol (707)823-1510

SIMPLY PURE BABY FOOD
RFD #3, Box 99, Bangor ME (800)426-7873

DIAPER SERVICES

Diaper services are the most intelligent, least wasteful solution to a huge disposable diaper problem. Diaper services are accredited, abide by health regulations, and don't throw away huge numbers of non-biodegradable, health-threatening waste. It is time to deal with waste in all senses of the word; it cannot be just thrown away. Unfortunately, diaper services often use polluting bleaches and detergents. The *Green Pages* looks forward to the development of natural and biodegradeable cleaners being regularly used in this industry.

ABC DIAPER SERVICE INC.
1800 2nd St, BERK (415)549-1133

CLOUD 9 DIAPER SERVICE
9970 Horn Rd., Sacramento (800)821-8264

DY-DEE WASH COTTON DIAPER SERVICE
1373 Lorie Ave., SSF (415)761-4445

MISSION DIAPER SERVICE
142 Stockton Ave., Santa Cruz (408)426-8286

DIAPERS

Using cotton diapers, rather than disposable diapers, saves space in landfills, avoids contaminating groundwater, and may even free your baby from diaper rash. Current research indicates that more than 18 billion disposable diapers, using more than 75,000 tons of petroleum-derived plastic and roughly 800,000 tons of tree pulp, are discarded, and create over 4.5 million tons of non-biodegradable waste each year.

The diapers listed here are not run-of-the-mill. All are super-durable, and last through many more washings than regular diapers. You can buy regular, less expensive diapers and wash them yourself or contract a diaper service.

BABY BUNZ & CO.
P. O. Box 1717-D, Sebastopol (707)829-5347
See display ad on this page.

BIOBOTTOMS
P.O. Box 6009, Petaluma (707)778-7945
See display ad, page 276

BUMKINS
6040 Bonnie Doon Road, Santa Cruz (408)459-6630

SUCH A BUSINESS
1 Rhode Island, SF (415)431-1703

TOYS

HEARTHSONG
P.O. Box B, Sebastopol (707)829-0900
See display ad on this page.

LOLLY AND COMPANY
6100 Fourth Ave., Seattle WA (206)762-6423

QUINBY'S
3411 California, SF (415)751-7727

SUCH A BUSINESS
1 Rhode Island, SF (415)431-1703

Tip: Always buy pesticide-free food, particularly for your children. It may cost a little more, but it's an investment in a long and healthylife.

CLOTHING

We are so accustomed to looking at some things in our lives that we don't even consider their relationship to the environment. Clothing is like that. Every shirt, skirt, sock, and hat you buy is made of resources, and uses energy and chemicals in its manufacturing process. One way you can save energy, money, and materials is by buying used clothing, and giving your cast-offs to thrift stores and charities. The *Green Pages* highly recommends that people take the interesting route of thrift store, garage sale, and flea market high-fashion.

When you do need to buy new clothes, buy from stores that donate to environmental groups, and that use natural fibers. Obviously anyone can go out and find a new clothing store. There can be a difference, however. Outback, for example, tithes a portion of its profits to environmental causes. In general, any company that carries natural fiber clothing has a leg-up on the ubiquitous polyester clothing purveyor.

Most fabrics that contain polyester, or are "permanent-press" or "wrinkle-resistant" have been fabricated with formaldehyde finishes that will not break down, even after numerous washes. Synthetic fabrics are also manufactured from petrochemicals, and never biodegrade.

Natural fibers include cotton, wool, silk, linen, and ramie (made from a nettle fiber grown in Asia). Rayon, though sounding like an artificial variant of nylon, is actually a cellulose product. It is manufactured from waste cotton, paper and wood pulp. Other reasons for not buying new clothing, even if made from natural fibers, is the fact that the growing techniques are usually chemical-intensive (Cotton and ramie in particular). Rayon also requires a petrochemical process to break the cellulose down into a usable form.

NEW CLOTHING

OUTBACK
1331 Eighth St., BERK (415)527-6886
See display ad on this page.
401 Bayshore Blvd., SF (415)285-5244
Willow Shopping Center, Concord (415)686-2270

WHOLE EARTH ACCESS
2990 Seventh St., BERK (415)845-3000
See display ad, page 291.

WOOLIES
413 Shrader, SF (415)221-8556

EDUCATION

There are a large number of unique environmental educational opportunities in the Bay Area — enough to have justified our dedicating a whole section of the *Bay Area Green Pages* to them (see the Education section, starting on page 171). There are people eager to show you anything from how to build a passive-solar-heated addition on your house to the workings of a salt marsh.

ELMWOOD INSTITUTE
P.O. Box 5765,BERK (415)845-4595
See display ad, this page

GAIA BOOKSTORE
1400 Shattuck Ave. #10, BERK (415)548-4172

MARINE SCIENCE INSTITUTE
P.O. Box 7142, Redwood City (415)364-2760

MERRITT COLLEGE - SELF RELIANT HOUSE
12500 Campus Dr., OAK (415)436-2405

OCEANIC SOCIETY EXPEDITIONS
Bldg. E, Rm. 257, Fort Mason , SF (415)441-1106

OWNER BUILDER CENTER CONSULTING
1250 Addison St., Ste. 209, BERK (415)848-6860
See display ad, page 278

WILDERNESS TRANSITIONS
(415)332-9558
See display ad, this page

FINANCIAL SERVICES

If you have money to invest, deciding where and how to invest it is as important as deciding what products and services to buy. Investing in a business supports that business. We need to use the power we have as investors to withdraw support from companies that are ruining the planet, and to support those that are taking green values seriously. We are used to looking at investments from a strictly rate-of-return (and usually short-term rate of return at that) point of view. What this perspective ignores is that part of investing is investing in the future. An irony which escapes many non-ethical investors is that they are investing (for future returns) in companies which are busy undermining the chances for that very future.

Socially-responsible, environmentally-sound investing does not have to mean poor returns. Both the Ariel Environmental Appreciation Fund (Calvert) and The Eco-Logical Trust did very well this year, and they have two of the cleanest environmental screens around. The criteria used by "socially responsible" investment funds vary widely. Some are purely subjective, while others have carefully outlined rules.

It can be very helpful to talk to a financial planner to help understand what your options are. All the financial planners and consultants listed here are members of the Bay Area Socially Responsible Investment Professionals. They are used to helping their clients make socially responsible investments, and discussing the types of criteria and goals the client may have. (see the article, *The Stockbrokers Smile*, p. 54)

Fact: The Savings and Loan debacle will cost Americans $500 billion over the next 30 years, and has almost disappeared from the public eye.

MARTY ACKERMAN
1 Alvarado Square, San Pablo (415)215-3030

BONNIE ALBION
Albion Financial Associates
2550 Ninth St., Suite 209B, BERK (415)486-8333

ERIC ARNOW
Northwestern Mutual Life
One Sansome St., Suite 1700, SF (415)956-8900 x259

BAY AREA SOCIALLY RESPONSIBLE
Investment Professionals
A Chapter of the Social Investment Forum
P.O. Box 26277, SF

WENDY BOLKER
Certified Personal Financial Planner
P.O. Box 2567, Santa Cruz (408)423-9407

JOSEPH ALAN BRAMAN, INC.
mutual fund
1315 Polk St. #301, SF (415)441-5141 x301

ANN M. BRENT
Dean Witter Reynolds (415) 955-6036

MICHELLE CHAN
East/West Securities Co.
120 Montgomery St. Suite 993, SF (415)397-3400

CLEAN YIELD ASSET MANAGEMENT
224 State St., Portsmouth, NH (603)436-0820

S. LOREN COLE
2550 Ninth St., #111, BERK (415)843-3080

NICHOLAS DEWAR, CPA
317 Noe St., SF (415)863-8485

PAUL B. DRESCHER
Certified Financial Planner
P.O. Box 1131, Capitola (408)462-3200

CAROL FELTON-MALNICK
Prudential-Bache Securities
525 University Ave., Palo Alto (415)328-0110

FRANKLIN RESEARCH & DEV. CORP.
65 K Gate 5 Road, Sausalito (415)332-5822

GIL FRIEND
Social Entrepreneurs Network Direct, Inc.
2118 Seventh St., BERK (415)548-7904

FRED FULD, CFP
New Alternatives Fund
3043 Clayton Road, Concord (415)686-9067

DAN GEIGER
Vanguard Public Foundation
14 Precita Ave., SF (415)285-2005

JOYCE G. HAROLD
Baraban Securities, Inc.
100 Hagan Burger Road, OAK (415)633-0433

DAWN HOLLERITH
Integrated Resources Equity Corp.
630 San Gallinas, San Rafael (415)381-2300
101 California St., SF (415)955-6893

KAYE HOXTER, CFP
1159 Clay St. #7, SF (415)928-1586

VAN KASPER & COMPANY
50 California St., Suite 2400, SF (415)391-5600

BONNIE LEVINE
33 Buchanan Dr. Sausalito (415)331-3820

HORACE R. MARKS
Greenleaf Financial Services
220 Montgomery St., SF (415)399-8848

NZINGA NYAGUA
370 Turk St., Suite 331, SF (415)826-8716

LINCOLN PAIN
2014 Los Angeles Ave., BERK (415)525-1230
See display ad on this page.

THE PARNASSUS FUND
244 California St., SF (415)362-3505

ENID PEARSON
Pearson Wood Associates, Inc.
127 2nd St., Suite 1, Los Altos (415)949-5393

PLANNED INVESTMENTS, INC.
700 Larkspur Landing Cr., Suite 120, Larkspur
(415)461-8558

ROGER PRITCHARD
Financial Alternatives
1514 McGee Ave., BERK (415)527-5604

PRATEK CORPORATION
235 Montgomery St., SF (415)956-3940

STOREY & GREEN ASSOCIATES
230 California St., Suite 500, SF (415)398-3950

PAUL TERRY & ASSOCIATES
1269 Rhode Island, SF (415)641-4456

WORKING ASSETS MONEY FUND
230 California St., SF (415)989-3200
See display ad on this page.
230 California St., SF (800)522-7759

FINANCIAL

Some Investment Criteria:

Calvert Social Investing Fund
Calvert Group
Seeks advice from local and national environmental groups; screens out major polluters and the nuclear industry.

Eco-Logic Trust '90
Merrill Lynch
(available from any major broker)
Mutual fund with a strict environmental screen, applied by Progressive Asset Management. Donates a portion of profits to the Environmental Federation of America.

New Alternatives Fund
Sector fund specializing in alternative energy. Excludes nuclear power.

Parnassus Fund
Vague environmental criteria, but screens out most egregious violators.

Pax World Fund
Invests in companies with "sound environmental policies andpractices," among other criteria. Advised by Clean Yieldnewsletter. Consults with local environmentalists, and screensout major polluters.

Progressive Environmental Fund
Progressive Asset Management
Sector fund investing in environmental services, with strict screen to exclude environmental services which also pollute.

Working Assets Money Fund
Fairly strict environmental criteria screen out EPA violators; fund looks for, but is not limited to, alternative energy and energy conservation firms.

A trip to the supermarket is like a trip to Las Vegas: bright colors, flashy displays, elaborate packaging, mellow music, and perfect glowing fruit dazzle and delight us. But this perfection comes at a high price, one greater than just the bar code readout. Chemical additives and preservatives mask the long distance and time between fresh-picked produce and the food relics we eat. The harried clerks, long lines, and production-line speed of "check-out" are incredibly remote from a connection to farmers and land.

Many consumers don't realize the environmental and social price tag they are actually paying to eat flawless or out-of-season produce. American lands and waterways are polluted with agricultural run-off, farm subsidies encourage unsustainable mass production, and dangerous pesticides threaten farm workers here and in the Third World. The EPA has determined that agriculture is the greatest non-point source of ground water sources in the U.S and has ranked pesticide residues as the third most important environmental factor for cancer risks in the country. (See the article on pesticides, p. 32)

This section of the Consumer Guide is an attempt to remedy this estrangement from our food supply. To address the energy costs and environmental pollution that result from transcontinental shipping of foods, we have compiled a list of local farmers, bakers, and manufacturers, and detailed in what grocery stores or farmers' markets their goods can be found. Local producers who grow or are using organic ingredients are given special attention as we feel their efforts particularly warrant your support.

Don't overlook the many remaining farms that operate in the area. The Bay Area has some of the finest agricultural soils in the United States although they are quickly being lost to development. Discover the area's bounty by following Farm Trail Maps for the ten counties. (See the Farm Trails listing on page 37). You can fight for preservation of the region's agriculture by joining local groups (see the Organization section).

The Bay Area Green Pages has made a decision to not include meat here regardless of how benignly the animals are raised. This decision is based not only on a concern for animal rights, but also on a human rights and environmental perspective. Two thirds of our grain exports to the Third World are used to feed the livestock that these countries export, while many of the people go hungry. The toll extracted from the health of Third World growers and the environmental damage done to these countries (and our own) is too high. Producing enough food to feed a meat-eater requires 4,200 gallons of water a day. To feed a vegetarian requires only 300 gallons. Tropical rain forests are being cleared in Latin America to raise cattle; a pound of hamburger represents 55 square feet of burned-off forest. Almost 4 billion tons of topsoil are lost each year in the U.S., primarily because of overgrazing of livestock. It also takes 39 times more energy to produce beef than soybeans having equal caloric value.

BAKERIES

These bakeries make a committment to the use of organic or minimally-processed ingredients. Some make contributions to their communities as well. Bakeries in general fulfill a greener aim by being local food processors. The less distance our food travels, and the more we know about its producers, the better off we, and the planet, are.

BOLINAS BAY BAKERY
Box 475-20 Wharf Rd., Bolinas (415)868-0211

LEAVEN & EARTH BAKERY
5427 College Ave., OAK (415)658-9836

TASSAJARA BREAD BAKERY
1000 Cole also
Fort Mason, SF (415)664-8947

THE BREAD WORKSHOP
1250 Addison St., BERK (415)649-9735

UPRISINGS BAKING COLLECTIVE
1210 Tenth St., BERK (415)527-4043

CATERING

FROM THE HEART CATERING
1-800-CREATED (415)386-7132

FARMER'S MARKETS

Farmer's markets are a great thing. Even the most jaded city type can enjoy the colors, noise and lively " bazaar" flavor of a farmer's market. Here you meet real farmers. County Agricultural Commissioners certify that the purveyors are indeed farmers. You can ask questions regarding pesticide use, and express your desire for organic farming techniques directly. Farmers peddle their wares quite effectively by offering tastes of their produce. Who can resist? It's cheap, too.

If your city does not have a farmer's market, lobby your local government to get one. If you do have a market, support it by shopping there regularly.

ALAMEDA FARMERS' MARKET
Center and Webster, Concord (415)798-7061
See display ad, page 281

ALEMANY FARMERS' MARKETS
SF (415)647-9423

BERKELEY FARMERS' MARKET
Tues.-Derby and Novia, BERK
Sat.-Center and Novia, BERK (415)548-2220

CLOVERDALE FARMERS' MARKET
Cloverdale (707)894-4623

Fact: All essential nutrients can be obtained from non-meat sources

DOWNTOWN FARMERS' MARKET FESTIVAL
4th @ B St., San Rafael
Thursdays (415)457-2266

DOWNTOWN NAPA FARMERS' MARKET
Comb St., Napa (707)257-0322

FARMERS' MARKET OF NAPA
Napa (707)965-3652

FELTON FARMERS' MARKET
Felton (408)335-9364

GILROY FARMERS' MARKET
5th @ Monterey, Gilroy
Thursdays 3-7, June-Oct. (408)842-6964

HEALDSBURG FARMERS' MARKET
900 Sunnyvale Dr., Healdsburg (707)433-6063

LIVERMORE FARMERS' MARKET
K and 2nd, Livermore (415)373-1795

LOS ALTOS FARMERS' MARKET
Los Altos (408)948-3366

MARIN COUNTY FARMERS MARKET
1114 Irwin St., San Rafael (415)456-3276

MORGAN HILL FARMERS' MARKET
(408)779-5130

NAPA FARMERS' MARKET
Napa (707)252-3972

NOVATO FARMERS' COUNTRY
1114 Irwin, Novato (415)456-3276

PALO ALTO FARMERS' MARKET
Gilman and Hamilton, Palo Alto
Saturday 8-12 (415)325-2088

PETALUMA FARMERS' MARKET
4th between Western and B, Petaluma
Saturday 2-5 (707)762-0344

PLEASANT HILL FARMERS' MARKET
Taylor Blvd. and Morello, Pleasant Hill
Saturday 8-1 (415)625-4333

REDWOOD CITY FARMERS' MARKET
609 Price Ave. #205, Redwood City (415)366-8401

RICHMOND FARMERS' MARKET
Richmond Civic Center,RICH
Friday 2-6 (415)724-2283

HEART OF THE CITY FARMERS' MARKET
1182 Market St. at 7th, SF (415)558-9455

SAN JOSE FARMERS' MARKET
San Jose (408)298-4303

SAN JOSE FARMERS' MARKET 2
647 S. King Rd., San Jose (408)926-5555

SANTA CRUZ FARMERS' MARKET
Santa Cruz (408)722-6483

SANTA ROSA FARMERS' MARKET
Santa Rosa (707)823-7023

SEBASTOPOL FARMERS' MARKET
1764 Cooper Rd., Sebastopol (707)823-0823

ST. HELENA FARMERS' MARKET
Napa (707)252-3972

VALLEJO FARMERS' MARKET
Vallejo (415)456-3276

WALNUT CREEK FARMERS' MARKET
Broadway and Lincoln,Walnut Creek
Sunday 8-1 (415)945-2940

WATSONVILLE FARMERS' MARKET
Watsonville (408)726-2521

Tip: Buy food in bulk, saving on expensive packaging, and on the food itself.

FOOD & PRODUCE

This section contains local retailers and wholesalers of products that offer alternatives to overprocessed supermarket fare. To be listed, groceries and distributors must offer some organic items as well. These organic items need not be third-party certified as do those in our "Organic food" category.

The third-party organic certification process (in California, the CCOF mark prevails) can be a lengthy and costly one, so many growers may instead abide by the Department of Health Services' regulations for organic production, which are less stringent than those of the CCOF, and then label produce " Grown in accordance with California Health Code 26569 11."

The use of the words "natural" and "no-spray" or "transitional" mean nothing in a legal sense. There is no verification process for these claims. This doesn't mean that the grower is being dishonest, but the buyer should beware. "No spray" typically means that the crops have not been sprayed with pesticides, yet makes no similar claim regarding the use of synthetic fertilizers or fungicides in the soil. "Pesticide free" probably applies only to the mature fruit or vegetable; nothing is claimed regarding the soil, or the immature plant's treatment. A "transitional" designation means the grower is in the process of obtaining third-party or state organic certification. She or he is adhering to organic guidelines but soils take awhile to recover from any synthetic method of farming. For specifics on organic labels see the article *So You Want To Go Organic*, page 36 .

Health food stores are also a good place to find natural soaps and vegetable washes which can help you clean off pesticide residue from any non-organic food you do obtain. A cautionary note: health food stores are also major outlets for vitamins and food supplements of questionable value. There are no quick fixes that can take the place of good nutrition.

AGACCESS
catalog of food goods (916)756-7177

BACK TO BASICS
74 E. Washington St., Petaluma (707)762-9289

EL CERRITO NATURAL GROCERY CO.
10367 San Pablo Ave., El Cerrito (415)526-1155

BERKELEY NATURAL GROCERY CO
1336 Gilman St., BERK (415)526-2456

BREAD OF LIFE
1690 S. Bascom Ave., Campbell (408)371-5000

BREAD OF LIFE
1690 S. Bascom Ave., Campbell (408)371-5000

BUFFALO WHOLE FOOD COMPANY
1058 Hyde St., SF (415)474-3053

COUNTRY VILLAGE NATURAL FOODS
3611 Union Ave., San Jose (408)377-1431

DEL TOMASO PRODUCE
4930 Telegraph Ave, OAK (415)655-2906

EDEN FOODS, INC.
701 Tecumseh Rd., Clinton MI (517)456-7424

ENVIROMINTS
11500 Pinehurst Wy. N.E. #411, Seattle WA (206)298-1230

BUFFALO WHOLE FOOD COMPANY
598 Castro St., SF (415)626-7038

FOR YOUR HEALTH NATURAL FOODS
1087 Foxworthy Ave., San Jose (408)267-0825

GOOD LIFE GROCERY
1524 20th, SF (415)282-9204

GOOD NATURE GROCERY
1359 N. Main St., Walnut Creek (415)939-5444

RIMA GREEN VALLEY
297 Page St., SF (415)431-7250

HEALING MOON HEALTH FOODS
523 Main St., Half Moon Bay (415)726-7881

HIGH HEALTH SHOPPE
2172 Chestnut, SF (415)921-1400

INNER SUNSET COMMUNITY FOOD STORE
1319 20th Ave., SF (415)664-5363

JAFFE BROS.
P.O.Box 636, Valley Center (619)749-1133

LAUREL NATURAL FOODS
777 Laurel, San Carlos (415)593-3274

LIVING TREE CENTER
P.O. Box 10082, BERK (415)420-1440

LLOYD'S NATURAL FOODS MARKET
39145 Fremont Hub, Fremont
(415)792-3000

MAD RIVER FARM COUNTY KITCHEN
P.O. Box 155, Arcata (707)822-7150

MOLLY STONE'S
100 Harbor St., Sausalito (415)331-6900

NATURAL FOODS
708 Ferry, Martinez (415)228-4747

NATURAL TEMPTATIONS
190-A Alamo Plaza, Alamo (415)820-0606

NOE VALLEY COMMUNITY FOOD STORE
1599 Sanchez, SF (415)824-8022

OTHER AVENUES
3930 Judah, SF (415)661-7475

REAL FOOD
5800 Shellmound Ave., Emeryville (415)420-8085
See display ad, page 281.

S.F. HEALTH FOOD STORE
Cosmetics, home products
333 Sutter St., SF (415)392-8477

SAN RAFAEL HEALTH FOODS
1132 4th St., San Rafael (415)457-0132

SANTA ROSA COMMUNITY FOODS
1215 Morgan St., Santa Rosa (707)546-1806

STAPLETON'S
415 River St., Santa Cruz (408)425-5888

SUNSHINE HEALTH FOOD
1 Embarcadero Ctr., SF (415)788-1380

THOM'S NATURAL FOODS
5843 Geary Blvd., SF (415)387-6367

TKO FARMS
101 East Manor Dr., Mill Valley (415)389-8137
See display ad, page 282.

TOTAL LIFE HEALTH FOODS
1757 Taraval, SF (415)681-5544

TRUE FOODS NATURAL MARKET
45 Camino Alto, Mill Valley (415)383-4515

FROZEN FOOD

CRYSTAL WAVE
P.O. Box 2298, Sebastopol (707)829-8336

Fact: One pound of beef requires 16 pounds of grain and soy to produce.

HERBS

Organically grown herbs are available in these shops.

AU NATUREL HEALTH FOOD STORE
2370 Market, SF (415)431-9963

LHASA KARNAK HERB CO.
2513 Telegraph Ave., BERK (415)548-0380

MOUNTAIN ROSE HERBS
P.O. Box 2000, Redway (800)879-3337

NATURE'S HERB CO.
1010 46th St., Emeryville (415)601-0700

NATURE'S SUNSHINE PRODUCTS
3557 Perada Dr., Walnut Creek (415)256-6851

RAINBOW GROCERY & GENERAL STORE
1899 Mission St., SF (415)863-0620

SAN FRANCISCO HERB COMPANY
250 14th St., SF (415)861-3018

THOM'S NATURAL FOODS
5843 Geary Blvd., SF (415)387-6367

ORGANIC FOODS

This is a list of growers and stores that sell products certified to be organic by a third-party certifier. Third-party certifcation is the most difficult for a grower to obtain, and assures that the produce or food is about as healthy a product as one could get. For a discussion of the third-party organic certification process and other certifiers, see the article *So You Want to Go Organic* starting on page 3. If your grocery doesn't sell organics that have been produced locally, you might want to visit an organic farm in your area. For a list of local organic farms that welcome visitors, see the list on page 32

ALMA'S PLACE
6154 Sawmill Rd., Paradise (916)872-9107

AMERIAN BROS. PRODUCE
335 E. Taylor, San Jose (408)293-7030

APPLE A DAY RANCH MARKET
4955 Deerwood Dr., Santa Rosa (707)525-1881

FOOD

ASSOCIATED COOPERATIVES
322 Harbour Way Suite 12, RICH (415)232-1111

BACK TO BASICS
74 E. Washington St., Petaluma (707)762-9289

BODEGA COFFEE COMPANY
P.O. Box 252, Kentfield (415)485-1111

BREAD OF LIFE
1690 S. Bascom Ave., Campbell (408)371-5000
See display ad, page 283

BUFFALO WHOLE FOOD COMPANY
1058 Hyde St., SF (415)474-3053

CALIFORNIA CERTIFIED ORG. FARMERS
P.O. Box 8136, Santa Cruz (408)423-2263
See display ad, page 281

CANYON LAKES MARKET
500 Bollinger Canyon Way, San Ramon (415)735-0801

COMMUNITY FOODS
2724 Soquel Ave., Santa Cruz (408)462-0458

CUPERTINO NATURAL FOODS
(408)253-1277

DAVID'S FINEST PRODUCE
P.O. Box 372, Stinson Beach (415)868-2024

DEL TOMASO PRODUCE
Providing the largest selection of organically grown produce in Oakland
4930 Telegraph Ave, OAK
(415)655-2906

EARTHBEAM NATURAL FOODS
1399 Broadway Ave., Burlingame (415)347-2058

FOOD FOR THOUGHT
7231-B Healdsburg Ave., Sebastopol (707)829-9801

GOOD EARTH NATURAL FOODS
123 Bolinas Ave., Fairfax (415)454-4633
See display ad, page ##

GRAINAISSANCE INC.
1580 62nd. St., Emeryville (415)547-7256

RIMA GREEN VALLEY
297 Page St., SF (415)431-7250

GREEN VALLEY ORGANIC SPECIALTIES
P.O. Box 2523, Sebastopol (707)823-6510

GREENLEAF PRODUCE
1980 Jerrold Ave., SF (415)647-2990

HAIGHT-FILLMORE WHOLE FOODS
501 Haight, SF (415)552-6077

INI INTERNATIONAL
309 Sheffield Ave., Mill Valley (415)383-1869

INNER SUNSET COMMUNITY FOOD STORE
1319 20th Ave., SF (415)664-5363

LINDA'S GARDEN
17262, Bodega Hwy., Bodega (707)876-3466

LIVING FOODS 2
224 Greenfield Ave., San Anselmo (415)457-4317
See display ad, page 283.

MAD RIVER FARM COUNTY KITCHEN
P.O. Box 155, Arcata (707)822-7150

MILL VALLEY ORGANIC & NATURAL FOOD
905 Sir Francis Drake Blvd.,Suite D, Kentfield
(415)456-5112

NOE VALLEY COMMUNITY FOOD STORE
1599 Sanchez, SF (415)824-8022

ORGANIC GROCERIES
2481 Guerneville Rd., Santa Rosa (707)528-3663

OTHER AVENUES FOOD STORE
3930 Judah, SF (415)661-7475

PLEASANT GROVE FARMS
PO Box636, Pleasant Grove (916)655-3391

PLEASANT HILL PRODUCE CO.
"The Good Foods Merchant"
2203 Morello Ave., Pleasant Hill
(415)798-5275

RAINBOW GROCERY & GENERAL STORE
1899 Mission St., SF (415)863-0620

REAL FOOD #7
5800 Shellmound Ave., Emeryville (415)420-8085
See display ad, page 281

RIVER VALLEY ORGANIC GARDEN
21040 Railroad Ave., Geyersville (707)857-3868

SANTA ROSA COMMUNITY FOODS
1215 Morgan St., Santa Rosa (707)546-1806

SEQUOIA FAMILY MARKET
5243 Hwy. Nine, Felton (408)335-9337

SONOMA GOLD ORGANICS
Organic Apples & Apple Product
(707)829-1121

STAPLETON'S
415 River St., Santa Cruz (408)425-5888

SUPER NATURAL FOODS
147, Corte Madera Town Center, Corte Madera
(415)924-7777

THE APPLE FARM BATES AND SCHMITT
18501 Greenwood Road, Philo (707)895-2333

THE FOOD BIN
1130 Mission, Santa Cruz (408)423-5526

THE SPROUT HOUSE
40 Railroad St., Great Barington (413)528-5200

TRUE NATURE FOODS
13070 Hwy. 9, Boulder Creek (408)338-2105

VERITABLE VEGETABLE, INC.
1600 Tennessee St., SF (415)821-1450

WHOLE FOODS MARKET
774 Emerson, Palo Alto (415)326-8666

WHOLE HERB COMPANY
P.O. Box 1085, Mill Valley (415)383-6485

WILD ROSE RANCH
308-B Center St., Healdsburg (707)433-7869

SOY PRODUCTS

VITASOY (U.S.A.) INC.
99 Park Lane, Brisbane (800)848-2769
See display ad on this page.

Fact: Eating less meat can help reduce the amount of meat imported from Central and South America, where increasing cattle production has greatly contributed to deforestation.

GARDEN

More than 69 million Americans gardened in 1988, according to the National Gardening Association. It is America's favorite recreational activity, according to a1987 Louis Harris poll. Unfortunately, most of that gardening is done with chemicals. Almost 400 lawn care products are sold in stores and through the mail. Residential lawns and gardens receive as much as 10 pounds of chemicals per acre each year, according to a 1980 National Academy of Sciences study, while soybean crops, for instance, generally receive only two. Only 120 of the 600 principal ingredients in those commercially available pesticides have been registered by the U.S. government so far.

The quest for cheaper, more ecologically-correct living can begin in the garden—if you have one. Here, you can practice chemical-free growing techniques, compost variations, low-water-use landscaping, indigenous plants, and proper maintenance. Here you can also nourish your connection with the earth and truly learn to appreciate its bounties.

In order to do either, however, you need outlets that supply non-chemical products, sound advice, and native plants. Peruse and frequent the following as you create your own garden of Eden, where you actually feel safe eating the fruit.

COMPOST

If you have only just started your own compost pile, you might not be able to wait two months to feed your plants with mother nature's most balanced food. You might be an apartment dweller with little room for a compost pile. If you are a serious gardener with a large plot to amend you will probably need some help. Whatever your problems, these organic compost suppliers can help.

Remember– if you don't compost yourself, you can probably find a friend who does, and would be happy to take contributions of organic materials.

AMERICAN SOIL PRODUCTS
2222 3rd St., BERK (415)540-8011
See display ad on this page.

EBMUD
2130 Adeline St., OAK (415)835-3000

ECOLOGY CENTER
2530 San Pablo, BERK (415)548-2220

RECYCLED WOOD PRODUCTS
3rd and Harrison St., BERK (415)525-4557

THE NATURAL GARDENING CO.
217 San Anselmo Ave.San Anselmo (415)456-5060

WEBB FARM SUPPLIES
5381 Old San Jose Rd., Soquel (408)475-1020

COMPOSTERS

These are the large, screened bins, or storage chutes that contain your compost pile and make your compost area neater and easier to manage. You can make your own composter, or simply pile the stuff in a corner. (See the article "Setting up a Home Compost Heap"on page 27.)

COMMON GROUND GARDEN SUPPLY
2225 El Camino Real, Palo Alto (415)328-6752

COMMUNITY COMPOSTERS
BERK (415)769-4020

KEMP COMPOST TUMBLER
160 Koser Rd., Lititz, PA (717)627-4566

RINGER'S THERMAL COMPOST BIN
Eden Prairie, MN (800)654-1047

FERTILIZER

Our good health is intrinsically dependent on the food we eat, and healthy food is derived from plants grown in fertile, organic soil.

AMERICAN SOIL PRODUCTS
2222 3rd St., BERK (415)540-8011
See display ad on this page.

BENNETT VALLEY FARM & COMPOST CO.
Sebastopol (707)823-7039

COMMON GROUND GARDEN SUPPLY
2225 El Camino Real, Palo Alto (415)328-6752

COUNTRY GARDEN NURSERY
1000 El Camino Real, SSF (415)583-8421

EBMUD
2130 Adeline St., OAK (415)835-3000
See display ad on this page.

ECOLOGY CENTER
2530 San Pablo, BERK (415)548-2220

FLOORCRAFT GARDEN CENTER
550 Bayshore Blvd., SF (415)824-1900

GOODMAN LUMBER AND NURSERY
445 Bayshore Blvd, SF (415)285-2800

LAS BAULINES NURSERY
Olema-Bolinas Rd. Star Route, Bolinas (415)868-0808

NORTH AMERICA ORGANIC
P.O. Box 70, RedwoodValley (707)485-7947

POTRERO GARDENS
1201 17th St., SF (415)861-8220

SLOAT GARDEN CENTER
327 3rd. Ave., SF (415)752-1614

SUNSET GARDEN SUPPLY
320 Alemany Blvd., SF (415)648-4242

THE URBAN FARMER STORE
2833 Vicente St., SF (415)661-2204

U-SAVE GARDEN CENTER
9317 MacArthur Blvd., OAK (415)578-3072

URBAN RESOURCES SYSTEMS, INC.
783 Buena Vista Ave. West, SF (415)621-3260

GARDEN SUPPLY

These suppliers offer at least one of the following: organic soil amendments, beneficial organisms for pest control, drip irrigation systems and advice, or non-toxic pest controls. Some offer organic seed and native plants as well. These companies may also be a good source for advice and tips on organic gardening techniques.

BENEFICIAL INSECTARY
14751 Oak Run Rd., Oak Run (916)472-3715

CLARKS HOME & GARDEN BLDG MATERIAL
23040 Clawiter Rd., Hayward (415)783-6366

CONNECTICUT ST. PLANT SUPPLIES
306 Connecticut, SF (415)821-4773

Fact: In 1990 there were over 300,000 known locations in the US containing hazardous substances.

ECOLOGY CENTER
2530 San Pablo, BERK (415)548-2220

GARDENER'S EDEN
P.O. Box 7307, SF (415)239-9191

GREEN GULCH FARM
1601 Shoreline Hwy., Sausalito (415)383-3134

GREENWOOD GARDENS
636 San Pablo, Pinole (415)724-5091

HARMONY FARM SUPPLY
P.O. Box 460 Dept NNE Graton (707)823-9125

HASENFLUG'S TOPSOIL & ROCK CO.
3232 San Pablo Dam Rd., San Pablo (415)232-4799

MOON MOUNTAIN WILDFLOWERS
P.O. Box 34, Morro Bay (805)772-2473

NORTH AMERICA ORGANIC
P.O. Box 70, Redwood Valley (707)485-7947

Fact: Only one percent of the produce in the U.S. is tested for pesticides, and these tests detect only one half of all pesticides.

SLOAT GARDEN CENTER
327 3rd. Ave., SF (415)752-1614

SMITH AND HAWKEN STORE
495 Miller Ave., Mill Valley (415)389-6061
See display ad on this page.

SUNSET GARDEN SUPPLY
320 Alemany Blvd., SF (415)648-4242

THE NATURAL GARDENING CO.
217 San Anselmo Ave,. San Anselmo (415)456-5060

THE NATURAL GARDENING COMPANY
217 Anselmo Ave., San Anselmo (415)456-5060

THE URBAN FARMER STORE
2833 Vicente St., SF (415)661-2204

URBAN ECOLOGY
P. O. Box 10144, BERK (415)549-1724

WEST COAST LADYBUG
P.O. Box 903, Gridley (916)534-0840
Fashion Island Shop. Ctr., S. Mateo (415)578-9200
401 Bayshore Blvd., SF (415)285-5244
Willow Shopping Center, Concord (415)686-2270

WHOLE EARTH ACCESS
2990 Seventh St., BERK (415)428-1600

GARDENERS

The gardeners in this section are committed to the use of at least two of the following: organic soil amendments, native or drought resistant plants, drip irrigation, and nontoxic pest management methods.

APPROPRIATE FLORA
3017 Wheeler, BERK (415)465-8491

ARBOR-VITAE
P.O. Box 24354, SF (415)822-5520

ATLAS LANDSCAPES/TREE CARE
1539 47th Ave., SF (415)566-5536

CAROLYN ATHERTON
25 Sirard Lane, San Rafael (415)457-1349

ENVIRONMENTAL ENHANCEMENT
2300 Market St. Suite 38, SF (415)469-2017

FOOTHILL COTTAGE GARDENS
13925 Sontad Rd., Grass Valley (916)272-4362

NON-TOXIC HOUSEHOLD WORKERS CO-OP
P.O. Box 22324, SF (415)681-6113
See display ad, page ##301

YOU BET FARMS
15595 You Bet Road, Grass Valley (916)273-5931

LANDSCAPE ARCHITECTS

4 DIMENSIONS LANDSCAPE DEVELOPMENT
4121 Culver, OAK (415)261-5820

DON JENSEN & ASSOCIATES
20 E. 5th St., Morgan Hill (408)778-2495

NATIVE LANDSCAPES
1021 University, San Jose (408)379-1323

LANDSCAPING

CAROLYN ATHERTON
25 Sirard Lane, San Rafael (415)457-1349

BRENDE & SHAPIRO
480 Vassar Ave., BERK (415)528-3906

NURIT BARUCH
317 Page St., SF (415)621-1262

BLOOM GARDENS
5855 Bernhard Ave., Richmond (415)234-5196

CALIFORNIA GARDENS
1171 Bay View Ave., OAK (415)654-2429
See display ad on this page.

EARTH MOODS
PO Box 8001, BERK (415)526-1140

ENVIRONMENTAL ENHANCEMENT
2300 Market St. Suite 38, SF (415)469-2017
See display ad on this page.

FIREWHEEL LANDSCAPING
350 East Todd Rd., Santa Rosa (707)585-3231

ALAN HOROBIN
San Rafael (415)456-6214

HUMMINGBIRD LANDSCAPING
1410 -14th Ave., SF (415)566-2844

INSIGHT LANDSCAPE SERVICES
306 Connecticut, SF (415)821-4773

PHIL JOHNSON
La Fayette (415)370-7614

LANDSCAPES BY MARC
558 Woodbine, San Rafael (415)472-6742

LARNER SEEDS
PO Box 407, Bolinas (415)868-9407

LUTSKO ASSOCIATES
San Fransico (415)391-0777

PHILIP MOYER
215 Chapman Road, Mill Valley (415)383-5874

NATIVE LANDSCAPES
1021 University, San Jose (408)379-1323

ROBERT'S LANDSCAPE SERVICE
P.O. Box 2223, Aptos (408)662-8681

TOM ROGALSKY
Ukiah (707)468-9160

ESTELLE SOLOMON
2363 Alva St. El Cerrito (415)235-7451

SUNRISE ORG. GARDEN & LANDSCAPING
Design,installation & maintenance (415)664-7782

TOMSASO
Native Landscaping, San Ramon (415)867-4123

GLORIA LUCIA VIANA
BERK (415)524-7679
See display ad on this page.

WAXMAN LANDSCAPING
2805 Eastman Lane, Petaluma (707)778-8529

WESTIN MILES DESIGN
17400 Monterey Rd., Morgan Hill (408)779-6686

NURSERIES

These nurseries carry drought-resistant or California native plants, seeds, or seedlings. They have exhibited expertise in regionally-appropriate landscaping techniques and supply.

ARBOR & ESPALIER COMPANY
201 Buena Vista Ave. East, SF (415)626-8880

BAY AREA SUCCULENTS
6556 Shattuck Ave., BERK (415)547-3564

BAY VIEW GARDENS
1201 Bay St., Santa Cruz (408)423-3656

BAYLANDS NURSERY
1103 Weeks St., E. Palo Alto (415)323-1645

BERKELEY HORTICULTURAL NURSERY
1310 McGee Ave., BERK (415)526-4704

C. H. BACCUS
900 Boynton Ave., San Jose (408)244-2923

CALIFORNIA FLORA
P.O. Box 3, Fulton (707)528-8813

CHRISTENSEN NURSERY
16000 Sanborn Rd., Saratoga (408)867-4181

CLYDE ROBIN SEED COMPANY
3670 Enterprise Ave., Hayward (415)785-0425

COUNTRY GARDEN NURSERY
1000 El Camino Real, SSF (415)583-8421

BOTANIC GARDEN-APRIL SALE
Tilden Regional Park, BERK (415)841-8732

FLOORCRAFT GARDEN CENTER
550 Bayshore Blvd., SF (415)824-1900

GLASS MOUNTAIN FOREST TREE NURSERY
P.O. Box 440, St. Helena (707)963-2372

HUMMINGBIRD GARDENS
P.O. Box 225, La Honda (415)747-0679

LARNER SEEDS
PO Box 407, Bolinas (415)868-9407
See display ad, page 288

LAS BAULINES NURSERY
Olema-Bolinas Rd., Star Route, Bolinas (415)868-0808

LIVING TREE CENTER
P.O. Box 10082, BERK (415)420-1440
See display ad, page 288

MAGIC GARDENS
729 Heinz Ave., BERK (415)644-1992

MOSTLY NATIVES
27215 Hwy 1 Box 258, Tomales Bay (707)878-2009

PACIFIC OPEN SPACE SEED COMPANY
P.O. Box 744, Petaluma (707)769-1213

REDWOOD NURSERY
2752 or 2800 El Rancho Dr., Santa Cruz (408)438-2844

SONOMA HORTICULTURAL NURSERY
3970 Azalea Ave., Sebastopol (707)823-6832

STRYBING ARBORETUM-MONTHLY SALE
9th Ave/Lincoln Way, Golden Gate Pk., SF (415)661-1316

SUNSET COAST NURSERY
P.O. Box 221, Watsonville (408)726-1672

THE NATURAL GARDENING COMPANY
217 Anselmo Ave., San Anselmo (415)456-5060

TIEDEMANN NURSERY
4835 Cherry Vale, Soquel (408)475-5163

TREE OF LIFE NURSERY
P.O. Box 736, San Juan Capistrano (714)728-0685

UC BOTANICAL GARDEN BIANNUAL SALE
Centennial Dr., BERK (415)642-0849

WESTERN HILLS RARE PLANTS
16250 Coleman Valley Road, Occidental (707)874-3731

WILDWOOD FARM
10300 Sonoma Hwy., Kenwood (707)833-1161

YA-KA-AMA
6215 Eastside Rd., Forestville (707)887-1541

YERBA BUENA NURSERY
19500 Skyline Blvd., Woodside (415)851-1668

Fact: Americans treat their gardens, lawns, and parks with 270 million pounds of pesticides annually. Fertilizer runoff, leaks from landfills, septic tanks, and underground storage tanks are major contributors to groundwater polution.

Fact: Plants grown in soil that is rich in humus (compost) are healthier and better prepared to withstand insect attack.

SEEDS

The businesses listed below offer either organic seeds or native plant seeds or both.

CONSERVASEED
P.O. Box 455, Rio Vista (916)775-1646

ECOLOGY CENTER
2530 San Pablo, BERK (415)548-2220

LARNER SEEDS
P.O.Box 407, Bolinas (415)868-9407

MOON MOUNTAIN WILDFLOWERS
Specializing in California Natives Since 1981
PO Box 34, Morro Bay
(805)772-2473

ROBINETT BULB FARM
7345 Healdsburg Ave., #9, Sebastopol (707)829-2729

S & S SEEDS
P.O.Box 1275, Carpenteria (805)684-0436

SUNSET GARDEN SUPPLY
320 Alemany Blvd., SF (415)648-4242

THEODORE PAYNE FOUNDATION
10459 Tuxford St. Sun Valley (818)768-1802

TREE SERVICE

BRENDE & SHAPIRO
480 Vassar Ave, BERK (415)528-3906
See display ad, page 287

Tip: Marigold, borage, dill or basil planted around and through the rows of tomatoes will ward off tomato hornworm. An effective old-fashioned spray for control of mealy bugs, spider mites, scales and aphids is Ced-o-flora, which contains hemlock oil.

GENERAL MERCHANDISE

This section contains the business that either encompass so many aspects of greener living that they warrant a catch-all category or they didn't really fit under any of our other categories. Many of these companies are excellent places to find new ideas, start your first greener living attempts, and get an understanding of the many alternatives that exist. Many operate through mail-order and would be happy to send you a catalog.

CATALOGS

Green and "earthsafe" catalogs are springing up all over. There are several very good catalogs which offer a wide range of products. There are also some which focus on specific product areas such as cosmetics or paper products.

Check that the catalog is printed on recycled paper (it should tell you what kind of paper, and what its recycled content is). How are they verifying the claims they are making? Are they selling products that you need to buy by mail order or could they be found locally? This issue is important because mail order is relatively wasteful of packaging, and often means that products are shipped a lot further than they would be through normal distribution channels. Direct distribution by manufacturers are an exception to this.

There are some things that are hard to find locally, especially if you don't live in an "eco-city." Educate yourself, andbuy the best products you can find. It is important that you shop carefully and support the products and companies that are part of the solution and not part of the problem.

GEN. MERCH

BEAUTY NATURALLY
Cosmetics (415)459-2826

BEAUTY WITHOUT CRUELTY
Personnal Care Products (415)382-7784

BEAUTY WITHOUT THE BEAST
Box 951, Boulder, CO (303)440-0188

EARTH CARE PAPER, INC.
P.O. Box 14140, Madison WI (608)256-5522

ECCO BELLA
Cleaning products,cosmetics & more (800)888-5320

ECO DESIGN CO.
Many good things (505)438-3448

ECO LOGIC
Environmentally safe products for home (714)338-5694

GARDENER'S SUPPLY
128 Intervale Rd., Burlington VT (802)863-4535

Great Lake Herb Co.
P.O. Box 6713, Minn. MN
Catalog – 600 kinds of herbs (612)722-1201

GREENER ALTERNATIVES
914 Mission St.,Suite A, Santa Cruz (408)423-0701

KIDSYSTEMS TOUGH TRAVELER LTD.
Children's goods (518)377-8526

LIVOS PLANT CHEMISTRY
Products for the environmentally safe home (800)621-2591

REAL GOODS TRADING CO.
966 Mazzoni St. , Ukiah (800)762-7325

SAFER PRODUCTS
Least toxic products for home,garden (800)447-2229

SAN FRANCISCO HERB COMPANY
250 14th St., SF (800)622-0768

SAVE ENERGY COMPANY
2410 Harrison St., SF (415)824-6010
See display ad on this page.

SELF-CARE CATALOGS
349, Healdsburg Ave., Healdsburg (707)431-1100

SEVENTH GENERATION
Many hard to find safe products. (800)456-1177

SIMMONS HOLISTICS
(800)533-6779
See display ad on this page.

THE LIVING SOURCE
Nontoxic products of all sorts (817)756-6341

THE SPROUT HOUSE
40 Railroad St., Great Barington, MA (413)528-5200

WALNUT ACRES
A catalog of organic foods (800)433-3998

GENERAL MERCHANDISE

BAUBIOLOGIE HARDWARE
207 16th St., Pacific Grove (408)372-8626

MITTIE CUETARA
Illustrations (415)864-8511
See display ad, page 289

EARTHSAKE
1844 Market St., SF (415)626-0722
See display ad on this page.

ENVIRONMENTAL CONCERNS
9051-E Mill Station Rd., Sebastopol (707)829-7957
See display ad, page ##

GREENER ALTERNATIVES
914 Mission St.,Suite A, Santa Cruz (408)423-0701
See display ad, page 20

GREENPEACE STORE
890 North Point, SF (415)474-1870

SAGE: THE ENVIRONMENTAL STORE
(415)485-5441

PENINSULA CONSERVATION CTR. STORE
2448 Watson Ct., Palo Alto (415)494-9301

SAMUEL HOMANS
1129 Fourth St., BERK (415)864-8511
See display ad on this page.

SIMMONS HOLISTICS
P.O. Box 3193, Chattanooga TN (800)533-6779

SOCIETE ANTIQUAIRE ET DES ARTS
1456 2nd Av #279 ., N.Y., NY (212)737-2350
See display ad on this page.

THE COHERENCY COMPANY
P.O. Box 553, Occidental (707)869-0956

THE HOMESTEAD
11 Glen Rd., San Anselmo (415)485-5441

WHITE DOLPHIN
218 F St., Eureka (707)445-2094

SHOPPING BAGS

BLUE RHUBARB, INC.
Old Creamery Road, Harmony (800)926-1017

NET-SAK
P.O. Box11065, Aspen CO 81612 (303)927-4135

SMITH AND HAWKEN STORE
495 Miller Ave., Mill Valley (415)389-6061

WHOLE EARTH ACCESS
BERK (415)845-3000
Concord 686-2270
SF 285-5244
San Mateo 578-9200
San Rafael 459-3533

WINDY CROW
P.O. Box 11566, Santa Rosa (707)546-4238

HAZARD ABATEMENT

Things have gotten so mucked up that we have developed a whole new industry of clean-up products and services to reduce the health risks and impacts of such simple things as breathing, drinking water or just living. It would obviously be better to avoid creating the problems in the first place. But the fact of widespread environmental dangers in our homes and workplaces has led many sensible people to turn to technological fixes.

There are now numerous air filters, water filters (for both drinking water and shower water), radon tests, environmental testing labs for the home, and home-toxin consultants stepping in to fill this new niche.

The *Green Pages* has listed just a few of the product resources available. Many catalogs, hardware stores, and general merchandise outlets carry air and water filters and home test kits.

We have included CFC recycling places for auto-air conditioning repair here because we believe CFCs are one of the greatest hazards humans and all life on the planet face.

AIR FILTERS

The best way to cut down on indoor air pollution is to remove its source, pliable plastics of all kinds, chemical-containing cleaners and personal care products, gas appliances, and synthetic materials in furniture. Obviously, removing all potential sources cand be very difficult, because just about all of the things we use give off pollutants. So air filters are a good idea, particularly for the chemically sensitive.

Filters are also useful to control unavoidable pollen, dust or mold. Indoor plants have been found to be very effective natural air filters. Spider plants, for instance, are known to remove urea formaldehyde, which is a toxin common in building materials.

There are four basic kinds of air filters:

Activated carbon filters need a fair amount of carbon to be effective (drug store models are somewhat inadequate). They can be impregnated with salts to increase their uptake of formaldehyde.

Mechanical filters use fine-fibered filters to trap particles like dust and pollen. Ask for brands that are not bonded with a polyvinyl acetate.

Electrostatic filters collect pollutants by electrical charge - these filters range in efficacy, need consistent cleaning with petrochemically based solvents, and give off positive ions.

Negative ion generators remove only particles from smoke and smog. They can't remove dust, pollen or toxic gases. They do, however, produce negative ions which appear to be beneficial. These generators might be best utilized in conjunction with a carbon filter.

AIR EXCHANGE INC.
1185 San Mateo Ave., San Bruno (415)871-2945

AIREOX RESEARCH CORPORATION
P.O. Box 8523, Riverside (714)689-2781

ALLERMED CORPORATION
31 Steel Rd., Wylie TX (214)442-4898

OPTIMUM INTERIORS
415 Trowbridge St., Santa Rosa (707)545-8838
See display ad on this page.

NONSCENTS
151 Coral Reef Ave., Half Moon Bay (415)728-9573

CFC RECYCLING

AIR COND. AND AUTO EL. OF FREMONT
37555 Central St., Fremont (415)793-2340

AIR CONDITIONING AUTOMOTIVE CLINIC
2035 Divisidero St., SF (415)563-2915

ALAMEDA TIRE, BRAKE & MUFFLER
1825 Webster St., ALA (415)521-1541

ALLIED RADIATOR AND AIR CONDITIONING
920 Gilman St., BERK (415)524-3860

AUTOHAUS - HAYWARD
21650 Mission Blvd., Hayward (415)881-1915

AUTOHAUS - SANTA CLARA
2555 Lafayette St., Santa Clara (408)986-1391

DON'S RADIATOR
647 23rd. St., RICH@cCo:Kings Auto Air/Heating, Inc.
2550 High St., OAK (415)534-1946

MADSEN'S AUTO AIR AND HEATING
1132 Meadow Ln., Concord (415)798-1131

THE FOAM SHOP
813 A St., San Rafael (415)453-3626

Fact: CFC's stay in the atmosphere for at least fifty years afer being released. Once it reaches the ozone layer and breaks down, each atom of chlorine can destroy millions of ozone molecules.

ENVIRONMENTAL TESTING

The companies listed below are of two types. The first are consultants or consulting labs that can determine potential problems, take samples if needed, and arrange for the testing of samples from your home or workplace to monitor levels of toxins. The others are state-certified laboratories that will accept residential, or small commercial samples of soil, water, or other things that may contain toxins. Basically, the non-certified companies are capable of advising you on the need for tests, and the labs do the testing. This process can be an expensive venture, but in many areas where groundwater may be contaminated, or in older neighborhoods which have had many years of lead paint use, it becomes an important one, especially if children are regularly present.

ADVANCED WATER SYSTEMS
P.O. Box 1775, Capitola (408)476-0515
WATER

ALPHA CHEMICAL AND BIOMEDICAL LABS
245 Kentucky Street, Petaluma (707)778-8607
AIR, WATER, SOIL, ASBESTOS-STATE CERTIFIED

CALCOAST ANALYTICAL LABS
P.O. Box 8702, Emeryville (415)652-2979
ASBESTOS, MERCURY, LEAD-STATE CERTIFIED

HOMESAFE ASSESSMENTS
4651 Mira Loma St., Castro Valley (415)537-8956
ASBESTOS (AND OVERSEEING OVERHAUL),RADON,LEAD PAINT,FORMALDEHYDE-STATE CERTIFIED

ITEK ENVIRO SERVICES
248 Riviera Circle, Larkspur (415)924-8534
ASBESTOS, FREON,LEAD,FORMALDEHYDE-STATE CERTIFIED

KELLCO ASBESTOS ANALYTICAL SERVICES
44814 Osgood Road, Fremont (415)659-9751
ASBESTOS-STATE CERTIFIED

KENNEDY/JENKS/CHILTON
674 Harrison Street, San Francisco (415)362-6065
WATER AND SOIL-STATE CERTIFIED

MED-TOX ASSOCIATES INC.
3440 Vincent Road, Pleasant Hill (415)930-9090
RADON, ASBESTOS, FREON, COLIFORM BACTERIA, FORMALDEHYDE, LEAD PAINT. ETC.-STATE CERTIFIED

PRECISION ANALYTICAL LABORATORY INC.
4136 Lakeside Drive, Richmond (415)222-3002
AIR AND SOIL-STATE CERTIFIED

SAFE ENVIRONMENTS
2512 Ninth St. #7, BERK (800)356-2663
See display ad, page 292.

SEQUOIA ANALYTICAL LABORATORY
680 Chesapeake Drive, Redwood City (415)364-9222
FREON, LEAD, FORMALDEHYDE, COLIFORM BACTERIA-STATE CERTIFIED

WATER FILTERS

DAN CLURMAN
5756 Claremont, OAK (415)547-2380
See display ad, page 338.

GETZ, ANDREW
5756 Claremont, OAK (415)547-2380
See display ad, page 338.

MULTI-PURE WATER FILTERS
James Baraz (415)465-2324

THE ENVIRONMENTAL NETWORK
2312 Castro St., SF (415)826-0195

Fact: Traces of more than 2000 different chemical pollutants have been found in public water supplies.
Tip: Boycott products that pollute, and don't pour those you have down the drain.
Fact: Front-loading washing machines use up to 40 percent less water than top-loading machines.
Fact: A dripping faucet wastes about 300 gallons of water a month. A running faucet averages 180 to 300 gallons every hour.
Tip: *Be aware* of the ways you waste or pollute water, and find ways to prevent it.

It is hard to believe that humans ever managed to live without our current health technologies. Diagnostic tools, drugs, surgery, and mechanical therapies save millions of lives yearly. Unfortunately, these efforts have put the cart before the horse. Most conventional western medicine emphasizes curing symptoms after a disease has developed, instead of preventing the illness in the first place. Too often, the manifestations of disease are eradicated, while the root cause, which may be emotional or spiritual, remains untouched. Anatomy becomes destiny because doctors see nothing else.

Alternative health care focuses on preventing disease, rather than masking it, and it does this by treating whole people, not body parts. It sees physical disorders as signs that something is wrong in the person's life in general. Rather than knocking out a problem after it arises, these therapies help prevent the problem by identifying the person's weak points, and strengthening them. They provide a comprehensive and subtle approach to the entire human being, not just the human body.

Since1980, the increase in costs of medical care in the U.S. has more than doubled to a whopping figure of $600 billion in 1989. If medical care alone were compared to the gross national products of other countries, the U.S. would rank as the ninth largest economy in the world. Yet the United States is also ranked as only the 17th-most effective nation in preventing infant mortality. The high cost of health care in this country literally cannot be afforded by many people. About 33.5 million, or 14 percent, of Americans are without medical insurance. Without health insurance, many can barely cover the cost of a doctor's visit. The whole system has

HEALTH

become impersonal, profit-directed, and inefficient. Too many people are falling through its cracks because they cannot pay the prices.

Preventive health care offers affordable and common-sense-based approaches to personal health. We offer a short list of local practioners as an introduction. There are many more, and you should consult your area's Yellow Pages for others.

A few cautionary notes: Alternative treatments bolster your general health and cut down the need for other forms of medicine. Neither the *Green Pages* staff, nor most alternative health providers, advocate avoiding the use of drugs, surgery, or other modern medicine when needed.

You may think that herbalists exclude the use of animal products, but this is not always the case. In Chinese tradition, doctors (not necessarily herbalists) prescribe substances that contain parts of animals. As a result of this demand, wild animals are being poached in great numbers. One example in California is the black bear. Poaching is making a rapid dent in population numbers, and if the trade is not stopped we may see regional extinction.

Not all herbal remedies or alternative-health products are benign. Antibiotics and steroids imported from Mexico are being sold outside of pharmacies, in local stores, and at flea markets. Chinese "herbal" medicines that are imported often contain poisons. The California Department of Health Services has issued warnings regarding the following imported products:

Chuifong Toukuwan – BB-sized pills prescribed for arthritis contain potent painkillers and occasionally toxic levels of cadmium and lead.

Koon Yick Hung Far Oil – can also cause convulsions, vomiting and even death.

Cinnabar Sedative Pills – contain mercury.

Niu Huang Xiao Yan Wan – contains arsenic disulfide which can result in liver and kidney damage.

Watson's Baby Water— prescribed to keep your baby quiet, contains chloroform.

ACUPRESSURE

ACUPRESSURE INSTITUTE
1533 Shattuck Ave., BERK (415)845-1059

VIVIAN BIAS
Hayward (415)581-1290

NANCY CARROLL
1205 Gravenstein Hwy S, Sebastopol (707)829-5629

JO ELLA CASKEY
Danville (415)837-2768

CENTER OF TRADITIONAL CHINESE MEDICINE
411 Primrose Rd., Burlingame (415)375-0425

PAMELA CLARK
OAK (415)652-5614

COLLEGE MEDICAL CLINIC
2051 Market, SF (415)863-3500

CANDACE COAR
(415)525-1801

CONNIE CRONIN
OAK (415)655-4650

VANESSA LILLIE
201 Miramar, Santa Cruz (408)423-8617

NANCY MARCHANT
PO Box 742, Mendocino (707)937-0220

ANTON PARADISE
BERK (415)644-1846

GLEE PASOL
OAK (415)763-3755

MARK PETTY, CMT
681 Oak Grove Ave., Menlo Park (415)323-4432

RUTH SCOLNICK
2602 Highland Ave., OAK (415)536-9814

AYNE SHORE
OAK (415)530-2023

IONA MARSAA TEEGUARDEN
366 California Ave., Palo Alto (415)328-1811

DEVON WORTZ
Corte Madera (415)924-7472

ARTEMAS YAFFE
524 Sunset Way, Redwood City (415)365-3248

ACUPUNCTURE

There are certification requirements for acupuncturists in California. The C.A. (Certified Acupuncturist) or D. O. M. (Doctor of Oriental Medicine) marks are two designations that insure some consistency with established practise. Many practitioners have other training certification from various colleges and institutes. Disposable needles are the rule, but check to make sure.

ACUPUNCTURE CLINIC
4444 Geary Blvd., Suite 305, SF (415)752-6680

ACUPUNCTURE COLLEGE OF S.F.
2051 Market St., SF (415)863-3500

ALDERMAN ACUPUNCTURE CLINIC
801 Brewster Ave., Redwood City (415)366-2165

AMERICAN COLLEGE OF TRADITIONAL CHINESE MEDICINE
clinic, 450 Connecticut (415)282-9603
school, 455 Arkansas, SF (415)282-7600

LUCIA CASTILLO
ALA (415)523-933645

CLASSICAL MEDICINE CENTER
360 Bryant, Palo Alto (415)324-8420

CONNIE CRONIN
OAK (415)655-4650

NG DANIEL MD
4444 Geary Blvd.Suite 102, SF (415)386-7169

KEN DENSMORE
San Rafael (415)485-9797

ROBERT DORSETT
BERK (415)849-0980

CHERYL DRAKE
48 Page St., SF (415)255-7300

JAY LASKO
BERK (415)644-0111

HIGGY LERNER
OAK (415)444-1501

LAI KIM MAN'S ACUPUNCTURE CLINIC
50 Skyline Plaza,Suite E, Daly City (415)994-6020

HOPE MCDONNEL
BERK (415)848-7400

PENINSULA ACUPUNCTURE CENTER
1425 Broadway #8, Burlingame (415)343-9911

NATHAN SCHULMAN
OAK (415)444-1501

THE ACUPUNCTURE CLINIC
2943 Broadway, Redwood City (415)369-5190

THE BANCROFT CENTER OF CHINESE MED.
1720 Bancroft Way, BERK (415)849-1809
See display ad, page 294

TURNING POINTS
2301 Broadway OAK (415)444-1501
See display ad, page 294

Fact: Each year about 5 million babies die before their first birthday for lack of clean water.

AIDS INFORMATION & PREVENTION

The AIDS epidemic is costing this country a great amount of money, talent and life. As insurance companies encounter the higher costs of providing for their AIDS-afflicted members (when they actually even cover their costs) they are passing these costs on to others through raised premiums. As fewer and fewer AIDs victims can find insurance or pay for their care, we are seeing growing neglect of these terminally ill people. The following groups are trying to prevent the spread of AIDS, educate the public about misconceptions, and offer a non-corporate, caring response to the problem.

AIDS FOUNDATION EDUCATION
333 Valencia, SF (415)864-4376

AIDS FOUNDATION OF SAN FRANCISCO
25 Van Ness, SF (415)864-5855

AIDS HOTLINE (800)367-2437

AIDS PROJECT
225 W 37th Ave., San Mateo (415)573-2588

PALO ALTO MEDICAL CLINIC
300 Homer Ave., Palo Alto (415)321-4121

EDUCATION

ACUPRESSURE INSTITUTE
1533 Shattuck Ave., BERK (415)845-1059

ACUPUNCTURE COLLEGE OF S.F.
2051 Market St., SF (415)863-3500

FIVE BRANCHES INSTITUTE
200 7th Ave., Santa Cruz (408)476-9424
See display ad, page 294.

THE ELMWOOD INSTITUTE
P.O.Box 5765, Berkeley (415)845-4595
See display ad, page 278

FAMILY PLANNING

Family Planning and Women's Health Clinics offer an understanding and open-minded medical response to family needs. They give women and families a better understanding of their options, more so than in most hospitals. Much of what is emphasized at these places is the notion of self-knowledge and self-control over one's own health.

BERKELEY WOMEN'S HEALTH COLLECTIVE
2908 Ellsworth St., BERK (415)843-6194

BUENA VISTA WOMEN'S SERVICES
2000 Van Ness, SF (415)771-5000

CHOICE MEDICAL GROUP
2280 Geary Blvd., SF (415)922-6656

FAMILY PLANNING CLINIC, SF GENERAL
1001 Potrero Ave, SF (415)648-6300

MISSION NEIGHBORHOOD HEALTH CENTER
240 Shotwell, SF (415)552-3870

PLANNED PARENTHOOD
SF, 815 Eddy St. Suite 300 (415)441-5454
San Mateo,2211 Palm Ave. (415)574-2622
Fremont, 39055 Hastings St., Suite204 (415)794-7566
Antioch, 1104 Buchanan (415)754-4550
Concord,1899 Clayton Rd. (415)680-0414
Daly City, 219 South Gate (415)991-3092
Fairfield, 1325-C Travis Blvd. (707)429-8855
Hayward,1453 B St. (415)538-6990
San Rafael,20 H St. (415)454-0476
Napa, 1735Jefferson St (707)252-8050
San Ramon, 130 Ryan Ct. (415)838-2108
OAK,482 West MacArthur Blvd. (415)652-7526
Richmond, 3050 Hilltop Mall Rd. (415)222-5290
San Jose,1691 The Alameda (408)287-7532
Vallejo, 990 Broadway (707)643-4545
Walnut Creek,1291 Oakland Blvd. (415)935-4066

PREGNANCY CONSULTATION CENTER
1801 Bush, SF (415)567-8757

WOMEN'S CLINIC
1650 Valencia, SF (415)285-7788

WOMEN'S HEALTH CENTER
582 Market St.,Suite100, SF (415)982-0707

HEALTH INSURANCE

Health insurance that promotes and pays for preventive health care is rare. The following company operates in a socially responsible and environmentally concerned manner, but does not specifically promote alternative health techniques.

HEALTH

Investments made by the company are only in projects like low-income housing, minority business development, and environmentally-sound community development. They provide unisex rates, domestic partner coverage regardless of marital status or gender, and other progressive features. Consider this short list of one a call for more responsible insurers. If you know of any please contact us.

CONSUMERS UNITED INSURANCE COMPANY
2100 M St.,N.W. Washington DC (800)255-4432

HEALTH PRODUCTS

BREAD OF LIFE
1690 S. Bascom Ave., Campbell (408)371-5000

HERBACEUTICALS
478 Post St. San Francisco (415)986-2607

HERBAL RESOURCES
3557 Perada Dr. Walnut Creek (415)930-9186
See display ad on this page.

WACHTER'S ORGANIC SEA PRODUCTS
360 Shaw Rd.South, SF (415)588-9567
See display ad on this page.

HERBALISTS

ACUPUNCTURE COLLEGE OF S.F.
2051 Market St., SF (415)863-3500

AMERICAN COLLEGE OF TRADITIONAL CHINESE MEDICINE
455 Arkansas, SF (415)282-7600

CLASSICAL MEDICINE CENTER
360 Bryant, Palo Alto (415)324-8420

LAI KIM MAN'S ACCUPUNCTURE CLINIC
50 Skyline Plaza,Suite E, Daly City (415)994-6020

PENINSULA ACUPUNCTURE CENTER
1425 Broadway #8, Burlingame (415)343-9911

CATHERINE ABBY RICH
201 Hawthorne St., Larkspur (415)924-5961

THE ACUPUNCTURE CLINIC
2943 Broadway, Redwood City (415)369-5190

HERBS, MEDICINAL

ALDERMAN ACUPUNCTURE CLINIC
801 Brewster Ave., Redwood City (415)366-2165

LHASA KARNAK HERB CO.
2513 Telegraph Ave., Berkeley (415)548-0380

MOUNTAIN ROSE HERBS
P.O. Box 2000, Redway (800)879-3337

NATURE'S HERB CO.
1010 46th, Emeryville (415)601-0700

SIMPLERS BOTANICAL CO.
Box 39, Forestville (707)887-2012

HOMEOPATHY

BOERICKE AND TAFEL
238 Circadian Way, Santa Rosa (800)876-9505

MICHAEL CARLSTON, M.D.
2801 Yulupa Ave.Ste. B, Santa Rosa (415)506-7000

DAVID FIELD, NDCA
1201 Noe, SF (415)282-9337

HAHNEMANN MEDICAL CLINIC
1918 Bonita Ave., BERK (415)849-1925

GREGORY MANTEUFFEL, M.D.
1109 Vicente, Suite 104, San Francisco (415)681-4440

NATURE'S SUNSHINE PRODUCTS
3557 Perada Dr., Walnut Creek (415)256-6851

RANDY NEUSTAEDTER, M.D.
360 Bryant, Palo Alto (415)324-8420

JONATHAN SHORE M.D.
900 s. Eliseo, Greenbrae (415)461-1981

CORY WEINSTEIN, M.D.
1109 Vicente St. #104, San Francisco (415)681-4440

MIDWIFERY

BAY AREA MIDWIFERY SERVICES
390 Laurel, Suite 300, SF (415)750-6422

BIRTH & BONDING EDUCATIONAL SVCS.
357 Vernon Street, OAK (415)839-0822

JAN SHELL DUSAL CNM
444 Rosalie Dr., Sonoma (707)938-5890

MARIA GREULICH, CN M
825 Pollard Rd., Los Gatos (408)866-4200

HEART & HANDS
2247 44th Ave., SF (415)753-5997

LABOR OF LOVE MIDWIFERY
333 Miller Ave., Mill Valley (415)381-8119

MARIN BIRTH CENTER
1314 Lincoln Ave., San Rafael (415)453-1813

MATERA BIRTH CENTER
515 South Dr., Mountain View (415)961-6346

MIDWIFERY CARE
Certified Nurse Midwife (415)337-8424

ESTES MILTON N MD
333 Miller Ave., Mill Valley (415)383-6622

KATHRYN NEWBURN CNM
1301 Sanchez Ave., Burlingame (415)347-6943

NOE VALLEY MIDWIFERY
223-A 29th St. SF (415)695-1290

REDWOOD MIDWIFERY SERVICE
1154 Montgomery Dr., Santa Rosa (707)579-4420

RENAISSANCE WOMAN MIDWIFERY
3106 Arizona St., OAK (415)530-4339

SONOMA COUNTY BIRTH NETWORK
13140 Frati Ln., Sebastopol (707)829-1131

ANN STAREGO, RN, CNM
1650 Valencia, SF (415)285-7788

WOMAN'S OBSTETRICS & MIDWIFERY
5309 College Ave., OAK (415)420-1200

WOMEN'S HEALTH ASSOCIATES
1240 S. Eliseo Dr., Greenbrae (415)461-7800

NATURAL VISION CARE

A very effective program of eye exercises that reduce the need for glasses.

DR. CARL HIRSCH
1613 Locust Walnut Creek (415)932-4362

DR. C. STEPHEN JOHNSON
2551 San Ramon Valley Blvd. San Ramon(415)743-1222

DR. STEWART MANN
1234 Cherry St. San Carlos (415)593-1661

DR. ROBERT MONETTA
2532 Ocean Ave. SF (415)239-2544

DR. MICHAEL ROY
675 Ignacio Walnut Creek (415)933-4700

NUTRITIONISTS

CATHERINE HEYMANN J RD
1828 El Camino Real, Burlingame (415)697-8602

JON D. KAISER, M.D.
3448 Sacramento, SF (415)922-8971

KOSHER NUTRITION PROJECT
320 15th Ave., SF (415)221-1025

JAN MCBRIDE, R.D.
Twentieth St., SF (415)824-8703

RITA SALVATORE, R.N.
623 12th Ave., SF (415)387-2536

THERESA A. TSINGIS, D.C.
220 Bush St., SF (415)397-9960

WHOLE BODY HEALTH
1110 Burligame Ave., Burlingame (415)343-3373

Tip: Recipes for non-toxic cleaning substutes can be found in *Cheaper and Better: Homemade Alternatives to Storebought Goods* by Nancy Birnes, Harper & Row.

Tip: To keep your name from being sold for use on most large mailing lists, writeto:
Mail Preference Service
Direct Marketing Association
6 E 43rd St., NY, NY 10017

HOUSEHOLD

Our homes are the environments over which we theoretically have the most control, yet our houses are often filled with suprisingly toxic and hazardous chemicals and substances. Increasing numbers of people are fallingvictim to a recently-recognized condition called "environmental illness." It appears that chronic exposure to low levels of toxins causes an allergic reaction, sometimes leading to complete intolerance of one's own house or office. It is not completely clear how this sensitivity develops, but it is clear that reducing the number of toxins in your home, or eliminating them altogether, greatly reduces your chances of developing this and may alleviate other unexplained health problems. Non-toxic cleaning-product alternatives are becoming abundant, and non-toxic cleaning services are a welcome development from these products. Natural fibers can replace polyvinyl shower curtains and bed linens.

Homes are an excellent arena for wise energy use. Our listings give guidance on energy conserving appliances - some miserly enough to work well with solar electiric systems. Used appliances are inherently energy-efficient because no manufacturing energy is spent creating yet another shiny, plastic-appointed big ticket item. Other energy saving devices — meters for light use, and compact fluorescent lighting for instance — are also important.

Avoid the purchase of any kind of furniture that has been manufactured with CFCs (in foam), halon-based fire extinguishers, and fire detectors or compact flourescents with radioactive elements. Ask salespeople about these issues and make sure they, and you, know what the manufacturing processes are and what elements the products you purchase contain. If you own any low-level radioactive items, contact your fire department for proper disposal.

APPLIANCES

Recent government standards require all new appliances to be 10 to 30 percent more efficient than models made in the past. These standards, it is estimated, will save consumers on the whole at least $28 billion over the new product's lifetime on appliances sold through the year 2000. Individual households can save $300 annually. Such savings not only easily pay back the slightly higher cost of buying energy-efficient appliances, but will also ease the economic and environmental ills caused by high energy production and use.

Appliances, heating, and cooling cost an average household more than $1000 per year. You can sharply reduce your own energy bill by using high-efficiency appliances and space conditioning equipment.

Stoves:

- Whenever possible, use smaller appliances. It's not worth making toast in a pizza oven.
- If you are pondering the idea of buying a new gas range, get one with an electronic ignition instead of one with a pilot light. This will cut fuel costs by 40 percent.
- Consider turning off the pilot and light the unit with a hand ignitor.
- A convection oven cooks food faster at a lower temperature through the use of a small fan which circulates the heat inside the oven.

• When preparing a meal, turn off your burners and oven a little early and let the residual heat finish the cooking.

Refrigerators

The problem with refrigerators is not only that they consume energy, but they rely on ozone-destroying chlorofluorocarbons as their coolant. Keep your refrigerator in good condition and have it serviced regularly. The energy-efficiency of refrigerators has improved considerably over the past fifteen years,. A typical new refigerator with automatic defrost and a top-mounted freezer uses about 1000 KW per year, whereas the typical model sold in 1973 used about 2000 KWper year.

• Keep your refrigerator between 38 and 42 degrees, and set the freezer between 0 and 5 degrees. Make sure the door gaskets are clean so the seal is tight, and replace gaskets when they begin to wear.

• If you are going to store some food which has just been cooked, let it cool down before putting it in the fridge. The-heat will consume a great deal of energy.

• Cleaning the coils on the back of the fridge will sustain the life of the motor and make it run more efficiently.

• If you are a single person, buy a smaller refrigerator.

Freeezers

The energy-efficiency of freezers has improved over the past decade, although at a slower rate than refrigerators.

Dishwashers

An energy-efficient dishwasher can consume less energy than doing dishes by hand, unless your dishes are done strictly in cold water. Runthe dishwasher only when you have a full load. When washing by hand, use a tub of water for rinsing as opposed to letting the faucet run.

Clothes Washers

• Washing your clothes in cold rather than hot water can, depending on your local energy costs, save you $100 to $200 a year.

• Washers use 32 to 59 gallons of water for each cycle, so make sure you have a full wash load.

• When washing, use cold water as much as possible. Nine tenths of the energy used washing clothes goes into heating the water.

The following are among the most energy efficient refrigerators:

TYPE	Brand	Model	Cubic ft.	Kwh/yr	Yearly Energy Cost
SINGLE DOOR, MANUAL DEFROST					
	Kenmore	86111*2	11.0	367	$28
	U-line	UR-110	11.6	434	$33
	Sanyo	SR 1057	11.6	489	$38
TOP FREEZER, PARTIAL AUTOMATIC DEFROST					
	Kenmore	83831	13.7	735	$57
	General Electric	TB13S	13.4	735	$57
	Hotpoint	CTA13C	13.4	735	$57
TOP FREEZER, AUTOMATIC DEFROST					
14.5-16.4 cubic feet					
	Frigidaire	FPI-16TE-0	16.0	766	$59
	Kelvinator	TPK160BN	16.0	766	$59
16.5-18.4 cubic feet					
	Gibson	RT17FU3A	16.6	766	$59
	Frigidaire	FPD-17TIF-0	16.6	766	$59
18.5-20.4 cubic feet					
	Frigidaire	FP-19TF-0	18.6	839	$65
	Gibson	RT19FU3A	18.6	839	$65
25-24.4 cubic feet					
	General Electric	TBX21KC	20.7	941	$73
	Hotpoint	CTX21KC	20.7	941	$73
SIDE-BY-SIDE FREEZER, AUTOMATIC DEFROST					
18.5-20.4 cubic feet					
	Amana	S120J	19.9	1032	$80
	Frigidaire	FPZ-19VF-0	19.9	1127	$87
20.5-22.4 cubic feet					
	Maytag	RSD22A	21.8	1117	$86
	Holland Dist.	NDNT2292	22.0	1117	$86
22.5-26.4 cubic feet					
	Frigidaire	FPCE-24VF-0	24.0	1117	$86
	Maytag	RSD 24A	23.8	1123	$86

The following are among the most energy efficient freezers

TYPE	Brand	Model	Cubic ft.	KW/yr	Yearly Energy Cost
UPRIGHT, MANUAL DEFROST					
13.5-17.4 cubic feet					
	Frigidaire	UFS16D-W3	15.8	585	$45
	Montgomery Ward	FFT46858	16.1	610	$47
17.5-21.4 cubic feet					
	Kenmore	82639	19.2	725	$56
	Whirlpool	EV190ESO	19.2	725	$56
UPRIGHT, AUTOMATIC DEFROST					
13.5-17.4 cubic feet					
	Amana	ESUF16D	16.2	871	67
	Kitchen Aid	KFF15MSY	14.9	918	71
CHEST, MANUAL DEFROST					
13.5-17.4 cubic feet					
	Home Products	HPC 16A	16.8	428	33
	Panasonic	NR 1705 FC	16.8	430	33
17.5-23.4 cubic feet					
	Panasonic	NR 2105 FC	20.7	519	40
	Home Products	HPC 20A	20.7	521	40

• Front loading washing machines use up to 40 percent less water than top-loading machines.
• Use shorter cycles for smaller loads.
• Too much detergent makes your washing machine work harder and use more energy.

Water Heaters

The energy-efficiency of a water heater is indicated by its Energy Factor (EF), an overall efficiency based on the use of 64 gallons of hot water per day. The average new gas water heater sold in 1988 had an EF of about 0.50. The average new electric water heater sold in 1988 had an EF of about 0.85.

All other things being equal, the smaller the water heater tank, the higher the EF. The water heater should, however, provide enough hot water at the busiest time of the day. For example, a household of two adults may never use more than 30 gallons of hot water in an hour, but a family of six may use as much as 60 gallons in an hour. The ability of a water heater to meet peak demands for hot water is indicated by its "first hour rating" (listed in the "rating " column below).

APPLIANCES, NEW

APPLIANCES, USED

Many fully functional appliances are thrown away. We often buy new ones because the old ones have gone out of style or we feel the need to buy something new. In many cases the manufacture of appliances takes more energy than the appliance will use in its lifetime. While this may not be true for air conditioners or refrigerators (appliances which stay on for long periods of time) it is certainly true for blenders, mixers, juicers and the like. Reuse is the most efficient form of recycling.

The following are among the most energy efficient dishwashers:

Brand	Model	KW/yr.	Energy cost/yr.
White-Westinghouse	SU200JX*2	501	$38
Caloric	DUS314-19	639	$49
Frigidair	DW180***1	639	$49

The following are among the most energy efficient washers:

TYPE	Brand	Model	KW/yr	Energy cost/yr.
COMPACT				
	General Electric	213JB 118	623	$50
	Gibson	213JB 24F	623	$50
	Kelvinator	213JB C340	623	$50
	Montgomery Ward	213JB 6507	623	$50
STANDARD FRONT LOADING				
	Gibson	WS 27 M6-V	291	$23
	White-Westinghouse	LT00L	291	$23
TOPLOADING				
	Montgomery-Ward	LNC6105A	651	$52
	Maytag	A212S	720	$58

The following are among the most energy efficient water heaters:

(30 gallons)	Model	Rating	Tank size	EF
GAS WATER HEATERS		(gal./hr.)	(gal.)	
Bradford-White	M-III-303T5CN-7	64	29	0.64
Rheem	21VR30-5	52	30	0.64
U.S. Water Heater	M-III-303T5CN-7	64	29	0.64
(40 gallons)				
American Appliance	N463V	86	40	0.72
Craftmaster	N463V	86	40	0.72
Mor-Flo	N463V	86	40	0.72
(50 gallons)				
Bock	50-HE-N	81	50	0.62
Bradford-White	M-III-503S5CN-7	81	50	0.62
Lochinvar	HEN050-ES	81	50	0.62
OIL WATER HEATERS				
Bock	32PP	131	32	0.63
Bradford-White	F-I-30SE50	120	50	0.62
Carlin	RCG-30	120	50	0.62
ELECTRIC RESISTANCE WATER HEATERS				
(40 gallons)				
Marathon	M30238	39	30	0.98
Rheem	81S*30DT	42	30	0.98
(50 gallons)				
Lowe's	26303	49	40	0.96
Marathon	MP40238	49	40	0.96
(60 gallons)				
Lowe's	26305	57	50	0.96
Rheem	81G*52*	57	50	0.96
(70 gallons)				
Rheem	81G-66D	73	66	0.95
Richmond	8*G66-*	73	66	0.95
(80 gallons)				
Rheem	80G-809D*	82	80	0.94
Richmond	8*G80-, H**G80*	82	80	0.94

D AND K APPLIANCE
1005 W. College Ave., Santa Rosa (707)575-1522

DYCOR APPLIANCES
1030 W. Washington Ave., Sunnyvale (408)245-7200

J. CASEBER WASHERS AND DRYERS
(415)548-4419

MACKALL APPLIANCE
249 California Dr., Palo Alto (415)321-2750

SOLEM'S APPLIANCES
25 North St., Healdsburg (707)433-4838

CFC-FREE PRODUCTS

We have a very short list here and we apologize for this. It is very difficult to find products or salespeople that publicize or know the CFC manufacture histories of their products. We will list a few guidelines, however, that should help in your search.

In general, when buying foam furniture make sure the foam was not blown using CFC propellants. Union Carbide makes three CFC alternative foams, marketed under the names Ultracel, Hyperlite,and Geolite. Union Carbide has declined publicizing which furniture carries them, unfortunately, but you can ask before you buy, and insist on finding out. Better yet, don't buy foam products at all and buy used furniture instead.

There are other types of ozone-destroying materials other than CFCs. One such substance is Halon, a common stable gas found in fire extinguishers. Make sure you buy halon-free extinguishers, and complain to companies that continue to manufacture halon-based equipment.

Although aerosol cans were banned from using CFCs many years ago, some replacement propellants cause up to sixteen percent of the ozone destruction in the atmosphere. One propellant in particular, 1,1,1-trichloroethane, made by Dow Chemical, PPG Industries, and Vulcan Chemical, is found in items like Elmer's Contact Cement, Scotch Spray Mount (art supply), Raid Fogger, and Kiwi shoe and leather products. Write to NRDC Publications, 40 W. 20th Street, New York NY,10011, for a list of others, called *Wanted:Public Enemy 1,1,1.* This chemical is sometimes listed on products, though there is no law saying it has to be, as methyl chloroform, Chloroethane, TCA, MCF, or just 1,1,1. Call manufacturers of any of your aerosol-based items and ask as a last resort. Better yet, use any of the non-toxic, non-aerosol alternatives that are available.

Do not buy electronic equipment cleaning sprays (almost entirely CFCs), or styrofoam "disposable" serving ware (cups, plates etc.) Petition your coffee stand to use paper cups, or better yet, bring your own non-disposable cup in a waterproof bag in your purse or briefcase.

For insulation needs in the home, use recycled cellulose fiber insulation if possible. Rigid foam insulation contains and is processed with many CFCs. Other less harmful alternatives include fiberglass, fiberboard and gypsum.

GREENER ALTERNATIVES
914 Mission St.,Suite A, Santa Cruz (408)423-0701

THE FOAM SHOP
813 A St., San Rafael (415)453-3626

CLEANING PRODUCTS

AFM ENTERPRISES
1140 Stacy Ct., Riverside (714)781-6860
See display ad, page 272.

BAU, INC.
P.O.Box 190, Alton NH 03809 (800)628-8113

BAUBIOLOGIE HARDWARE
207 16th St., Pacific Grove (408)372-8626

BROOKSIDE SOAP CO.
P.O. Box 55638, Seattle WA (206)363-3701

CLEAN ENVIRONMENTS INC.
P.O. Box 17621, Boulder CO (303)494-1770

DR. BRONNER'S PURE CASTILLE SOAP
All-One-God-Faith Inc.
P. O. Box 28, Escondido (619)745-7069

ECCO BELLA
free catalog w/ many good products (800)888-5320

ECO LOGIC CATALOG
Products for environmentally safe homes (714)338-5694

GRANNY'S OLD FASHIONED PRODUCTS
P.O.Box 256, Arcadia (818)577-1825
See display ad, page 300

GREEN CONSULTANCY
P.O. Box 20098, Santa Barbara (805)563-1817

HEALTHY KLEANER
P.O. Box 4656, Boulder (303)444-3440
See display ad, page 300

JERRICK
18149 Ventura Blvd. #149, Tarzana (818)344-5214

LIFE TREE PRODUCTS
P.O. Box 120, Sebastopol (707)577-0324
See display ad, page 288

LIFELINE NATURAL SOAPS
P. O. Box 531, Fairfax (415)457-9024

NONTOXIC ALTERNATIVES
13 Ramona Dr., Orinda (415)376-6998

SAFER PRODUCTS
Catalog (800)447-2229

SHAKLEE PRODUCTS DISTRIBUTOR
124 Avenida Drive, BERK (415)843-8544

SOLUTIONS TO INDOOR POLLUTION
10565 Caminito Bayon, San Diego (619)271-6082

THE LIVING SOURCE
Catalog of nontoxic home products (817)756-6341

WACHTER'S ORGANIC SEA PRODUCTS
360 Shaw Rd.South, SF (415)588-9567
Fashion Island Shop., Ctr.,S. Mateo (415)578-9200
Fashion Island Shop. Ctr.,S.Mateo (415)578-9200
401 Bayshore Blvd., SF (415)285-5244
Willow Shopping Center,Concord (415)686-2270

WHOLE EARTH ACCESS
2990 Seventh St., BERK (415)845-3000

CLEANING SERVICE

NON-TOXIC HOUSEHOLD WORKERS CO-OP
P.O. Box 22324, SF (415)681-6113
See display ad on this page.

SNOW WHITE ASSOCIATES
7 Mabry Way, San Rafael (415)472-2909
See display ad on this page.

ENERGY-SAVING DEVICES

These stores and companies offer a wide-range of energy-saving devices including water heaters, boilers, rechargeable batteries, motion sensor switches, and low-energy lighting.

ALL ENERGY SERVICES
2489 Mission, SF (415)334-0124

BASTIN CONTROLS
1801-D Clement Ave, ALA (415)521-4977
See display ad on this page.

BRAYER LIGHTING COMPANY
1177 Harrison, SF (415)255-1793

EARTHSAKE
1844 Market St., SF (415)626-0722

ENERGY COMPLIANCE SYSTEMS
2881 Hemlock, Suite 2, San Jose (408)244-1220

ENERSAVE INC
2525 Van Ness Ave., SF (415)346-3779

GREENER ALTERNATIVES
914 Mission St.,Suite A, Santa Cruz (408)423-0701

MAXIMUM TECHNOLOGY
80 Industrial Way, Brisbane (415)468-2560

REAL GOODS NEWS
966 Mazzoni St., Ukiah (800)762-7325
See display ad, page 289

RENEWABLE RESOURCE SYSTEMS INC
3000 Sand Hill Rd., Menlo Park (415)854-7793

SAVE ENERGY COMPANY
2410 Harrison St., SF (415)824-6010

SOLAIRE OF CALIFORNIA
3020 Bridgeway #342, Sausalito (415)331-5293

WHITE ELECTRIC COMPANY
1511 San Pablo Ave., BERK (415)845-8534

HOUSEHOLD

LIGHTING

Lighting is an excellent area in which to save energy. Typical home lighting that uses incandescent bulbs is the most wasteful; ninety percent of the energy used in these bulbs goes to heating the coil – all wasted energy. There are other options that offer considerable energy savings. The following companies regularly stock, and sell these options:

Fluorescent lighting -including compact flourescent lighting

Fluorescent fixtures can be five times as efficient as incandescent lighting, and may last up to twenty times longer than ordinary bulbs.

The recent development of compact fluorescent bulbs allows the homeowner to screw fluorescent bulbs into ordinary sockets. Most are in the 60-75 watt range and last up to ten times as long, thus justifying their higher cost for initial purchase ($12 - $25). They have also been designed to give off a light color similar to warmer incandescent light, and they don't flicker or hum. These bulbs use 90 percent less energy, and, depending on local rates, may save you up to forty-five dollars per bulb in energy savings. The bulbs last from 7,500 -100,000 hours compared to the 750-1000 hour lives of incandescent bulbs. These bulbs should not be used in enclosed fixtures, on dimmer circuits, or outside. Also, ask the salesperson whether there is any radioactive material in the bulbs, and avoid the brands that contain them. (See the article "The Changing of the Bulb," on page 17). Many hardware stores carry compact fluorescent bulbs, and you should request them if they do not.

High Intensity Discharge (HID) Lighting

These lights typically use a halogen gas to prevent the darkening of an incandescent bulb, thus using less electricity for more light. They require special fixtures and can get very hot. They are often best used in an office setting, where the electricity expenses are high enough to warrant the conversion. (See the Business-Guide Lighting section.)

Full Spectrum Lighting

There is growing opinion that lighting that is not full-spectrum is unhealthy and difficult to work under. Some companies have developed full-spectrum lights to address the problem.

ACTIVE TECHNOLOGY
2518 MacArthur Blvd., OAK (415)482-8025

C. MARKUS
301 Jefferson, OAK (415)832-6522

ECOLIGHTS
Nuclear Free America
325 E. 25th St., Baltimore MD (301)235-3575

FOX HARDWARE
564 Sixth St., SF (415)777-4400

GREENER ALTERNATIVES
914 Mission St.,Suite A, Santa Cruz (408)423-0701

JERRICK
18149 Ventura Blvd. #149, Tarzana (818)344-5214

ORCHARD SUPPLY HARDWARE
East Bay

OSRAM COMPACT FLOURESCENT
Nuclear Free America
325 E. 25th St., Baltimore MD (301)235-3575

REAL GOODS NEWS
966 Mazzoni St., Ukiah (800)762-7325
See display ad, page 289.

SAVE ENERGY COMPANY
2410 Harrison St., SF (415)824-6010

THE HOMESTEAD
11 Glen Rd., San Anselmo (415)485-5441

WHITE ELECTRIC CO., INC.
The Light Bulb Place1511 San Pablo Ave., BERK
(415)845-8535
Willow Shopping Center, Concord (415)686-2270

WHOLE EARTH ACCESS
2990 Seventh St., BERK (415)845-3000

NATURAL FIBERS

BIOBOTTOMS
P.O. Box 6009, Petaluma (707)778-7945

GREENER ALTERNATIVES
914 Mission St.,Suite A, Santa Cruz (408)423-0701

PAINT, NON-TOXIC

AFM ENTERPRISES
1140 Stacy Ct., Riverside (714)781-6860

ARMSTRONG PAINTING & WATERPROOFING
774 Harrison St., SF (415)777-1234

AURO ORGANIC PAINTS
PO Box 857, Davis (916)753-3104

SAGE: THE ENVIRONMENTAL STORE
(415)485-5441

LIVOS PLANT CHEMISTRY
Catalog of nontoxic home products (800)621-2591

MILLER PAINT
317 S.E. Grana, Portland (503)233-4491

PACE CHEM INDUSTRIES, INC.
779 S. La Grange Ave, Newbury Park (805)499-2911

SINAN COMPANY
P.O. Box 857, Davis (916)753-3104

THE COHERENCY COMPANY
P.O. Box 553, Occidental (707)869-0956

THE LIVING SOURCE
3500 Mac Arthur Drive, Waco, TN (817)756-6341

RECYCLING SYSTEMS

These are the bins and storage systems for your household or office's recycling convenience. The Green Pages is sorry there are so few, but some that were encountered were rejected for their poor design. One was simply a garbage can with a can crusher attached, and another was a rolling vegetable bin upon which you had to neatly stack newspaper, and carefully arrange about fifteen glass or aluminum containers in a precarious front-loading system of plastic. The following companies have devoted much more thought to the problem. Write to them for a catalog, or description.

ANTON PHILLIPP LAB
P.O. Box 613, Charlevoix (800)748-0563

DIVERSIFIED RECYCLING SYSTEMS
5606 N. Hiwy. 169, New Hope MN 55428 (612)536-6828

Fact: Between 1950 and 1988, Americans doubled the amount of waste they throw away, from an average of two pounds per person per day to four. Other industrial countries, such as Japan and West Germany, generate only half that amount.

MEDIA & COMMUNICATIONS

Getting and staying informed is the first step towards becoming an effective activist; knowledge is the precursor to power. Of course, nothing is effortless. Keeping informed takes the same kind of personal commitment as does shopping for safe food and recyclable containers. But it has great rewards. Becoming aware of what's happening, and knowing the truth about the condition of our planet is a difficult thing to bear sometimes. Through these alternative media, however, we can also find out how people are solving problems. This section contains listings of TV and radio stations, news-services and networks that do a good job of covering environmental issues. (See the Publications heading for print resources.)

INFORMATION SERVICES

ECO-NET
3228 Sacramento St., SF (415)923-0900
See display ad on this page.

ENVIRONMENT NEWS SERVICE
3505 W 15th Ave., Vancouver,BC (900)446-4761
See display ad, page 304.

GREENLINE
758 Gilman St., BERK (415)526-3900 x76
See display ad, page 305.

RADIO STATIONS

KALW — 91.7 FM
2905 21st St., SF (415)695-5740
See display ad on this page.

KALX — 90.7 FM
2311 Bowditch, BERK (415)642-1111
See display ad, page 304

MEDIA & COMM.

Information

Learn the Facts.
It takes more than recycled paper cups to change the world. It takes a willingness to educate ourselves, to make others aware of the issues so vital to our survival. "Global consciousness" is spread through the ideas we communicate. Since 1949, KPFA has presented environmental programming that provokes, informs, and educates.

Listen in to *The Morning Show*, weekdays at 7:00 am; *Brainstorm*, weekdays at noon; and to *Prime Time*, weeknights at 7:00 pm for some of the wide ranging discussions and documentaries concerning political, social, and environmental issues. We also ask that you tell your friends.

Inform Yourself. Educate Others.

fm 94.1

KPFA fm94

LISTENER SPONSORED PACIFICA RADIO
2207 SHATTUCK AVENUE • BERKELEY, CA 94704 • (415) 848-6767

ENS

Environment News Sevice
The Environment News Service is an independent publishing news agency with a network of international correspondents.

We offer daily and weekly feeds of breaking environmental news of the world.

Why wait to see it somewhere else?

Via computer, as the **E-Line**
Via fax, as the **E-Sheet**
Via Air Mail
Via audio-text as the **GreenLine News**

1-900-446-4761 ext.70

Contact: News Director
Enviromental News Service
3505 W. 15th Avenue
Vancouver, B.C. V6R 2Z3
CANADA
Voice: 1-604-732-4000
Fax: 1-604-732-4400
InterNet: a13@mindlink.uucp
EcoNet: web:ens

Available worldwide!

KALX 90.7 FM

Alternative music, public affairs, news, sports and special programming. Community radio for the Bay Area.

Now on Viacom Cable San Rafael.

Call 642-1111 to obtain a programming schedule.

University of California at Berkeley Radio

MEDIA & COMM.

KCSM — 91.1 FM
1700 West Hillsdale Blvd., San Mateo (415)574-6586
See display ad, page 308.

KFJC — 89.7 FM
12345 El Monte Road, Los Altos Hills (415)949-7260
See display ad, page 306.

KPFA — 94.1 FM
2207 Shattuck Ave., BERK (415)848-6767
See display ad, page 304.

KQED — 88.5 FM
500 Eighth St., SF (415)553-2108

KSJS — 90.7 FM
San Jose State University (408)924-4549
See display ad, page 306.

KUSF — 90.3 FM
2130 Fulton St., SF (415)386-5873
See display ad, page 305.

KUSP — 88.9 & 90.3 FM
P.O. Box 423, Santa Cruz (408)476-2800

KZSC — 88.1 FM
University of California, Santa Cruz (408)459-2811

KZSU — 90.1 FM
P.O. Box B, Stanford (415)725-4868
See display ad, page 306.

TELECOMMUNICATIONS

WORKING ASSETS LONG DISTANCE
230 California St., SF (800)522-7759
See display ad on this page.

TELEVISON STATIONS

KCSM
1700 West Hillsdale Blvd., San Mateo (415)574-6586

KQEC
500 Eighth St., SF (415)864-2000

KQED
500 Eighth St., SF (415)553-2108

KTEH
100 Skyport Dr., San Jose (408)437-5454

VIDEO PRODUCTION

GREEN TV
1125 Hayes St., SF (415)255-4797
See display ad on this page.

MEDIA & COMM.

OFFICE

Most of us end up spending a third of our lives at work. It's very important that we all bring our ecological standards to the office. Not only could you be spreading knowledge to your associates, but you could also be helping your employer. All businesses have a lot to gain by becoming environmentally sound.

Businesses waste 1.4 billion pounds of paper each year. Office recycling is very necessary AND very simple to set up. Order a how-to booklet, consult one of the advisors listed below, or contact your city's commercial recycling efforts (the conservation corps comes straight to your office to get you started–see the Organization Index, under Recycling).

Some businesses resist recycling for fear it is too complicated or costly. It is not. For example, after Pacific Bell began recycling programs at all their western offices, they saved $140,000 on disposal fees, the equivalent of 1700 acres of trees, and boosted their corporate image by keeping up with the times.

So if you have trouble starting recycling at your office, stay with it — don't give up! Don't hesitate to ask for help from local groups and city recycling offices.

If you are responsible for ordering office supplies, utilize your position. If your supplier is not offering recycled paper and other sound products, suggest that they do, and failing that, file a complaint or take your business elsewhere. Insist on recycled paper, recycled products and a recycling system. Promote nontoxic office supplies, green plants, natural fibers, full-spectrum lights. If you feel your workplace is making you sick call Cal-OSHA (Occupational Safety and Health - see County Government Section, under Workplace).

ADVISORS

The following consultants will advise your office on how to set up office recycling systems, reduce energy, and purchase greener office supplies. Nearly every city now has an office recycling pickup service, and some county departments address business needs.(Also see the Public Agency section and the Business guide's Recycled Products section.

EARTH TALK
P. O. Box 426, Larkspur (415)924-3822

WORKING WITH NATURE
12, OAK Knoll Ave., San Anselmo (415)457-7656

PAPER, RECYCLED RETAILERS

Buy recycled. Purchase the products with the most post-consumer waste percentages (post-consumer waste is that paper recovered through collection of USED paper, not mill leftovers). Besides these retailers, consult the Business Guide's Paper Recycled heading for wholesalers.

BRIGHT IDEAS
3450 Golden Gate Wy, Lafayette (415)283-5199
See display ad, page 311.

CONSERVATREE PAPER CO.
10 Lombard St., Ste. 250, SF (415)433-1000

ECOLOGY CENTER
2530 San Pablo, BERK (415)548-2220

EVERGREEN RECYCLED PAPERS
1938 Old Middlefield Way, Mountain View (415)961-8928

GREENER ALTERNATIVES
914 Mission St.,Suite A, Santa Cruz (408)423-0701

J.C. PAPER CO.
1920 Army St., SF (415)285-6220

JAMES RIVER CORPORATION
300 Lakeside Dr, OAK (415)874-3458

KET ENTERPRISES
Santa Cruz (408)335-3216

NATIONWIDE PAPERS
345 Schwerin, SF (415)586-9160

OFFICE CLUB
1631 Challenge Dr., Concord (415)682-2582
See display ad on this page.

STATLER TISSUE
300 Middlesex Ave., Medford MA (617)395-7770

PAPER, RECYCLED WHOLESALERS

Businesses buy huge amounts of paper. There is no reason to use virgin stock paper when there are so many recycled papers available for all kinds of uses, in many grades and colors, for everything from book stock to xerographic needs. If you are concerned that recycled paper does not convey a certain quality image, let your customers know why you have made the switch, and you may find they actually commend your actions. Let your paper suppliers know what you need. Ask them for unbleached supplies. Most in-house and letter paper needs do not require perfectly bleached white paper. Try to use paper with the most post-consumer recycled content available. It is only when we do this that we are actually diverting paper from the landfills. If all the businesses in the Bay Area demanded and used paper with a minimum ten percent post-consumer content it would make a huge difference. (See the article "Greening Your Business" on page 54.)

A.C. PAPER AND SUPPLY
1321 Seventh St., BERK (415)527-0841

ALTE SHULE, U.S.A.
704 E. Palace Ave., Santa Fe NM (505)983-2593

APPLETON PAPERS, INC.
825 E. Wisconsin Ave., Appleton WI (414)735-8810

ATLANTIC RECYCLED PAPER COMPANY
P.O.Box11021, Baltimore MD (800)323-2811
See display ad, page 309.

BANDELIER DESIGNS
1570 Pacheco, Suite E7, Santa Fe NM (505)986-1600

BUTLER PAPER COMPANY
180 West Hill Pl., Brisbane (415)468-1100

CARTER RICE COMPANIES
273 Summer St., Boston MA (617)542-6400

CONSERVATREE PAPER CO.
10 Lombard St., #250, SF (415)433-1000
See display ad, page 312.

CROSS POINT PAPER CO.
1295 Bandana Blvd. N., St. Paul MN (612)644-3644

DANCING TREE RECYCLED PAPER PROD.
2253 Glen Ave., BERK (415)486-1616

EARTH CARE PAPER, INC.
P.O. Box 14140, Madison WI (608)256-5522

ECO-VISION
2215 Market, SF (415)864-2316

FORT HOWARD PAPER
2379 Norwood Rd, Livermore CA (415)447-5796

FUTURE FIBRES GROUP, INC.
4380 Redwood Hwy., #C-21, San Rafael(800)447-3111

GREENER ALTERNATIVES
914 Mission St.,Suite A, Santa Cruz (408)423-0701

J. C. PAPER CO.
650 Brennan Ave., San Jose (800)527-2737

JAMES RIVER CORPORATION
300 Lakeside Dr #1461, OAK (415)874-3458
See display ad, page 311

KIRK PAPER COMPANY
6001 East Randolph St., Los Angeles (213)728-8117

OFFICE CLUB
1631 Challenge Dr., Concord (415)682-2582
See display ad, page 309

PAPER, INC.
8041-2 Queenair Dr., Gaithersburg MD (301)948-6300

PERFORMANCE COMPUTER FORMS
21673 Cedar Ave., Lakeville, MN (612)469-1400

RECYCLED PAPER COMPANY
100% Post-consumer Paper
185 Corey Rd., Boston, MA 02146 (617)277-9901

SCRIBECRAFT
P. O. Box 77, Newfield,NY (607)564-7572

SEABOARD PAPER COMPANY
336 Oyster Point Blvd., SSF (800)832-2311

SIMPSON PAPER COMPANY
600 Market St., SF (415)391-8140

THE PAPER PROJECT
P.O. Box 12, Arcata (415)540-1127

ZELLERBACH
245 S. Spruce Ave., SSF (800)468-0200
See display ad, page 312.

RECYCLED PRODUCTS

If you're not buying recycled products, you're not recycling. High-quality sources of recycled paper products, such as toilet paper, and paper towels are becoming increasingly available. Other office items like brooms and recycling bins that are made of recycled plastic can be found here. Call these places to find out what they carry. Make the most important link of the recycling chain a high priority. Some city governments, (e.g. the City of San Jose) can advise you on sources as well. (See the Public Agency Section.)

BAY WEST JANITORIAL SUPPLIES
P.O. Box 2867 So., SF (415)873-0160

BIRITE FOODSERVICE DISTRIBUTORS
201 Alabama St., SF (415)621-6909

BUSINESS ECOLOGY PRODUCTS
P.O. Box 946, Martinez
(415)687-1202

CLEAN MASTER
1780 3rd St., SF
cleaning and food service supplies (415)495-2900

COMPETITIVE PAPER SALES
151 Second St.,OAK
food service products (415)832-6714

DATACOM INC.
Evolution Recyled Paper Products
201 Gilbraltar Rd., Horsham PA
(215)443-5700

EARTH CARE PAPER, INC.
P.O. Box 14140, Madison WI (608)256-5522

FLEXEL-ATLANTA
19961 Merrit Drive, Copertino (415)828-0603

FORMS NET
10915 Merritt St., Castroville (408)688-6964

RECYCLED TONER CARTRIDGES
Unconditionally guaranteed
...INFOTEK...
free trial
(415)921-7135

MONAHAN PAPER
175 Second St., OAK (415)835-4670
See display ad, page 348.

POPE & TALBOT
P.O. Box 8171 Portland,OR (503)228-9161

PURE PAPER
1232 Masonic Ave, BERK (415)524-8079

TURN AROUND PRODUCTS OF MARTINEZ
1258 Pine St., Martinez (415)372-6974

RECYCLING SYSTEMS

Recycling systems for offices and businesses can be elaborate or suprisingly simple. Call your city's solid waste or recycling office to see if they provide business customers with recycling bins.

ANTON PHILLIPP LAB
P.O. Box 613, Charlevoix (800)748-0563

DIVERSIFIED RECYCLING SYSTEMS
5606 N. Hwy. 169, New Hope, MN 55428 (612)536-6828

REHRIG PACIFIC CO.
4010 E. 26th St., Los Angeles (213)262-5145

WINDSOR BARREL WORKS
P.O. Box 47, Kempton, PA (215)756-4344

PRINTERS

ALONZO PRINTING
1094 San Mateo Ave.,SSF (415)873-0522
See display ad, page 347.

BLACK LIGHTNING,INC.
RR 1-87, Depot Road, Hartland, VT (800)252-2599

BRIGHT IDEAS PRESS
3450 Golden Gate Wy, Lafayette (415)284-5350

BRIGHT IDEAS COPY SHOP
99 Brookwood Road, Orinda (415)253-0260

GRAFFIK NATWICKS
240 Golden Gate Ave., SF (415)441-0878

NORCO-BILLINGS PRINTING
1152-A Marina Blvd., San Leandro (415)351-8288

SUNCHEMICAL
14300 Catalina St.,San Leandro (415)351-7661

PRINTERS' SUPPLY

AMERICAN INK PRODUCTS COMPANY
630 E. Tenth St., OAK (415)268-0825

SUNCHEMICAL GENERAL PRINTING INK
14300 Catalina St. San Leandro (415)351-7661

ZELLERBACH
245 S. Spruce Ave., SSF (800)468-0200

PAPER

Recycled paper plays a crucial role in our environmental future. When you consider that paper takes up as much as half of our landfill space; that the average American uses at least one 100-foot tree per year to meet his or her wood and paper needs, and that America is cutting down more trees per year than any country in the world — twice as much as Brazil — it becomes clear that we must use recycled paper.

Recycled paper can be made from scraps of virgin paper left over from the paper-making process and from recycled wastepaper fromenvelope makers, printing shops, homes, businesses and paper mills. The term "post-consumer waste" refers to the amount of recycled content that was actually used by consumers. The paper you buy should contain at least some fiber that originated through recycling efforts. In addition to making a commitment on the purchasing end of recycling efforts, the use and patronage of post-consumer waste paper minimizes the production of toxics like dioxin (a deadly carcinogen) that are used in the bleaching process for virgin paper products.

Recycled paper is made in many styles, many of which are indistinguishable from non-recycled types in appearance, texture, and performance. There is recycled photocopying paper, stationery, envelopes, gift wrap, paper towels, and so on.

Don't forget that the purchase and use of recycled paper AND the committment to then recycle this paper play an equal role in conservation.

(For a full discussion of recycled paper, please see the article "Creating Markets for Recycled Paper," p. 69.)

PAPER PRODUCTS, RECYCLED

These listings contain companies that offer recycled products other than printing paper or writing paper. You can find disposable food service items, envelopes, and other products that enable you to "buy recycled".

BRUSH DANCE
218 Cleveland Ct., Mill Valley (415)389-6228

DANCING TREE RECYCLED PAPER PROD.
2253 Glen Ave., BERK (415)486-1616
See display ad on this page.

MONAHAN PAPER
175 Second St., OAK (415)835-4670

PAPER CHOICE
303 E.Sixth Ave.Vancouver, B.C. (604)873-5700

WALDECK'S
526 Washington St., SF (415)981-3381
863 E. St. Francis Blvd. (415)578-9200
Willow Shopping Center, Concord (415)686-2270

WHOLE EARTH ACCESS
2990 Seventh St. BERK (415)845-3000

PAPER PRODUCTS, UNBLEACHED

The bleaching of paper, from copier paper to toilet paper, is a process that releases tons of carcinogenic dioxins into our rivers and environment all over the country. This is an unnecessary result of an unintelligent aesthetic. Consumers should buy unbleached paper goods and unbleached paper for their offices. Many of these companies offer unbleached coffee filters. Others carry products like paper towels and toilet paper that is unbleached. Check the General Merchandise category for more multi-product vendors, as many of the companies listed there also carry unbleached paper items. Check the Paper, Recycled headingfor sources of office paper that are unbleached.

COFFEE, TEA & SPICE
1630 Haight, SF (415)861-3953

GREENER ALTERNATIVES
914 Mission St.,Suite A, Santa Cruz (408)423-0701

PAPER

KET ENTERPRISES
Santa Cruz (408)335-3216

PEET'S COFFEE
2156 Chestnut, SF (415)931-8302

PURE PAPER
1232 Masonic Ave, BERK (415)524-8079
See display ad on this page.

SPINELLI COFFEE COMPANY
3966 24th St., SF (415)550-7416

THE CASTRO BEAN
415 Castro, SF (415)552-0787

PurePaper
Unbleached Coffee Filters

Great tasting coffee without chlorine bleach or dioxin

Available at natural and fine food stores, or call PurePaper at (415) 524-8079

PAPER, RECYCLED

Buy recycled. Purchase the products with the most post-consumer waste percentages (post-consumer waste is that paper recovered from community and business collection of USED paper, not mill leftovers). Besides these retailers, consult the Business Guide's - Paper, Recycled heading for wholesalers.

BRIGHT IDEAS
3450 Golden Gate Wy, Lafayette (415)283-5199
See display ad, page 311.

CONSERVATREE PAPER CO.
10 Lombard St., Ste. 250, SF (415)433-1000

DATACOM INC.
Evolution Recycled Paper Products
201 Gilbraltar Rd., Horsham PA
(215)443-5700

ECOLOGY CENTER
2530 San Pablo, BERK (415)548-2220

EVERGREEN RECYCLED PAPERS
1938 Old Middlefield Way, Mountain View (415)961-8928

Zellerbach
A Mead Company
THE ONLY SOURCE YOU NEED TO KNOW FOR RECYCLED PAPER

A complete line of:
•Text•
•Cover•
•Coated•
•Writing•
Recycled paper

Complete advisory sample department
(800) 468-0204

Dedicated to providing the widest range of recycled choices in the market

GET REAL!

If you're in the market for *real* recycled paper, come to the source:
CONSERVATREE PAPER COMPANY.

Our commitment to our customers is to furnish the highest quality recycled paper and greatest variety at the lowest prices in the United States.

Experience the difference that commitment can make.
When you want environmentally sound paper contact the country's most experienced and knowledgeable recycled paper source: **Conservatree Paper Company.**

Conservatree Paper Company
Since 1976... The Leader in Quality Recycled Paper
415-433-1000 x 24

Conservatree currently only markets recycled paper on a wholesale basis — minimum order $200.

GREENER ALTERNATIVES
914 Mission St.,Suite A, Santa Cruz (408)423-0701

J.C. PAPER CO.
1920 Army St., SF (415)285-6220

JAMES RIVER CORPORATION
300 Lakeside Dr, OAK (415)874-3458
See display ad, page 311.

KET ENTERPRISES
Santa Cruz (408)335-3216

NATIONWIDE PAPERS
345 Schwerin, SF (415)586-9160

OFFICE CLUB
1631 Challenge Dr., Concord (415)682-2582

STATLER TISSUE
300 Middlesex Ave., Medford, MA (617)395-7770

STATIONERY

THE JOHN ROSSI COMPANY
259 Washburn Rd., Briarcliff,NY (914)941-1752

BRUSH DANCE
218 Cleveland Court, Mill Valley (415)389-6228

PERSONAL CARE

It is ironic that a whole industry built upon the notion of personal health care is actually very damaging to the environment. The companies and products listed here represent alternatives in personal care (cosmetics, soaps, lotions, feminine products, shaving products, and-dental care). These companies have addressed the major problems of over-packaging, animal testing, non-biodegradeability, unrecycled and unrecycleable packaging, and in some cases products which are actually harmful to the user. No product overcomes all of these hurdles; the most natural shampoo inevitably comes in a plastic bottle.

ALLERGY RELIEF SHOP
2932 Middlebrook Pike, Knoxville, TN (800)678-2028

AMERICAN MERFLUAN

AUROMERE
1291-G Weber St., Pomona (800)735-4691
See display ad on this page.

BARE ESSENTIALS COSMETICS
809 University Ave., Los Gatos (408)354-8853

BEAUTY NATURALLY
Catalog (415)459-2826

BEAUTY WITHOUT CRUELTY
Catalog of Cosmetics (415)382-7784

CHENTI PRODUCTS, INC.
21093 Forbes Ave., Hayward (415)887-0546

COLUMBIA COSMETICS MFG., INC.
1661 Timothy Dr., San Leandro (415)562-5900

COMMON SCENTS
3920A 24th St., SF (415)826-1019

COSMETICS
2479 Edison Way, Menlo Park (415)364-6343

COUNTRY VILLAGE NATURAL FOODS
Cosmetics, body care
3611 Union Ave., San Jose (408)377-1431

DP LABORATORIES
2724 Third Ave, San Diego (619)233-7213

DR. BRONNER'S PURE CASTILLE SOAP
P. O. Box 28, Escondido (619)745-7069

GOOD LIFE DISCOUNT NUTRITION
Body, health care products
437 South Kiely Blvd., San Jose (408)247-7814

GRANNY'S OLD FASHIONED PRODUCTS
for the body and home
P.O.Box 256, Arcadia (818)577-1825
See display ad, page 300

HERBACEUTICALS
478 Post St. , SF (415)986-2607

IDA GRAE COSMETICS-NATURES COLORS
424 Laverne Ave., Mill Valley (415)388-6101

JASON NATURAL COSMETICS
8468 Warner Dr., Culver City (213)838-7543

NATURAL LIFE FOODS
Health & beauty Products
2638 Pleasant Hill Rd., Pleasant Hill (415)932-1292

NATURE'S GATE HERBAL COSMETICS
9183 Kelvin Ave., Chatsworth (800)327-2012

NATURE'S OWN COSMETICS
San Lorenzo (415)278-4600

OCEANS OF LOTIONS
842 Cole St., SF (415)566-2326

SAN RAFAEL HEALTH FOODS
Beauty w/ cruelty products
1132 4th St., San Rafael (415)457-0132

SHATOIYA- BODY CARE
13935 Dry Creek Rd., Auburn (916)878-2411

THE BODY SHOP
1341 Seventh St., BERK (415)644-3557

THE LIVING SOURCE
Catalog of natural, earthwise products (817)756-6341

PERSONAL CARE

THOM'S NATURAL FOODS
Cosmetics and more
5843 Geary Blvd., SF (415)387-6367

WACHTER'S ORGANIC SEA PRODUCTS
360 Shaw Rd.South, SF (415)588-9567
Willow Shopping Center, Concord (415)686-2270

WHOLE EARTH ACCESS
2990 Seventh St., BERK (415)845-3000

ZIA COSMETICS
300 Brannan, SF (415)543-7546

CAL BEN FIVE STAR SOAP
9828 Pearmain St., OAK (415)638-7091

Fact: Homeowners annually use between 20 and 30 million pounds of chemical pesticides in their gardens.

Imagine this: you are just visiting this planet for the first time. You have taken a little time to find out which things are poisons so as to try and avoid contact with them. While you are at the supermarket you notice the natives buying toxic chemicals (usually neurotoxins) and then taking them home and spreading them all over the place, frequently in their kitchens.

There is a better way. One of the first things we need to do is not be so frightened of bugs. Then, when we get over our entophobia, we should learn to distinguish between beneficial insects and pests, and perhaps even notice the ways the beneficial insects help us and how amazing insects are.

"The-only-good-bug- is- a-dead-bug-gimme-that-spray" approach is based on ignorance; ignorance of insects and ignorance of the effects of insecticides on humans. The business and products in this section are based on the philosophy that it's not worth killing people to get rid of the bugs. The health of people and pets is a central concern of the biological pest control techniques used by the businesses listed here.

There were fewer than twenty pests known to be resistant to available pesticides in 1950; most pests in California today are resistant to at least one chemical. You can help naturally-occurring biological control by introducing beneficial insects, bacteria which attack pest insects, and by using insect-repellent plants, crop rotation, and enhancement of soil fertility (i.e., use compost).

For difficult pest problems use natural botanical pesticides with careful monitoring.

For more information, please refer to the article on integrated pest management on page 26.

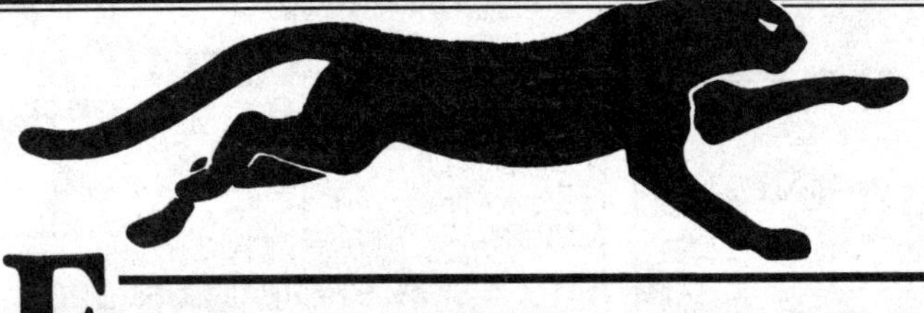

PEST CONTROL

BIOSYS
1057 East Meadow Cr., Palo Alto (415)856-9500

BENEFICIAL INSECTARY
14751 Oak Run Rd., Oak Run (916)472-3715

BIO INTEGRAL RESOURCE CENTER
P.O. Box 7414 BERK (415)524-2567

BIO-CONTROL CO.
P.O. Box 247, Cedar Ridge (916)589-5227

CHRIS POEHLMANN
Consultation on biological termite control (415)763-7576

CONNECTICUT ST. PLANT SUPPLIES
306 Connecticut, SF (415)821-4773

ELECTRACAT
5588 N. Palm Ave., Fresno (209)432-2707
See display ad, page 314.

ECOLOGY CENTER
2530 San Pablo, BERK (415)548-2220

ELECTRACAT
1315 Olympia Ave., Campbell (408)377-9866

ERLANDER'S NATURAL PRODUCTS
2279 N. Lake, Altadena (800)562-8873

FLEA CONTROL @ R VALUE/WEST
150 N Sta Bonita Ave.Ste 300, Arcadia (818)446-4888

DRAX ANT CONTROL @ R VALUE/WEST
150 N Sta. Bonita Ave.Ste300, Arcadia (818)446-4888

SEABRIGHT LABORATORIES
4026 Harlan St., Emeryville (415)655-3126

WEST COAST LADYBUG
P.O. Box 903, Gridley (916)534-0840

PEST CONTROL, STRUCTURAL

LIVE OAK STRUCTURAL
801 Camelia St. Suite B, BERK (415)524-7101

Tip: The best way to control household pests is by simply keeping them out. Check the exterior of your house carefully and seal all holes - even the tiniest ones.

PUBLICATIONS

Where would we be without the printed word? The following publications can keep you informed about environmental issues in the Bay Area and around the world. (Please see the Review section for other important books and source materials.)

BOOKS

CHRONICLE BOOKS
275 Fifth St., SF (415)777-7240

EARTHWORKS PRESS
P.O. Box 251400 Shattuck Ave., BERK (415)841-5866

EBMUD
2130 Adeline St., OAK (415)835-3000
See display ad, page 316.

FOGHORN PRESS
P.O. Box 77845, SF (415)641-5777

NEW SOCIETY PUBLISHERS
P.O. Box 582, Santa Cruz (408)458-1191

NON-TOXIC, NATURAL AND EARTHWISE
P.O. Box 279, Forest Hills (415)488-4614
See display ad on this page.

SOLANO PRESS
Box 773, Point Arena 95468 (707)884-4508
See display ad, page 316.

BOOKSTORES

ALICORN IRISH BOOKS
3428 Balboa, SF (415)221-5522

PUBLICATIONS

PUBLICATIONS

BLACK OAK BOOKS
1491 Shattuck Ave., BERK (415)486-0698

BOOKLOVER'S HAVEN
433 Georgia, Vallejo (707)557-4190

ECOLOGY CENTER
2530 San Pablo, BERK (415)548-2220
See display ad, page 317

FOREVER AFTER BOOKS
1475 Haight, SF (415)431-8299

GAIA BOOKSTORE
1400 Shattuck Ave. #10, BERK (415)548-4172
See display ad, page 315.

GATEWAYS
825 Pacific Ave., Santa Cruz (408)429-9600

HEYDAY BOOKS
P.O.Box 9145, BERK (415)549-3564

OLD WIVES' TALES
1009 Valencia, SF (415)821-4675

ROOKS & BECKORDS
222 Polk, SF (415)771-7909

SHAMBHALA BOOKSELLERS
2482 Telegraph, BERK (415)848-8443

SIERRA CLUB STORE
730 Polk St., SF (415)923-5600
See display ad, page 315.

JOURNALS

EARTH ISLAND INSTITUTE
300 Broadway, Ste. 28, SF (415)788-3666
See display ad, page 316.

LIFEWEB
P.O. Box 20803, San Jose (408)997-6067
See display ad, page 318

LIVING TREE JOURNAL
P.O. Box 10082, BERK (415)420-1440

ORGANIC FOOD MATTERS
Comm. for Sustainable Agriculture
P.O. Box 1300, Colfax (916)346-2777

MAGAZINES

FOLIO
2207 Shattuck Ave., BERK (415)848-6767

MOTHER JONES
1663 Mission St., SF (415)558-8881

OPTIONS MAGAZINE
11 Bolinas Rd.,Fairfax (415)485-5329
See display ad, page 318.

OWNER BUILDER CLASS CATALOG
1250 Addison St., Ste. 209, BERK (415)848-6860

NEWSLETTERS

CALIFORNIA CERTIFIED ORG. FARMERS
P.O. Box 8136, Santa Cruz (408)423-2263

PUBLICATIONS

FOGHORN PRESS
P.O. Box 77845, SF (415)641-5777

GREEN CONSENSUS
777 Valencia St., SF (415)255-2940

LOMA PRIETAN
2448 Watson Ct., Palo Alto (415)494-9901

REDWOOD NEEDLES
825 College Ave., Santa Rosa (707)544-7651

S. F. LEAGUE OF URBAN GARDENERS
2540 Newhall St., SF (415)468-0110

SONOMA SIERRAN
P.O. Box 466, Santa Rosa (707)544-7651

THE EARTHWISE CONSUMER
P.O. Box 279, Forest Knolls (415)488-4614

PUBLISHERS

GREEN MEDIA GROUP
P.O. Box 11314, BERK (415)521-0884
See display ad on this page.

PARALAX PRESS
P.O. Box 7355, BERK (415)548-3721

RECREATION

Travel and tourism often bring out the worst in us. We tend to use lots of energy and waste heaps of resources when we're on vacation. This is often marketed to us as luxury. We try to get away from the stresses of modern life, but often end up more tired after our vacation. In response to this, there has been an explosive growth in "natural" holidays and offerings of wilderness trips. Delicate environments have been degraded for a quick buck and a not-so-cheap-thrill. You cannot stop being an environmentalist when you go on vacation, especially if you are visiting some fragile ecosystem.

The businesses listed in this section are included for the following reasons: they encourage local and low-impact recreation, they have an environmental education component, and they donate money to environmental causes. Foreign travel programs are also sensitive to the culture of the visited country, and the impact of tourism on it.

EXPEDITIONS

BACKROADS BICYCLE TOURING
1516 Fifth St., BERK (415)527-1555

GLOBAL EXCHANGE
2141 Mission, Rm. 202, SF (415)255-7296

HIMALAYA TREKKING. & WILDERNESS. EXPEDITIONS
1900 Eighth St., BERK (415)540-8040

INNERASIA EXPEDITIONS
2627 Lombard, SF (415)922-0448

MARIAH
P.O. Box 248, Point Richmond (415)233-2303

MOUNTAIN TRAVEL
6420 Fairmont Ave., El Cerrito (415)527-8100

MOUNTAINEERING BASICS
2570 Dell Ave., Mountain View (415)965-3822

OCEANIC SOCIETY EXPEDITIONS
Bldg. E, Rm. 257, Fort Mason, SF (415)441-1106

SOBEK'S EXCEPTIONAL ADVENTURES
P.O. Box 1089, Angels Camp (800)777-7939

VISION QUEST
Sausalito (415)331-5380

OUTINGS

SIERRA CLUB STORE
730 Polk St., SF (415)776-2211

WINE COUNTRY WAGONS
510 Kenilworth Ave., Kenwood (707)833-2724

TRAVEL

AYH TRAVEL SERVICES
425 Divisadero #306, SF (415)863-9939

CENTER FOR US-USSR INITIATIVES
3268 Sacramento, SF (415)346-1875

HOWARD MILLER
100 California St., SF (415)956-7800

RIVER TRAVEL CENTER
P.O. Box 6-R, Point Arena (800)882-7238

WHITE MAGIC UNLIMITED
317 MillerAve., Mill Valley (415)381-8889

WILDERNESS TRAVEL
801-AZ Allston Way, BERK (415)548-0420

WORKING ASSETS TRAVEL SERVICE
230 California St., SF (800)522-7759

RECYCLING

This section was formed by consulting all the local and county groups that have made it their job to monitor the recycling facilities in their areas. We have made every attempt to make this a complete list. Sadly, you may find some of these businesses have closed down before you call them. Recycling is a volatile industry, and the prices recyclers can get for their materials fluctuate wildly.

Here are a few tips on the nuts and bolts of household recycling, and what you can and cannot do with your recyclable waste.

Glass: Recyclers reject entire truckloads of glass if there are rocks,dirt, or even one piece of ceramic material in the lot. Glass recycling is a high-temperature process in which any contamination ruins the ability of theplant to make pure glass. Do not throw a coffee cup, or a Grolsch Beer top in with your glass containers. The process can accept, however, those thin rings of plastic or tin soda lids attached to bottles without major distress. The process has problems with the lead around wine bottles, so you should remove these. Jar glass is fine. Window glass is recycled with a different process than container glass, and therefore few recyclers take it because collected quantities are too low to warrant a run.

Aluminum: Recyclers love it. Some cat and dog food cans are aluminum, not tin, and recyclable with your other aluminum.

Plastic: Most recyclers only take PET soda bottles. Manufacturers claim that all kinds of plastics are recyclable, but the fact is that there are no recycling plants close enough to the Bay Area to make recycling plastic economical. The manufacturers have marked their products with the universally-used recycling symbol, and have coded their products 1 to 7 to enable separation into different types of plastic, but have failed to develop any way to re-use their containers.

Tin: Few recycling centers take tin cans because the reprocessing is expensive and the resale value low. The cans are recycled primarily for their steel, not their tin. Recyclers request that labels be taken off before they'll accept loads of cans. Much of the steel that is recycled winds up in a copper smelting process, and not as a new can.

Paper: If the paper looks white, or does not contain a dye, it can be recycled as high-grade white paper. If you've photocopied solid black onto a white sheet, reject it. Glossy paper can be recycled, but at present there are only a few U.S. mills processing it. Glossy paper has a high clay content that must be removed in the recyckling process. The occasional junk-mail glossy insert in your mixed paper collection (cardboard cereal boxes, junk mail, etc) won't gum up the system too much, but you should separate out magazines and large quantities of glossy paper.

Miscellaneous: Don't worry about staples in paper. Pull off the clear windows of envelopes if your material is high grade white paper. Don't worry about them if you are collecting mixed paper.

Jar lids can be recycled with your tin, but companies hate the plastic coated baby food lids, or pure plastic lids. Throw them away — and avoid buying them if you can

Call and write the companies that continue to use stupid packaging. Let's get milk back in glass bottles, and soda out of plastic. Encourage intelligence.

In addition to the following places that accept motor oil for recycling, some oil changing businesses take used motor oil as well. They are required by law to recycle it anyway.

Most oil recyclers charge a per-gallon fee. Many cities are beginning to address the problem by applying for storage tank permits for their drop-off sites. (For more information, please see the chapter on recycling beginning on page 59. Refer to the main index to find non-profit organizations and public agencies that deal with recycling issues.)

LISTING CODES:

B = Buy -back center
C = Curbside pickup
D = Dropoff point

RECYCLING

ALAMEDA

CFC RECYCLING

AIR CONDITIONING AND AUTO ELECTRIC OF FREMONT
37555 Central St., Fremont (415)793-2340

ALAMEDA-OAKLAND TIRE, BRAKE, & MUFFLER
1825 Webster St., Alameda (415)521-1541

ALLIED RADIATOR AND AIR CONDITIONING
920 Gilman St., Berkeley (415)524-3860

AUTOHAUS - DUBLIN
6058 Dougherty Road, Dublin (415)873-3013

AUTOHAUS - HAYWARD
21650 Mission Blvd., Hayward (415)881-1915

CHEMICALS FOR RESEARCH & INDUSTRY,INC.
2928 Poplar Street, Oakland (415)893-8257
SOLVENTS-CALL FOR DETAILS OF WHAT THEY TAKE.

KING'S AUTO AIR HEATING INC
2550 High St., Oakland (415)534-1946

KINGS AUTO AIR/HEATING, INC.
2550 High St., Oakland (415)534-1946

SSC AIR CONDITIONING CO.
8541 Amelia St., Oakland (415)430-1600

RECYCLING

ACTION RECYCLING
727 Industrial Pkwy, Hayward (415)537-6075
B: NEWS, MIX, WHITE, COMP, COLOR, GLASS, REDEMP

ALLIED METALS
3426 Peralta St., Oakland (415)547-2408
B: SCRAP

APPLING & SON - BERKELEY
1941 San Pablo, Berkeley (415)653-5460
B:GLASS, SCRAP, REDEMPTION.
D: NEWS, CARD, MIXED AND WHITE PAPER

APPLING & SON - OAKLAND
5800 Bancroft Ave.
10111 E. 14th Ave, Oakland (415)653-5460
B: GLASS, SCRAP, REDEMP, ALUM
D: NEWS, CARD

ARATA CO.
6565 Smith Ave., Newark (415)795-7228

AARON METALS CO.
750 105th Ave, OAK (415)569-6767
See display ad on this page.

BAY CITY PAPER STOCK CO.
2615 Davis St., San Leandro (415)638-4327
B: GLASS, ALUM, NEWS, CARD, COMP,PET

CIRCO RECYCLERS
6565 Smith Ave., Newark (415)791-6980
B: ALUM, GLASS

COGIDO RECYCLING SERVICES INC.
850 77th Ave, OAK (415)444-1871
B: ALUM, GLASS, COMP, WHITE, CARD, MIX, NEWS, PET
C: WHITE, COMP

COMMUNITY BUSINESS FOUNDATION
Oakland (415)540-4724

COMMUNITY CONSERVATION CTRS.
669 Gilman Street, Berkeley (415)524-5355
B: NEWS, CARD, GLASS, WINE, ALUM, SCRAP
D: MIX, WHITE, PHONEBK, TIN

COMMUNITY CONSERVATION CTRS.-MLK
MLK and Dwight Wy., Berkeley (415)524-5355
D: NEWS, CARD, MIX, GLASS, ALUM, TIN

COMMUNITY RECYCLING CTRS.
3550 Adeline Street, Oakland (415)653-2104
B: NEWS, GLASS, ALUM, PET, CARD
D: TIN

CONTAINER RECOVERY CORP.
29995 Ahern Ave., Union City (415)471-4776
B: REDEMP, SCRAP

EAST BAY CONSERVATION CORPS
120 Linden Street, Oakland (415)891-3800

EAST BAY DEPOT
1027 60th St., Oakland (415)547-6470
D:MATERIALS USEFUL FOR CHILDRENS' PROJECTS, STYROFOAM PELLETS, BUTTONS

GALLAGHER AND BURK
7100 Mountain Blvd., Oakland (415)635-5200
CONCRETE AND ASPHALT (FEE).

GOODWILL INDUSTRIES - PLEASANTON
3550 Bernal Ave., Pleasanton
B: REDEMPTION

GOODWILL INDUSTRIES - RECYCLING
1301 30th Ave., Oakland (415)534-6666
B: REDEMPTION

LAKESIDE NON-FERROUS METALS
4th and Madison, Oakland (415)444-5494

LIVERMORE AUTO SALV. AND RECYC.
101 N. Greenville Rd., Livermore (415)443-0310
B: SCRAP, ALUM

MERRITT COLLEGE RECYCLING
3600 Joaquin Miller Road, Oakland (415)530-4818
D: GLASS, ALUM, NEWS, COMP, CARD, MIX, WHITE, MAGS, PET

NATIONAL RECYCLING
1312 Kirkham St., Oakland (415)268-1022
D: NEWS, CARD, WHITE, COMPUTER AND COLORED PAPER

NO. OAKLAND RECYCLING CENTER
1027 60th St., Oakland (415)547-6470
D: NEWS, GLASS, ALUM, TIN, SCRAP, CARD, MIX, WHITE, MAGS, PHONEBK, OTHER BOOKS, PLASTICS

NORTHERN CALIFORNIA PULP & PAPER
2085 Wayne Ave., San Leandro (415)483-3255
D: NEWS, WHITE, MAGS, COLORED, COMPUTER PAPER

OWENS-BROCKWAY
6150 Stoneridge Mall Rd., Ste. 375, Pleasanton (415)463-7950

PAPER RECOVERY OF NORTHERN CA.
25670 Nickle Pl., Hayward (415)785-7311
B: COMP

RECYCLING AND RES. RECOVERY
3110 Busch Rd., Pleasanton (415)846-2042
B: REDEMP
D: NEWS, CARD

TRI-CED COMM. RECYCLING
33300 Central Ave., Union City (415)471-3850
B: NEWS, GLASS, ALUM, CARD, COMP, TIN; D: PET

UNIVERSITY VILLAGE RECYCLING CENTER
1125 Eighth St., Albany
(415)526-8505

D: NEWS, CARD, MIX, GLASS, MAGS, WINE, ALUM, TIN

URBAN ORE BUILDING MATERIALS
1325 Sixth, BERK (415)526-7080
See display ad on this page.

URBAN ORE FLEA MARKET
1231 2nd St. at Gilman, Berkeley (415)526-9467
REUSABLE HOUSEHOLD GOODS, FURNITURE, APPLIANCES, BLDG. MATERIALS, SCRAP, GLASS, ART OBJECTS

CAR BATTERIES

AMERICAN BATTERY
22851 Sutro St., Hayward (415)881-5122

B & T COMPANY
1034 40th Ave., Oakland (415)261-6051

Fact: Recycling one ton of paper saves approximately 17 trees, and prevents 60 pounds of pollution from being released into the atmosphere.

RECYCLING

CURBSIDE PICKUP

ALLSTAGE
1001 San Leandro, OAK (415)638-8323
C: GLASS, ALUM, TIN, NEWS, MIX, WHITE, COMP, SCRAP, WASHING MACHINES

DAVE'S RECYCLING
Oakland (415)339-2037
C: ALUM, GLASS, NEWS, PAPER, CARD, MIX, PHONEBK, MAGS, SCRAP

EAST BAY DISPOSAL CO.
Fremont & Newark (415)797-1235
C: NEWS, GLASS, ALUM, TIN, PET

ECOLOGY CENTER - CURBSIDE RECYCLING
Berkeley (415)644-3822

KARL'S RECYCLING
Rock, Glen, N.Oak, Pdmnt, Mtclr, Dimond, Oakland (415)635-1727
C: NEWS, ALUM, GLASS, CARD, MIX, COMP, WHITE, MAGS, PHONEBKS, PET

LIVERMORE/DUBLIN DISPOSAL SERVICE
6175 South Front Road, Livermore (415)447-1300
C: GLASS, ALUM, TIN, NEWS, PET

OAKLAND SCAVENGER
Emeryville (415)430-8509
C: GLASS, ALUM, TIN, NEWS, PET

RECOVERY RECYCLING
Berkeley (415)523-7853

RRR-CURBSIDE
3110 Busch Road, Pleasanton (415)846-2042
C: NEWS, GLASS, ALUM, TIN.

WASTE MANAGEMENT
Albany (415)524-9283
C: NEWS, GLASS, ALUM, TIN

WASTE MANAGEMENT OF DANVILLE
Danville (415)935-8900 x23
C: GLASS, ALUM, NEWS, PET

MOTOR OIL

CAMPUS AUTO CARE
1752 Shattuck Ave., Berkeley (415)845-8828

DAVIS STREET TRANSFER STATION
2615 Davis St., San Leandro (415)638-2303

EAST BAY DISPOSAL - MOTOR OIL
Durham Road, Fremont & Newark (415)797-0440

EL CERRITO RECYCLING CTR.-MOTOR OIL
7501 Schmidt Ln., El Cerrito (415)527-6077

EMERYVILLE UNION 76
1400 Powell St., Emeryville (415)653-2251

EVERGREEN OIL INC.
P. O. Box 248, Newark (415)795-4400
WASTE OIL RECOVERY, RECYCLING

FIRE STATION #1
4550 East Ave., Livermore (415)373-5450

LARRY'S UNION SERVICE STATION
3374 Grand Ave., Oakland (415)832-9605

TRI-CITY ECOLOGY CENTER
Fremont (415)745-2725
CURBSIDE

UNIVERSITY AVE. TEXACO SERVICE
6th and University, Berkeley (415)843-4910

WHITMORE'S AUTO SERVICE
1701 Buena Vista Ave., Alameda (415)522-3388

CONTRA COSTA

CFC RECYCLING

AUTOHAUS-CONCORD
2575 Monument Blvd., Concord (415)825-7111

DON'S RADIATOR
647 23rd. St., Richmond
AUTO AIR CONDITIONER

MADSEN'S AUTO AIR AND HEATING
1132 Meadow Ln., Concord (415)798-1131

MADSENS AUTO, AIR, AND HEATING
1132 Meadow Lane, Concord (415)798-1131

RECYCLING

ACME LANDFILL & TRANSFER STATION
950 Waterbird, Martinez (415)228-7099
MATTRESSES, BOX SPRINGS, WOOD, CONCRETE, METALS

ANTIOCH WASTE FIBER RECOVERY
Brentwood Exit, Hwy. 4
Behind Bonanza, Antioch (415)625-9372
UNTREATED WOOD AND PRUNINGS

BLAKE ENTERPRISES-LASER CARTRIDGES
2154A North Main St., Walnut Creek (415)939-1718
See display ad, page 348.

BRIGHT IDEAS
3450 Golden Gate Wy, Lafayette (415)284-5350

DELTA SCRAP AND SALVAGE CO.
Route 1, Box 73, Oakley (415)754-1474
B: GLASS, ALUM, TIN, SCRAP, PET

EL CERRITO RECYCLING CENTER
10890 San Pablo Ave., El Cerrito (415)527-6077
D: NEWS, CARD, WHITE, COMP, MAGS, PHONEBK, GLASS, WINE, ALUM, SCRAP, OIL, PET, CAR BATTERIES

ENCORE! ENV. CONTAINER REUSE
860 South 19th St., Richmond (415)234-5670
D: U.S. PREMIUM WINE BOTTLES

LMC METALS
600 S. Fourth St., Richmond (415)236-0606
B: ALUM, SCRAP

MANY HANDS
P.O. Box 1487, Pittsburg (415)432-1171
B: REDEMP
D: NEWS, CARD, WHITE, COMP, GLASS, TIN, HDPE

MT. DIABLO PAPER STOCK CO.
4080 Mallard Dr., Concord (415)682-9113
B: NEWS, CARD, MIX, WHITE, COMP, GLASS, ALUM, PET
D: STYRENE

PINOLE VALLEY H.S.
2900 Pinole Valley Rd., Pinole (415)758-4664
D: NEWS, GLASS, ALUM, STYRENE

PLEASANT HILL BAYSHORE DISPOSAL
441 North Buchanan Circle, Pacheco (415)685-4711
See display ad, page 322.

SAN RAMON VALLEY H.S.
140 Love Lane, Danville (415)837-0533
D: NEWS, GLASS, ALUM

ST. AGNES ELEMENTARY SCHOOL
3886 Chestnut Ave., Concord (415)689-3990
D: NEWS

VALLEY WASTE MANAGEMENT
1801 Oakland Blvd. Suite 250, Walnut Creek (415)935-8900
PLEASE SEE WASTE MANAGEMENT OF ALAMO OR SAN RAMON.
See display ad, page 322.

W.C. COSTA SANITARY LANDFILL
end of Parr Blvd., Richmond (415)262-1600
D: NEWS, CARD, GLASS, ALUM

WASTE FIBRE RECOVERY
(415)525-4557
WOOD

CAR BATTERIES

EL CERRITO BATTERIES
10890 San Pablo, El Cerrito (415)547-6470

CURBSIDE PICKUP

CONTRA COSTA RES. RECOVERY
615 Escobar St., Martinez (415)372-0625
C:NEWS, CARD, GLASS, ALUM, TIN, SCRAP,PET, OIL, CAR BATTERIES

EL CERRITO CURBSIDE RECYCLING
10890 San Pablo Ave., El Cerrito (415)527-6077
C: NEWS, GLASS, ALUM, TIN, PET

ORINDA/MORAGA DISPOSAL
1 Northwood Dr., Orinda (415)254-2844
C: GLASS, ALUM, NEWS

PITTSBURG DISPOSAL
Pittsburg (415)432-6262
C: GLASS, ALUM, NEWS, PET

PORT COSTA RECYCLES
P.O. Box 35, Port Costa (415)787-2579
C: NEWS, CARD, MIX, GLASS, ALUM, TIN, PET

RICHMOND SANITARY SERVICE
205 41st Street, Richmond (415)236-8000
C: NEWS, GLASS, ALUM, TIN,PET, HDPE

WASTE MANAGEMENT OF ALAMO
Alamo (415)935-8900 x23
C: NEWS, GLASS, ALUM, PET

WASTE MANAGEMENT OF SAN RAMON
San Ramon (415)935-8900 x23
C: GLASS, ALUM, NEWS, PET

MOTOR OIL

BAY CITY DISPOSAL
2615 Davis St., San Leandro (415)638-4327

BOB LEE CHEVRON
4295 Clayton Rd., Concord (415)680-9766

CARDOZA CHEVRON
400 Parker Ave., Rodeo (415)799-4430

COMPLETE AUTO SERVICE
3831 Shopping Heights, Pittsburg (415)439-2081

MORAGA EXXON
530 Moraga Rd., Moraga (415)376-0692

PHIL MADRUGA'S CHEVRON
2295 Morello Ave., Pleasant Hill (415)685-1975

PLEASANT HILL BAYSHORE
411 N. Buchanan Circle, Pacheco (415)685-1975
See display ad, page 322.

RON'S SHELL
3630 Alhambra Ave., Martinez (415)228-7115

STEVE'S UNION SERVICE
3160 Carlson Blvd., El Cerrito (415)526-9227

JACK RICE CHEVRON
1500 Palos Verdes Mall, Walnut Creek (415)933-5851

MARIN

RECYCLING

SHORELINE DISPOSAL SERVICE
Bolinas School, Bolinas (800)862-4659
D: NEWS, CARD, GLASS, ALUM, PET

CURBSIDE PICKUP

MARIN RECYCLING AND RESOURCE RECOVERY
Marin County (415)453-1404
C: NEWS, MIX, PHONEBK, MAGS, GLASS, ALUM, TIN, PET

NOVATO DISPOSAL SERVICE
P. O. Box 768, Novato (415)897-4177
C: NEWS, GLASS, ALUM, TIN, PET

SAN ANSELMO GARBAGE DISPOSAL
80 Greenfield Ave., San Anselmo (415)453-4610
C: NEWS, CARD, GLASS, TIN, ALUM, PET

MOTOR OIL

MARIN RECYCLING—MOTOR OIL
535 Jacoby St., San Rafael (415)453-1404

REDWOOD SANITARY LANDFILL
Redwood Hwy, Novato (415)892-2851

NAPA

CFC RECYCLING

AUTOHAUS - NAPA
480 Soscol Ave., Napa (707)252-1141

RECYCLING

REYNOLD'S ALUMINUM RECYCLING CTR.
4000 Belaire Plaza, Napa (800)228-2525
B: ALUM CANS AND SCRAP.

UPPER VALLEY DISPOSAL SERVICE
Clover Flat Landfill Site, St. Helena (707)963-7988
B: ALUM, GLASS, ETC.
MOTOR OIL

VALLEY RECYCLING
874 Jackson, Napa (707)258-8301
B: ALUM. GLASS.
PLASTIC, SCRAP ALUMINUM, BRASS, COPPER.

CURBSIDE PICKUP

NAPA GARBAGE & RECYCLING
400 Clay, Napa (707)255-0462

MOTOR OIL

NAPA GARBAGE & MOTOR OIL
400 Clay, Napa (707)255-0462

UPPER VALLEY DISPOSAL - MOTOR OIL
Clover Flat Disposal Site', St. Helena (707)963-7988
CAN ACCEPT OIL IN QUANTITIES OF FIVE GALLONS OR LESS.

SAN FRANCISCO

CFC RECYCLING

AIR CONDITIONING AUTO CLINIC
2035 Divisadero, San Francisco (415)563-2915

AIR CONDITIONING AUTOMOTIVE CLINIC
2035 Divisidero St., San Francisco (415)563-2915

AUTO AIR CONDITIONING (2)
650 Divisadero, San Francisco (415)923-1576
AUTO AIR CONDITIONING
JACOB RANDLISH

RECYCLING

AUTHOR ENTERPRISES INT.
811 Oak, San Francisco (415)255-9366
C: CARD, NEWS, COMP, ALUM, GLASS

BAY AREA METALS
3201 Third St., San Francisco (415)826-3432
B: METALS, WIRE

BAYSHORE BUYBACK
Tunnel Ave., San Francisco (415)330-1407
B: GLASS, ALUM, NEWS
D: BI-METAL, PET

CIRCOSTA IRON AND METAL CO. INC.
1801 Evans Ave., San Francisco (415)282-8568
B: ALL METALS

DUMP SEVENTY-ONE SALVAGE & RECYCLING
1930 Oakdale Ave., San Francisco (415)282-1491
B: WOOD, TIN , IRON

HAIGHT-ASHBURY NEIGHBORHOOD COUNCIL
780 Frederick, San Francisco (415)753-0932
D: NEWS, CARD, COMP, MIX, WHITE, MAGS, GLASS, WINE, ALUM, REDEMP, BI-METAL, PET

MANY HAPPY RETURNS
701 Amador, San Francisco (415)285-0669
B: REDEMP
D: NEWS, HDPE

METAL RECYCLING INC.
1438 Donner Ave., San Francisco (415)822-9096

MOBILE RECYCLING UNIT
Powell St., San Francisco (415)285-0669
B: GLASS, ALUM, NEWS, OFFICE PAPER
D: BI-METAL, PET, DONATION

NORCAL WASTE SYSTEMS, INC.
5 Thomas Mellon Circle, Ste. 266, SF (415)330-1000
See display ad on this page.

PAPER RUSH CO., INC.
2372 Jerrold Ave., San Francisco (415)282-2344
B: NEWS, OFFICE, CARD

RELIANCE PAPER
2150 Folsom, San Francisco (415)553-6644
B: ALUM, OFFICE

RICHMOND ENVIRONMENT
249 Anza, San Francisco (415)387-3117
B: GLASS, ALUM, NEWS, OFFICE
D: BI-METAL, TIN, CARD, PET

SAN FRANCISCO COMMUNITY RECYCLERS
701 Amador at Cargo, San Francisco (415)285-0669
LOCATIONS: 1. MARKET AT DOLORES 2. MASON AT CHESTNUT 3.GEARY AT WEBSTER. SEE ORGANIZATION LISTING FOR DETAILS.

SAN FRANCISCO METAL
99 Mississippi, San Francisco (415)863-3508
B: NON-FERROUS METALS

SF RECYCLING COORDINATOR
City Hall, Room 271, San Francisco (415)554-6193
FOR INFORMATION ONLY

WASTE PAPER INC.
998 Indiana St., San Francisco (415)550-7500
B: NEWS, OFFICE PAPER, CARD

WATERFRONT RESOURCES
250 China Basin, San Francisco (415)543-3939
B: NEWS
D: PHONEBKS, MAGS

WEST COAST SALVAGE
1900 17th St., San Francisco (415)621-6200
B: GLASS, ALUM, NEWS
D: BI-METAL, PET, DONATION

WEST COAST SALVAGE AND RECYCLING
350 Rhode Island, SF (415)621-3840
GLASS, ALUM, PLASTIC, PAPER, COMERCIAL PICKUP

CURBSIDE PICKUP

RAINBOW RECYCLING SERVICE
519 Guerrero #2, San Francisco (415)863-8455
C: NEWS, GLASS, ALUM, PET

SF FIRE FIGHTERS TOY PROGRAM
San Francisco (415)777-0440
C: ALUM, COMP

MOTOR OIL

ARMY CIRCLE SERVICE MOBIL PRODUCTS
2831 Army St., SF (415)826-8855

CHAN'S AUTO REPAIR
301 Claremont Rd., San Francisco (415)759-9770

CHESTER'S SERVICE STATION
1699 Pine St., San Francisco (415)474-4843

GEARY BOULEVARD CHEVRON SERVICE
6000 Geary Blvd., San Francisco (415)386-9604

HYDE STREET STATION
168 Hyde St., San Francisco (415)673-9455

MAHMOUD ARCO GAS STATION
302 Potrero, San Francisco (415)626-1129

POTRERO CHEVRON
16 th and Potrero, San Francisco (415)626-1129

POTRERO UNOCAL
401 Potrero Ave., San Francisco (415)863-6043

S.S JEREMIAH O'BRIEN
Pier 3, East Fort Mason, San Francisco (415)441-3101
URGES RESIDENTS TO USE OTHER OIL RECYCLERS

SANITARY FILL
501 Tunnel Ave., San Francisco (415)468-2440

VALENCIA UNOCAL SERVICE
1298 Valencia St., San Francisco (415)824-5622

SAN MATEO

CFC RECYCLING

AUTOHAUS - SO. SAN FRANCISCO
1166 San Mateo Ave., South San Francisco(415)873-3013

CFC RECYCLING

A1 RADIATOR & AIR CONDITIONING
317 So. Norfolk St., San Mateo (415)347-0731

AUTO AIR CONDITIONING
777 Industrial Rd., San Mateo (415)595-3736

RECYCLING

BAYSHORE SALVAGE
Tunnel and Beatty Rd., Brisbane (415)467-0567
B: GLASS, ALUM, NEWS, CARD, PET

BROWNING FERRIS INDUSTRIES (BFI)
375 Quarry Rd., Belmont (415)637-1411
B: GLASS, ALUM, NEWS, CARD, COMP, PET
D: TIN, WHITE, HDPE
See display ad on this page.

BLUE LINE TRANSFER STATION
180 Oyster Point, South San Francisco (415)589-5511
B: GLASS, ALUM, PET
D: CARD, OIL

GOLDEN WEST
555 Quarry Rd., Belmont (415)591-6224
B: GLASS, ALUM, CARD, COMP, WHITE
D: NEWS

LOS ALTOS GARBAGE CO.
1285 Pear Ave., Mountain View (415)961-8010
D: GLASS, ALUM, NEWS

OX MOUNTAIN
13210 Hwy. 92, Half Moon Bay (415)726-1819
D: GLASS, ALUM, NEWS, OIL

PACIFIC LIONS CLUB
P.O. Box 1132, Pacifica (415)359-1815
D: GLASS, ALUM, NEWS

PALO ALTO SANITATION CO.
2380 East Embarcadero Rd., Palo Alto (415)329-2495
D: GLASS, ALUM, TIN, NEWS, CARD, SCRAP, OIL, PET, WHITE, COMP, CAR BATTERIES

REYNOLDS ALUMINUM RECYCLING CENTER
633 Bayshore Blvd., Brisbane (415)467-9798
B: ALUM, COPPER

SAN BRUNO BUYBACK CENTER
Skycrest Shopping Center, San Bruno (415)583-8536
B: REDEMP
D: NEWS

STANFORD RECYCLING CENTER
Stanford Campus, Pampas Ln., Stanford (415)723-0919
D: NEWS, GLASS, ALUM, CARD, COMP

YOUTH ENTERPRISES RECYCLING
3641 Haven Ave., Menlo Park (415)364-3333
B: NEWS, GLASS, ALUM, CARD, SCRAP, COMP
D: TIN

Fact: At least eighty percent of what we consume is recyclable, but only ten percent is ever actually recycled.

CURBSIDE PICKUP

BROWNING FERRIS INDUSTRIES
375 Quarry Rd., Belmont (415)637-1411
C: PET

CITY OF PALO ALTO CURBSIDE RECYCLING
P.O. Box 10250, Palo Alto (415)329-2495
C: NEWS, GLASS, ALUM, TIN, OIL, PET, CARD, SCRAP, COMP

COASTSIDE SCAVENGERS
P.O. Box 1099, Pacifica (415)355-9000
C: GLASS, ALUM, NEWS, CARD, PET

COASTSIDE SCAVENGERS-EL GRANADA
El Granada (415)355-9000
C: GLASS, ALUM, NEWS, CARD, PET

SAN BRUNO GARBAGE CO.
101 Tanforan, San Bruno (415)583-8536
C: GLASS, ALUM, TIN, NEWS, PET

SAN FRANCISCO SCAVENGERS
P. O. Box 348, South San Francisco (415)589-4020
C: GLASS, ALUM, TIN, NEWS, OIL, PET

SOUTH SAN FRANCISCO SCAVENGERS
69 S. Linden Ave., South San Francisco (415)589-4020
C: GLASS, ALUM, TIN, NEWS, PET, OIL

SUNSET SCAVENGER
Tunnel Ave./ Beatty Rd., San Francisco (415)330-2872
C: GLASS, ALUM, TIN, PET, NEWS, CARD, MIX, MAGS, JUNK MAIL, PHONEBK

MOTOR OIL

ADAM'S REDWOOD SHELL
690 Veterans Blvd., Redwood City (415)369-6675

KEN LORD CHEVRON
375 N. El Cabrillo, Half Moon Bay (415)726-4055

LADERA CHEVRON
104 La Mesa Dr., Menlo Park (415)854-4504

MILLBRAE UNION 76
5 El Camino Real, Millbrae (415)692-8222

OX MOUNTAIN-MOTOR OIL
13210 Hwy. 92, Half Moon Bay (415)726-1819

PALO ALTO MOTOR OIL RECYCLING
2380 East Embarcadero Rd., Palo Alto (415)329-2495

PLAZA UNION
1500 Woodside Rd., Redwood City (415)368-4237

SANFORD FIRESTONE
705 Hickey Blvd., Pacifica (415)355-1157

SOUTH SAN FRANCISCO SCAVENGERS-MOTOR OIL
69 S. Linden Ave., South San Francisco (415)589-4020

RECYCLING

SANTA CLARA

CFC RECYCLING

AUTOHAUS - SANTA CLARA
2555 Lafayette St., Santa Clara (408)986-1391

RECYCLING

A-1 RECYCLING CIRCUS
150 Howson St., Gilroy (408)842-0288
B: NEWS, GLASS, ALUM, CARD, SCRAP, COMP, PET

ALL RECYCLING
175 Lewis Rd.,bldg 25, San Jose (408)629-4061
B: NEWS, GLASS, ALUM, TIN, SCRAP, COMP, PET, MAGS

ARATA WESTERN INC.
388 E. Alma, San Jose (408)297-1022
B: NEWS, CARD, COMP

BAY AREA RECYCLING
700 E. McGlincy Ln., Campbell (408)371-5834
B: NEWS, GLASS, ALUM, SCRAP, COMP, PET
D: CARD

CITY METALS AND SALVAGE
11665 Berryessa Rd., San Jose (408)452-0777
B: ALUM, SCRAP

CITY OF SUNNYVALE RECYCLING
PO Box 3707, Sunnyvale (408)730-7262
D: NEWS, GLASS, ALUM, TIN, CARD, OIL, COMP, PET

COASTAL FIBERS
1045 Commercial Ct., San Jose (408)453-1960
B: NEWS, GLASS, ALUM, CARD, COMP, PET
D: MAGS

CUPERTINO COMMUNITY RECYCLING CENTER
De Anza College Parking Lot
Cupertino (408)255-8033
D: NEWS, GLASS, ALUM, TIN, CARD, OIL, COMP, PET

FOOTHILL DISPOSAL CO.
935 Terra Bella Ave., Mountain View (415)967-3034
B: GLASS, ALUM, PET
D: NEWS, TIN, OIL

LOS ALTOS CARDBOARD RECYCLING
Loyola Fire Station, 765 Fremont Ave., Los Altos (415)948-1491
D: CARDBOARD

LOS ALTOS CARDBOARD ALMOND AVE
Sequoia Fire Station,10 Almond Ave., Los Altos (415)948-1491
D: CARDBOARD

MISSION TRAIL GARBAGE CO.
735 Reed St., Santa Clara (408)727-5365
D: NEWS, GLASS, ALUM

NORTH CENTRAL DISPOSAL
4835 Newell Creek Rd., Ben Lomond (408)336-8610
D: ALUM, GLASS, NEWS, CARD, OIL, METALS, MATTRESSES, BATTERIES

SAN JOSE METALS
1032 N. Tenth St., San Jose (408)293-4032
B: ALUM, SCRAP

SARATOGA RECYCLING CENTER
19700 Allendale, Saratoga (408)867-4005
D: NEWS, GLASS, ALUM, CARD, COMP, PET

SCOTT'S VALLEY HOST LIONS CLUB
Kings Village Rd., Scotts Valley
D: NEWS, CARD

TOWN OF LOS GATOS RECYCLING CENTER
41 Miles Ave., Los Gatos (408)354-6809
D: NEWS, GLASS, ALUM, CARD, OIL, SCRAP

TRINIE MARTIN RECYCLING
8565 1/2 Monterey St., Gilroy (408)842-2565
B: NEWS, GLASS, ALUM, SCRAP, COMP, PET

VALLEY WOMEN'S CLUB
Hwy 9 and Fillmore St., Ben Lomond (408)338-6578
B: REDEMP

WATSONVILLE METALS CO.
213 Dias Ln., Watsonville (408)728-1551
B: MOST METALS, BALED CARDBOARD SOME COLLECTION

CAR BATTERIES

BATTERY EXCHANGE
193 Barnard Ave. #3, San Jose (408)286-8039

BAYLAND BATTERY CORPORATION
800 Faulstich Court, San Jose (408)453-3522

MONTGOMERY WARD AUTOMOTIVE CENTER
879 Blossom Hill Rd., San Jose (408)224-2357

SAN JOSE BATTERY
670 Stockton Ave., San Jose (408)947-1726

SEARS AUTOMOTIVE CENTER
2180 Tully Rd., San Jose (408)238-1122

CURBSIDE PICKUP

CITY OF LOS ALTOS CURBSIDE RECYCLING
City Hall, 1 N. San Antonio Rd, Los Altos (415) 1491 x222
C: NEWS, GLASS, ALUM, TIN, OIL

CITY OF MOUNTAIN VIEW CURBSIDE RECYCLING
231 N. Whisman, Mountain View (415)966-6329
C: NEWS, GLASS, ALUM, TIN, OIL, PET

CITY OF SUNNYVALE CURBSIDE RECYCLING
PO Box 3707, Sunnyvale (408)730-7262
C: NEWS, GLASS, ALUM, TIN, OIL, PET

RECYCLE AMERICA
715 Comstock, Santa Clara (408)423-2022
C: ALUM, GLASS, NEWS, CARD, OIL

SAN JOSE CITY CURBSIDE PROGRAM
777 N. 1st, San Jose (408)748-1616
C: NEWS, GLASS, ALUM, TIN, COMP

SANTA CLARA RECYCLING INC.
P.O. Box 4838, Santa Clara (408)727-3044
C: NEWS, GLASS, ALUM, PET, OIL

MOTOR OIL

BLOSSOM HILL ROAD/SNELL CHEVRON
Blossom Hill Rd./ Snell Chevron, San Jose (408)22 1560

BRITISH PETROLEUM
Blossom Hill Rd./ Almaden Exp., San Jose (408)26 7621

C AND M SERVICE
1098 W. San Carlos, San Jose (408)298-5385

CUPERTINO RECYCLING CENTER-MOTOR OIL
De Anza College Parking Lot, Cupertino (408)255-8033

FOOTHILL DISPOSAL-MOTOR OIL
935 Terra Bella Ave., Mountain View (415)967-3034
CURBSIDE OR DROP-OFF

LOS GATOS RECYCLING CENTER-MOTOR OIL
41 Miles Ave., Los Gatos (408)354-6809

LOU'S TOWN AND COUNTRY EXXON
425 S. Winchester Blvd., San Jose (408)296-7503

RECYCLE AMERICA-MOTOR OIL
715 Comstock, Santa Clara (408)423-2022

SANTA CRUZ

RECYCLING

AL PARIS RECYCLING
1111 River St., Santa Cruz (408)459-9619
C: WHITE PAPER, GLASS, ALUM, CARD
D: WHITE PAPER, GLASS, ALUM, CARD

APTOS NEIGHBORS, NATIVIDAD GROUP HOMES
Aptos (408)688-9898
B: ALUM, GLASS, NEWS

CABRILLO COLLEGE
6200 Soquel Dr.
(Gym's North Parking Lot), Aptos (408)476-1201
D: NEWSPAPER

CALIFORNIA GREY BEARS
2710 Chanticleer Ave., Santa Cruz (408)479-1055
D: NEWS, ALUM, GLASS, CARD, BROWN PAPER BAGS, PET, WHITE, COMP AND COLORED PAPER.

CITY OF SANTA CRUZ RECYCLING PROGRAM
605 Dimeo Lane, Santa Cruz (408)429-3657
D: ALUM, NEWS, GLASS, CARD, OIL, PET, MATTRESSES, BATTERIES

COUCH DISTRIBUTING
104 Lee Rd, Watsonville (408)724-0649
B: ALUM

D AND D RECYCLING
P.O. Box 1777, Freedom (408)722-3597
B: ALUM, GLASS, SCRAP

HEDRICK DISTRIBUTORS
210 Encinal St., Santa Cruz (408)427-3773
D: WASTE FUEL, OIL, BATTERIES

MONTEREY BAY RECYCLING
Santa Cruz (408)426-0112

ROUNDUP RECYCLING
3820 Soquel Dr., Soquel (408)462-6701
B: ALUM, GLASS, CARD, OFFICE PAPER, NON-FERROUS METALS
D: ALUM, GLASS, CARD, OFFICE PAPER, NON-FERROUS METALS

SANTA CRUZ AND CENTRAL COUNTY GARBAGE
Buena Vista Dr., Watsonville (408)688-7250
D: ALUM, GLASS, NEWS, CARD, BATTERIES, METALS, MATTRESSES, OIL, PET

STATE STEEL COMPANY
56 Porter Dr., Pajaro (408)724-7111
B: SCRAP

CURBSIDE PICKUP

CAPITOLA GARBAGE CO.
Capitola (408)476-9288
C: ALUM, GLASS, NEWS
COMMERCIAL: CARD

CITY OF SANTA CRUZ CURBSIDE RECYCLING
809 Center St., Rm. 101, Santa Cruz (408)429-3666
C: ALUM, NEWS, GLASS, CARD, OIL, TIN, PET

HEDRICK DISTRIBUTORS-CURBSIDE
210 Encinal St., Santa Cruz (408)427-3773
C: WASTE FUEL, OIL, BATTERIES

RICK'S NEWS
P.O. Box 711, Freedom (408)728-5915
C: NEWS, OFFICE PAPER

Fact: Every hour, Americans go through 2.5 million plastic beverage bottles. We also throw out 2.5 billion styrofoam cups each year.

SOLANO

RECYCLING

C & C METALS
11320 Dismantle Ct., Rancho Cordova (916)635-5600

PACIFIC RIM RECYCLING
433 Industrial Wy., Benicia (415)372-8524

VACAVILLE SANITARY RECYCLING
302 N. First Street, Vacaville (707)448-2945
B: ALUM, GLASS, ETC. EXCEPT NEWSPAPER
D: NEWSPAPER

VACAVILLE SANITARY-RECYCLING
855 1/2 Davis Street, Dixon (707)448-2945
B: GLASS, ALUM, ETC.
D: NEWSPAPER

VALCROE RECYCLING CENTER
38 Sheridan, Vallejo (707)645-8258
B: ALUM. GLASS, ETC.
MOTOR OIL

MOTOR OIL

VALCROE RECYCLING - MOTOR OIL
38 Sheridan, Vallejo (707)645-8258

SONOMA

CFC RECYCLING

NOVATO AIR CONDITIONING
904 Vallejo Ave., Novato (707)897-6139

RECYCLING

BECOMING INDEPENDENT
980 Hopper Ave., Santa Rosa (707)527-5904
CORK, EMPLOYS THE DISADVANTAGED AND DISABLED

GARBAGE REINCARNATIONS' RECYCLING CENTER
3899 Santa Rosa Ave., Santa Rosa (707)584-8666
B: GLASS, ALUM,OFFICE PAPER
D: BROWN BAGS, MOTOR OIL, CRDBRD, NEWS.
RECYCLING DEPOT; COMMERCIAL OFFICE PAPER PICK-UP, MOBILE DEPOTS THROUGHOUT COUNTY, SALES OF SHREDDED PACKAGING PAPER AND RECYCLED PAPER STOCK.

LARRY'S RECYCLING CENTER
7085 Gravenstein Hwy., Cotati (707)795-2319

PETALUMA JUNK CO.
519 Lakeville, Petaluma (707)778-7432
B: ALUM, BRASS, COPPER, IRON, AUTO RADIATORS.

PETALUMA LANDFILL REUSE
Depot 403 Mecham Road, Petaluma (707)795-3660
RE-USE, SALVAGE.

PETALUMA RECYCLING CENTER
315 Second St., Petaluma (707)763-4761
B: GLASS, ALUM, NEWS, CARD
D: GLASS, ALUM, NEWS, CARD

RECYCLE AMERICA, SANTA ROSA
3400 Standish Ave., Santa Rosa (707)585-0291
B: ALUM, GLASS, OFFICE PAPER.
D: NEWS, CARD, COMPUTER PAPER, PET, MAGAZINES, MILK JUGS (HPDE), TIN.

CAR BATTERIES

BATTERY RECYCLING
P.O. Box 1965, Sebastopol (707)829-2950

CURBSIDE PICKUP

RECYCLE AMERICA - CURBSIDE
3400 Standish Ave., Santa Rosa (707)585-0291

MOTOR OIL

GARBAGE REINCARNATIONS-MOTOR OIL
3899 Santa Rosa Ave., Santa Rosa (707)584-8666

PETALUMA LANDFILL MOTOR OIL
Depot 403 Mecham Road, Petaluma (707)795-3660

Americans eat one of every three meals away from home, and this average is rising. Your decisions about where and what to eat can support restaurants with a commitment to using organic food, and which serve vegetarian menus. The businesses listed below meet one or both of these criteria.

RESTAURANTS, FAST FOOD

DHARMA'S NATURAL FAST FOODS
4250 Capitola Road, Capitola (408)462-1717

RESTAURANTS, ORGANIC

CAFE LATTE
100 Bush, SF (415)989-2233

FAZ RESTAURANT AND BAR
132 Bush, SF (415)362-4484

FLEA ST. CAFE
3607 Alameda de Las Pulgas, Menlo Pk (415)854-1226

GOVINDA'S
86 Carl St., SF (415)753-9703

GREEN'S AT FORT MASON
Fort Mason, Bldg. A, SF (415)771-6222

HOBEES 4
740 Front St. #100, Santa Cruz (408)458-1212

LATE FOR THE TRAIN
150 Middlefield Road, Menlo Park (415)321-6124

MUDD'S RESTAURANT
10 Boardwalk, San Ramon (415)837-9387

STARS RESTAURANT
150 Redwood Alley, SF (415)861-7827

WATERFRONT RESTAURANT
Pier 7,Embarcadero, SF (415)391-2696

POT AND PAN RESTAURANT
1243 9th Ave., SF (415)665-2833

SILVER MOON
2307 Clement St., SF (415)386-7852

SUN N' SOIL
2700 Stevens Crk Blvd., CUP (408)257-8887

THE GANGES
775 Frederick, SF (415)661-7290

THE RED CRANE
1115 Clement St., SF (415)751-7226

THE TURNING EARTH
13 Columbus, SF (415)391-6307

TORTILLA FLATS
4724 Soquel Dr., Soquel (408)476-1754

VEGI FOOD 2
2083 Vine St., BERK (415)548-5244

RESTAURANTS, VEGETARIAN

ADULIS ETHIOPIAN FOOD
2519 Durant Plaza, Berkeley (415)843-4990

ALL YOU KNEAD
1466 Haight St., SF (415)552-4550

AMAZING GRACE VEGETARIAN REST.
216 Church St., SF (415)626-6411

ANANDA-FUARA
3050 Taraval, SF (415)564-6766

DIAMOND STREET RESTAURANT
737 Diamond, SF (415)285-6988

GOLDEN CARROT
1621 West Imola, Napa (707)224-3117

GOVINDA'S
86 Carl St., SF (415)753-9703

HARVEST MOON
339 Judah St., SF (415)664-3044

HOBEES 4
740 Front St. #100, Santa Cruz (408)458-1212
See display ad on this page.

LONG LIFE VEGGIE HOUSE
2129 University Ave., BERK (415)845-6072

LOTUS GARDEN
532 Grant St., SF (415)397-0707

LUCILLE'S
38 Miller Ave., Mill Valley (415)381-4613

MACROBIOTIC GROCERY
1050 Fourthieth St., OAK (415)653-6510

MILLY'S
1613 Fourth St., San Rafael (415)459-1601

The solar revolution has gone and come. A significant number of the technologies which flowered in the energy-expensive seventies have survived the subsidized oil and governmental hostility of the eighties.

Solar electric applications have dramatically improved efficiency while prices have dropped. There are solar water and space heating systems which have been in place and working well for the last ten years, in spite of what Mobil Oil says about practical solar energy being decades away.

Solar electric (photovoltaic) systems are commercially available for off-grid and portable applications such as portable radios, phones, and cabins in the woods or water pumps in the middle of fields. For such applications, solar is already the best alternative.

For city dwellers in the Bay Area, solar water and space heating are environmentally and economically sensible options.Our planet has been running on solar energy for the past four billion years. There's no reason it can't continue to do so.

SOLAR COOKER

These companies either offer designs for easy-to-construct solar cooking devices or sell them.

KERR ENTERPRISES
P. O. Box 27417, Tempe (602)968-3068

SOLAR BOX COOKERS INTERNATIONAL
1724 Eleventh St., Sacramento (916)444-6616

SOLARCHEF™ SOLAR OVEN
now cook with the sun
Syncronos Design, Incorporated
P.O. Box 10657Albuquerque NM
(505)897-1440

SOLAR ENERGY

ADVANCED ENERGY ENGINEERING
837 San Anselmo Dr., San Anselmo (415)456-6925

ADVANCED SOLAR INSTALLATIONS
P.O. Box 4098, Mountain View (415)968-2383

ALTERNATIVE ENERGY COMPANY
1437 Imola Ave., Napa (707)255-6116

ALTERNATIVE ENERGY ENGINEERING
PO Box 339, Redway (707)923-2277

APPI
1527 Commercial Way, Santa Cruz (408)476-6363

APPLIED SOLAR PRODUCTS
4580 Almaden Expwy., San Jose (408)269-3050

BRANDALICK ENERGY SERVICE
1543 Cavendish Ave., Santa Rosa (707)795-8535

CALIFORNIA MICRO UTILITY
Design, Feaseability Studies, Software, Energy Monitoring, Product Development.
1195 Park Ave., Suite 202A, Emeryville
(415)658-1273

DELUCCHI PLUMBING HEATING & SOLAR
1670 15th St., SF (415)431-1656

FOUR SEASONS DESIGN AND REMODELING
26 Medway Rd., San Rafael (415)459-6216

FULL CIRCLE SOLAR
200 7th Ave., Suite 111, Santa Cruz (408)479-7613

GLOBAL LIGHT AND POWER
55 New Montgomery, Suite 424, SF (415)495-0494

GOLDEN WEST ENERGY CONTROL
4592 E. 2nd Unit A, Benicia (707)746-5479

HELIODYNE INC
4910 Seaport Ave., RICH (415)237-9614

HELIOS SOLAR INC.
1755 E. Bayshore Rd., Redwood City (415)367-8439

INTEX II ENTERPRISE, INC.
P.O. Box 5734, SSF (415)952-6472

M C SOLAR ENGINEERING
1096 Rembrandt Dr., Sunnyvale (408)720-0352

PACIFIC ENERGY TECHNOLOGIES
P.O. Box 34211, SF (415)468-8074

PHOTOCOMM INCORPORATED
Solar Energy for the 90's
Photo Voltaic Systems
930 Idaho-Maryland Rd., Grass Valley
(800)544-6466

POCO
3345 Keller St., Santa Clara (408)970-0680

RELIABLE ENERGY PRODUCTS
Specializing in solar pool and water heating and freeze damage
...since 1979.... P.O. Box 3706, Napa
(707)252-2002

SAN FRANCISCO PLUMBING CO.
2345 Wawona, SF (415)664-2345

CAZZATO & ASSOCIATES
(408)425-7797
See display ad on this page.

SEVENTH SUN
1611 9th St., BERK (415)524-5821

SMITH AND SUN
612 Euclid Ave., BERK (415)525-9126

SOLAR CRAFT
448 Ignacio Blvd.,Suite 342, Novato (415)382-7717

SOLAR DEPOT
61 Paul Dr., San Rafael (415)499-1333

SOLAR SELF-HELP, INCORPORATED
965-D Detroit Ave., Concord (415)680-4343

SOLAR SERVICE SONOMA MECH. INC.
P.O. Box 750926, Petaluma (707)763-8000

SOLAR WORKS
17192 Fitzpatrick Ave., Occidental (707)874-2998

SUN LIGHT & POWER CO.
Installation & Service since1976.
Engineering, Design; Active & Passive Systems.
1035 Folger Ave., BERK
(415)845-2997

SUNTEK
8940, OAK Grove Ave., Sebastopol (707)823-9322

U.S. WINDPOWER
500 Sansome, SF (415)455-6012

SOLAR HEATING

ALCOR SOLAR POOL HEATING SYSTEMS
P.O. Box 21244, Concord (415)825-5658

AMERICAN SOLAR NETWORK
12811 Bexhill Ct., Herndon VA (703)620-2242

DIABLO SOLAR SERVICE
5021 Blum Rd., Martinas (408)729-6383

DORIUS AND MACCARTHY
P.O. Box 649, San Anselmo (415)453-3910

HELIOS SOLAR, INC.
1755 Bayshore Rd., Redwood City (415)367-8439

RADIANT ENERGY SYSTEMS
1818 Harmon St., BERK (415)654-1486

SEALED AIR CORPORATION
3433 Arden Rd., Hayward (415)887-8090

SUN PIRATE
22 Varda Landing Rd., Sausalito (415)332-7246

SUNWRIGHT INC.
1501 Feiton Rd., Healdsburg (707)433-3693

THE POOL SCENE, INC.
2 Bridge Ave., San Anselmo (415)454-5554

SOLAR ELECTRIC SYSTEMS

ACTIVE TECHNOLOGY
2518 MacArthur Blvd., OAK (415)482-8025

ECHO ENERGY PRODUCTS
219 Van Ness St., Santa Cruz (408)423-2429

FULL CIRCLE SOLAR
Research and Development...passive solar design & portable solar electric systems
200 7th Ave., Suite 111, Santa Cruz
(408)479-7613

SOLAR

HOLLY SOLAR PRODUCTS
2013 Bodega Ave., Petaluma (707)763-6173

SOLAR ELECTRIC ENGINEERING
The SOLAR/ELECTRIC CAR (707)586-1987

SOLAR ELECTRIC SPECIALTIES CO.
P.O. Box 537, Willits (707)459-9496

SOLAR-INVERTERS

WEST MARINE PRODUCTS
500 West Ridge Dr., Watsonville (408)728-2700

WEST MARINE PRODUCTS2
608 Dubuque,SSF (415)873-4044

THRIFT STORES & FLEA MARKETS

When you buy second-hand goods, you save the energy used in manfacturing and transporting new goods, as well as the raw materials. So when you get the urge to go shopping, go to second-hand stores rather than department stores. Explore the oddities and singular items at local thrift stores and flea markets, and have fun practically guiltlessly.

ALAMEDA

FLEA MARKETS

BERKELEY FLEA MARKET
Ashby Bart Brk (415)644-0744

COLISEUM SWAP MEET
5401 Coliseum Way, Oak (415)534-0325

FLEA SHOP THE
601 E 14th, SnLdro (415)568-1087

LANEY FLEA MARKET
7th & Fallon, Oak (415)769-7266

THRIFT STORES

ALTA BATES SHOWCASE
5615 College Av., Oak (415)653-1527

AMERICAN CANCER SOCIETY
3241 Grand Av., Oak (415)452-2201

VESPER SOCIETY THRIFT SHOP
378 Davis., SnLdro (415)483-7977

BEE HIVE THRIFT SHOP
1840 San Pablo Av., Oak (415)465-2018

BUENAS VIDAS THRIFT SHOP
2350 First Lvrmr (415)447-3329
4341 Railroad, Pls (415)846-5776

CELLAR THRIFT SHOP
2345 Channing Wy, Brk (415)644-3262

CEREBRAL PALSY CENTER THRIFT SHOP
320 38th St., Oak (415)658-6851

CLAUSEN HOUSE THRIFT SHOP
5000 Telegraph Av., Brk (415)653-4831

DRESS BEST FOR LESS
3820 Piedmont Av., Oak (415)658-8525

EAGLES NEST THE
37581 Niles Blvd., Fmt (415)792-6532

ECONOMY CORNER NO. 2
2440 Telegraph Av., Oak (415)451-0266

ELEGANT SCARECROW THE
3336 Grand Av., Oak (415)444-2656

ENCORE THRIFT SHOP - OAKLAND EAST BAY SYMPHONY
545 19TH, Oak (415)465-9788

FANTASY JUNCTION
108 B St., Hay (415)886-0922

FRED FINCH THRIFT STORE
3800 Coolidge Av., Oak (415)482-4093

GOODWILL INDUSTRIES
1220 Broadway, Oak (415)834-6123
14750 E 14th St., SnLdro (415)352-6966
893 West A St., Hay (415)785-3967

GRANDMA'S ATTIC
354 42nd St., Oak (415)654-3764

HOPE HOSPICE THRIFT SHOP
3037 Hopyard Rd., Pls (415)426-8988

JOSEPHENUM THRIFT SHOP
1910 Encinal Av., Almda (415)865-3077

JUNIOR LEAGUE OF OAKLAND
536 15th St., Oak (415)451-2704

JUST FOR YOU THRIFT SHOP
1800 Telegraph Av., Oak (415)839-8009

LADIES AID - RETARDED CHILDREN THRIFT STORE
22809 Mission Blvd., Hay (415)582-1922

MISS TOTT'S THRIFT SHOP
7114 E 14th St., Oak (415)638-3813

MY SISTER'S CLOSET
4050 Peralta Blvd., Fmt (415)795-0131
2205 First, Livmr (415)455-4446
3622 Old Santa Rita Rd Plstn (415)463-9626

NEARLY NEW SHOP
2140 University Av., Brk (415)841-3954

NEW TO YOU
2480 Shattuck Av., Brk (415)540-7778
16330 E 14th., SnLdro (415)276-5767

OAKLAND PENIEL MISSION
811 Washington Oak (415)452-3758
3550 E 14th St., Oak (415)532-5144

ONCE UPON A TIME
5332 College Av., Oak (415)653-2924

ORT OPPORTUNITY SHOP
5809 College Av., Oak (415)653-6212

PURPLE HEART THRIFT SHOP
825 A St., Hay (415)581-8094

ST VINCENT DE PAUL SOCIETY
2315 Lincoln Av., Almda (415)865-1109
2009 San Pablo Av., Brk (415)841-1504
2272 San Pablo Av., Oak (415)834-4647
3777 Decoto Rd.., Fmt (415)792-3711
22331 Mission Blvd.., Hay (415)582-0204

SALVATION ARMY THE
1382 Solano Av., Alby (415)524-5100
7th & Franklin., Oak (415)451-4514
2179 E 14th SnLdro (415)895-0602
40733 Chape Wy., Fmt (415)639-0637
22500 Foothill Blvd., Hay (415)538-2698
35201 Newark Blvd., Nwrk (415)791-9198
2491 First St., Lvrmr (415)447-2303

SECOND HAND ROSE
1111 Solano Av., Alby (415)527-7742

SIMS BUDGET STORE
7011 Hamilton, Oak (415)638-9148

SLASH CLOTHING
2840 College Av., Brk (415)841-7803

SLENDER POCKETS
3807 MacArthur Blvd., Oak (415)530-0606

TAKE-A-PEEK THRIFT SHOP
823 Rincon Av., Lvrmr (415)449-1664

THE SHOP
536 15th St., Oak (415)451-2704

THRIFT CENTER THRIFT STORE
1311 Bockman Rd., SLz (415)481-1930
40645 Fremont Bl., Fmt (415)490-0485
29498 Mission Bl., Hay (415)881-0222

THRIFT TOWN
16160 E 14th, SnLdro (415)278-1766

THRIFT VILLAGE
20812 Hesperian Blvd., Hay (415)783-0101

UHURU FURNITURE ETC
2700 Park Blvd., Oak (415)763-3342
2680 Union, Oak (415)268-0994

VALLEY MEMORIAL HOSPITAL THRIFT
1111 East Stanley Blvd., Lvrmr (415)373-4069

VALUE THRIFT CENTER
22683 Main, Hay (415)582-2404

WOOD'S ENTERPRISES
1444 23rd Av., Oak (415)533-1651

CONTRA COSTA

THRIFT STORES

AMERICAN CANCER SOCIETY
1389 N Main St., WC (415)944-1991
10313 San Pablo Av., EC (415)527-1469

ASSISTANCE LEAGUE OF DIABLO VALLEY
3521 Golden Gate Wy., Laf (415)527-1469

CLAUSEN HOUSE THRIFT SHOP
3616 Willow Pass Rd., Cncd (415)680-0878

CLOTHESLINE THE
3490 Clayton Rd, Cncd (415)676-8066

CONTRA COSTA ASSOC. FOR RETARDED CITIZENS
14237 San Pablo Av., SP (415)235-1536

DOTTY JO'S THRIFT SHOP
1992 23rd, SP (415)236-7203

FAMILY SAVERS THRIFT SHOP INC
2265 Contra Costa Bl PH (415)680-6535
1835 Colfax, Cncd (415)827-2667

GARMENT EXCHANGE
250 Pacific Av., Rodeo (415)799-7680

GARRET THRIFT SHOP OF JOHN MUIR MED CTR
1530 3rd Av., WC (415)932-9474

GILDED CAGE THRIFT SHOP
3439 Chestnut Av., Cncd (415)686-0324

GOODWILL INDUSTRIES
12915 San Pablo Av., Rich (415)236-2422
2665 Pleasant Hill Rd., PH (415)935-7414
3495 Clayton Rd., Cncd (415)676-8140

HALL CLOSET
827 Las Juntas, Mrtnz (415)228-0230

HOLY TRINITY THRIFT SHOP
555 37th Rich (415)236-6225

HOSPICE THRIFT SHOP
1550 Olympic Bl WC (415)947-1064

KIDS RECYCLE
1944 Monument Blvd., Cncd (415)827-4482

KNIGHTSEN TRADING POST
Delta & Knightsen Rds, Kngtsn (415)625-3838

LADY MARIAN THRIFT SHOP
3669 Desanie Cir., Pit (415)458-9726

LEFTOVERS THRIFT SHOP
2333 Boulevard Cir., WC (415)930-9393

MT DIABLO REHABILITATION CENTER
1539 N Main St., WC (415)935-4211
2 E 5th., Pit (415)439-8300

MY SISTER'S CLOSET
7009 Village Pkwy., Dbln (415)829-4773

NEW TO YOU
3219 McDonald Av., Rich (415)234-4972

NIFTY THRIFT
261 Lafayette Cir., Laf (415)284-5237

PURPLE HEART VETERANS SERVICES
2158 Solano Wy., Cncd (415)825-8960

RAG DOLL THE
1109 Bont Ln., W.C. (415)937-2344

RAGS AND RICHES
110072 San Pablo Av., EC (415)232-2024

ST VINCENT DE PAUL
1025 13th., Rich (415)236-5521
650 El Portal Shopping Ctr., SP (415)620-0203
390 Central Av., Ptsbg (415)439-5060
3325 N Main PH (415)934-5063
Hwy 4 & Sunset Dr., Brtwd (415)634-0473

SALVATION ARMY THE
1430 Cutting Blvd., Rich (415)234-5181
1806 Linda Dr., PH (415)685-6900
1900 A Ant (415)778-9813

SECOND FIDDLE THE
2103 San Pablo Av., Pnle (415)724-7948

THRIFT STATION THE
123 N Hartz Av., Dan (415)820-1988

THRIFT TOWN
3645 San Pablo Dam Rd., ElSbt (415)222-8696

TURNABOUT SHOP THE
10052 San Pablo Av., EC (415)525-7844

VALLEY MEM. HOSPITAL THRIFT SHOPS
7387 Village Pkwy., Dbln (415)829-8822

VALUE VILLAGE THRIFT STORE
14237 San Pablo Av., SP (415)235-1536

WAY SIDE INN THRIFT SHOP & COSTUME RENTAL
3521 Golden Gate Wy., Laf (415)284-4781

MARIN

FLEA MARKETS

LITTLE FLEA MARKET
175 Belvedere, SR (415)457-9359

MARIN CITY FLEA MARKET
147 Donahue, MC (415)331-6752

MARIN CITY FLEA MARKET
630 Drake Av., MC (415)332-1441

THRIFT STORES

AMER. CANCER SOC. DISCOVERY SHOP
928 Grant Av., Nov (415)898-1149
801 A Street, SR (415)456-0340
31 Sunnyside Av., MVly (415)389-9518

BARGAIN BOX SUNNY HILLS
508 Irwin St , SR (415)459-2396

CATCH-ALL
541 San Anselmo Av., SAnslmo (415)456-1206

CHURCH OF THE REDEEMER THRIFT SHOP
121 Knight Dr, SR (415)456-0509

FAMILY STORE THE
448A Miller Av., MVly (415)388-9596

GOOD STUFF
1942 4th SR (415)456-4227

GOODWILL STORES
809 Lincoln Av., SR (415)456-5273

HOSPICE HODGEPODGE THRIFT & GIFT SHOP
1541 4th, SR (415)459-4686

HUMANE SOCIETY THRIFT SHOP
360 San Anselmo Av., SAnslmo (415)459-9929

MARIN GENERAL HOSPITAL THRIFT SHOP
1115 3rd, SR (415)456-1430

MT CARMEL SALVAGE SHOP
45 Lovell Av., MVly (415)388-4332

ONE MORE TIME
1821 4th, SR (415)459-6137

ST ANSELM'S RUMMAGE RACK
43 Mariposa Av., SAnslmo (415)453-1226

ST PATRICK'S THRIFT SHOP
457 Magnolia Av., Lrkspr (415)924-5393

ST VINCENT DE PAUL SOCIETY
881 Grant Av., Nov (415)897-7011

SALVATION ARMY THE
4th & Mary, SR (415)454-7201

SAUSALITO SALVAGE SHOP
19 Princess, Sau (415)332-3471

SECOND BANANA THE
405 Sycamore Av., MVly (415)381-5437

SEEK AND FIND THRIFT SHOP
1815 4th, SR (415)456-7353

THE DEPOT AT ST VINCENT'S SCHOOL
4900 Highway 101, SR (415)479-0366

TIBURON THRIFT SHOP
1101 Tiburon Blvd, Tibrn (415)789-9891

WEST MARIN COMMUNITY THRIFT SHOP
11431 State Route 1, PtReys (415)663-9227

NAPA

FLEA MARKETS

FLEA MARKET & AUCTION
303 Kelly Rd., Napa (707)255-9428

THRIFT STORES

ACORN THRIFT SHOP
657 Main, SH (707)963-7489

COMMUNITY PROJECTS INC
715 Franklin, Napa (707)226-7585

CONSIGNMENT SHOPPE
1924 Yajome, Napa (707)255-8911

GOODWILL STORE
2412 Jefferson, Napa (707)257-6610

SALVATION ARMY
1234 3rd, Napa (707)224-8220
200 Lytton Springs Rd, Hldbrg (707)224-0630

SAN FRANCISCO

THRIFT STORES

ANOTHER WAY THRIFT
1661 Fillmore, SF (415)931-1713

ATTIC SHOP
1040 Hyde, SF (415)474-3498

BARGAIN MART
1823 Divisadero, SF (415)921-7380

BEARLY-NEW THRIFT SHOP
1752 Fillmore, SF (415)921-2789

CATHEDRAL SCHOOL SHOP
1036 Hyde, SF (415)776-6630

COMMUNITY THRIFT STORE
625 Valencia, SF (415)861-4910

GOODWILL INDUSTRIES,SAN FRANCISCO
980 Howard, SF (415)362-0778

IZALCO SECOND HAND STORE
2708 24th St., SF (415)695-0327

JO JO'S THRIFT SHOP
4416 3rd St., SF (415)695-1278

MORE FOR LESS
407 Valencia, SF (415)863-6279

NEXT-TO-NEW SHOP
2226 Fillmore, SF (415)567-1627

OCEAN AVENUE THRIFT SHOP
1619 Ocean Av.,e., SF (415)239-8766

PURPLE HEART VETERANS THRIFT SHOP
1855 Mission, SF (415)621-2581

REPEAT PERFORMANCE
SF Symphony Thrift
2223 Fillmore, SF (415)563-3123

SECONDS TO GO
2252 Fillmore, SF (415)563-7806

ST. ANTHONY DISTRIBUTION CENTER
111 Golden Gate Av.,e., SF (415)241-8300

ST. FRANCIS CHURCH THRIFT SHOP
Ocean Av.,e.& San Fernando Wy., SF (415)587-1082

ST. VINCENT DE PAUL SOCIETY
1745 Folsom, SF (415)863-3315

THE STREET SHOP
2011 Divisadero, SF (415)931-4382

THIRD HAND STORE
1839 Divisadero, SF (415)567-7332

THRIFT TOWN
2101 Mission, SF (415)861-1132

TOWN SCHOOL CLOTHES CLOSET
3325 Sacramento, SF (415)929-8019

VICKIE'S STREET SHOP BOUTIQUE
603 Haight, SF (415)552-5997

VICTORIAN HOUSE THRIFT SHOP
2318 Filmore, SF (415)923-3237

SAN MATEO

THRIFT STORES

ALTERNATIVE THRIFT SHOP
522 Plaza Alhambra, ElGrda (415)726-9911

AMERICAN CANCER SOCIETY
1410 Broadway, Burl (415)343-9100
746 Santa Cruz Av., MenloPk (415)325-8939

BARGAIN BOUTIQUE-CHILDREN'S HOME SOCIETY
31 North B, SMto (415)344-4208

BEST FOR LESS
3247 Middlefield Rd, MnloPk (415)366-4148

DISCOVERY SHOP- JUNIOR LEAGUE INC
2432 Broadway, RC (415)363-2238

FAMILY TREE THE
1589 Laurel, SC (415)592-6150

GOODWILL INDUSTRIES
33 Southgate Av., DC (415)992-1720

JANNETTE'S COLLECTABLES
726 Villa, MVw (415)967-8684

ORT OPPORTUNITY SHOP
437 San Mateo Av., SB (415)871-7170

PENINSULA HUMANE SOCIETY
56 42nd Av., SMto (415)345-1024

PESCADERO THRIFT SHOP
356 Stage Rd., Pscdro (415)879-0114

PRIMROSE HOUSE OF VALUE
307 Primrose Rd., Burl (415)343-1920

RAGS & RICHES
715 South B, SMto (415)348-3590

ROUND ROBIN
124 Second St., LAltos (415)948-0905

ST VINCENT DE PAUL SOCIETY
344 Grand Av., SSF (415)589-8445
113 South B, SMto (415)347-5101
831 Main, RC (415)369-5898

SALVATION ARMY THE
650 El Camino Rl, Blmnt (415)591-5499

SAVERS
2840 Geneva Av., DC (415)468-0646

THIS'N THAT SHOP
1300 5th Av., Blmnt (415)591-6166

THRIFT CITY
852 Walnut, RC (415)367-0646

SANTA CLARA

FLEA MARKETS

FLEA MARKET CAPITOL
Capitol Expwy & Monterey Rd., SJ (408)225-5800

FLEA MARKET INC
12000 Berryessa Rd., SJ (408)453-1110

MORGAN HILL FLEA MARKET
140 E Main Av., MorganHill (408)779-3809

THRIFT STORES

AMERICAN CANCER SOCIETY
1346 Lincoln Av., SJ (408)294-3982
7432 Monterey, Gilroy (408)847-4409

BARGAIN BOX (CHILDREN'S HEALTH COUNCIL)
318 Cambridge Av., PA (415)326-0458

COMMUNITY THRIFT CENTER
420 S 1st, SJ (408)288-9837

CORINTHIAN THRIFT SHOP
1405 Hacienda Av., Cml (408)866-7502

ECHO SHOP OF ST ANDREWS CHURCH
14477 Big Basin Wy., Stg (408)867-3995

HAPPY DRAGON THRIFT SHOP
245 West Main, LGtos (408)354-4072

MEDICAL MISSION SISTERS
465 California Av., PA (415)321-8616

MORGAN CENTER SECOND SEASON SHOP
380 Cambridge Av., PA (415)323-6827

ST VINCENT DE PAUL
1290 S 1st, SJ (408)993-9502
747 1st St., Gilroy (408)842-5524

SALVATION ARMY THE
1494 Halford Av., SClara (408)249-1715

SARA'S CONSIGNMENT FASHION BOUTIQUE
448 N San Pedro, SJ (408)298-1100

SAVERS
60 Dempsey Rd, Milpts (408)263-8338

THRIFT BOX - CHILDREN'S HOSPITAL
1362 Lincoln Av., SJ (408)294-4490

THRIFT CITY
1795 Alum Rock Av., SJ (408)259-7683

TURNABOUT SHOP
2335 El Camino Rl., PA (415)321-9853

SANTA CRUZ

FLEA MARKETS

RED BARN FLEA MARKET
1000 Highway 101, Aro (408)726-3101

SKYVIEW FLEA MARKET
2260 Soquel Dr., SCz (408)462-4442

THE OLD CORRAL FLEA MARKET
16130 Highway 156, Salinas (408)663-3611

THRIFT SHOPS

AWARE
1515 Commercial Wy, SCz (408)476-0644

DAISY THE (FAMILY SERVICE ASSOC.)
817 Front, SCz (408)426-1399

HUB THE
1054 Amesti Rd., SCz (408)761-3722

ONCE AROUND LIGHTLY
871 41st Av., SCz (408)462-3796

ST JOHN THE BAPTIST
101 Capitola Av., Cap (408)475-0957

ST VINCENT DE PAUL SOCIETY
3912 Portola Dr., SCz (408)462-1647
12860 Hwy 9, BC (408)338-2422
135 E Lake Av., Wat (408)722-0420

SALVATION ARMY
322 Front, SCz (408)429-8118

THRIFT CENTER THRIFT STORE
559 Main, Wat (408)761-0291

UNITED CEREBRAL PALSY
504 Front, SCz (408)429-6974

YWCA
367 E Lake Av., Wat (408)728-3766

SOLANO COUNTY

FLEA MARKETS

BROADWAY SHOPPING OUTLET
710 Broadway, Vljo (707)557-0566

THRIFT SHOPS

AMERICAN CANCER SOCIETY
3712 Sonoma Blvd., Vljo (707)643-6606

GOODWILL INDUSTRIES
1039 Tennessee, Vljo (707)643-3624

FATHER JOHN'S THRIFT SHOP
822 Marin, Vljo (707)642-5525

SALVATION ARMY THE
2136 Sacramento, Vljo (707)643-3433

SECOND CHANCE THRIFT SHOP
640 1st, Ben (707)745-4565

SUTTER SOLANO MED CTR THRIFT SHOP
1011 Broadway, Vljo (707)648-0566

VALUE CENTER
418 Georgia, Vljo (707)557-0238

SONOMA COUNTY

FLEA MARKETS

MIDGLEY'S COUNTRY FLEA MARKET
2200 Gravenstein Hwy S Sbtpl (707)823-7874

THRIFT STORES

AMERICAN CANCER SOCIETY
2300 Mendocino Av., SRs (707)526-3349

AS IS SHOP
28 W Spain, Snma (707)996-8923

CHURCH MOUSE
10 Boyes Blvd., (707)938-9839

GOODWILL INDUSTRIES
3535 Industrial Dr., SRs (707)545-2492
172 Petaluma Blvd., Pet (707)778-7485

NEW TO YOU SHOP
5 W 6th, SRs (707)546-9484

PRICELESS THRIFT SHOP
1110 Petaluma Hill Rd., SRs (707)528-2724

RESCUE MISSION THRIFT STORE
4056 Sebastopol Rd., SRs (707)523-3152

SACK'S THRIFT AVENUE
322 Western Av., Pet 765-2228

ST JOHN'S USED TREASURES
201 Center, Hbg (707)433-8533

ST VINCENT DE PAUL SOCIETY
9869 Main, MRio (707)865-1339

SALVATION ARMY THE
Lytton Springs Rd., Hbg (707)433-7404

WELFARE LEAGUE SHOP
126 4th., SRs (707)542-7480

Transportation in the Bay Area has become a nightmare — one that our whole way of life is wrapped up in. A big part of the problem is that many of us live so far from work. The congestion at rush hour (which now seems to extend through almost the entire day) is horrendous as a result. Yet we continue to get in our cars and face it every day rather than deal with buses, which are even slower, or trains, because the system simply isn't extensive enough, and we have to drive to get to a BART station anyway.

We remain dependent on the automobile, even though we are all quite aware of the list of heavy tolls their use takes on the environment — the hydrocarbons that are the main contributor to the thickening smog, the carbon dioxide that contributes to global warming, the lead that poisons our children, and the importing of millions of gallons of oil which seems to necessitate the occasional disastrous oil spill on the high seas.

Because cars are so central to our way of life, the Green Pages staff felt that we could not follow our original inclination to leave them out entirely. We argued about this at length, and in the end decided that we would print a list of the most efficient cars, but we would not solicit or accept advertising from any car companies. We also have listed some fuel-saving devices, and one company that sells a kit you can use to convert your small car to electric power.

Here you will also find bicycling information (ride a bike — they are the best form of transportation), garages that will service you auto air conditioner without releasing the CFCs into the atmosphere, and extensive information about mass transit.

Most fuel efficient cars:	Engine size	Transmission	MPG	Fuel cost/yr.
Geo Metro XFI	1L, 3cyl	5spd	53/58	$259
Geo Metro LSI	1L, 3cyl	5spd	46/50	296
Geo Metro	1L, 3cyl	5spd	46/50	296
Suzuki Swift	1L, 3cyl	5spd	46/50	296
Honda Civic CRX HF	1.5L, 4cyl	5spd	43/49	320
Daihatsu Charade	1L,3cyl	5spd	38/42	362
Volkswagon Jetta	1.6L, 4cyl	5spd	37/43	365
Ford Escort	1.9L, 4cyl	4spd	32/42	387
Isuzu Stylus	1.6L, 4cyl	5spd	32/37	402
Honda Civic CRX	1.5L, 4cyl	5spd	32/35	418
Pontiac LeMans	1.6L, 4cyl	5spd	31/40	420
Suzuki Samarai	1.3L, 4cyl	5spd	28/29	475

Fact – Automobiles use close to two-thirds of all oil consumed in the U.S.

AUTOMOBILE - AIR-CONDITONING REPAIR

AIR COND.& AUTO ELECT. OF FREMONT
37555 Central St., Fremont (415)793-2340

AIR CONDITIONING AUTOMOTIVE CLINIC
2035 Divisidero St., SF (415)563-2915

ALAMEDA TIRE, BRAKE & MUFFLER
1825 Webster St., ALA (415)521-1541

ALLIED RADIATOR AND AIR CONDITNG
920 Gilman St., BERK (415)524-3860

AUTOHAUS - HAYWARD
21650 Mission Blvd., Hayward (415)881-1915

AUTOHAUS - SANTA CLARA
2555 Lafayette St., Santa Clara (408)986-1391

DON'S RADIATOR
647 23rd. St., RICH

KINGS AUTO AIR/HEATING, INC.
2550 High St., OAK (415)534-1946

MADSEN'S AUTO AIR AND HEATING
1132 Meadow Ln., Concord (415)798-1131

AUTOMOBILE - FUEL-SAVING DEVICES

AUTOMOBILE FUEL SAVERS ASSOC.
5337 College Ave Dept 512, OAK (415)601-5095
See display ad, page 335.

AUTOMOBILE - ELECTRIC CARS

SOLAR ELECTRIC ENGINEERING
175 Cascade Ct., Rohnert Park (707) 586-1987

BICYCLES

BIKES ON BART

NON-COMMUTE HOURS
800 Madison St., OAK (415)464-7127

CALTRANS/COMMUTER BICYCLE SHUTTLE
Takes bikes from MacArthur BART to SF Terminal (415)464-0669

CITY CYCLE
3001 Steiner, SF (415)346-2242

SEAL ROCK CYCLES
6350 Geary Blvd., SF (415)387-3152

MASS TRANSIT

AC TRANSIT
1600 Franklin St., Oakland (415)839-2882
East Bay and transbay buses. Local, transbay, express, youth, senior and disabled monthly passes. Ticket books for discounts without an expiration date. AC/BART and Muni passes. Buses run from Crockett south to Fremont.

AMTRAK
(800)872-7245
(800) USA-RAIL. The train.

BART
800 Madison St., Oakland (415)464-6000
Bay Area Rapid Transit. Our inter-urban rail system, including: BART Express Buses, BARTlink (Vallejo to El Cerrito Del Norte BART), and Traveler's Transit (Richmond-San Rafael). Excellent San Francisco & East Bay BART & BUSES guide.
For information on connecting buses from BART (all in Area Code 415):

Antioch/Pittsburg	*754-BART*
Fremont/Union CITY	*793-BART*
Hayward/San Leandro	*783-BART*
Concord/Lafayette/Walnut Creek	*933-BART*
Livermore/Pleasanton	*462-BART*
Berkeley/Oakland/Orinda	*465-BART*
El Cerrito/Richmond	*236-BART*
San Francisco/ Daly City	*788-BART*
S. San Francisco/San Bruno	*873-BART*
TTY 839-2220	

BERKELEY TRIP
2033 Center St., Berkeley (415)644-7665
Clearinghouse of mass transit info. Passes, tickets, commute and bicycle maps. They'll get you going.

CALTRAIN
930 Montgomery St.
5th Floor, San Francisco (800)558-8661
Rail service from S.F. to the peninsula & San Jose.

CALTRANS/COMMUTER BICYCLE SHUTTLE
Dept. of Transportation
Toll Bridge Region/ P.O. Box 7310, SF (415)464-0669
Provide van service which takes bikes from MacArthur BART to Transbay Terminal during commute hours.

COMMUNITY TRANSIT SYSTEM
1705 Enterprise Drive, Fairfield (707)422-6378
The Community Transit System encompasses:VacAv.,ille City Coach, Fairfield Flier, Benlcia Bay Connection, North Bay Shuttle, and DART- a shuttle for disabled people.

COUNTY CONNECTION
2477 Arnold Industrial Way, Concord (415)676-7500
1990 N. California Blvd., Walnut Creek
Contra Costa. Serves Central CCC, has commute card, 40-ride card, 20-ride card for Cal Commuter Express, Monthly Cal Commuter.

EAST BAY PENINSULA COMMUTE CLUB
(415)654-5778
Membership organization providing bus transportation to mid-Peninsula area.

GOLDEN GATE TRANSIT
1011 Andersen Drive, San Rafael (415)332-6600
Ticket books, ferry service. Serves S.F.'s financial district to Santa Rosa,Marin County,and Sonoma County, from the Bay to the Coast.

GREEN TORTOISE
P.O. Box 24459, SF (415)285-2441
Alternative bus trips around country and to national parks-so you don't haveto drive.

GREYHOUND LINES
155 Fremont St., Transbay Terminal, SF (415)558-6789
Trailways Bus Line is also at this number. From 408 area - 297-8890. From 707 area - 226-1856.
Grey hound/Trailways-our national mass transit bus system. There are many bus stations in our area, and around the country. Call for the nearest.

HEALDSBURG TRANSIT
126 Mathson, Healdsburg (707)431-3309

MUNI
949 Presidio Ave., SF (415)673-6864
San Francisco bus service -Monthly Pass, Monthly Sticker for AC Transbay Pass, Senior/Disabled Pass. Publishes a timetable pamphlet for all lines. Call 673-MUNI.

PETALUMA TRANSIT
67 Magnolia St., Petaluma (707)778-4460

SAMTRANS
945 California Drive
P.O. Box 957, Burlingame

South San Mateo	*(415) 367-1500*
Central San Mateo Call	*(415) 348-8858*
North San Mateo	*(415) 871-2200*
Half Moon Bay, Coastside	*(415)726-5541.*
San Francisco	*(415) 761-7000*
Hayward Area	*(415) 582-1500*
TTY	*(415) 343-3222*

Bus service in San Mateo County with connections to Daly City, Hayward, SF Airport, Cal Train, Santa Clara Transit, Transbay Terminal. Discount ride pass, senior/disabled,and youth passes.

SANTA CLARA COUNTY TRANSPORTATION AGENCY
1555 Berger Dr., San Jose (408)287-4210
Call for free trip planning.

SANTA CRUZ METROPOLITAN TRANSIT
920 Pacific Ave.,
Marketing Dept., Santa Cruz (408)425-8600
Santa Cruz bus service.

SANTA ROSA CITY BUS
P.O. Box 1678, Santa Rosa (707)524-5306

SEBASTOPOL TRANSIT
P.O. Box 1776, Sebastopol (800)345-7433

SONOMA COUNTY TRANSIT
355 W. Robles, Santa Rosa (800)345-7433
800-576-RIDE

TRI-DELTA TRANSIT
801 Wilbur, Antioch (415)754-4040
Antioch, Pittsburg, W. Pittsburg

UNION CITY TRANSIT
705 Bradford Street, Union City (415)471-1411

VALLEJO BARTLINK
1850 Broadway, Vallejo (707)648-4666
Provides service between BART and Vallejo. Monthly Pass, 10-Ride Punch Card. You can also call 236-BART for info and schedules.

VALLEJO TRANSIT
1850 Broadway, Vallejo (415)648-4669
Bus Service, Vallejo Bartlink

VINE (NAPA CITY BUS)
P.O. Box 660, Napa (707)255-7631

WESTCAT (WESTERN CONTRA COSTA COUNTY)
953-B San Pablo Ave., Pinole (415)724-7993
Fixed route-Hercules and Pinole; Dial-a-Ride for elderly and disabled covers Crockett, Port Costa, Rodeo, Hercules, and Pinole.

WHEELS (DUBLIN, LIVERMORE, PLEASANTON)
87 Fenton Street, Suite 200, Livermore (415)455-5414

MASS TRANSIT-AIRPORTERS

ABC TRANSPORTATION
P.O. Box 691, Soquel (408)462-5312

AIRPORT COMMUTER
11 Airport Blvd., SSF (415)759-5463

AIRPORT CONNECTION
100 B Produce Ave., SSF (415)841-0150
Van service between Berkeley Hotels and SF Airport, Door-to door service Av.,ailable.

AIRPORTER
923 Folsom, SF (415)673-2433
Express bus service from several locations in downtown S.F.

BAY AREA BUS SERVICE
(415)444-4200
Regular service from the Oakland Hyatt Hotel to S.F. and Oakland airports.

BAY AREA SHUTTLE
18 Chateau Ct., SSF (415)873-7771

BAYPORTER EXPRESS
100 N. Hill Drive #29, Brisbane (415)467-1800

CAPITOLA LIMOUSINE CO.
2535 Seventh Ave., Santa Cruz (408)476-5466

EXPRESS AIRPORT SHUTTLE
851 SpringfieldDr., Campbell (408)378-6270

LARKSPUR LIMOUSINE
P.O. Box 745, Larkspur (415)927-4422

MARIN AIRPORTER
300 Larkspur Landing Circle, Larkspur (415)461-4222

NAVE'S LIMOUSINE SERVICE
351 Airport Rd., Novato (415)485-1764

OVER THE HILL TRANSPORTATION SERVICE
(408)462-4598

SAMTRANS AIRPORT SERVICE
945 Calikfornia Dr., Burlingame (415)761-7000
Buses from SF Transbay Terminal to SFO.

SANTA CRUZ AIRPORTER
224 Walnut Ave., Santa Cruz (408)423-1214

SAUSALITO LIMOUSINE
P.O. Box 9738, San Rafael (415)332-7811

SOUTH BAY AIRPORT SHUTTLE
840 E. Campbell Ave., Campbell (408)559-9477

SUPER SHUTTLE
700 16th St., SF (415)558-8500

YELLOW AIRPORT SHUTTLE
1200 Mississippi, SF (415)282-7433
Only 24 hr. shuttle in area

MASS TRANSIT-FERRIES

GOLDEN GATE FERRY
(415)453-2100
Ferry service from Sausalito and Larkspur to SF Ferry Building.

RED AND WHITE FLEET
Pier 41, Fisherman's Wharf, SF (415)546-2628
Commute ferry service from East Bay (Jack London Square, and Alameda), San Francisco,Tiburon, Vallejo. Mid-day and weekend excursions to Alameda, Oakland, Fisherman's Wharf, Sausalito, Tiburon, Napa and Sonoma Valley, and Alcatraz. Charters available.

MASS TRANSIT-VAN POOLS

AMCAR VANPOOLS
2833 Mankas Blvd., Fairfield (707)425-7877

BERKELEY TRIP/COMMUTE STORE
2033 Center St., Berkeley (415)644-7665
Free car/vanpool matching service for commuters to and from Berkeley.

BISHOP RANCH TRANSPORTATION CENTER
2682 Bishop Dr, San Ramon (415)830-0101

HALF- COST CAR POOL
2720 Martin Luther King Way, BERK (415)845-1769

PEOPLE RIDESHARING SYSTEMS
2200 Wood, BERK (415)839-3444

RIDES FOR BAY AREA COMMUTERS
(415)861-7665
Free carpool/vanpool matching service for commuters within ten-county Bay Area.

RIDES FOR BAY AREA COMMUTERS, INC.-BERK
2033 Center, BERK (415)644-7665

RIDES FOR BAY AREA COMMUTERS,INC.-SF
60 Spear, SF (415)861-7665

SHARE-A-RIDE
701 Ocean, Santa Cruz (408)429-7665
Free car-pool matching. Luxury vanpool commuting.

SOLANO RIDESHARE
1961 Walters Ct., Fairfield (707)447-7665

VP SHUTTLE
4791 Teakwood Dr., Oakley (415)625-4193

VPSI (VAN POOL SERVICES INC.)
400 Oyster Point Blvd., SSF (415)952-8774

WATER

Water is life, and California is a dry place. We live lifestyles based on wasting water. Rather than cut back, we continue to build new dams to the ruination of dwindling wilderness resources.

Most of the water we waste does not add to our standard of living in any way . We could easily get by without having our houses surrounded by non-native plants which require watering in the dry season. We use high-quality drinking water for low-quality uses like cleaning our driveways, washing our cars, and flushing our toilets. Meanwhile, many of our freshwater sources are threatened by pesticide runoff, toxic dumps and airborne pollutants.

There are simple ways we can save water around home and work. First, do a water use audit. Where does the water go in your house? What are the uses? Who are the big users? Where would it be easiest to cut down?

Replace your toilets with water saving ones, or if your landlord is replacing the toilet, get him or her to use a water saving model. Even better, get a composting toilet. Install low-flow shower heads and water meters.

If you have a garden or a lawn, look into getting a grey water recycling system. Essentially, this is a storage tank for water diverted from your bathtub, washing machine, or hot water spigot (to save that precious water when you are heating up the stream). The State Health Department is trying to make these systems illegal (as health hazards), even though they are designed to rid the water of bacteria. You do have to avoid pouring toxins like household bleach down the drain when using a greywater system, and you should use biodegradable soaps. But we should all be doing this anyway.

If you are concerned about the quality of your drinking water, there are good water filtration systems which remove most of the pollutants you can expect.

There are also home water testing kits with which you can check for common pollutants you suspect might be in your drinking water.

WATER CONSERVING DEVICES

The companies listed below are either manufacturers or sellers of low-flush toilets, low-flow shower heads, or water meters. Many of these items can be found in any hardware store.

Many counties and cities in California have recently passed laws requiring the installation of low-flush toilets in all new housing. If you ever replace your toilet you should buy a low flush model, but if you plan to maintain your six- to seven-gallon water hog, at least place a toilet dam in it. Low-flow shower heads and faucets are another easy water and money saving item to install.

CAL-STEAM SUPPLY
777 Mariposa, SF (415)861-3071

COMPOSTING TOILET SYSTEMS
1211 Bergen Rd., Newport,WA (509)447-3708

EARTHSAKE
1844 Market St., SF (415)626-0722

EBMUD
2130 Adeline St., OAK (415)835-3000

R.D. KINCAID, INC.
615 7th St., SF (415)861-9400

SAVE ENERGY COMPANY
2410 Harrison St., SF (415)824-6010

Fact: One gallon of motor oil can contaminate up to 750,000 gallons of drinking water.

WATER DISTILLERS

Combined with active carbon filters (see Water Filters below) water distillers give the most consistent and pure water you can get. Water is distilled through a boiling and recondensing process that eliminates all impurities and particles. Volatile chemicals can be removed through an additional carbon filter. Though they use energy to operate, they require no replacement parts.

Disadvantages: Distillers require active involvement on the part of the user. Some have to be filled by hand, turned off or on, and they all must be cleaned regularly. They require energy to run. More water is wasted than distilled. They also make the water *too* clean, as distillation removes trace minerals that are important to your health.

MCCRACKEN SOLAR COMPANY
329 West Carlos, Alturas (916)233-3175

WATER FILTERS

There are several kinds of water filters. Not all of them are represented in this list, but the following descriptions should give you the information you need to ask pertinent questions of your supplier. No purification device is 100 percent effective at removing all contaminants, but these options can improve your supply. Drinking water is not the only source of water contamination; skin absorption while bathing is a concern, and you may want to look into shower head filters as well. In addition to the companies in our short list, consult the General Merchandise section of the guide for other companies that address these water filtering issues. Check your local hardware store too, as many carry both water filters and home water-testing kits.

Activated Carbon Filters

These filters remove volatile chemicals such as chlorine, solvents and pesticides. Make sure your unit has a good supply of carbon for effectiveness; cheaper models may not have sufficient amounts. A carbon block is preferable to carbon granules.

Disadvantages: Once fully saturated with contaminants, the filter must be changed. This maintenance, and the cost of new filters deter some people. Ask for a double system of sediment filter and carbon filter in one unit, which cuts the cost of filter replacements.

Reverse Osmosis

These filters use both carbon and a thin membrane. The membrane prevents the passage of all particulate matter, including radioactive particles. The filter gives the volatile chemical-filtering advantages of activated carbon. These are more elaborate, three-part units that use a storage tank (needed because of the time it takes for water to pass the membrane).

Disadvantages: Some membrane materials cannot tolerate chlorine, and thus need a pre-filter of carbon. Others require the presence of chlorine to remain bacteria-free. The units are made mostly of plastic, which can be a problem for chemically-sensitive people. The filters decline in efficiency over time, and in conjunction with any pH, temperature or water pressure changes. All parts of the filter need to be replaced periodically. The units waste about twenty percent of the water that is filtered (it goes down the drain). The membranes themselves are very susceptible to bacterial breakdown, and require either chlorine (or a chlorinated water supply) or other means of killing bacteria at the membrane section.

DAN CLURMAN
5756 Claremont, OAK (415)547-2380
See display ad, page 337.

ENVIRONMENTAL PURIFICATION SYSTEMS
The Only 100% Stainless Steel Drinking Water Filter
P.O. Box 191, Concord
(800)829-2129

GETZ, ANDREW
5756 Claremont, OAK (415)547-2380
See display ad, page 337.

MULTI-PURE WATER FILTERS
James Baraz (415)465-2324

THE ENVIRONMENTAL NETWORK
2312 Castro St., SF (415)826-0195

WIND POWER

After a good bit of discussion, we decided to place the wind power category in the Business to Business Guide (see p. 350) because the latest wind generators are high-capacity machines, more appropriate to large-scale applications such as office parks than to private residences.

We don't mean to discourage you from looking into wind power, however. If you are interested, and live in an appropriately windy spot, consider forming a cooperative relationship with others in your neighborhood, and get a single windmill to provide power to all of your houses. Ridgetops in the Bay Area are generally windy enough to make it worthwhile.

WINE

Listed below are wineries and wine brokers who insist upon organically grown grapes. Enjoy.

BELLROSE VINEYARD
435 West Dry Creek Rd. (707)433-1637

FREY VINEYARDS
14000 Tomki Rd., Redwood Valley (800)345-3739

NEVADA COUNTY WINE GUILD
11372 Winter Moon Bay, Nevada City (916)265-3662

OLSON WINERY INC.
3620 Road B, Redwood Valley (707)485-0323
See display ad on this page.

THE ORGANIC WINE COMPANY
54 Genoa Place SF (415)433-0167
See display ad on this page.

HALLCREST VINEYARDS
379, Felton Empire Rd. (408)335-4441

WHAT DO YOU KNOW THAT WE DON'T?

Now that you've carefully perused the first edition of the *Bay Area Green Pages*, you've probably noticed that your rabbit's favorite brand of organic pet food is conspicuously missing, or that we have committed some other grievous error of omission.

We depend on our readers to help us make this a good resource, so we would like to hear from you. (Of course, we are also willing to hear compliments on our thoroughness.)

Write: Green Media Group,
P.O. Box 11314, Berkeley, CA 94701

OLSON
Winery
Dedicated to making the finest wines in Mendocino County from organically farmed vineyards. Visit our tasting room & picnic area.
3620 Road B
Redwood Valley, CA 95470
(707) 485-0323

IMPORTERS OF FINE FRENCH ORGANIC WINES

Véronique Raskin-Denille, President
415.433.0167

Section Eight

BUSINESS TO BUSINESS

It is true that businesses are by far the biggest consumers in our society, and the purchasing (and service contracting) decisions they make have a proportionally large effect on the environment. They are the major consumers of paper products, energy resources, and virtually all raw materials.

But business does not exist in a void, pillaging the planet all by itself for obscure reasons all its own. We all help out. As consumers we encourage irresponsible corporate behavior through indifference to the provenance of the goods we buy. Most of us are employed by a business of one kind or another. Regardless of whether that work is in a chemical factory or a public relations firm, we all share responsibility for the commercial activities we participate in, directly or indirectly. Business will only adopt an environmental ethic when the society that supports it does.

At present, there are only a few resources available for the businesses that are ready to take the green plunge, making this section of the book necessarily brief. But among the few we foundwere some real pearls.

The Business to Business Guide contains many of the same classifications as you will find in the Consumer Guide, the difference being that the businesses listed here are either wholesalers or specialize in services to business. Some of the categories (cartography, for instance) are unique to this section.

Use the Table of Contents opposite to find the category in which you are interested. The index to the Consumer Guide, pages 267-270, also contains subject references to the Business to Business Guide and is the quickest way to find the specific item or service you are seeking.

BUSINESS CONSULTANTS

The first of these companies specializes in assisting nonprofit organizations with their computer needs. The second provides business advice and assistance to worker and housing cooperatives, and promotes the democratic control of economic and housing opportunities. The Green Pages looks forward to hearing about more innovative business consultancies.

NOE VALLEY COMPUTERS,INC.
(assist non-profit organizations)
85 29th St., SF (415)282-3000

URBAN ALTERNATIVES
(415)644-8336

BUSINESS LOCATIONS

BERKELEY OFFICE OF ECONOMIC DEV.
2180 Milvia Fifth Floor, BERK (415)644-6309
See display ad on this page.

CARTOGRAPHY

EUREKA CARTOGRAPHY
2161 Shattuck Ave., BERK (415)-845-MAPS
See display ad on this page.

CFC RECYCLING EQUIP.

These companies manufacture equipment for recycling CFC's in commercial refrigeration and auto air conditioning repair applications. The Bay Area needs many more locations for the recycling and capture of the CFC's that are such a menace to the planet. (See the article on p. 118.)

AUTOHAUS HEADQUARTERS
1925 Republic Ave., San Leandro (415)357-6210

JOHNSON SUPPLY CO. OF OAKLAND
2140 Livingston Ave., Oakland (415)534-6755

THERMAFLO
3640 Main Street, Springfield (800)848-4232
MANUFACTURERS OF PORTABLE CFC RECOVERY EQUIPMENT FOR REFRIGERATION REPAIR CONTRACTORS, SAFETY VALVES, PUNCTURE VALVE SEALS, AND MORE.

WESTERN LUBRICANTS
1130 Pacific St., Union City (415)487-4991

CLEANING PRODUCTS

Conventional janitorial cleaning products contain some of the worst chemicals for the environment, and can easily be replaced with non-toxic and environmentally-safe alternatives. There are many non-toxic and completely biodegradeable cleaners and janitorial supplies available for all sorts of applications. The businesses listed in this section can supply you with industrial cleaners, institutional soaps, shampoos, abrasives, and polishes.

BIO-CARE
2105 Bascomb, Campbell (408)559-7500

NUTRISOURCE
P.O. Box 81106, Seattle, WA (206)467-7190

LIFE TREE PRODUCTS
P.O. Box 120, Sebastopol (707)577-0324

SIERRA NAT'L FOODS - ECOVER PRODUCTS
440 Valley Dr., Brisbane (415)468-8800

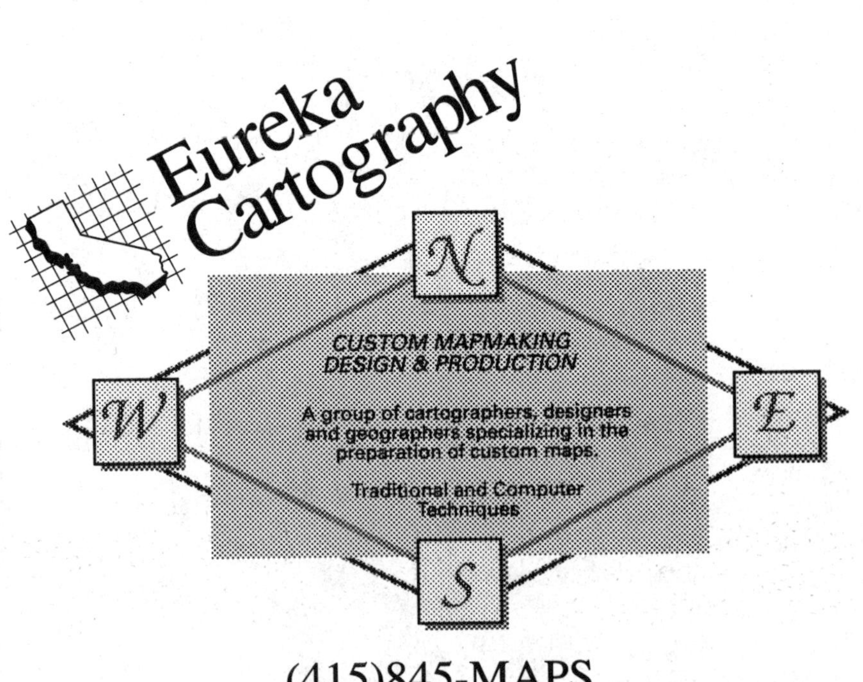

ENERGY

The following companies specialize in Title 24 compliance and energy-efficiency design in office and multi-unit buildings. There are many such consultants in the Bay Area. Insist on using the most energy-efficient techniques for your buildings and retrofits. For companies that also incorporate non-toxic considerations in their, please see Building consultants in the Consumer Guide Section.

BUILDING RESOURCES MANAGEMENT CORP
1750 112th. St., Bellevue WA 98004 (415)446-1949

GABEL. ASSOCIATES
1818 Harmon St., BERK (415)428-0803

GOLDEN WEST ENERGY CONTROL
4592 E. 2nd Unit A, Benicia (707)746-5479

INQUIRING SYSTEMS, INC.
3111 Deakin St., BERK (415)843-3135

ENVIRONMENTAL CONSULTANTS

These consultants have expertise in diverse areas of restoration, office design, energy consulting, building technique, and recycling. Since they have knowledge of numerous fields, they should be able to help you directly, or refer you to someone else to meet your business' or project's particular needs.

BROWN, VENCE ASSOCIATES
120 Montgomery, SF (415)434-0900

DAVID AMMEE- OPEN SPACE MANAGEMENT
1314 Curtis St., BERK (415)526-9257

ECOSERVE
690 Garcia Ave., Pittsburg (415)439-5766

GREGORY HOUSE FARMS
P.O. Box 1615, Davis (916)753-3361

MARIANNE HEGEMAN
7538 Stockton Ave., El Cerrito (415)525-8811

PHILIP WILLIAMS & ASSOCIATES
Pier 35, The Embarcadero, SF (415)981-8363

SUSAN ENGLEBRY
P.O. Box 843, Bolinas (415)868-0256

FOOD — PRODUCE TESTING

This fee-for-service company independently tests produce for certain pesticides and provides third-party certification of pesticide-free status.

NUTRICLEAN
1611 Telegraph Ave., Suite 1111, OAK (415)832-1415

FOOD WHOLESALERS

These companies sell organic or natural food products wholesale. They can supply your grocery, cafeteria or catering supply needs.

ALMA'S PLACE
6154 Sawmill Rd.,Paradise (916)872-9107

ANTOLINI FARM
4575 Paul Sweet Rd. Santa Cruz (408)476-1648

APPLESEED RANCH
1834 High School Rd., Sebastopol (707)823-4408

APPLEWOOD RANCH
6551 Glen Haven Rd., Soquel (408)476-4321

ASSOCIATED COOPERATIVES
322 Harbour Way Suite 12, Richmond (415)232-1111

B & B ORGANIC FARM
P.O. Box 309, Leggett (707)984-8420

BETA BROKERAGE
170 West Lake St., Watsonville (408)728-0414

CACHE CREEK ORGANIC FARM & NURSERY
P.O. Box 85, Rumsey (916)796-3521

CASCADIAN FARM
Star Route Box 568, Concrete, WA (206)853-8175

CRYSTAL WAVE
P.O. Box 2298, Sebastopol (707)829-8336

DEER GARDEN FOODS
2230 Soquel Ave., Santa Cruz (408)462-6715

EDEN FOODS, INC.
701 Tecumseh Rd., Clinton MI (517)456-7424

FALCON TRADING COMPANY INC.
1055 17th Ave., Santa Cruz (408)462-1280

FARMVEST, INC.
P.O. Box 201, Clayton (415)672-8843

FOWLER BROTHERS
P.O. Box 2324 San Rafael (415)459-3406

FROM THE RAINFOREST
8 East 12th St., NY, NY (212)645-7177

GARDEN OF EATIN'
5300 Santa Monica Blvd., Los Angeles (213)462-5406

GIUSTO'S SPECIALTY FOODS, INC.
241 East Harris Ave., SSF (415)873-6566

GREENE & HEMLY INC.
11275 State Hwy.160, Courtland (916)775-1379

GREENLEAF PRODUCE
1980 Jerrold Ave., SF (415)647-2990

GREGORY HOUSE FARMS
P.O. Box 1615, Davis (916)753-3361

GROSSI FARMS
6652 Petaluma Hill Road, Santa Rosa (707)664-7723

H-S FARMING COMPANY
P.O. Box 724, Healdsburg (707)838-4570

INI INTERNATIONAL
309 Sheffield Ave., Mill Valley (415)383-1869

JACOB'S FARMS
P.O. Box 508, Pescadero (415)879-0580

KNOLL ORGANIC FARMS
PO Box 65-H, Brentwood (415)634-5959

KONA KAI FARMS
1824 5th St., BERK (415)486-8334

KOZLOWSKI FARMS
5566 Gravenstein Hwy., Forestville (707)887-1587

LITTLE BEAR ORGANIC FOODS
15200 Sunset Blvd.#202, Pacific Palisades (213)454-4542

LUNDBERG FAMILY FARMS
P.O. Box 369, Richvale (916)882-4551

MADE IN NATURE
7 Mt. Lassen Dr., Ste A120, San Rafael (415)296-7046

MCFADDEN FARM
sun dried, organic tomatoes (800)544-8230

NEW NATIVE/GREENSWARD
1255 Hames Rd., Aptos (408)728-4136

OCEAN ORGANIC PRODUCE COMPANY
P.O. Box 570, Moss Landing (408)722-1166

ORGANIC PRODUCE CO.
P.O. Box 14443, Santa Rosa (707)579-9297

PACIFIC SOYBEAN AND GRAIN
495 De Haro St., SF (495)863-0867

PERLINGER NATURAL PRODUCTS
238 Petaluma Ave., Sebastopol (707)829-8363

PLANT SCIENCES, INC.
514 Calabasas Rd., Watsonville (408)728-7771

PLEASANT GROVE FARMS
PO Box636, Pleasant Grove (916)655-3391

PLEASANT HILL PRODUCE COMPANY
2203 Morello Ave., Pleasant Hill (415)798-5275

QUONG HOP AND COMPANY
161 Beacon St. South, SF (415)761-2022

S & S PRODUCE CO.
P.O. Box 570, Moss Landing (408)761-0881

SANTA CRUZ NATURAL
P.O. Box 1510, Freedom (408)728-0515

SIERRA NATURAL FOODS
440 Valley Dr., Brisbane (415)468-8800

SIMPLY PURE BABY FOOD
RFD #3, Box 99, Bangor ME (800)426-7873

SOYFOODS UNLIMITED
14670 Doolittle, San Leandro

STAR VALLEY FARM
2985 Mix Canyon Road, Vacaville (707)448-5303

SWANTON BERRY FARMS
5221 Coast Road, Santa Cruz (408)426-9614

TASSAJARA BREAD BAKERY
5700 3rd St., SF (415)822-5770

THE APPLE FARM BATES AND SCHMITT
18501 Greenwood Road, Philo (707)895-2333

TKO FARMS
101 E. Manor Dr., Mill Valley (415)389-0507
See display ad, page 282.

VALLE VERDE FARMS
11397 Green Valley Road, Sebastopol (707)823-0156

VAN DYKE RANCH
7665 Crews Road, Gilroy (408)842-5423

VERITABLE VEGETABLE, INC.
1600 Tennessee St., SF (415)821-1450
See display ad on this page.

VITASOY (U.S.A.) INC.
99 Park Lane, Brisbane (800)848-2769
See display ad, page 284.

WEBB'S ORGANIC FARM
5381 Old San Jose Road, Soquel (408)475-1020

WESTERN SKIES RANCH
1156 West Ave., Santa Rosa (707)545-8079

WHOLEFOOD EXPRESS
3134 Jacobs Ave., Eureka (707)445-3185

WILD ROSE RANCH
308-B Center St., Healdsburg (707)433-7869

WILDWOOD NATURAL FOODS
135 Bolinas Rd., Fairfax (415)459-8636

HERB WHOLESALERS

GREAT LAKE HERB CO.
P.O. Box 6713, Minneapolis MN (612)722-1201

MCFADDEN FARM
16000 Powerhouse Rd., Potter Valley (800)544-8230

MOUNTAIN ROSE HERBS
P.O. Box 2000, Redway (800)879-3337

TRADITIONAL MEDICINALS
4515 Ross Rd., Sebastopol (707)823-8911

LIGHTING

The first company specializes in energy-efficient lighting for offices including the use of compact fluorescents, hi-intensity discharge fixtures, and the retrofitting of existing fluorescent fixtures with energy-efficient ballasts and hi-efficiency optical reflectors. The second company supplies compact fluorescents. It will soon carry balanced spectrum lights for a healthier ambience in the office. Please see the Lighting section in the Consumer Guidefor more suppliers.

COMMERCIAL ENERGY LIGHTING INC.
731 Mandana Blvd., OAK (415)839-7638

GREENER ALTERNATIVES
914 Mission St.,Suite A, Santa Cruz (408)423-0701
See display ad, page 290.

MARKETING

BUTTERFIELD ASSOCIATES.
1339 61st St., Emeryville (415)428-1991
See display ad on this page.

NURSERIES

These companies specialize in the wholesale supply of large quantities of drought-resistant plants or California natives for use in restoration projects or commerical landscaping. They can help you design a beautiful and environmentally-conscious landscape for both meanings of "company image". See also the Garden section of the Consumer Guide.

APPLETON FORESTRY
1369 Tilton Rd.Sebastopol (707)823-3776

CALIFORNIA FLORA
P.O. Box 3, Fulton (707)528-8813

CIRCUIT RIDER PRODUCTIONS
9619 Old Redwood Rd., Windsor (707)838-6641

CLYDE ROBIN SEED COMPANY
3670 Enterprise Ave., Hayward (415)785-0425

MOSTLY NATIVES
27215 Hwy 1 Box 258, Tomales Bay (707)878-2009

PACIFIC COAST SEED
7074 D Commerce Cr., Pleasanton (415)463-1188

SIMEON'S ORGANIC PLANT FOOD
5685 Holland Loop Rd., Cave Junction (503)592-4753

TREE OF LIFE NURSERY
P.O. Box 736, San Juan Capistrano (714)728-0685

WAPUMNE
3807 Mount Pleasant Rd., Lincoln (916)645-9737

Fact: 25 billion styrofoam cups are thrown away every year in the U.S.

PACKAGING

Packaging is one of the major sources of waste paper and plastics. Even when new, it is one step away from being garbage. If packaging is a necessity, use recycled and recyclable materials. The manufacturer and/or shipper has the responsibility to use a design and materials which can be easily recycled and to know how this can be done. More important is to use recycled materials in your packaging The suppliers in this section have the materials and systems to help you do it.

ALASKA PLASTIC SALES
939 East Eleventh St., OAK (415)832-5441
See display ad, page 346.

ALEXANDER FRUIT & TRADING CO.
510 Hwy. 128, Geyserville (707)433-1944

EVERETT GRAPHICS
P. O. Box 8095, OAK (415)654-5652 x3214
See display ad on this page.

FLEXEL-ATLANTA
19961 Merrit Drive, Copertino (415)828-0603
See display ad, page 346.

GREENER ALTERNATIVES
914 Mission St.,Suite A, Santa Cruz (408)423-0701
See display ad, page 290.

ZELLERBACH
245 S. Spruce Ave., SSF (800)468-0200
See display ad, page 346.

PACKING

Don't shipout anything that contains packing material that you wouldn't know how, and where, to recycle yourself. The number of places that recycle styrofoam peanuts is very small. The best packing alternative that we have found is EcoPack's product that uses 100 percent post-consumer-waste, recycled paper that is unbleached, reusable and recyclable. How intelligent! The Alexander Fruit & Trading Co. in Geyserville, CA has begun shipping its fruit in real popcorn – completely biodegradable and most certainly stale. We really must get away from styrofoam peanuts for two reasons: They are amajor source of CFCs - one polystyrene cup (you figure the peanut equivalency) contains one billion-billion CFC molecules that last 150 years, and continuously breakdown the protective ozone layer during their lifetimes, Wildlife, especially birds, think these items are food, consume them and then starve because their stomachs can no longer accommodate real food.

ECOPACK INDUSTRIES, INC.
7859 S 180th StSpringbrook B3, Kent WA (206)768-9600

ZELLERBACH
245 S. Spruce Ave., SSF (800)468-0200
See display ad on this page.

PAPER, RECYCLED

Businesses buy huge amounts of paper. There is no reason to use virgin stock paper when there are so many recycled papers available of every grade and color, and for everything from book stock to xerographic needs. Let your paper suppliers know what you need. Ask them for unbleached supplies. Most in-house and letter paper needs do not require perfectly bleached white paper. Try to use paper with the most post-consumer recycled content available. It is only when we do this that we are actually diverting paper from the landfills. If all the businesses in the Bay Area demanded and used paper with a minimum ten percent post-consumer content it would make a huge difference. Using recycled paper can have it's benefits in public relations as well. People are in favor of positive action for the environment. So let your customers know you've switched to recycled paper and why. It might just improve business. (See article "Greening Your Business" on page 54.)

A.C. PAPER AND SUPPLY
1321 Seventh St., BERK (415)527-0841

ALTE SHULE, U.S.A.
704 E. Palace Ave., Santa Fe NM (505)983-2593

APPLETON PAPERS, INC.
825 E. Wisconsin Ave., Appleton WI (414)735-8810

ATLANTIC RECYCLED PAPER COMPANY
P.O.Box11021, Baltimore MD (800)323-2811
See display ad, page 309.

BANDELIER DESIGNS
1570 Pacheco, Suite E7, Santa Fe NM (505)986-1600

BRIGHT IDEAS
3450 Golden Gate Wy, Lafayette (415)283-5199
See display ad, page 311.

BUTLER PAPER COMPANY
180 West Hill Pl., Brisbane (415)468-1100

CARTER RICE COMPANIES
273 Summer St., Boston MA (617)542-6400

CONSERVATREE PAPER CO.
10 Lombard St., #250, SF (415)433-1000
See display ad, page 312.

CROSS POINT PAPER CO.
1295 Bandana Blvd. N., St. Paul MN (612)644-3644

DANCING TREE RECYCLED PAPER PROD.
2253 Glen Ave., BERK (415)486-1616
See display ad, page 311.

EARTH CARE PAPER, INC.
P.O. Box 14140, Madison WI (608)256-5522

ECO-VISION
2215 Market, SF (415)864-2316

FORT HOWARD PAPER
2379 Norwood Rd, Livermore (415)447-5796

FUTURE FIBRES GROUP, INC.
4380 Redwood Hwy., #C-21, San Rafael(800)447-3111

GREENER ALTERNATIVES
914 Mission St.,Suite A, Santa Cruz (408)423-0701
See display ad, page 290.

J. C. PAPER CO.
650 Brennan Ave., San Jose (800)527-2737

JAMES RIVER CORPORATION
300 Lakeside Dr #1461, OAK (415)874-3458
See display ad, page 311

KIRK PAPER COMPANY
6001 East Randolph St., Los Angeles (213)728-8117

OFFICE CLUB
1631 Challenge Dr., Concord (415)682-2582
See display ad, page 309.

PAPER, INC.
8041-2 Queenair Dr., Gaithersburg MD (301)948-6300

PERFORMANCE COMPUTER FORMS
21673 Cedar Ave., Lakeville, MN (612)469-1400

RECYCLED PAPER COMPANY
100% Post-consumer Paper
185 Corey Rd., Boston,MA 02146 (617)277-9901

SCRIBECRAFT
P. O. Box 77, Newfield,NY (607)564-7572

SEABOARD PAPER COMPANY
336 Oyster Point Blvd., SSF (800)832-2311

SIMPSON PAPER COMPANY
600 Market St., SF (415)391-8140

THE PAPER PROJECT
P.O. Box 12, Arcata (415)540-1127

ZELLERBACH
245 S. Spruce Ave., SSF (800)468-0200
See display ad, page 312.

PEST CONTROL

These companies provide biological and non-toxic pest control products for wholesale producers of organic food, and other farming operations. For office pest control please see the Pest Control heading in the Consumer Guide.

BENEFICIAL INSECTARY
14751 Oak Run Rd., Oak Run (916)472-3715

ELECTRACAT
5588 N. Palm Ave., Fresno (209)432-2707
See display ad, page 314.

NATURAL PEST CONTROL
8864 Little Creek Dr., Orangevale (916)726-0855

PRINTERS

Make sure your printer stocks recycled paper and uses less-toxic, soy-based inks. The companies below have made a commitment to the use of recycled paper, and some have been regularly using less toxic inks - with great sucess. Ask your regular printer to use soy ink which contains fewer petrochemical products than regular ink, and is based on soy oils. (We have used it for the printing of this book). Printing, in general, is a toxic business; solvents and chemicals are used in great amounts. We are happy to report that the industry is beginning to replace these items with substitutes that do not destroy the ozone and worker health.

ALONZO PRINTING
1094 San Mateo Ave.,SSF (415)873-0522
See display ad on this page.

BLACK LIGHTNING,INC.
RR 1-87, Depot Road, Hartland, VT (800)252-2599

BRIGHT IDEAS PRESS
3450 Golden Gate Wy, Lafayette (415)284-5350

BRIGHT IDEAS COPY SHOP
99 Brookwood Road, Orinda (415)253-0260

GRAFFIK NATWICKS
240 Golden Gate Ave., SF (415)441-0878

NORCO-BILLINGS PRINTING
1152-A Marina Blvd., San Leandro (415)351-8288

SUNCHEMICAL
14300 Catalina St.
San Leandro (415)351-7661

PRINTERS' SUPPLY

AMERICAN INK PRODUCTS COMPANY
630 E. Tenth St., OAK (415)268-0825

ZELLERBACH
245 S. Spruce Ave., SSF (800)468-0200

RECYCLED PRODUCTS

If you're not buying recycled products, you're not recycling. High-quality sources of recycled paper products, such as toilet paper and paper towels, are becoming increasingly available. Other office items like brooms and recycling bins, made of recycled plastic, can alsobe found here. Call these places to find out what they carry. Make the most important link of the recycling chain a high priority in your business. Some city governments, (e.g. the City of San Jose) can advise you on sources as well — see the Public Agency section of this book (p. 231).

BAY WEST JANITORIAL SUPPLIES
P.O. Box 2867 So., SF (415)873-0160

BIRITE FOODSERVICE DISTRIBUTORS
201 Alabama St., SF (415)621-6909

BUSINESS ECOLOGY PRODUCTS
P.O. Box 946, Martinez
(415)687-1202

CLEAN MASTER
1780 3rd St., SF
cleaning and food service supplies (415)495-2900

COMPETITIVE PAPER SALES
food service products (415)832-6714

DATACOM INC.
Evolution Recyled Paper Products
201 Gilbraltar Rd., Horsham PA
(215)443-5700

EARTH CARE PAPER, INC.
P.O. Box 14140, Madison WI (608)256-5522

FLEXEL-ATLANTA
19961 Merrit Drive, Copertino (415)828-0603
See display ad, page 346.

FORMS NET
Santa Cruz (408)688-6964

RECYCLED TONER CARTRIDGES
Unconditionally guaranteed
... INFOTEK ... **free trial**
(415)921-7135

MONAHAN PAPER
175 Second St., OAK (415)835-4670
See display ad on this page.

POPE & TALBOT
P.O. Box 8171 Portland,OR (503)228-9161

PURE PAPER
1232 Masonic Ave, BERK (415)524-8079
See display ad, page 312.

TURN AROUND PRODUCTS OF MARTINEZ
1258 Pine St., Martinez (415)372-6974

RECYCLERS

BY COUNTY

Setting up a recycling program at your business can save you money. Normally you pay to have your trash hauled away but, if it is appropriately sorted, there are many companies that will pick it up for free or even pay you for the privilege. Some of these recycling companies have consultants to help you design your recycling program, and resource materials to help educate employees. Many companies have set up very successful recycling programs which have saved them money and improved company morale. See the article "Greening Your Business" on page 54. Businesses should also try to buy recycled goods whenever possible, whether as raw materials (recycled plastic feedstock, for instance) or as finished products for office use. Make sure you are not making the problem worse by increasing the demand for our scarce resources (see also Packaging,Recycled Products and Recycled Paper).

ALAMEDA

AARON METALS CO.
750 105th Ave, Oakland (415)569-6767
See display ad, page 320.

AMERICAN INK PRODUCTS COMPANY
630 E. Tenth St., Oakland (415)268-0825
LASER PRINTER TONER CARTRIDGES

BAY CITY PAPER STOCK
2615 Davis St., San Leandro (415)638-2303
PAPER

BAY POLYMER CORPORATION
44530 Grimmer Blvd., Fremont (415)490-1791
PLASTIC

COMMUNITY RECYCLING CTRS.-COMMERCIAL
3550 Adeline, Oakland (415)653-2104
WHITE PAPER

INDEPENDENT PAPER STOCK-COMMERCIAL
3300 Wood St., Oakland (415)654-3787
WHITE, COMP

INDUSTRIAL PALLET COMPANY
PO Box 161, Union City (415)489-4050
PALLETS

LASER-LIFE TECHNOLOGIES
6711 Serria Ct, Ste C, Dublin (415)829-9215
LASER PRINTER TONER CARTRIDGES

MISSION PALLET COMPANY
2100 Dennison St., Oakland (415)261-0994
PALLETS

OAKLAND PLASTICS SALES
9733 San Leandro Street, Oakland (415)562-6023
PLASTIC

OWENS-BROCKWAY
6150 Stoneridge Mall Road, Ste 375, Pleasanton (415)463-7950
GLASS

PAPER RECYCLING OF NORTHERN CALIF.
25670 Nickel Place, Hayward (415)785-7311
PAPER

SMURFIT RECYCLING COMPANY
3300 Wood St., Oakland (415)654-3787
B: NEWS, CARD, MIX, MAGS, WHITE, COMP

CONTRA COSTA

AMERICAN LASER CHARGERS
1301 Franquette Ave., Ste C, Concord (415)827-9220
LASER PRINTER TONER CARTRIDGES

BAY AREA PALLET CORPORATION
P.O. Box 6102, Concord (415)757-3524
PALLETS

BLAKE ENTERPRISES
2154A North Main St., Walnut Creek (415)939-1718
LASER PRINTER TONER CARTRIDGES
See display ad, page 348.

CONTRA COSTA RES. RECOVERY-COMMERCIAL
615 Escobar St., Martinez (415)372-0625
CARD, WHITE, COMP

RICHMOND SANITARY- COMMERCIAL RECYCLING
205 41st St., Richmond (415)262-1600
GLASS, CARD, WHITE PAPER

SAN FRANCISCO

BERG MILL SUPPLY
440 Davis Ct., San Francisco (415)398-5222
PAPER

GOLDEN GATE DISPOSAL
900 Seventh St., San Francisco (415)553-2935
GLASS, ALUM (FROM FOOD BUSINESSES), OFFICE RECYCLING

LASER CARTRIDGE NETWORK
2120 E. Howell, Suite 409, Anaheim (800)333-2505
LASER PRINTER TONER CARTRIDGES

LASERTONE
23 Sienna Way, San Rafael (415)485-5440
LASER PRINTER TONER CARTRIDGES

METAL RECYCLING INC.
1438 Donner Ave. SF (415)822-9096

NORCAL WASTE SYSTEMS, INC.
5 Thomas Mellon Cr., Suite 266, SF (415)330-1000
See display ad, page 324

PAPER RUSH CO., INC.
2372 Jerrold Ave., SF (415)282-2344

RAINBOW RECYCLING SERVICE-COMMERCIAL
49 Linda, San Francisco (415)863-8455
NEWS, GLASS, ALUM, PET

UNITED WAY
410 Bush St., San Francisco (415)872-6736
OFFICE EQUIPMENT

WEYERHAEUSER PAPER
925 Toland, San Francisco (415)821-0900
OFFICE PAPER

SAN MATEO

BUFFALO INCORPORATED
535 Middlefield Rd, Ste 245, Menlo Park (415)326-3103
LASER PRINTER TONER CARTRIDGES

CALIFORNIA SOLVENT RECYCLING
P.O.Box 50728, Palo Alto (415)856-7577
SOLVENT. ALL CLIENTS MUST BE REGISTERED WITH THE EPA AS A HAZARDOUS WASTE GENERATOR.

PORTA PULPER
800 S. Amphlett Blvd., San Mateo (415)344-6336
DOES CONFIDENTIAL PAPER DESTRUCTION AND RECYCLES

SUNSET SCAVENGER - COMMERCIAL
Tunnel Avenue & Beatty Road, Brisbane (415)330-2872
CARD, SOME WHITE, NEWS (HIGH VOLUME).

SANTA CLARA

A AND A PALLET RECYCLE
800 Spreckles Ave., Alviso (408)263-7878
PALLETS

AC PALLET SERVICE
1370 Taylor St., Alviso (408)263-4864
PALLETS

AMERICAN TRANSITECH
1917 Old Middlefield Wy, Ste 5, Mountain View (415)968-4397
LASER PRINTER TONER CARTRIDGES

C AND M DIVERSIFIED
330 N. Montgomery St., San Jose (408)294-5185
USED BUILDINGS MATERIALS, FROM DOORS TO TOILETS

COOKE RESOURCES COMPANY
PO Box 1144, San Jose (408)238-6096
RECOVERY OF SILVER FROM PHOTOGRAPHIC SOLUTIONS

GUADALUPE RUBBISH DISPOSAL COMPANY
P.O. Box 20957, San Jose (408)268-1670
ALUM, PAPER, GLASS, CARD, PET

HARRIS RECYCLING
787 E Norht King Rd., Milpitas (408)259-2290
PALLETS

IN-HOUSE RECYCLING
996 Minnesota Ave., Ste. 204, San Jose (408)292-4372
PAPER

LARSON PALLET COMPANY
1000 Yosemite Dr., Milpitas (408)946-4971
PALLETS

LASER RECOVERY SERVICE
25670 Nickle Pl., Hayward (415)785-7311
LASER PRINTER TONER CARTRIDGES

MARKOVITS AND FOX
1633 Old Oakland Rd., San Jose (408)453-7888
B: GLASS, ALUM,SCRAP, PET

MOBILE SHREDDING
349 Laurelwood Rd., Santa Clara (408)748-0508
PAPER

RECYCLING SPECIALISTS
1720 Old Bayshore Hwy, San Jose (408)437-7553
GLASS, ALUM, NEWS, COMP, PET

SAINT FRANCIS CABRINI CHURCH
15333 Woodard Rd., San Jose (408)377-3090
PAPER

SALVATION ARMY
702 West Taylor St., San Jose (408)298-7600
OFFICE EQUIPMENT

SAN JOSE CONSERVATION CORPS
2650-A Senter Rd., San Jose (408)998-5884

SAN JOSE RECYCLING & TRANSFER STATION
291 Barnard Ave., San Jose (408)294-7007

SAN JOSE RECYCLING II
1675 Pomona Ave., San Jose (408)294-1370
ALUM, PAPER, PET, GLASS, CARD

SAN JOSE TALLOW COMPANY
PO Box 610116, San Jose (408)452-8777
GREASE

SECOND HARVEST FOOD BANK
297 Commercial St., San Jose (408)286-9170

SECURITY SHRED AND BALE
Watsonville (408)722-4101
CONFIDENTIAL PAPERS AND FILES

SECURITY SHREDDING
PO Box 3304, San Jose (408)452-5996
PAPER

SMURFIT
2005 De La Cruz, Ste.102, Santa Clara (408)727-7576
BUSINESS - NEWSPAPER, MAG.

SOUTH BAY RECYCLING
Santa Clara (408)980-9900
VARIOUS BUSINESS RECYCLING SERVICES

TECHNALLOY INCORPORATED
101 Metro Dr., Ste. 259, San Jose (408)437-1995
SCRAP METALS

U-SAVE BROKER PALLET REPAIR
752 N. Tenth St., San Jose (408)295-4840
PALLETS

YU AI KAI
565 N. Fifth St., San Jose (408)294-2505
PAPER

ZANKER ROAD RESOURCE MANAGEMENT
705 Los Esteros Rd., San Jose (408)263-2383

SANTA CRUZ

C AND M SALVAGE
Watsonville (408)761-2390
CARD, OFFICE PAPER

CARDBOARD SEEKERS
101 Graham Ave., Santa Cruz (408)426-1748
CARD, GLASS, COMPUTER AND LEDGER PAPER, ALUM

CITY OF SANTA CRUZ-COMMERCIAL RECYCLING
Santa Cruz (408)429-3490
CARDBOARD, GLASS

INDUSTRIAL STAK-RITE
Watsonville (408)724-3905
CARDBOARD, OFFICE PAPER

SOLANO

ATLAS METAL & IRON CORP.
2490 Cement Hill Road, Fairfield (707)437-4808
B: ALUM, NON-FERROUS METALS, GLASS.

SONOMA

INDUSTRIAL CASTING
P. O. Box 3935, Santa Rosa (707)585-0511
GLASS, PAPER, ALUMINUM, ETC.

WEST COAST SCRAP CO.
20439 Old Redwood Hwy., Windsor (707)838-9731

RECYCLING SYSTEMS

Recycling systems for offices and businesses can be elaborate or suprisingly simple. Call your city's solid waste or recycling office to see if they provide business customers with recycling bins.

ANTON PHILLIPP LAB
P.O. Box 613, Charlevoix (800)748-0563

DIVERSIFIED RECYCLING SYSTEMS
5606 N. Hwy. 169, New Hope MN 55428 (612)536-6828

REHRIG PACIFIC CO.
4010 E. 26th St., Los Angeles (213)262-5145

WINDSOR BARREL WORKS
P.O. Box 47, Kempton PA (215)756-4344

RESTORATION

Restoration has really only become an industry since the passage of environmental laws requiring governmental agencies and corporations to mitigate for the environmental destruction they cause. Restoration techniques employ replanting or engineering to re-create or enhance a certain habitat.

Much research needs to be done in order to separate out those companies who engage in restoration thoroughly, ethically, and with the most scientific and current techniques available, from those that have jumped on a lucrative bandwagon with little regard for efficacy in restoration or long-term viability of their projects.

The following (very short) list is just a start. These companies impressed us, but by no means are they only acceptable companies out there.

PACIFIC OPEN-SPACE, INC.
P.O. Box 744, Petaluma (707)769-1213

PT. REYES NAT'L SEASHORE REST.
P.O. Box 447, Pt. Reyes Station (415)663-1171

THE HABITAT RESTORATION GROUP
6001 Butler Ln. Ste. 1, Scotts Valley (408)438-4102

SHOPPING BAGS

Only in the United States does virtually every store give the consumer a shopping bag every time he or she makes a purchase. The environmental price tag of this custom is staggering. If you are a retailer, have reusable bags for sale at a reasonable price near the counter. Print your store's name on the bags. They also make great promotional items for other sorts of businesses.

INDUSTRIAL WEAR
P.O.Box 1349, Studio City (818)508-7324

NET-SAK
PO Box11065, Aspen CO 81612 (303)927-4135
See display ad on this page.

WIND POWER

Twenty years ago, proponents of wind power had a vision of a society powered by a windmill on every roof. The vision and the technology have changed a bit since then. Today's wind turbines are powerful generators, many capable of generating in excess of 100 kilowatts in a good wind — much more electricity than one home would ever require. These machines are much more suitable for larger purposes, such as generating power for an industrial park or a whole neighborhood. For this reason, we have included wind power as part of the Business to Business Guide, rather than in the main Consumer Guide. Wind power is essentially a form of solar power, and like other kinds of solar power, it is cheap, clean, and indefinitely renewable. The following companies manufacture various kinds of wind generators.

AERMOTOR WINDMILL CORPORATION
P.O. Box 5110, San Angelo, TX 76902 (915) 658-2795

BERGEY WINDPOWER CO.
2001 Priestley Ave., Norman OK 73069 (405) 364-2078

NORTHERN POWER SYSTEMS
1 N. Wind RTd., Moretown, VT 05660 (802) 486-2955

SOUTHWEST WINDPOWER
P.O. Box 22178, Flagstaff, AZ 86004 (602) 256-54908

THERMAX CORPORATION
P.O. Box 3128, Burlington VT 05401 (802) 658-1098

U.S. WINDPOWER
6952 Preston Ave., Lvrmr, CA 94550 (415)455-6012

WIND ENERGY DEVELOPMENT CO.
4302 Arnold Dr., Wchta Falls, TX 76302(817) 569-3338

WIND HARVEST COMPANY
80 Lincoln Dr., Ventura, CA 93001 (213) 457-1563

The Index Explained

This index lists items alphabetically and by subject. The letters in parentheses tell you to which kind of listing the page numbers refer.

(BB) a classification in the Business-to-Business Guide
(CG) a classification in the Consumer Guide
(E) a listing in the Education section
(O) a listing in the Organizations section
(PA) a listing in the Public Agencies section
(Pk) a listing in the Parks section
(R) a title in the Reviews section